AF753548

EARTH'S NATURAL HAZARDS

Understanding Natural Disasters and Catastrophes

THIRD EDITION

Ingrid A. Ukstins • David M. Best

Kendall Hunt
publishing company

Cover image of the Taylor Complex fire, which burned more than 1,300,000 acres in Alaska in July 2004.

Cover photo source: United States Forest Service

Kendall Hunt
publishing company

www.kendallhunt.com
Send all inquiries to:
4050 Westmark Drive
Dubuque, IA 52004-1840

Copyright © 2010, 2019, 2022 by Kendall/Hunt Publishing Company

Text and KHQ ISBN 978-1-7924-6180-4
Text ISBN 978-1-7924-6181-1

All rights reserved. No part of this publication may be reproduced, stored in a retrieval system, or transmitted, in any form or by any means, electronic, mechanical, photocopying, recording, or otherwise, without the prior written permission of the copyright owner.

Published in the United States of America

Dedication

To Parker James, Jaxon Dean and their Mimi—my continuing inspirations.

D.M.B.

To my parents, Marion and Peter, who have always told me I could be anything I wanted to be and, more importantly, helped me to believe it was true.

I.A.U.

Brief Contents

Contents

Preface

The topic of natural hazards and disasters has always been of interest to the general populace. Each day on Earth brings some type of hazardous event that is part of the dynamic nature of the planet. This results in changes in the appearance of the landscape, and oftentimes the creation of a crisis for humans who are affected by the event. The broad range of disasters covered in this text addresses these events, from earthquakes, floods, and volcanoes to the biological hazards that often result indirectly from these events.

Normal geological processes have been studied and taught for more than two hundred years. The basics of river flow, landslides, and volcanoes were recognized early in the study of processes occurring on Earth, but geologists have developed a more detailed understanding of many others through the use of improved technology. By using satellite data and surface observations we are able to better explain how natural processes such as increased rainfall and river flow can generate floods in a given area. Monitoring movement in the subsurface gives insights into possible earthquake activity or volcanic eruptions. Examples of major disasters are numerous. During the past 30 years the eruption of Mount St. Helens in southwestern Washington, widespread flooding of the Midwest in the United States, the tsunami in South Asia, and earthquakes in Japan and the southwestern Pacific Ocean. Recent hurricanes in the Atlantic Ocean and typhoons in the Pacific have inflicted major damage and loss of life in those regions. These are just a few examples of geologically recent occurrences of life-changing events.

Featured Themes

The dynamic nature of the material covered in this text is presented through the use of carefully selected images and artwork that reduce the amount of visual overload that can occur in such a course. Pictures are worth a thousand words, as we oftentimes hear; we have attempted to do this without offering too many images. Additional images are available through the selection of the Images option provided by general web searches. Questions for Thought are at the end of each chapter as well as several websites along with brief lists of selected references and reading material.

Topic Organization

Following an introduction to hazards in general and the development of the Earth system, volcanoes and earthquakes, which are many times interrelated, are presented. Discussions of tsunami and mass movements follow earthquakes, as the movement of land and water is often a result of earthquakes. The discussion of objects from outer space provides information about hazards that are extraterrestrial.

Almost 75% of Earth's surface is covered by water. Several chapters address the role that the atmosphere and water play in the creation of natural hazards. A general introduction to global climate, followed by streams and oceans, explains how the presence of water is a key force on Earth. The uncontrolled presence of cyclonic storms completes the material that addresses the role of water in disasters. Wildfires, often the result of too little water in regions, have become an increasing problem in our country as well as in other parts of the globe. New topics provide material about biological hazards and the environment.

Acknowledgments

The impetus for the third edition of this text was provided by Paul Carty, Director of Publishing Partnerships of Kendall Hunt Publishing Company. His recognition of the need for an updated version is greatly appreciated. He also coordinated the addition of Ingrid as a second author. This edition is the result of the superb efforts of Beth Trowbridge, Senior Publishing Specialist, who oversaw the myriad details of its production. Her persistence and attention to detail are without equal.

Throughout the writing of this book and the previous edition David often called on a cadre of colleagues to listen to ideas and provide him the impetus necessary to complete the work. The assistance of the following colleagues and professionals has proven to help him in the writing and presentation of material: Sylvester Allred, James Best, Charles Denton, Lee Drickamer, Wendell Duffield, Duncan Foley, and Julie McCormick. Special thanks go to Darrell Boomgaarden, who granted an interview relating his personal experiences during the Great Alaskan earthquake of March 1964. The immeasurable efforts of Anne Luty and her capable staff at Composure Graphics added very significantly to our final result.

Finally, David wishes to express sincere thanks to his wife Mary, who provided continued encouragement whenever challenges arose. Her persistence allowed the work to come to fruition. In addition, her review of several chapters and her assistance with polishing the text made the material read much more clearly.

We also wish to express appreciation to the following reviewers who provided constructive and insightful comments.

Martin Acaster
Portland Community College

Christine Aide
Southwest Missouri State University

Stephen T. Allard
Winona State University

Steven Altaner
University of Illinois

Dr. Thomas Bicki
Emerson College

Paul Bierman
University of Vermont

Susan Bilek
New Mexico Tech

Amy Bloom
Illinois State University

Jon Boothroyd
University of Rhode Island

Dr. Patricia Cashman
University of Nevada, Reno

Heather K. Conley
Illinois State University

Winton Cornell
University of Tulsa

Juliet Crider
Western Washington University

Charles Denton
U. S. Forest Service (retired)

David Dinter
University of Utah

Wendell Duffield
U. S. Geological Survey (retired)

Todd Feeley
Montana State University

Duncan Foley
Pacific Lutheran University

John Foster
California State University, Fullerton

Kevin Furlong
Penn State University

David Gillespie
Washington University, St. Louis

Luis Gonzalez
University of Kansas

Michael Hamburger
Indiana University, Bloomington

Linda Hand
College of San Mateo

Keith Henderson
Villanova University

Simon Katterhorn
University of Idaho, Moscow

Robert Kay
Cornell University

Chris Kent
Community College of Spokane

Attila Kilinc
University of Cincinnati

Jean P. Kowal
University of Wisconsin, Whitewater

Stephen D. Lewis
California State University, Fresno

Lawrence Malinconico
Lafayette College

Doug McKeever
Whatcom Community College

Sue Morgan
Utah State University

Ken G. Sutton
Portland Community College

Harold Tobin
University of Wisconsin, Madison

James Wittke
Northern Arizona University

We hope you find the text interesting and engaging and ask that you make us aware of any errors, corrections, or additions you might have. The process of assembling a text is always difficult and oversights do occur. Thank you for your interest.

David M. Best
Flagstaff, Arizona

Ingrid Ukstins
University of Auckland
November 2021

About the Authors

David M. Best, PhD, grew up in North Carolina, where he graduated from the University of North Carolina at Chapel Hill with a BS in mathematics. He completed his MS in geology and then served in the U.S. Navy, including service in Vietnam and the Panama Canal Zone. After returning to UNC-Chapel Hill to obtain his PhD in geology, he began his teaching career at Northern Arizona University in 1978. During his 34 years on the faculty at NAU, he taught classes in introductory and physical geology, geophysics, statistical methods, and the geology of Arizona. He has also served in various administrative roles, including department chair, associate dean, and dean of the College of Arts and Sciences. After returning to the teaching ranks in 2003, David began teaching geologic hazards along with the geocommunications course required of geology majors. For 15 summers David also team-taught a biology and geology field course in the national parks until his retirement in 2010. He leads tours throughout the Southwest and maintains his interests and skills in cooking, woodworking, and studying the interconnection between geology and wines.

Ingrid A. Ukstins, PhD, grew up in Deptford, New Jersey, US. She graduated cum laude from Mount Holyoke College with a BA, where she majored in geology and minored in anthropology. Her MSc degree is from University of California, Davis, on the formation and deposition of pyroclastic rocks in the East Greenland flood basalts, North Atlantic Large Igneous Province. After her MSc, she spent a year working for an oil company in Houston, Texas, before completing a PhD in 2003 at Royal Holloway, University of London, on the eruption and emplacement mechanisms of silicic pyroclastic volcanism in Yemen and Ethiopia associated with the formation of the Red Sea and Gulf of Aden. She had a year of postdoctoral research at the Danish Lithosphere Center in Copenhagen, Denmark, and spent 15 years on the faculty in the Department of Earth and Environmental Sciences at the University of Iowa. She is currently a faculty member in the School of Environment, University of Auckland, New Zealand, where she teaches classes in natural disasters, geology, and planetary science. Her research on volcanology, geochemistry, and planetary geology has taken her to Australia, Chile, China, Greenland, Hawaii, Iceland, Kiritimati, New Zealand, and Vanuatu, among other places. In her free time she enjoys learning to stand-up paddleboard (mostly by falling off) and drinking coffee.

Living With Earth's Natural Hazards

1

Source: Ingrid Ukstins.

Mount Shasta in northern California is the largest volcano in the region. Although the population is relatively sparse in the surrounding area, this volcano has the potential to erupt and produce massive landslides and lahars. Such an event occurred about 11,000 years ago, covering the landscape with rock and water from the melting of glaciers on the summit.

© Asianet-Pakistan/Shutterstock.com.

Figure 1.1a **Heavy monsoon rains in Karachi, Pakistan, in August 2020 overwhelmed the city's water drainage systems, causing massive flooding and resulting in the deaths of at least 41 people.**

Does it seem like Earth is a dangerous place to live? A hazard by definition is "a possible source of danger." Among some of Earth's natural hazards are hurricanes, tornadoes, earthquakes, tsunamis, floods, mudslides, wildfires, and volcanic eruptions (**Figures 1.1a–c**) When these possible events do occur and impact people, they become disasters, such as the Australian bushfires in 2019–2020, flash flooding in Indonesia in 2020 that triggered landslides and killed 66 people, and the coronavirus COVID-19 pandemic that swept the planet in 2020. These all resulted in widespread destruction and loss of life (**Figures 1.2a–b**).

These scenes of unprecedented devastation have been etched into our minds and have made us increasingly aware of our human vulnerability to the awesome power of nature. However, these events are the result of natural processes involving physical, chemical, and biological mechanisms and forces. When the processes occur, they become hazards for people and create disasters.

Worldwide we are facing natural disasters that have an increasingly large impact on people. Data from the United Nations Office for the Coordination of Humanitarian Affairs outlined the global effects of natural disasters for 2016. In that year the number of deaths caused by the documented 342 disasters was 8,733, making it the second lowest in the past decade. The number of people affected by these natural disasters was 564 million and the economic damage was estimated to be $154 billion (US dollars).

Hydrological disasters (floods) and meteorological disasters (cyclonic storms) accounted for 80% of the events, while climatological (drought) and geophysical events (earthquakes) comprised the remaining 20%.

Source: NASA.

Figure 1.1b **NASA astronaut Chris Cassidy, aboard the International Space Station, photographed Hurricane Genevieve as it hit Mexico's Baja California peninsula in August of 2020.**

© DanielFreyr/Shutterstock.com.

Figure 1.1c **Just 40 kilometers (25 mi) away from Iceland's capital city of Reykjavik, Mt Fagradalsfall—beautiful valley mountain in Icelandic—started erupting on March 19, 2021, after earthquakes began in the region in December 2020.**

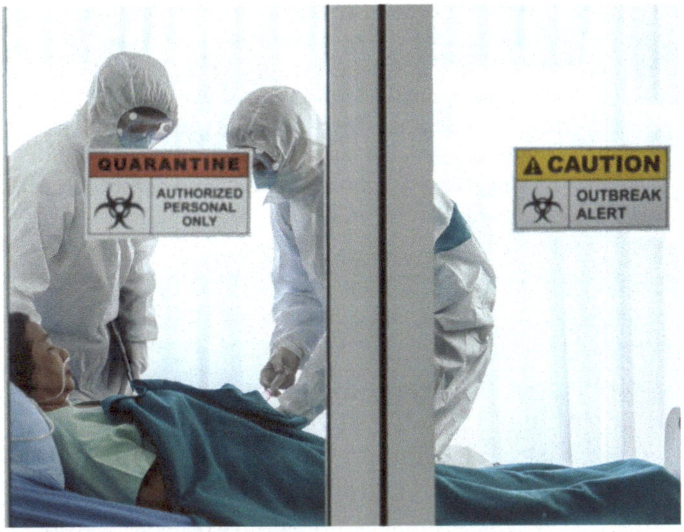

© Mongkolchon Akesin/Shutterstock.com.

Figure 1.2a The COVID-19 pandemic has killed millions of people worldwide and led to unprecedented challenges to public health, food and goods supply systems, and social and economic well-being.

Source: NOAA.

Figure 1.2b The Southeastern United States was hit by a deadly tornado outbreak over Easter Sunday and Monday, April 12–13, 2020.

China, the United States, India, Indonesia, and the Philippines are the top five countries most frequently affected by natural disasters. Coincidently, the first four of these are also the four most populated countries on Earth. Natural disasters kill on average 60,000 people per year and are responsible for 0.1% of global deaths. While global deaths may be low in many years, low-frequency high-impact events can cause devastating impacts: the 2004 Indian Ocean earthquake and tsunami killed an estimated 227,898 people; Cyclone Nargis struck Myanmar in 2008, killing more than 138,000 people; and the 2010 Haiti earthquake in Port-au-Prince killed more than 100,000 people. Such large numbers may appear abstract and difficult to conceptualize, but they are a harsh reality for many families who have lost loved ones, had their homes destroyed, or have watched their investments and economic future be destroyed by a natural disaster.

What can we do? Natural hazards are natural phenomena, but their transition into disasters or catastrophes is often a result of the organization, distribution, and behavior of our society (Figure 1.3). Natural disasters know no political boundaries. Smoke from forest fires can make air unbreathable in neighboring nations, ash plumes from volcanoes can disrupt air traffic across the globe, and emerging diseases and global warming impact the entire planet. Whether an extreme hazard event becomes a natural disaster depends on our ability to predict, prepare, and mitigate. Most decision makers and policymakers agree that the key to reducing the vulnerability and risk of human populations to natural hazards is the integration of disaster preparedness, **mitigation**, and prevention measures into future policy development. Yet funding patterns, an indicator of real priorities, show that disaster relief, not reduction or prevention, tops the list of disaster-management funding. This leads to continued losses and suffering after we recover until a disaster strikes again.

mitigation
The act of making less severe or intense; measures taken to reduce adverse impacts on humans or the environment.

© Kendall Hunt Publishing Company.

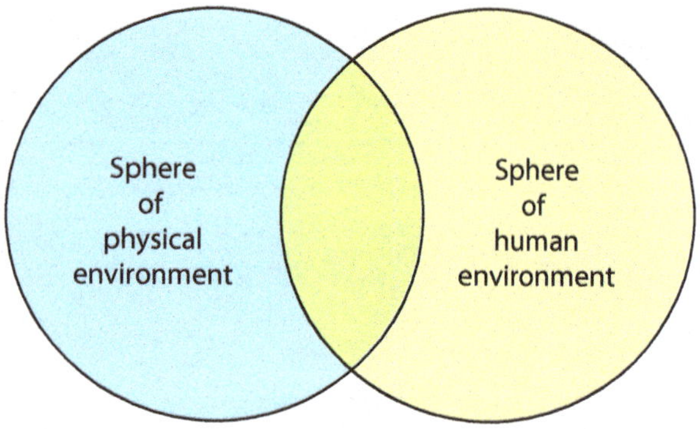

Figure 1.3 Natural hazards represent the potential interaction between humans and extreme natural events—it is the likelihood of the event, not the event itself. The overlapping area of physical environment and human environment represents conflicting natural forces at odds with the interests of human use. When an extreme event occurs, it causes negative impact to the vulnerable human population and constitutes a disaster (after Tobin and Montz, 2009).

Fortunately, public awareness of the losses in human lives and property from natural disasters is starting to grow. We still have a long way to go if we are to focus on hazard preparedness and prevention. Better information and public education about the natural environment and Earth processes are essential to reducing losses. Therefore, the focus of this book is to help develop an understanding of Earth's natural processes through the study of relationships between people and their environment. The information on natural hazards presented in this book is of practical value to people in making choices of where to live and understanding what hazards might be present so they may make the most informed decisions.

From Natural Hazards, to Disasters, to Catastrophes: Interactions With Natural Processes

natural hazard

An event or phenomenon that could have a negative impact on people and their property resulting from natural processes in the Earth's environment.

natural disaster

The loss of life, injuries, or property damage as a result of a natural event or process, usually within a more local geographic area.

natural catastrophe

A massive natural disaster often affecting a large region and requiring significant amounts of time and money for recovery.

In describing our interaction with natural processes we use the terms natural hazard, disaster, and catastrophe. A **natural hazard** is a potential event that could have a negative impact on people and their property resulting from natural processes in the Earth's environment. Some of these events include the possibility of occurrence of earthquakes, hurricanes, tsunamis, floods, volcanic eruptions, droughts, landslides, and coastal erosion. These events are the result of natural Earth processes that have been operating over the lifetime of our planet. They they will continue to happen, regardless of whether or not humans are exposed to them. Only when these processes threaten humans and their property do they become hazards, and if they occur they can become a disaster. A **natural disaster** is the loss of life, injuries, or property damage as a result of a natural event or process. A natural disaster could be as small as an individual's house damaged in a wildfire with losses handled through a private insurance carrier or a local community devastated by flooding that could require assistance from government agencies. A **natural catastrophe** is considered a massive disaster often affecting a larger region and requiring significant amounts of time and money for recovery. Communities along the Gulf of Mexico and in the Caribbean are prone to hurricanes—creating a hazardous condition. When a hurricane makes landfall, a disaster almost inevitable given the high populations in these areas and the number of people who will be impacted. New Orleans,

Louisiana, USA, was a city of almost 500,000 people, located along the edge of the Gulf of Mexico and built almost entirely below sea level, protected only by levees and floodwalls. In 2005, Hurricane Katrina hit New Orleans, flooding 80% of the city, causing $70 billion dollars (US) in damage, and killing more than 1,800 residents—a clear demonstration of how hazard can evolve into disaster and catastrophe.

Types and Characteristics of Natural Hazards

Natural hazards, and the disasters or catastrophes that result from them, come in a variety of forms. Natural hazards of terrestrial origin can be divided into three different categories: geologic, atmospheric (or hydro-meteorological), and environmental (or biological) (Table 1.1). These originate from the flow of energy and matter contained on or in our planet. Examples include geologic events such as earthquakes and volcanic eruptions, atmospheric events like hurricanes and tornadoes, and environmental events like wildfires and diseases. Natural hazards of extraterrestrial origin include the possibility of asteroid or comet impacts and solar-related events such as geomagnetic storms.

Hazards can also be categorized as **rapid-onset hazards** which expend their energy very quickly, such as volcanic eruptions, earthquakes, floods, landslides, thunderstorms, and lightning. These hazards can develop with little warning and strike rapidly. These are the hazards we hear about the most because of their violent effects and real-time media coverage. In contrast, **slow-onset hazards**, such as drought, insect infestations, disease epidemics, and climate change take years to develop and are usually neglected by the media as not newsworthy until long after the effects have started.

rapid-onset hazard

A hazard that develops with little warning and strikes rapidly. They expend their energy very quickly, such as volcanic eruptions, earthquakes, floods, landslides, thunderstorms, and lightning.

slow-onset hazard

A hazard that takes years to develop such as drought, insect infestations, disease epidemics, and global warming and climate change.

Table 1.1 Categories of Natural Hazards and Examples

TERRESTRIAL HAZARDS		
Geologic	**Atmospheric**	**Environmental**
• Volcanic eruptions • Earthquakes • Landslides • Land subsidence • Tsunamis • Coastal erosion • Floods	• Global warming • Tropical cyclones • Storms • Tornadoes • Droughts • Lightning • Blizzards	• Wildfires • Biological diseases • Insect infestations • Environmental pollution
EXTRATERRESTRIAL HAZARDS		
• Impacts from asteroids and comets • Solar flares • Geomagnetic storms		

Other types of hazards include man-made *technological hazards* that originate in accidental or intentional human activity, such as oil and chemical spills, building fires, plane crashes, and acts of terrorism. *Anthropogenic hazards* are also caused by humans, but they affect our environment and ecosystem, which eventually affects us in the long run. These include things such as pollution and deforestation that lead to global warming, ozone layer destruction, increased magnitude of storms and landslides, and other effects. We will explore some of the anthropogenic hazards that are linked to natural hazards and disasters.

Often, several hazards and processes can be linked to a single event and occur at the same time or be triggered by the main event. These multiple events must be taken into account in preparing in advance for a hazard. For example, along with strong winds, hurricanes are associated with intense precipitation that can cause flooding, erosion along the coast, and landslides on inland hill slopes. Volcanic eruptions can cause lahars (mudflows) and flooding, and if the volcano erupts or collapses into the ocean, it can cause a tsunami.

Natural hazards also frequently occur in a series, meaning they follow an initial onset. Each subsequent hazard acts upon the effects of the preceding event. For instance, a hurricane can down enormous stands of timber, an event that could result in a disastrous outbreak of detrimental insect populations and lead to increased wildfire risk in future years from the volume of dead wood. Likewise, during long periods of drought, wildfires that burn standing or downed timbers can denude hillsides of vegetation and thereby trigger soil erosion, flooding, and landslides.

Severe Weather

© Minerva Studio/Shutterstock.com.

Figure 1.4 **The tornado that hit Moore, Oklahoma, in 2013 was 27 km long, 2 km wide (17 by 1.2 mi) and had winds over 320 km per hour (200 mph). It was classified as an EF5; of about 1,000 tornadoes to hit the United States each year, only one on average reaches EF5 status.**

Severe weather events such as tornadoes, hurricanes, storms, and heat waves strike many areas of the United States each year (**Figure 1.4**). Due to changes in population demographics and the establishment of more complex weather-sensitive infrastructures, the United States is more vulnerable to severe weather events today than in the past. In 2017 the estimated cost of severe weather disasters in the United States was $300 billion.

Over the past 30 years, coastal population growth has quadrupled, along with accompanying property and infrastructure development, to the extent that more than 75 million people now reside along hurricane-prone coastlines. At least 1,464 people lost their lives in Hurricane Katrina (**Figure 1.5**, on page 8) and in the subsequent floods. The storm was also responsible for an estimated $70 billion (2005 US dollars) in damage, making it the costliest natural catastrophe in our history.

BOX 1.1 Case Study: Natural Hazards in the United States.

The extraordinary natural, climatic, and geographic diversity of the United States of America exposes people to a wide range of natural hazards. Natural hazards such as earthquakes, volcanic eruptions, wildfires, storms, and drought strike nearly every part of the United States, exacting an unacceptable toll on life, property, natural resources, and economic well-being (**Figure 1.1.1**). Each year, natural hazards cause hundreds to thousands of deaths and cost billions of dollars in disaster aid, disruption of commerce, and destruction of homes and critical infrastructure. The cost of major disaster response and recovery continues to rise, and property damage from natural hazards events doubles or triples each decade. The following discussion is a short summary of some common natural hazards in the United States intended to help illustrate the need for understanding these events and reducing their occurrence and effects.

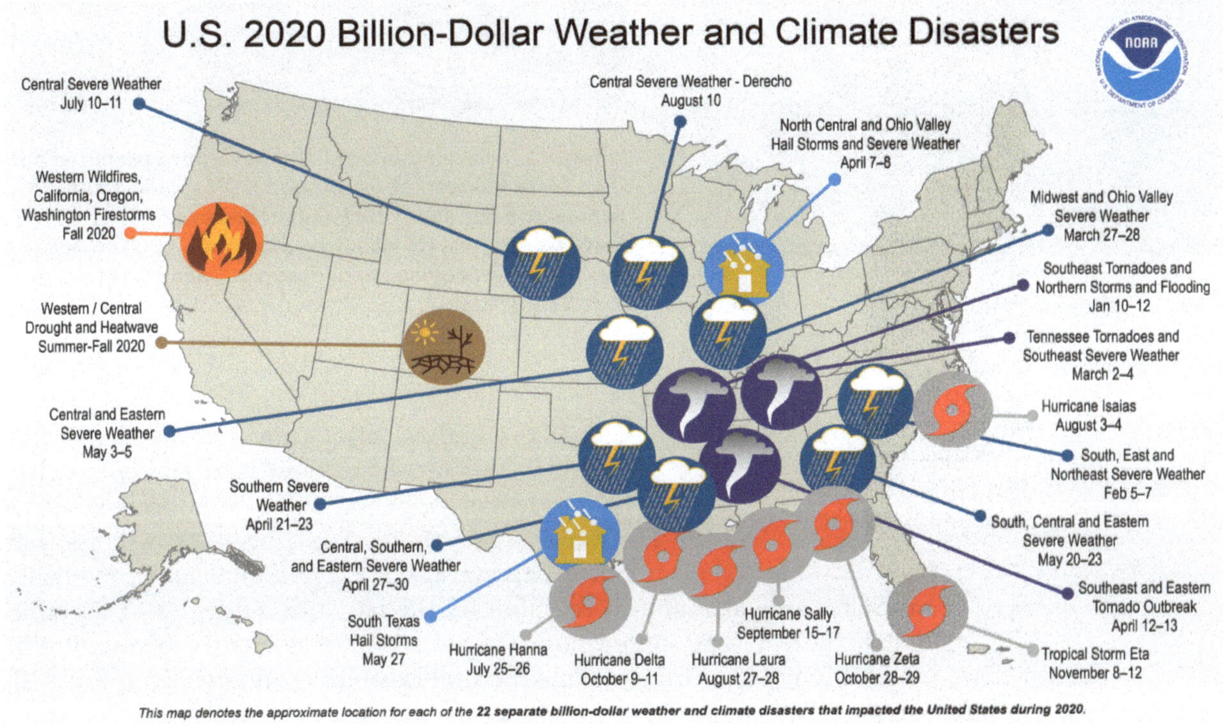

Source: FEMA.

Figure 1.1.1 In 2020, 22 separate billion-dollar weather and climate disasters occurred in the United States.

Tornadoes are more common in the United States than anywhere else in the world, with an average of 1,000 reported annually nationwide, resulting in an average of 80 deaths and over 1,500 injuries every year. The tornadoes that struck the Oklahoma City area on May 3, 1999, were some of the most devastating in history, destroying over 2,500 structures and causing over $1 billion in damage (**Figure 1.6**).

Rapid-onset hazards such as hurricanes and tornadoes strike with deadly force, but slow-onset hazards can also be deadly. Such is the case with heat waves. In July 1995, a heat wave in Chicago killed 739 people. From June to mid-August 2003 a severe heat wave struck continental Europe, killing more than 30,000 people.

Figure 1.5 Hurricane Katrina near peak strength on August 28, 2005. This developed into the most catastrophic storm to hit the United States in more than 80 years.

Figure 1.6 Several supercell thunderstorms produced more than 70 tornadoes that struck central Oklahoma, southern Kansas, and northern Texas, United States, on May 3, 1999. Forty-six people were killed and more than 8,000 homes and businesses were damaged or destroyed, at a cost of more than $1.2 billion.

Wildfires

Globally, wildfires, also known as wildland fires, engulfing millions of acres of land and encroaching on residential neighborhoods, have become all too familiar scenes (Figure 1.7). Since 2000 wildfires have burned an average of 6.5 million acres in the United States annually, and in recent years the nation has spent over $2 billion annually for fire suppression. This does not include costs to communities and individuals in terms of structural losses or economic disruptions. The year 2020 was the largest wildfire season for the United States, where record-breaking heat waves across the Western United States collided with lightning strikes from storms, setting off more than 650 fires just on August 24 alone (Figure 1.8). Aerosol from the smoke traveled across North America, reducing visibility and impacting human health across the continent. By the end of the fire season, almost 10,000 fires burned more than 4 million acres, including 5 of the top 10 largest wildfires in California's history. At their peak, there were around 3,400 firefighters plus personnel fighting the blazes, which ultimately killed 31 people.

Drought

Drought is a complex, slow-onset, nonstructural-impact natural hazard that affects more people in the United States than any other hazard, with annual losses estimated at $6 to $8 billion. Compared to all natural hazards, droughts are considered the leading cause of economic losses. In the year 2000, the Western United States entered what scientists believe is a megadrought—a prolonged drought period lasting 2 decades or longer, triggered by a combination of a naturally dry cycle and human-caused

© StratosBril/Shutterstock.com.

Figure 1.7 **Thousands of wildfires—9,639 in total—burned more than 4% of California's land area in 2020, including the world's first "gigafire," burning over 100 million acres. The fires destroyed over 10,000 structures and cost more than $12 billion (2020 US dollars), including $2 billion in fire suppression costs.**

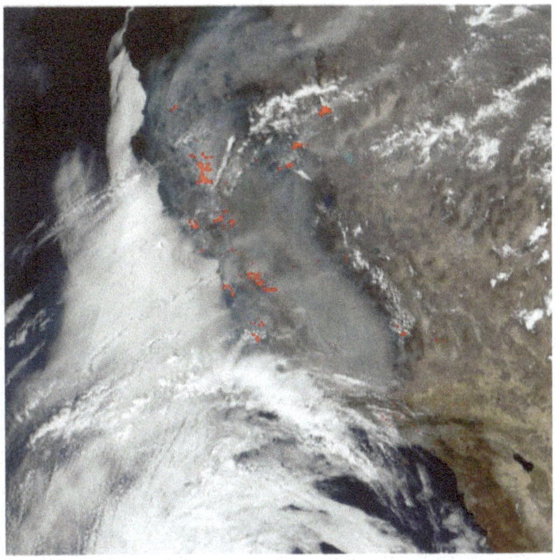

Source: NASA.

Figure 1.8 **August 2020 after lightning strikes ignited more than 650 wildfires.**

climate change. More than 60% of the Western United States is currently under severe, extreme, or exceptional drought conditions. Dr. Williams from Lamont-Doherty Earth Observatory of Columbia University recently published in *Science* a study of tree ring data from the Western United States that indicates that the last megadrought period was between the years 850 and 1600, and by comparing current conditions to past records, they estimate that 46% of the present drought's severity is due to climate change. The current megadrought is the second-driest period in 1,200 years of drought record, and may be just the start of the worst drought in modern history.

Earthquakes

Each year the United States experiences thousands of earthquakes. An average of seven earthquakes with a magnitude of 6.0 or greater (enough to cause major damage) occur each year. About 75 million people in 39 United States states face significant risk from earthquakes, and earthquakes remain one of the nation's most significant natural hazard threats. The last major earthquake to strike a large urban area was the Northridge, California, earthquake that occurred on January 17, 1994 (Figure 1.9). Fifty-seven people died as a result of the earthquake, and over 7,000 were injured. In addition, the earthquake caused an estimated $44 billion in damage and $49 billion in economic loss, making it one of the costliest natural disasters in United States history.

FEMA News Photo.

Figure 1.9 **The January 17, 1994, Northridge, California, earthquake. Numerous highways and approximately 114,000 residential and commercial structures were damaged and 57 deaths were attributed to the earthquake. Damage costs were estimated at $44 billion (US dollars).**

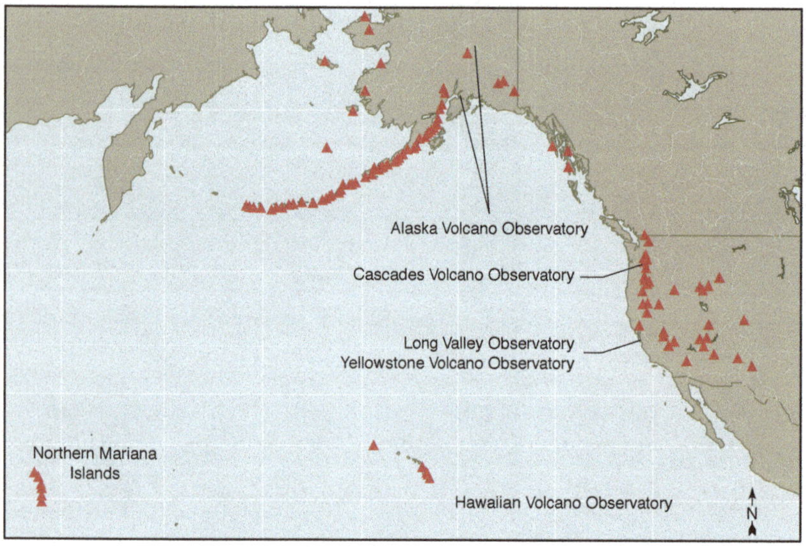

© Kendall Hunt Publishing Company.

Figure 1.10 The United States Geological Survey (USGS) is responsible for monitoring the nation's 170 active volcanoes (red triangles) for signs of unrest and for issuing timely warnings of hazardous activity to government officials and the public. This responsibility is carried out by scientists at the five volcano observatories operated by the USGS Volcano Hazards Program and also by state and university collaborators.

Source: USGS.

Figure 1.11 Mount St. Helens erupting at 8:32 a.m. on May 8, 1980, after a magnitude 5.1 earthquake triggered a collapse of the bulging upper part of the volcanic edifice. Fifty-seven people were killed or are still missing, including Dr. David Johnston, a geologist who was stationed at an observation point that morning.

Volcanoes

The United States is among the most volcanically active countries in the world with about 200 active or dormant volcanoes (Figure 1.10). During the past century, volcanoes erupted in Washington, Oregon, California, Alaska, and Hawaii, devastating thousands of square kilometers and causing substantial economic and societal disruption and loss of life. Since 1980, 45 eruptions and 15 cases of notable volcanic unrest have occurred at 33 volcanoes, producing lava flows, debris avalanches, and explosive blasts that have invaded communities, swept people to their deaths, choked major rivers, destroyed bridges, and devastated huge tracts of timber forests. Volcanic ash plumes ejected into the atmosphere can be a costly and serious danger to aircraft. A Boeing 747 sustained $80 million in damages when it encountered ash from Mount Redoubt in Alaska during a 1989 eruption (Figure 1.11).

Floods

Floods are the most frequent natural disaster in the United States and can be caused by hurricanes, weather systems, and snowmelt (Figure 1.12). Failure of levees and dams and inadequate drainage in urban areas can also result in flooding. Nearly 75% of federal disaster declarations are related to flooding, which, on average, kills about 140 people each year and causes $6 billion in property damage. In 1993, flooding in the Mississippi Basin resulted in an estimated $12 to $16 billion in damages across nine states in the Midwestern United States. An increase in population and development

FEMA, photo by Andrea Booher.

Figure 1.12 **Volunteers filled more than 300,000 sandbags in 1 day in order to build levees along the Red River in Fargo, North Dakota, as the river crested at 12.5 m (41 ft) in late March 2009.**

FEMA, photo by Adam DuBrowa.

Figure 1.13 **Roads in Cougar, Washington, were destroyed in February 2009 by landslides and mudslides produced by a series of several winter storms that came off the Pacific Ocean.**

in river floodplains and an increase in heavy rain events over the past 50 years have gradually increased the economic losses from floods.

Landslides

Landslide hazards occur and cause damage in all 50 states (Figure 1.13). Severe storms, earthquakes, volcanic activity, coastal wave attacks, and wildfires can cause widespread slope instability. Each year landslides in the United States cause $1 to $2 billion US dollars in damage and more than 25 fatalities. The May 1980 eruption of Mount St. Helens caused the largest landslide in history—large enough to fill 250 million dump trucks. Human activities and population expansion, however, are major factors in increased landslide damage and costs. Wildfires also contribute to increased landslide and mudslide activity as they reduce the vegetation that holds the surface in place.

Disease Epidemics

The spread of COVID-19 through the United States and across the world was the defining feature of 2020. COVID-19 is the disease caused by the severe acute respiratory syndrome SARS-CoV-2. United States government officials knew that a severe pandemic was possible and even plausible, yet when it came time to respond to the pandemic many parts of the United States were unnecessarily underprepared. The first reported case of COVID-19 in the United States was on January 20, 2020, from a man who had recently returned to his home in Washington State from visiting his family in Wuhan, China, which was later identified as the origin of the novel infectious coronavirus. On January 30, 2020, the World Health Organization warned that "all countries should be prepared for containment, including active surveillance, early detection, isolation and case management, contact tracing and prevention of onward spread" of the disease. The pandemic has led to a dramatic loss of life worldwide, and while analysis of individual country response and impact is ongoing, the effects are likely to be felt for years to decades.

Human Population Growth: Moving Toward Disaster

Complicating all natural hazards issues is the rapid growth of our world population (Figure 1.14), along with the goal of having a better standard of living which is heavily tied to natural resource use. Remember that a natural hazard only becomes a disaster when it affects people where they live or work. The increase we see compared to past natural disaster occurrences is due mostly to expansion of our population into hazardous areas which is occurring at a fast rate. The biggest population increases have been in cities and their suburbs (Figure 1.15). In the United States between 1950 and 2018, the urban population grew from 64% of the total population to 81%. The worldwide urban rate is only 54%, making the United States a very urbanized country. Densely populated cities can be easy targets for natural catastrophes as witnessed in New Orleans in 2005. Mexico City, with over 20 million people, is a megacity in a region of earthquake and volcanic hazards. The Southern and Western regions of the United States (areas prone to drought, wildfires, hurricanes, earthquakes, and mudslides) are expected to grow by 32 and 51%, respectively, by the year 2050.

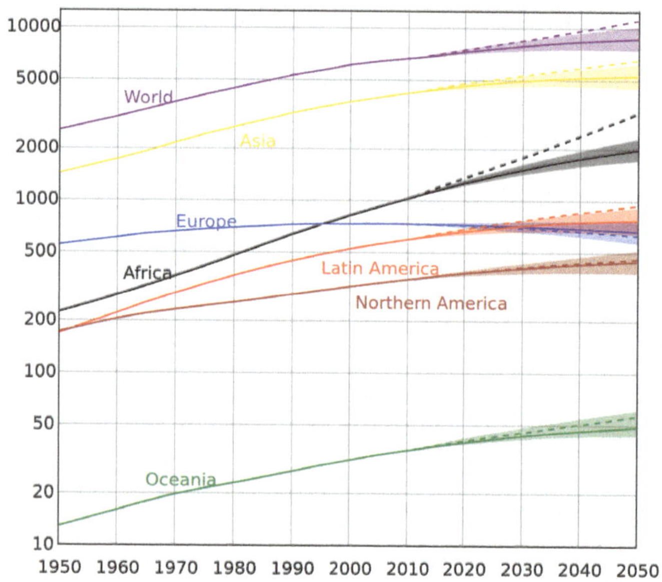

United Nations Population Center.

Figure 1.14 **Population growth curve showing different parts of the world. North America and Europe maintain a fairly constant growth while Asia and Africa have much higher growth rates.**

© JupiterImages Corporation.

Figure 1.15 **Cities throughout the world have grown rapidly in the past 50 years, using many more resources than in the past.**

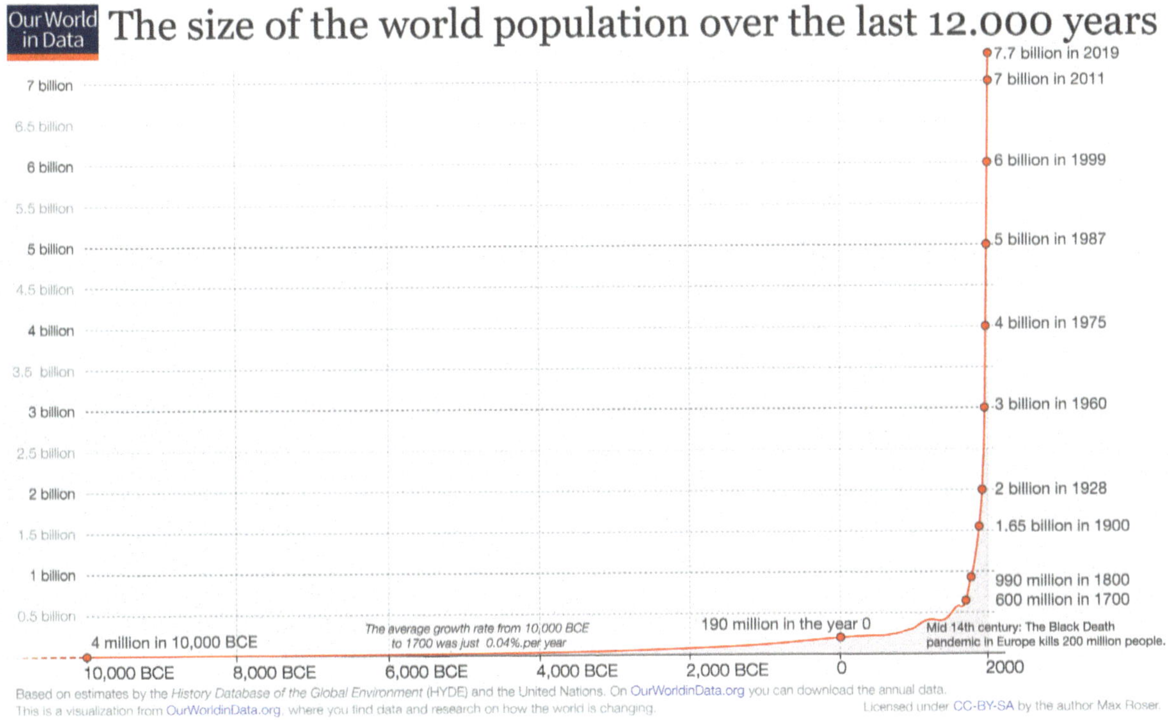

Source: Max Roser, Hannah Ritchie, and Esteban Ortiz-Ospina (2013)—"World Population Growth." Published online at OurWorldInData.org. Retrieved from https://ourworldindata.org/world-population-growth [Online Resource].

Figure 1.16 **The world population today is 1,860 times the size of what it was 12,000 years ago, and all of this growth happened very recently. Around the year 1800, the world population was about 1 billion, and has increased sevenfold since then.**

The increase in our world population by over 5 billion people during the past 200 years is without precedent in human history (**Figure 1.16**). The world population is estimated to have been only about 5 million people 10,000 years ago, but the nearly flat population curve began to rise about 8,000 years ago when agriculture and domestication of animals began to replace a hunter-gatherer culture. For thousands of years people lived in sparsely settled rural areas and growth rates were still very low, so population increased slowly but steadily to about 400 million by the Middle Ages (1100 to 1500 C.E. or common era). A sharp drop in population occurred when the Black Death struck Europe in the mid-1300s, but by 1700 the world population rebounded to about 650 million. Since that time, **exponential growth** in world population reached the world's first billion by the early 1800s, 2 billion by 1930, 3 billion by 1960, and 6 billion by 2000 (**Figure 1.17**). It is estimated that the world population will reach 9.8 billion by 2050.

exponential growth
Growth in which some quantity, such as population size, increases by a constant percentage of the whole during each year or other time period; when the increase in quantity over time is plotted, this type of growth yields a curve shaped like the letter J.

The rapid increase in human population is causing serious shortages of resources, including oil, food, and water, and is heavily degrading the environment. Some resources are renewable, such as water, whereas many others such as fuels (oil, natural gas, coal) and minerals (ores for copper, iron, etc.) are not. From population and natural resource data it becomes clear that it is impossible in the long run to support exponential population growth with our finite resources. Scientists are worried that it will be impossible to supply resources and a high-quality environment for the billions of people who will be added to the planet in the 21st century. Some scientists suggest that the present population is already above our planet's **carrying capacity**, which is the maximum number of people the Earth can hold without causing environmental degradation that reduces the ability of the planet to support the population.

carrying capacity
The maximum population size that can be regularly sustained by an environment.

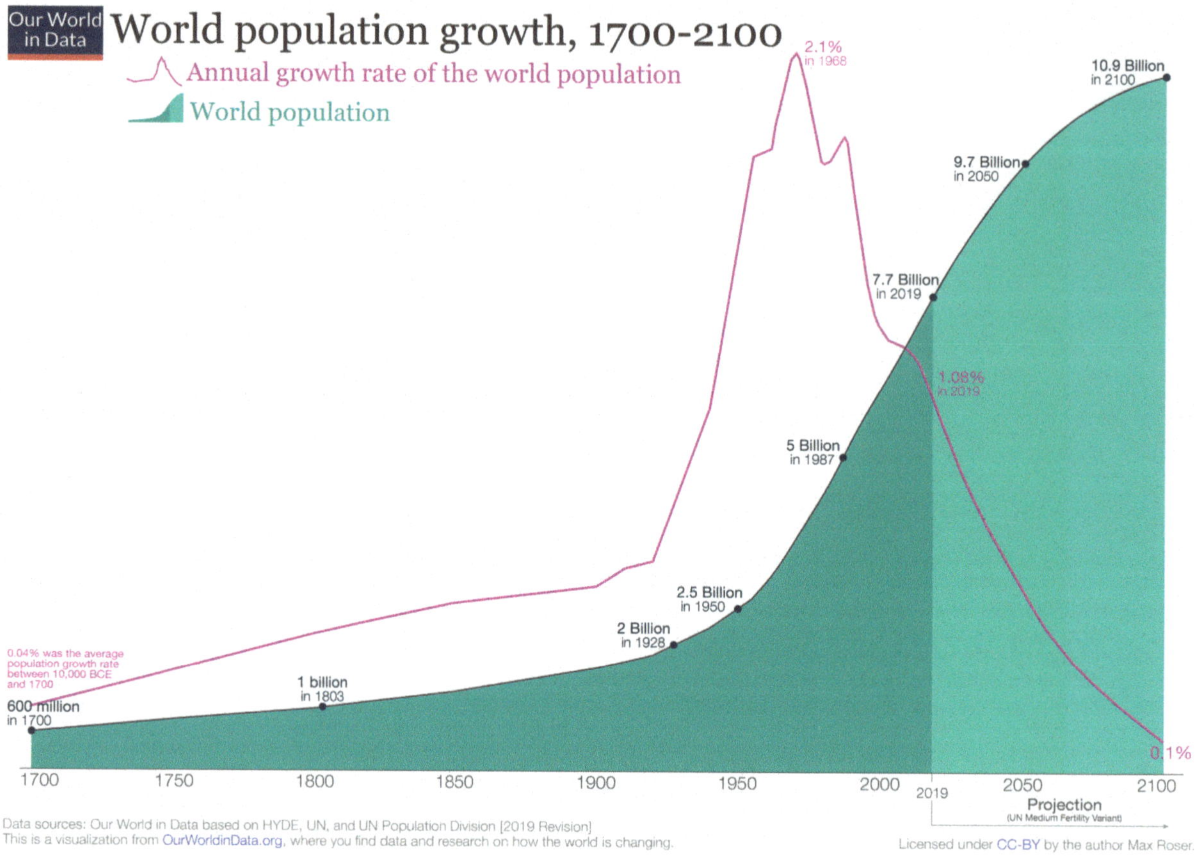

Source: Max Roser, Hannah Ritchie, and Esteban Ortiz-Ospina (2013)—"World Population Growth." Published online at OurWorldInData.org. Retrieved from https://ourworldindata.org/world-population-growth [Online Resource].

Figure 1.17 **Over the last 100 years, global population has more than quadrupled. We are living through the steepest increase in the population curve. Every year 140 million people are born and 58 million people die, adding 82 million more people to the world. Peak population growth of 2.1% was reached in 1968, and has declined since then. As population growth continues to decline, the curve representing world population gets less steep. If UN predictions of future population in the year 2100 are accurate, the world population will have increased by more than tenfold over the span of 250 years.**

Compounding the problem is the fact that technologically advanced societies consume greater quantities of resources that are only partially deemed essential, with the remaining consumed for convenience or luxury. Therefore, to reach our goal of achieving a higher standard of living, our resource consumption increases substantially per person, not to mention the vast quantities of wastes generated. Human links to environmental degradation, due to our thirst for using more resources, can greatly increase the risks of natural hazards. For example, deforestation for wood production and land development, which may be critical for economic development, can lead to more landslides and flooding. Therefore, population growth and resource consumption leads to more loss of life and property from individual hazard events, as well as more hazards or higher-magnitude hazards due to resource development.

One of the obvious ways to reduce natural disasters and catastrophes would be population control, for which there is no easy answer. Left unabated, some scientists predict that population growth will take care of itself through wars (mostly over resources such as land, fuel, minerals, water, etc.) and other catastrophes such as famine, disease, and ecosystem collapses. Others believe we will find better ways to control population through increased education, improved regulation of resources and space, and being environmentally friendly.

Impacts From Natural Hazards: Human Fatalities, Economic Losses, and Environmental Damage

Human Fatalities

Increasing world populations in urban areas and coastal regions has resulted in more people living in hazardous areas. As a result, more than 255 million people globally were affected by natural disasters each year between 1994 and 2003. During the same period, these disasters claimed an average of 58,000 lives annually, with a range of 10,000 to 123,000.

Table 1.2 Five Deadliest Natural Disasters, 2000 to 2022

Date	Event	Region	Overall Losses	Fatalities
December 26, 2004	Earthquake, tsunami	South Asia		230,000
January 12, 2010	Earthquake	Haiti	$14 billion	160,000
May 2, 2008	Cyclone	Myramar	$4 billion	138,000
May 12, 2008	Earthquake	Sichuan, China	$29 billion	88,000
October 8, 2005	Earthquake	Kashmir, Pakistan	$5 billion	>87,000

The five deadliest disasters since the year 2000 are listed in Table 1.2 and the 10 deadliest historical disasters are listed in Table 1.3. Because many of these events occurred in areas that had inadequate communication and reporting capabilities, the fatality numbers are often not precise and can vary. Notice that as deadly as the 2004 Asian disaster was, it was not the deadliest in human history. Also notice that the greatest disasters occurred where human population density is high in a belt of poorer countries running from China and Bangladesh through India, northwestward into Iran and Turkey. Population growth, urbanization, and the inability of poor populations to escape from the vicious cycle of poverty make it more likely that there will be a continued, and increased, number of people who are vulnerable to natural hazards.

Table 1.3 Ten Deadliest Natural Disasters

Rank	Event	Location	Date	Death Toll (Estimate)
1	1931 Yellow River flood	Yellow River, China	Summer 1931	850,000–4,000,000
2	1887 Yellow River flood	Yellow River, China	September–October 1887	900,000–2,000,000
3	1970 Bhola cyclone	Ganges Delta, East Pakistan	November 13, 1970	500,000–1,000,000
4	1938 Huang He flood	China	1938	500,000–900,000
5	Shaanxi earthquake	Shaanxi Province, China	January 23, 1556	830,000
6	1839 India cyclone	Coringa, India	November 25, 1839	300,000+
7	1642 Kaifeng flood	Kaifeng, Henan Province, China	1642	300,000
8	2004 Indian Ocean earthquake/tsunami	Indian Ocean	December 26, 2004	225,000–275,000
9	Tangshan earthquake	Tangshan, China	July 28, 1976	242,000*
10	1138 Aleppo earthquake	Syria	1138	230,000

* Official government figure. Estimated death toll as high as 655,000.

Economic Losses

The deaths and injuries caused by natural disasters are what grab our immediate attention, but there are economic losses as well that are increasing at a staggering rate (Figure 1.18). Since 1980 there have been hundreds of natural disasters in the world resulting in over $500 billion US in damage and thousands of deaths. The economic cost associated with natural disasters has increased fourteenfold since the 1950s.

The list of most expensive events is dominated by hurricane storm events along coastal regions (Table 1.4). Tables 1.2 and 1.4 show that poorer countries have significantly greater number of deaths, whereas wealthier countries have fewer deaths but much larger economic loss. In both developed and undeveloped countries population growth once again has led to the increases, with more people living in more dangerous places. However, in developed countries the number of deaths has not increased due to better prediction, forecasting, and warning systems, as well as safer buildings. Disasters cause the loss of property and lives wherever they occur, but more highly developed countries usually lose more property while poorer countries always lose more lives. Scientific predictions and evidence indicate that global climate change will increase the number of extreme events, creating more frequent and intensified natural hazards such as floods, storms, hurricanes, and droughts.

Source: Swiss Re Institute.

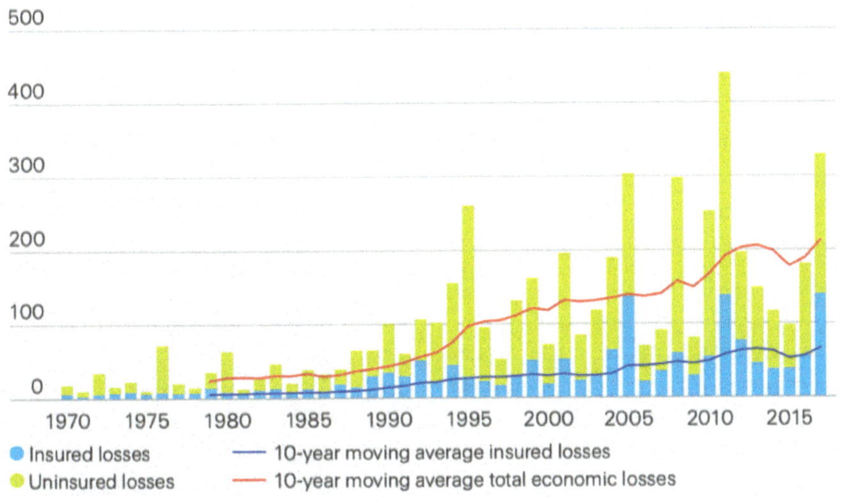

Figure 1.18 Global total and insured losses.

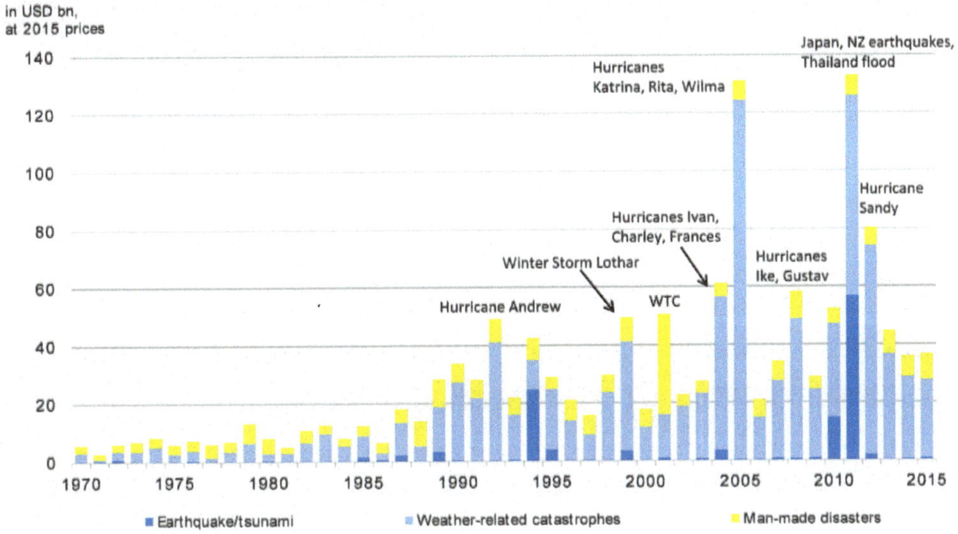

Table 1.4 **10 Costliest Natural Disasters**

Date	Loss Event	Region	Cost ($billions)*	Fatalities
2011	Tohoku earthquake and tsunami	Japan	$424	19,774
1995	Great Hanshin earthquake	Japan	$345	6,434
2008	Sichuan earthquake	China	$180	87,587
2005	Hurricane Katrina	USA	$167	1,836
2017	Hurricane Harvey	USA	$133	107
1988–1989	North American drought	USA and Canada	$116	4,800 to 17,000
2017	Hurricane Maria	North America	$98	8,498
2012	Hurricane Sandy	North America	$79	233
2019–2020	Bushfires	Australia	$72	451
2017	Hurricane Irma	North America	$69	134

* In 2021 US dollars.

Environmental Damage

© 1082492116/Shutterstock.com.

Figure 1.19 **Invasive species such as the zebra mussel can be spread during natural disasters.**

Natural hazards not only affect humans, but also affect and shape the environment and biodiversity. In August 1992 Hurricane Andrew had a strong economic effect not only on the city of Homestead, Florida, but also on the ecosystem of the nearby Everglades and other coastal waters. Despite the destruction in Homestead, timber losses during the winter storms of 1993–1994 cost insurers more than the total payout from Hurricane Andrew. The drought in 1988 in the Midwest and the 1993 floods both had impacts on riverine ecosystems as well as downstream coastal ecologies of the Gulf of Mexico, changing water salinity and adding toxins and other pollutants washed out upstream by the floods. The floods also accelerated the spread of invasive species, like the Zebra mussels, in the Mississippi watershed (Figure 1.19). These examples show that strategies for protecting renewable resources and biodiversity must be taken into account when looking at the effects of natural hazards.

Human Vulnerability and Natural Hazards

Each year natural hazards are responsible for causing significant loss of life and staggering amounts of property damage, and they continue to adversely affect millions of people every day. Statistics published by the International Strategy for Disaster Reduction, a subsidiary of the United Nations, show that there is an increasing trend of natural disasters, especially over the last several decades. However, scientists don't see Earth as becoming more violent; instead, because of population and land management trends in our society, humans are increasingly becoming more *vulnerable* to the natural hazards, and more people are being impacted by these events when they do occur. A summary of some reasons for the increase in disaster and catastrophe losses includes:

- Global population growth. In 1960, for example, there were 3 billion people living on Earth, and by 2000 there were 6 billion—a doubling in only 40 years. In 2011 the Earth's population reached 7 billion people.
- Increased settlement and industrialization of hazardous lands subject to floods, landslides, hurricanes, earthquakes, wildfires, volcanoes, and other hazards.
- Concentration of population and values in conurbations (when towns expand sufficiently that their urban areas join up with each other) and the emergence of numerous megacities—even in hazardous regions (e.g., metropolitan Tokyo, with more than 37 million inhabitants).

- The increased demand for natural resources and rising standard of living in nearly all countries of the world produces growing accumulations of wealth and the need for increased insurance. Losses thus escalate in the event of a disaster or catastrophe.
- The vulnerability of modern societies and technologies in the form of structural engineering, lifelines (water, sewer, electrical, etc.), transportation services, and other networks that are both fragile and costly to repair when damaged by natural hazards. Such effects are mostly due to lack of scientific understanding or education and lack of awareness of the hazards.
- Global changes in environmental conditions, climate change, water scarcity, and loss of biodiversity, mostly due to human activities and interactions with natural hazards.
- Globalization of the world economy now makes us all vulnerable to disasters wherever they occur, for example, by shutting off supplies of goods.

Therefore, natural hazard events are only classified as disasters or catastrophes when people or their properties are adversely affected. An earthquake in the Gobi Desert or gales in the Antarctic are not natural disasters if they do not have any impact on people. Conversely, a natural event that is considered to be a normal scale event in some places, such as the natural flooding of a stream, can quickly become a disaster in a region that is densely populated and poorly prepared.

Studying Natural Hazards: Fundamental Principles for Understanding Natural Processes

People are usually surprised to learn that every day the Earth experiences earthquakes, volcanic eruptions, landslides, floods, fires, and storms. Natural disasters are often perceived as being out of our hands with little causal relationship to human activities. However, a growing body of evidence points to the effects of human behavior on the global natural environment and on the possibility that certain types of natural disasters, such as floods, may be increasing as a direct consequence of human activity. Geologic events that cause natural hazards and disasters, which the geologic record shows have been continuously shaping our planet for hundreds of millions to billions of years, are not problems to be solved but are a critical part of how the Earth functions. Ecosystems and individual species have evolved to coexist with these hazards. Natural hazards become disasters only when natural forces collide with people, structures, or other property. Understanding how the Earth works and our relationship with it is the first step in reducing the impacts of natural hazards. The following concepts serve as a basic framework for our study and understanding of different natural hazards discussed in later chapters.

The Earth System and the Human Connection

Earth itself is a complex and dynamic planet, always in motion, and not static or unchanging. Earth's major components of land, water, air, and life do not exist as isolated entities, but are interconnected to form a unified dynamic whole. Even though we can study each component individually, each is related to the continuously interacting entity that we call the *Earth system*. We observe the most obvious connection between Earth's components when natural disasters strike. Events such as destructive volcanic eruptions, disastrous landslides, devastating earthquakes, tsunamis, hurricanes, and floods affect people in obvious ways. Even though we cannot prevent these natural events from happening, the more we study and understand how they operate, the more we will be able to predict them and mitigate their effects. Scientists who study the interconnections of Earth's components use an interdisciplinary approach known as **Earth system science**, which aims to study our planet as a system composed of numerous interconnecting subsystems governed by natural laws.

Earth system science
Study of our planet as a system composed of numerous interconnecting subsystems governed by natural laws.

A system is any assemblage or combination of interacting components that form a complex whole. A system is a group of any size composed of interacting parts within which we are interested in studying the flow of energy and matter. In our home we can talk about the heating system, electrical system, or plumbing system. Each functions as an independent unit, transferring material and energy from one place to another, and each is driven by a force that makes the system operate. A good example of a system is an automobile with its various components (or subsystems), such as the engine, transmission, steering, brakes, and radiator, that all function together as a whole when energy is supplied (gasoline or electricity). The many different component parts mutually adjust to function as a whole, with changes in one part bringing changes in the others. For example, if the oil pump in the car malfunctions and causes less oil lubricant to enter the cylinders, then increased friction will cause more heat to be generated in the engine, in turn causing the cooling system to adjust to the need for additional heat removal.

The overall nature of the Earth system is similar to an open system because it exchanges both energy and matter with its surroundings. Earth receives energy from the Sun and radiates it back into space, along with energy that leaves the Earth's interior. However, most of the Earth's materials such as rocks and water are neither gained nor lost to space but are continuously transferred and recycled within the Earth system. Thus, we most often view Earth as a self-contained **closed system** with regard to matter, even though some meteors fall to Earth and small amounts of gas escape into space. The solar energy that enters this closed system causes air and water to move and flow in patterns that can be hazards, such as hurricanes and wind storms, while heat energy from within the Earth causes motions that result in earthquakes, volcanic eruptions, and drifting continents.

closed system
A system in which no matter or energy can leave or enter from the outside.

Human Activities and Hazards

Natural systems tend toward balance among opposing factors or forces. The interactions of these energy sources produce and maintain a balanced state called **dynamic equilibrium**. A change in one part of the system will tend to become balanced by a change or changes in another part of the system to maintain overall equilibrium. Therefore, this is not a static equilibrium but a dynamic one, and human activities can cause or accelerate changes in natural systems (**Figure 1.20**).

dynamic equilibrium
The state in which the action of multiple forces produces a steady balance, resulting in no change over time.

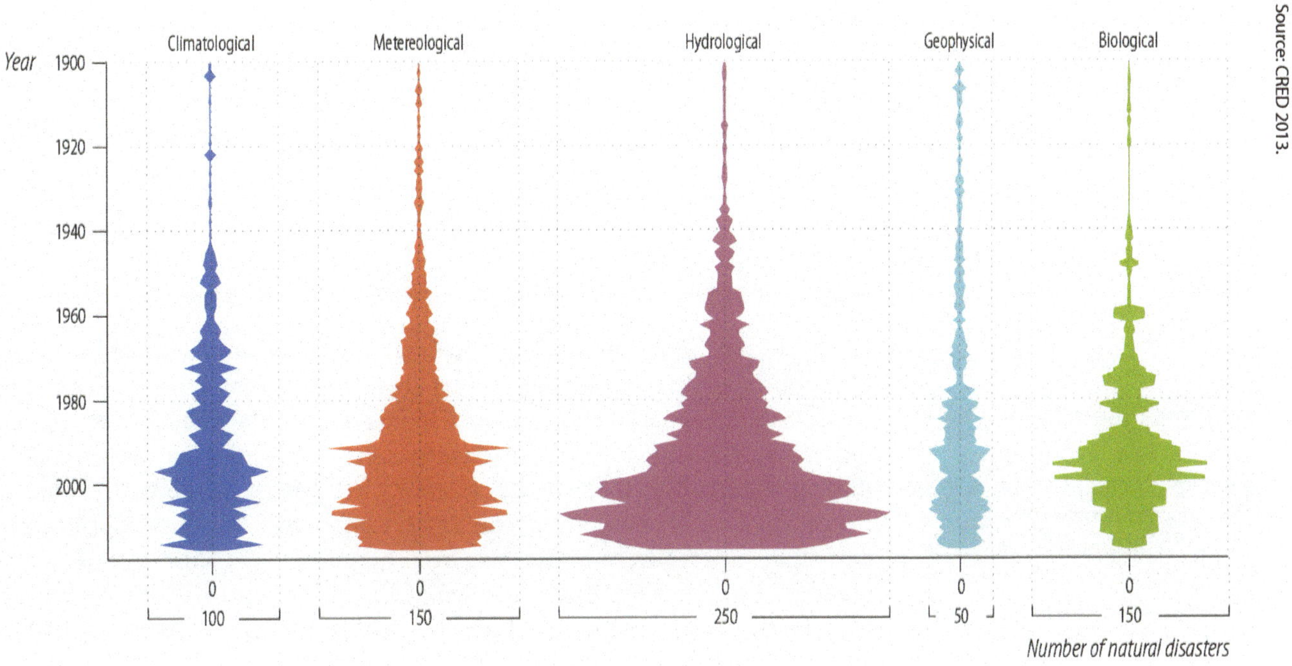

Figure 1.20 Between 1901 and 1910 there were 82 recorded natural disasters, whereas between 2003 and 2012 there were more than 4,000. The main types of disasters shown here include extreme temperatures and droughts (climatological), storms (meteorological), floods (hydrological), earthquakes (geological), and epidemics (biological).

The Earth's land surface is an important resource for people, plants, and the other animals with which we share the planet. Humans use land for housing, transportation, agriculture, manufacturing, mineral and energy production, deposition of waste products, and aesthetics. The increase in human population has increased the demand for land use, such as for urban and agricultural purposes. This has led to an increasing rate of human-induced change to our planet's surface. Human activity, including agriculture, urbanization, and mining, is now at the level where we move as much soil and rocks on an annual basis as do natural processes such as river transport. It appears that human activities are the most significant processes shaping the surface, which in turn upsets the dynamic equilibrium of Earth's systems and produces undesired hazards. For example, changes such as altering the steepness of hillslopes and removing vegetation can lead to unwanted increases in landslide and flooding events. Deforestation can lead to soil erosion and increase landslides and flooding.

The impact of humans on the environment is broadly proportional to the size of the population as well as technological advances. For example, air pollution can lead to the slow onset of hazards. The exhaust from one automobile pollutes only the air in its immediate vicinity, which can be dispersed through the atmosphere with negligible global impact. However, the collective exhaust from millions of automobiles has a global impact. Burning of fossil fuels over the past 250 years of increasing industrialization has caused measurable increases in atmospheric pollutants, such as the greenhouse gas CO_2, leading to global climate change. Next to human population growth, climate change has the greatest impact on the environment. Global warming will cause sea levels to rise as glaciers melt, leading to increased coastal erosion. Warming of the oceans will produce a warmer atmosphere and increase the **frequency** and severity of weather-related thunderstorms, tornadoes, and hurricanes. Changes in climate patterns can affect

frequency
Occurrence of specific events. Used in interpreting the past record of events to predict occurrences of that event in the future.

food-producing areas and increase precipitation in some areas producing flooding and landslides, while expanding desert areas in others causing droughts. Many of these changes could lead to population shifts, which could initiate wars or social and political upheavals. We must not forget that humans are part of the Earth system and that our activities can produce changes with potential consequences for us and our planet.

Time Perspectives: Geological and Historical Records

geologic time scale

A relative time scale based upon fossil content. Geological time is divided into eons, eras, periods, and epochs.

Geology is the scientific study of the Earth and its history. One of the important concepts that sets geology apart from most other sciences is the consideration of vast amounts of time. Geologists use the **geologic time scale** in describing the timing and relationships between events that have occurred during the history of the Earth (**Figure 1.21**). Many of the natural hazard processes have been operating on Earth long before humans populated the planet. Geologically, humans have been here for a very short period of time compared to the 4.6 billion year age of Earth. The most primitive human-type fossil remains discovered are 4.4 million years old. Modern humans (*Homo sapiens*) evolved about 200,000 years ago and migrated out of their African birthplace as recently as 80 thousand years ago to occupy Europe and beyond. On a geologic time scale, this is a very short time for humans to populate the Earth, less than 0.06% of the age of the Earth. Needless to say, humans have had an enormous impact on the surface of the Earth—an impact that is far out of proportion to the length of time we have occupied the planet.

The geologic record is full of examples of disasters that humans have never witnessed during historical times, such as, asteroid collisions large enough to cause the extinction of the dinosaurs, and violent caldera eruptions such as that of Yellowstone volcano, United States. Knowing that these hazards happened in the past means they certainly will happen again, and we should not be surprised when they do. Studying the events of the geologic past is vital to understanding the frequency with which we can expect hazardous events to occur.

The Scientific Method in Studying Earth's Hazards

scientific method

A systematic way of studying and learning about a problem by developing knowledge through making empirical observations, proposing hypotheses to explain those observations, and testing those hypotheses in valid and reliable ways.

hypothesis

A supposition or proposed explanation made on the basis of limited evidence as a starting point for further investigation.

Natural hazards, as the term implies, are a natural phenomenon and so can be studied using scientific principles. The Earth system is governed by natural laws that provide the keys to understanding the landscapes we live on and the processes that formed them. Our understanding of natural processes operating on Earth today and in the past has been influenced by the use of the scientific method for studying natural phenomenon.

The **scientific method** is based on logical analysis of collected data and observations to solve problems. From initial observations, scientists develop a tentative explanation to explain the cause, or "why," of the phenomenon being studied, which is known as a **hypothesis**. To see if the hypothesis is an explanation of the observations, it must be tested with new observations and experiments. A stated hypothesis should always be testable since the science of discovery evolves through continual testing with new observations

Era	Period	Epoch	Approximate Ages (in millions of years)
Cenozoic	Quaternary	Holocene	0
		Pleistocene	3
	Neogene	Pliocene	5
		Miocene	23
	Paleogene	Oligocene	33
		Eocene	56
		Paleocene	66
Mesozoic	Cretaceous	Late Early	144
	Jurassic	Late Middle Early	208
	Triassic	Late Middle Early	245
Paleozoic	Permian	Late Early	286
	Carboniferous Pennsylvanian	Late Middle Early	320
	Mississippian	Late Early	360
	Devonian	Late Middle Early	408
	Silurian	Late Middle Early	438
	Ordovician	Late Middle Early	505
	Cambrian	Late Middle Early	545
Precambrian	Locally divided into Early, Middle, and Late		3,900 +

Figure 1.21 The Geologic time scale.

and experiments. Alternate hypotheses are usually developed to test other potential explanations for the observed phenomenon, which are also tested. If the observations and experiments are inconsistent with a hypothesis, then they are rejected as an explanation for the phenomenon, or the hypothesis is revised and new tests are developed. If a hypothesis continues to be supported by continued testing over a long period of time, it becomes known as a **theory**. A theory, then, is an explanation for some natural phenomenon that has a large body of supporting evidence and has been tested for validity through the scientific method.

theory
A comprehensive explanation of a given set of data that has been repeatedly confirmed by observation and experimentation and has gained general acceptance within the scientific community.

Continued testing of a theory that is irrefutably correct leads to the statement of a **law** (or **principle**). Most disciplines have laws, such as those in biology, chemistry, economics, geology, and physics.

law (or principle)
The unvarying sequence of a set of naturally occurring events.

Hazard Mitigation and Management: Reducing the Effects From Natural Hazards

The effects of natural disasters will continue to increase unless concerted efforts are made to reduce them. At stake is our physical, economic, and social well-being. But these are increasingly at risk as the world population grows and concentrates in hazard-prone areas. Large numbers of buildings, critical facilities, and lifelines (utilities) remain or are still being constructed in areas that are vulnerable to natural hazards. Disasters and catastrophes from Earth's natural hazards cannot always be prevented, although some are avoidable with foresight and planning. Most of the planet's seemingly deadly impact natural events are largely the result of people making bad choices concerning where to live, deliberate placement in harm's way, or failure to see and/or act on warning signs. The main goal of preventing or avoiding a natural disaster or catastrophe is to understand and assess the natural processes at work that lead to natural hazards. Human awareness through education is a vital step toward knowing and understanding the risks involved.

Before a Natural Hazard Strikes

When disasters strike, we find that areas that had properly prepared for the hazard suffered fewer losses and recovered more quickly than those that did not have preparedness programs in place. Being prepared requires seven steps: (1) hazard process research and development, (2) hazard identification, (3) risk assessment, (4) risk communication, (5) mitigation, (6) prediction, and (7) preparedness, described as follows:

1. Hazard process research and development. The science activities dedicated to improving understanding of the underlying processes and dynamics of each type of hazard. This includes fundamental and applied research on geological, meteorological, epidemiological, and fire hazards; development and application of remote-sensing technologies, software models, infrastructure models, and organizational and social behavior models; emergency medical techniques; and many other science disciplines applicable to all facets of disasters and disaster management.

Knowledge of how hazards operate is the fundamental first step in understanding how to reduce impacts to humans and the environment. Basic research leads to fundamental breakthroughs in understanding natural processes, such as the theory of plate tectonics that greatly improved our knowledge of volcanic and earthquake hazards. Long-term scientific studies of coupled ocean–atmosphere oscillations enabled NOAA in 1997 to predict a powerful El Niño during 1997–1998 that produced large rainfall-induced landslide events in California. NOAA and USGS worked with the California Office of Emergency Services to prepare landslide hazard maps showing former landslide deposits and debris-flow source areas along with dangerous rainfall thresholds for the San Francisco Bay area. This prediction allowed individuals and emergency responders to prepare for the extreme hazard event. No loss of life occurred from landslides in this area, as compared to 25 deaths in comparable storms from an El Niño event in 1982.

2. Hazard identification. Determining which hazards threaten a given area. This includes understanding an area's history of hazard events and the range of severity of those events. The continuous study of the nation's active faults, seismic risks, and volcanoes is included in this category, as are efforts to understand the dynamics of hurricanes, tornadoes, floods, droughts, and other extreme weather events.

Since natural disasters are repetitive events caused by natural processes, we can identify potential hazards and assess their probability of occurring in the area. Some events may not occur more than once in a person's lifetime or even for several generations. Also, many events such as volcanic eruptions—with hundreds to thousands of years elapsing between eruptions—are not on the same timeline as humans (average life expectancy of 72.6 years), thus we often do not see the danger. Therefore we become complacent by not knowing there is a hazard until disaster strikes. It is important, therefore, to take our working knowledge of natural hazards (from hazard process research) and identify their potential occurrences. Also known as *hazard assessment*, hazard identification consists of determining *where* and *when* hazardous events occurred in the past and how they affected the area.

Hazards vary widely in their power, and the extent to which hazards impact an area is partly a function of their **magnitude**, the energy released, and the interval between occurrences, or frequency. All other factors being equal, larger-magnitude events occur less frequently but cause more damage than those of a smaller magnitude. The **recurrence interval** is how frequently, on average, an event of a certain magnitude occurs. This information is important and useful to planners and public officials responsible for making decisions in the event of a possible disaster.

magnitude

The size or scale of an event, such as an earthquake.

recurrence interval

A statistical expression of the average time between events equaling or exceeding a given magnitude. It also refers to the average time interval, usually in years, between the occurrence of events of a given magnitude or larger.

Geologists are well trained to consider the long- and short-term history of an area and assess natural hazards. Geologists understand that if a destructive event occurred in this area before, then it could possibly happen again. Therefore, hazard assessment studies usually try to: (1) quantify the recurrence of each hazard; (2) increase the amount of detail available about the location, frequency, and magnitude of the physical and biological effects on maps; and (3) ensure that assessments consider possible links between hazards (for example, the triggering of floods by hurricanes or an increase in wildfires during a drought).

3. Risk assessment. Determining the impact of a hazard or hazard event on a given area. This includes advanced scientific modeling to estimate loss of life, threat to public health, structural damage, environmental damage, and economic disruption that could result from specific hazard event scenarios. Risk assessment takes place both before and during disaster events.

A risk assessment tells us something about the loss expected from each hazard, which elements specifically are at risk (population, property, natural resources, environment), and how easily the various elements can be damaged by each hazard. Risk assessments include estimates of the overlapping vulnerability of new technologies and the way and extent to which human actions may influence the risk. Depending on the vulnerability of individuals or a community to a hazard of a certain magnitude, risk is essentially a hazard considered in terms of its recurrence interval and expected costs. Thus the greater the hazard, the shorter its recurrence interval; the more vulnerability, then the greater the risk. The higher the risk, the more critical that the hazard-specific vulnerabilities should be targeted by mitigation and preparedness efforts. However, if there are no vulnerabilities there will be no risk—for example, an earthquake occurring in a desert where no one lives.

4. Risk communication. Public outreach, communication, and warning at every stage of hazard management. Risk communication includes raising public awareness and instigating behavioral change in the areas of mitigation and preparedness; the deployment of stable, reliable, and effective warning systems; and the development of effective messaging for inducing favorable community response to mitigation, preparedness, and warning communications.

Once a risk has been identified, people who live and work in hazard-prone areas need to know the nature and probable impact of natural hazards and what they can do to protect themselves before as well as during and after a natural disaster occurs. This information is conveyed through schools, special television programs, publications, workshops, and many other channels. The challenge here is to make sure that accurate information is presented by highly credible sources and understood by the public, leading to preventative and precautionary actions. Federal agencies, working with state and local governments, usually try to promote the importance of disaster prevention and mitigation programs.

5. Mitigation. These are sustained actions taken to reduce or eliminate the long-term risk to human life and property from hazards based on hazard identification and risk assessment. Examples of mitigation actions include planning and zoning to manage development in hazard zones, stormwater management, fire fuel reduction, acquisition and relocation of flood-prone structures, seismic retrofit of bridges and buildings, installation of hurricane straps, construction of tornado-safe rooms, and flood-proofing of commercial structures.

Once the risk for each hazard is known, mitigation efforts attempt to prevent hazards from developing into disasters or to reduce the effects of disasters when they occur. Thus they are designed to reduce our vulnerability to a natural hazard. The mitigation phase differs from the other phases of response and recovery because it focuses on long-term measures for reducing or eliminating risk. From one disaster to another, we are now seeing that the benefits of mitigation before an event greatly affect the losses after the event, and that we need to move more in this direction to reduce the impacts of hazards. Like the adage "an ounce of prevention is worth a pound of cure," *mitigation is the most cost-efficient method for reducing the impact of natural hazards*. However, progress toward this end has been slow. The implementation of mitigation strategies can also be considered a part of the recovery phase if applied after a disaster occurs; however, any actions that reduce or eliminate risk over time are still considered mitigation efforts.

Mitigation measures can be divided into structural methods (engineering solutions) or non-structural methods (land use management) depending on the type of hazard risk. *Structural methods* include building codes that specify materials, construction techniques that minimize damage to structures during a hazard event, and construction of large-scale structural interventions such as flood levees and seawalls. Historically, we have mostly employed large engineering projects in order to try to control natural processes for the protection of lives and property. *Nonstructural methods* include restrictive zoning, relocation, and land abandonment. Zoning laws can greatly reduce loss of lives and structures if properly implemented and enforced. Zoning can limit population density in potentially hazardous areas, as well as specify the type of structures permitted in a hazardous area. However, in many hazard-prone areas, housing developments, utilities, schools, fuel storage tanks, and hospitals still sprawl across the land (for example, across active fault lines in earthquake-prone parts of California). This construction occurred because the hazard was not realized accurately at the time, or the potential risk was glossed over. Now, corrective measures are too late and will have to wait until the recovery phase after an event, so in the meantime structural measures are usually employed.

In the long run, the best solution may be to abandon the use of certain hazardous areas and adapt them for public use by constructing parks and athletic fields on river floodplains (**Figures 1.22a–b**). Unfortunately, these actions are usually taken after disasters have struck, but still are examples of proactive land-use management. The designation of public recreation, wildlife, or wilderness areas with minimal permanent structures is a prudent use of land in hazardous areas. This method greatly reduces the risk to people, avoids the expense of evacuations before an event, avoids emergency response expenses, and minimizes reconstruction costs afterward. An added benefit of this approach is that it affords our growing population of people beautiful recreation areas and needed open spaces.

Source: David M. Best.

Figure 1.22a **Indian Bend Wash Greenbelt in Scottsdale, Arizona, lies within the drainage of an area that floods every year from thunderstorm activity.**

Source: David M. Best.

Figure 1.22b **Minimal cleanup is required following a storm to return the area to its normal use and appearance.**

Land-use planning methods are an obvious and cost-effective way of avoiding many disasters or catastrophes. However, it is sometimes difficult to impose land-use restrictions because areas are already heavily populated, or because people want to live as close as possible to scenic areas such as coasts, rivers, and mountains. Citizens often resent being told they cannot live in a certain area and oppose attempts at restrictions because they feel it infringes on their property rights, and attempts at regulations in the public interest often are met with political and legal opposition.

6. Prediction. Predicting, detecting, and monitoring the onset of a hazard event. Federal agencies utilize weather forecast models, earthquake and volcano monitoring systems, remote-sensing applications, and other scientific techniques and devices to gather as much information as possible about the what, when, and where of a potential hazard, as well as the severity of each threat.

Predicting the time, place, and severity of a hazard event saves lives and reduces losses when it is followed by timely and effective communication of warnings to the public. Some disaster events can be predicted fairly accurately, such as the arrival of coastal hurricanes, tsunamis, certain storms, and some river floods; some precursor events have led to better volcanic eruption predictions. Our capability to predict natural hazards has grown considerably as a result of scientific and technological advances in recent decades. However, variability still exists from hazard to hazard, and we still need to monitor and observe hazards and possible precursors to close the gaps and translate predictions into easily understood warnings that can be communicated quickly to the public.

7. Preparedness. The advance capacity to respond to the consequences of a hazard event. This means having plans in place concerning what to do and where to go if a warning is received or a hazard is observed. Communities, businesses, schools, public facilities, families, and individuals should have preparedness plans.

Emergency plans must be prepared and tested many times over before a natural hazard strikes. Even the most comprehensive and well-constructed contingency plans should be tested and evaluated through field exercises and study so that planning activity voids can be filled. Although planning for small, frequent natural hazards seems challenging and expensive at the time, planning for infrequent disasters is more difficult but equally important.

During a Natural Disaster Event

Even while a natural hazard is in progress, actions can be taken to reduce the impact; the response phase takes place during a natural hazard event.

8. Response. The act of responding to a hazard event. Hazard response activities include evacuation, damage assessment, public health risk assessment, search and rescue, fire suppression, flood control, and emergency medical response. Each of these response activities relies heavily on information and communication technologies.

During the initial response following the occurrence of a hazard first responders arrive and perform their acts of saving lives and stabilizing the injured. A great deal of training goes into preparing all the various emergency services that are required to operate in a coordinated, efficient manner. Search and rescue teams move into high gear in order to reach injured people in the shortest amount of time. The role of individual citizens is also important as they can provide assistance with many tasks that do not require specific technical knowledge. Sandbagging swollen rivers requires the help of many people, but a smaller number of technical assistants are needed to oversee the task.

After a Natural Disaster Event

As the emergency response period following a disaster wanes, individuals and communities enter a long period of recovery, redevelopment, and restoration.

9. Recovery. Activities designed to restore normalcy to the community in the aftermath of a hazard event. Recovery activities include restoring power lines, removing debris, draining floodwater, rebuilding, and providing economic assistance programs for disaster victims. As with response, the recovery process relies heavily on the availability of up-to-date data and information about the various community sectors, and on the technology to obtain and communicate that information.

The primary goal of the recovery phase is to restore an area affected by a disaster to its pre-disaster condition. Once the initial emergencies are dealt with, a more calculated response is used to make decisions that involve the clean-up and possible reconstruction of damaged or destroyed property. Undoubtedly infrastructure such as electrical distribution, water systems, and transportation need to be restored to their normal operational levels (Figure 1.23). Federal and state financial aid are often available to help complete these needs.

Disasters are sometimes seen as opportunities to rebuild a community or relocate it so that it will not suffer great losses in the future from a repeat performance of a natural hazard. The challenge is that there must be a strong commitment from citizens and the local government of the affected area to rebuild in such a manner that the best hazard mitigation methods are employed.

FEMA, photo by Mike Moore.

Figure 1.23 Union Pacific Railroad workers repairing an area washed out by flooding near a street crossing on Galveston Island, Texas. Tidal surge and hurricane force winds brought by Hurricane Ike in September 2008 damaged infrastructure across the region.

However, the window of opportunity for change is very narrow following a disaster. Following the devastation of New Orleans, Louisiana, United States, by Hurricane Katrina in 2005, scientists studying the circumstances surrounding the catastrophe suggested it was a good time to relocate the city instead of rebuilding it. New Orleans is built on a floodplain area of the Mississippi River, and levees were constructed along the river to protect it from flooding. The levees stopped the floods as well as the deposition of sediments that built up the delta to begin with. This resulted in the city's land sinking below sea level, which required the building of more levees around the entire city and pumping of water to keep the city dry. In the meantime, the Mississippi River—which is the lifeblood of the city—naturally changes its course over time across the delta and tends to move westward away from the city. To prevent this from happening, the US Army Corps of Engineers spends billions of dollars in attempts to prevent this natural event from happening and thus keep the river flowing toward the city. The city continues to sink below sea level. Higher levees must be built and massive pumps must keep operating to keep it dry. Is it time to move to higher ground? The citizens and local government decided to stay and rebuild in hopes that another massive hurricane is not on the horizon. Studies have shown that major hurricanes hit the central Gulf Coast region every 100 to 200 years, a time span long enough to not be a concern to most people.

Lessons and Challenges

Today we hear more about natural hazards and have seen more vividly the negative effects these hazards have on people, property, and the environment. We cannot totally avoid natural hazards, but how we perceive and react to them determines our destiny. We can act to minimize and reduce their impacts. After all, natural hazards do not become disasters or catastrophes unless the people and communities they touch are not prepared to deal with them. The previously described strategy of reducing the impacts from inevitable natural hazards can be strengthened as we learn more about natural processes and work together to mitigate their hazards. We still have a long way to go if we are to focus on preparedness and prevention rather than the usual bandage solutions.

Our general approach to disaster management has remained reactive, focusing on relief, followed by recovery and reconstruction. Prevention planning or community preparedness has been poorly funded and has not been a major policy priority. Relief remains media-friendly, action oriented, easy to quantify (for example, the tons of food distributed, number of family shelters shipped), and readily accountable to donors as concrete actions in response to a disaster. With the increase in magnitude of disaster impacts, mostly in poorer developing countries, concern is rising over inadequate preparedness and prevention. Natural disasters create serious setbacks to the economic development of countries. This has been proven time and time again, particularly in the last 20 years with the devastation caused by Hurricane Mitch in Central America; the Yangtze River floods in China; earthquakes in Turkey, Iran, and Indonesia; the Japanese earthquake and tsunami; and Hurricanes Katrina and Sandy in the United States. All of these events diverted possible economic development funds toward reconstruction.

One of the biggest challenges is to change our culture from one of continually responding to and recovering from a natural disaster to one of learning to live with natural hazards through mitigation efforts, which are more

© jon lyall/Shutterstock.com.

Figure 1.24 **Policymakers in the New Zealand parliament in the Beehive in Wellington are responsible for developing strategies for dealing with natural hazards from disease outbreaks and pandemics, climate change, volcanoes, earthquakes, damaging winds, tsunamis, floods, wildfires, extreme weather events, landslides, storm surges, and coastal erosion, all of which have the potential to impact human life, property, or both.**

cost-efficient. The United States Geological Survey and the World Bank calculated the worldwide economic losses from natural disasters in the 1990s could have been reduced by $280 billion if $40 billion had been invested in disaster preparedness, mitigation, and prevention strategies. Unfortunately, many policies concerning hazards are driven by politics and special interests.

Policymakers must understand that mitigating hazards is prudent both politically and economically (Figure 1.24). Developers, companies, and local governments often allow or encourage people to move into known hazardous areas. Developers, real estate companies, and some corporations are reluctant to admit to the presence of hazards for fear of reducing the value of land and scaring off potential clients. So the adage "buyer beware" applies here. Many local governments consider any news about hazards to be bad for growth and economy and resist restrictive zoning for fear of losing their tax base. Even developers and private individuals view restrictive zoning as infringement on their property rights and doing as they please with their property. Having good land-use practices can lead to an improved quality of life, new recreation opportunities using open-space lands, and a stronger tax base in the long run.

Instead of mitigating hazards, natural disasters continue to kill and inflict human suffering as well as destroy property, economic productivity, and natural resources. Financial assets are diverted from much needed investments in our future—such as research, education, and the reduction of crime, disease, and poverty. Inaction today regarding natural hazards compromises our safety, economic growth, and environmental quality for future generations. Natural disasters greatly jeopardize our goal of sustainable development, which means meeting the needs of the present without compromising the ability of future generations to meet their needs. Sustainable development is usually understood to require (1) economic growth, (2) protection of the environment, and (3) sustainable use of ecological systems. However, a fourth criterion of equal importance is that sustainable development must be resilient to the natural variability of the Earth and the solar system. By variability, we mean natural hazards events such as earthquakes, tsunamis, hurricanes, and other events that brutally interrupt our societies and ecosystems. These events have not merely punctuated Earth's history as much as they have defined it, and so understanding them is the first step in our success in coping with them.

We have opportunities to prevent or reduce future natural disasters rather than just picking up the pieces after a hazard unleashes its energy upon us. Just like the adage, "an ounce of prevention is worth a pound of cure," we know that a habit of a good diet and exercise is a first defense against heart disease and stems off the higher cost of medical expenses associated with hospitalization or surgery. The need is the same for us as individuals to develop good habits about how we live and build on this planet. It is hoped that this book will help readers to understand natural hazards and what might occur in the event of a disaster, and to make sound decisions about them.

Key Terms

carrying capacity *(page 13)*
closed system *(page 20)*
dynamic equilibrium *(page 20)*
Earth system science *(page 20)*
exponential growth *(page 13)*
frequency *(page 21)*
geologic time scale *(page 22)*
hypothesis *(page 22)*
law (or principle) *(page 24)*
magnitude *(page 25)*
mitigation *(page 3)*
natural catastrophe *(page 4)*
natural disaster *(page 2, 4)*
natural hazard *(page 4)*
rapid-onset hazard *(page 5)*
recurrence interval *(page 25)*
scientific method *(page 22)*
slow-onset hazard *(page 5)*
theory *(page 24)*

Summary

Natural hazards are the result of natural processes involving physical, chemical, and biological mechanisms and forces that affect Earth's surface, and they have been operating continuously for millions of years. These events become hazards to humans when people live or work where these processes occur and become affected by them. Each year, natural hazards become disasters or catastrophes that are responsible for thousands of deaths and billions of dollars in disaster aid, disruption of commerce, and destruction of homes and critical infrastructure.

Hazards can also be categorized as rapid-onset hazards, which expend their energy very quickly, such as volcanic eruptions, earthquakes, floods, landslides, tornadoes, and lightning. These hazards can develop with little warning and strike rapidly. In contrast, slow-onset hazards, such as drought, insect infestations, disease epidemics, global warming, and climate change, take years to develop. Rapid population growth is forcing development in hazardous areas and has increased the demand for natural resources such as land, water, petroleum, metals, and timber.

Living with natural hazards requires us to keep some basic principles in mind. The dynamic nature of the Earth produces many types of hazards. We must recognize where these events can occur and alter our lifestyles in order to co-exist with nature. The effects of natural disasters will continue to increase as the population expands into more inhospitable regions on Earth.

References and Suggested Readings

Jones, L. (2018). *The big ones—how natural disasters have shaped us (and what we can do about them).* Doubleday.

Kieffer, S. W. (2013). *The Dynamics of Disaster.* New York: W. W. Norton & Company.

Kostigen, T. M. (2014). *Natural Geographic extreme weather survival guide—Understand, prepare, survive, recover.* National Geographic Society.

Parfit, M. (1998). Living with natural hazards. *National Geographic, 194*(1): 2–39.

Park Williams, A., Cook, E. R., Smerdon, J.E., Cook, I., Abatzoglou, J. T., Bolles, K., Baek, S. H., Badger, A. M. & Livneh, B. (2020). Large contribution from anthropogenic warming to an emerging North American megadrought. *Science, 368*(6844): 314–318.

Patton, S. (2013). *Natural disasters.* Greenhaven Press.

Smith, K. (2001). *Environmental Hazards: Assessing risk and reducing disaster* (3rd ed.). Routledge.

Tobin, G. A., & Burrell, E. M. (1997). *Natural Hazards: Explanation and Integration.* The Guilford Press.

Websites for Further Reference

https://www.census.gov

https://www.earthobservatory.nasa.gov

https://www.fema.gov

https://www.gdacs.org/

https://www.nationalgeographic.com

https://www.prb.org/data

https://www.usgs.gov

https://reliefweb.int

https://www.ready.gov

Questions for Thought

1. Explain the differences between natural hazards, disasters, and catastrophes.
2. What natural hazards are most common in the United States?
3. What are the nine hazard reduction and disaster management activities?
4. Contrast and explain the general nature of disaster losses in developed countries versus poor countries.
5. What is the difference between scientific hypotheses and scientific theories?
6. What is the world's population today, and how do growth rates compare to historical trends?
7. Explain the concept of carrying capacity for a species of this planet.
8. What is the difference between rapid-onset and slow-onset hazards? Give an example of each type of hazard.
9. Explain the concept of a system and how humans are a part of the Earth's system.
10. How do human activities upset the dynamic equilibrium of the Earth's system?

The Dynamic Earth System

2

Source: NASA/GeoEye, Inc. Copyright 2010. All rights reserved.

Earth has a wide range of environments that allow life to exist. The interaction of water, atmosphere, and land generate natural hazards that affect people and the planet.

Source: NASA.

Figure 2.1 View of the Great Lakes region of the United States, where the spheres on Earth interact constantly.

We live on a planet that is complex and dynamic, and has been in a continuous state of change since its origin about 4.6 billion years ago. The present-day landscapes and features we observe and live with are still changing as the result of interactions among Earth's many internal and external components. In studying the Earth, it is clear that the planet can be viewed as a system of interacting parts or spheres (or layers). The atmosphere, hydrosphere and cryosphere, biosphere, and geosphere, and all of their components, can be defined separately, but each is continuously interacting as a whole called the **Earth system** (Figure 2.1). Geologists attempt to understand the past, present, and future behavior of the whole Earth system by taking a planetary approach to the study of global change.

Earth system

Composed of the geosphere, hydrosphere, atmosphere, and biosphere, and all of their components, continuously interacting as a whole.

The Earth's surface is a major boundary in the Earth system where the four spheres intersect, where energy is transferred, and therefore, where we find volcanoes, landslides, earthquakes, severe weather, and long-term climate changes that impact us. The dynamic Earth system is powered by energy from two sources, one external and the other internal. Energy from the Sun drives the external processes that occur in the atmosphere, hydrosphere, and at Earth's surface of the geosphere. Processes such as weather, climate, ocean circulation, and erosion of the land are driven by the Sun's energy. Earth's interior portion of the geosphere is the second source of energy. Heat remaining from when the planet formed, along with heat produced continuously from decay of radioactive elements, powers the internal processes to produce volcanoes, earthquakes, and mountain ranges. This constant motion of the Earth's spheres will continue as long as the energy sources remain.

Earth's Spheres

Atmosphere

atmosphere

The air surrounding the Earth, from sea level to outer space.

Earth is surrounded by a thin envelope of gases called the **atmosphere**. The gases consist of a mixture of primarily nitrogen (78%), oxygen (21%) and minor amounts of other gases such as argon (less than 1%), carbon dioxide (0.035%), and water vapor. These gases are held near Earth by gravity but thin rapidly with altitude. Almost 99% of the atmosphere is concentrated

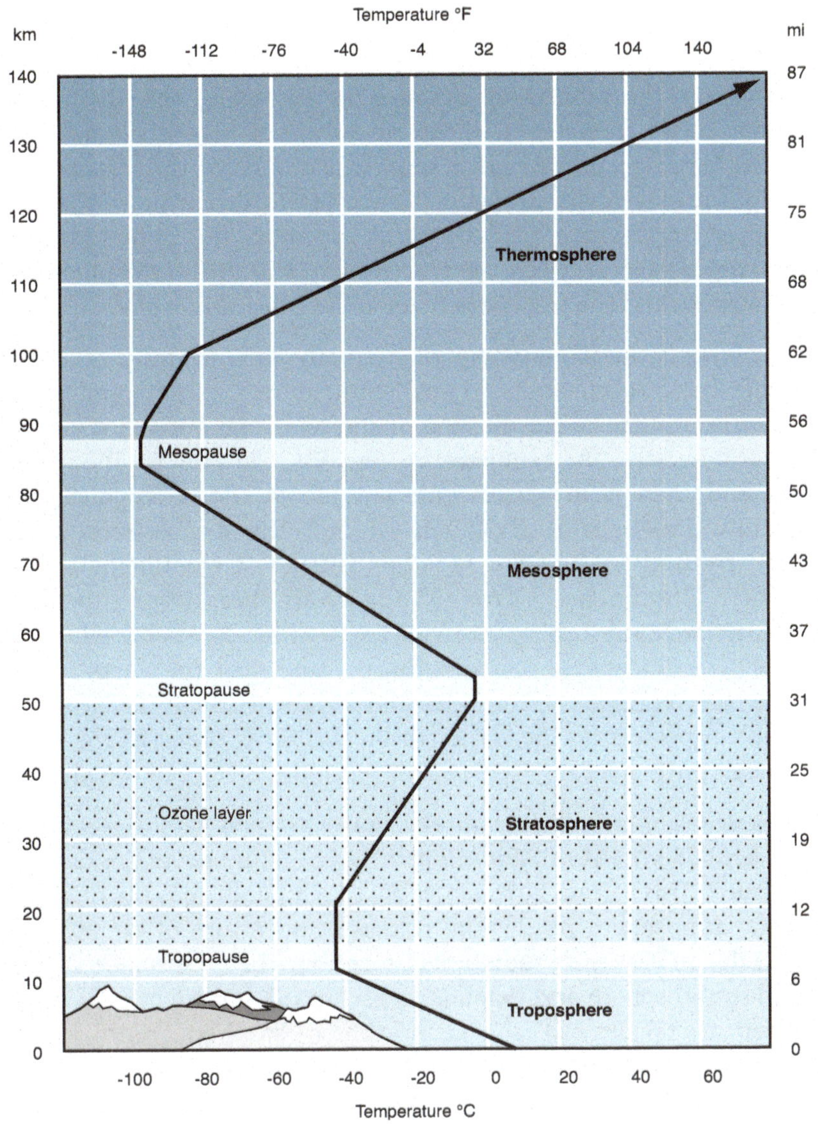

Source: NOAA.

Figure 2.2 Earth's atmosphere consists of four layers. The troposphere has the greatest effect on activity at the surface. The solid line shows the average temperature; note the variation of temperature with increased altitude. The ozone layer shields the surface from incoming ultraviolet radiation.

within 30 km (19 mi) of Earth's surface and forms an insignificant fraction of the planet's mass (less than 0.01%). Despite its modest dimensions, the atmosphere plays an integral role in planet dynamics. It supports life for both animals that need oxygen and plants that need carbon dioxide. It supports life indirectly by controlling climate, acting as a filter, and being a blanket that helps retain heat at night and shield us from the Sun's intense heat and dangerous ultraviolet radiation. The atmosphere is also responsible for transporting heat and water vapor from place to place on Earth, with solar heat being the driving force of atmospheric circulation. Energy is continuously being exchanged between the atmosphere and the surface. This interchange influences the different effects of weather and climate. The atmosphere is divided into four layers (Figure 2.2) in which different factors control the temperature.

Layers of the Atmosphere

The **troposphere** is the first layer above the surface and contains half of the Earth's atmosphere and about 75% of the total mass. This is where all plants and animals live and breathe. Virtually all of the water vapor and clouds exist

troposphere
The lowest part of the atmosphere that is in contact with the surface of the Earth. It ranges in altitude from the surface up to 10 to 12 km (6 to 7.5 mi). This is where nearly all weather takes place.

in this layer and almost all weather occurs here. The air is very well mixed and the temperature decreases with altitude. Air in the troposphere is heated primarily from the ground up because the surface of the Earth absorbs energy and heats up faster than the air does. The heat is spread through the troposphere because the air is slightly unstable. At the tropopause the steady decline in temperature with altitude ceases abruptly and the cold air of the upper troposphere is too dense to rise above this point. As a result, little mixing occurs between the troposphere and the stratosphere above. The tropopause and the troposphere form the lower atmosphere.

stratosphere

The second lowest level of the atmosphere, above the troposphere, which extends to about 55 km (34 mi) above the Earth's surface and contains the ozone layer.

The next layer is the **stratosphere** where many jet aircraft fly because it is very stable. It extends to about 55 km (34 mi) above the Earth and this layer plus the troposphere make up 99% of the total mass of the atmosphere. In the stratosphere, the temperature remains constant to about 20 km (12 mi) and then it increases with altitude to the stratopause (where temperature levels again) at 50 km (31 mi). The stratosphere is heated primarily from above by solar radiation instead of from below as in the troposphere. The presence of ozone (O_3) causes the increasing temperature in the stratosphere by absorbing ultraviolet rays which causes the temperature in the upper stratosphere to reach those found near the Earth's surface. Ozone is more concentrated around an altitude of 25 km (15.5 mi) and is important in protecting life on Earth by absorbing much of the harmful high-energy ultraviolet rays before they reach the surface of the planet.

Ozone concentrations decline in the upper portion of the stratosphere, and, at about 55 km (34 mi) above Earth, temperature decreases once again with altitude. This second zone of declining temperature is the mesosphere where the atmosphere reaches its coldest temperature of about −100 °C (−150 °F) due to very little radiation absorption. The mesosphere starts just above the stratosphere and extends to 85 km (53 mi) high until reaching the mesopause (Figure 2.3). The gases in the mesosphere are thick enough to slow down meteorites hurtling into the atmosphere, where they burn up by friction, leaving fiery trails, "shooting stars," in the night sky. The regions of the stratosphere and the mesosphere, along with the stratopause and mesopause, are called the middle atmosphere.

Source: NASA Expedition 22 crew.

Figure 2.3 **The space shuttle Endeavour floats above Earth's atmosphere—the orange layer is the troposphere, the whitish layer is the stratosphere, and the blue layer is the mesosphere.**

The thermosphere starts just above the mesosphere and extends to 600 km high (373 mi). The thermosphere contains auroras and is where the space shuttle orbits. The temperature increases with altitude due to the Sun's energy and can reach as high as 2482 °C (4500 °F). The air is very thin in the thermosphere but small changes in solar energy can cause large changes in temperature. The thermosphere also includes the region of the atmosphere called the ionosphere, which is filled with charged particles. The thermosphere layer is known as the upper atmosphere.

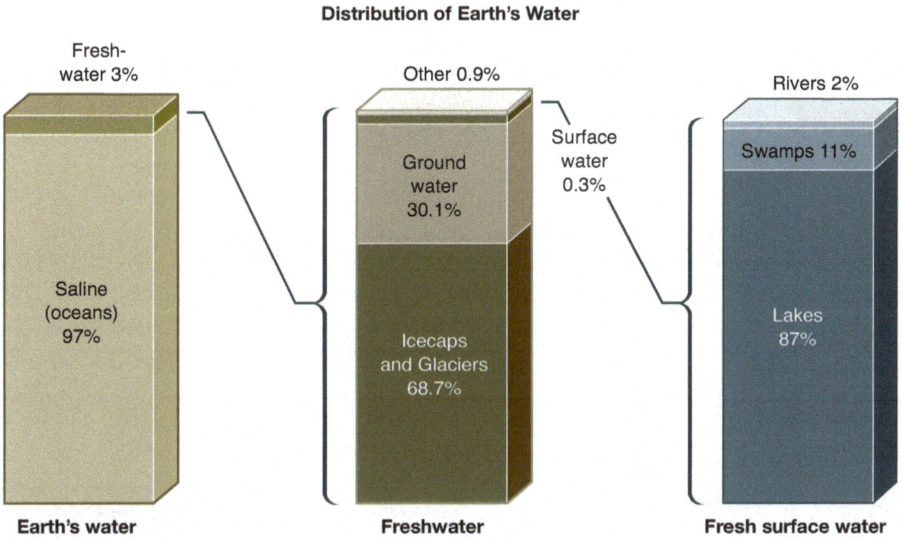

© Kendall Hunt Publishing Company.

Figure 2.4 Most of Earth's water is found in the oceans. About two-thirds of the relatively small amount of freshwater is confined to cold environments.

Hydrosphere and Cryosphere

The **hydrosphere** and **cryosphere** include all of Earth's water in liquid and solid form which continually circulate among the oceans, continents, and atmosphere. The most prominent feature of the hydrosphere is the global ocean, which covers 71% of the Earth's surface and contains about 97% of its water (Figure 2.4). Ocean currents transfer heat from areas of high heat input like the equator to areas of low heat input like the poles, helping to keep the planet at an equilibrium temperature. The remaining 3% of Earth's water is found as freshwater in the hydrosphere and cryosphere and is in ice caps, glaciers, lakes, streams, and groundwater. This freshwater is responsible for creating many varied landforms found on the surface of our planet. Water evaporates from the oceans and moves through the atmosphere, precipitating as rain and snow, and returns to the oceans in streams, groundwater, and glaciers. This movement of water over Earth's surface weathers, erodes, and transports rock material and deposits it as sediment thus constantly modifying the landscape.

hydrosphere
The part of the Earth composed of water including clouds, oceans, seas, lakes, rivers, underground water supplies, and atmospheric water vapor.

cryosphere
The frozen water part of the Earth system, including ice caps and glaciers.

Biosphere

The **biosphere** represents all life on Earth that exists in a thin layer in the uppermost geosphere, the hydrosphere, and the lower atmosphere (Figure 2.5). Plants and animals depend on the physical environment for the basics of life and are affected by Earth's environment. Organisms breathe air, need water, and live in a relatively narrow temperature range. Land organisms depend on soil, which is part of the geosphere. Ultimately, the biosphere strongly influences the other spheres. Essentially, the present atmosphere has been produced by chemical activity of the biosphere, mainly through photosynthesis by plants. Organisms control some of the composition of the oceans; for example, many organisms extract calcium carbonate from seawater for their bones and shells, and when they die, this material settles to the seafloor and accumulates as sediment that lithifies to limestone. The biosphere contains Earth's natural resources of coal, oil, and natural gas, and many of the rocks in the Earth's crust can have their origins traced back to some form of biological activity.

biosphere
The living and dead organisms found near the Earth's surface in parts of the lithosphere, atmosphere, and hydrosphere. The part of the global carbon cycle that includes living organisms and biogenic organic matter.

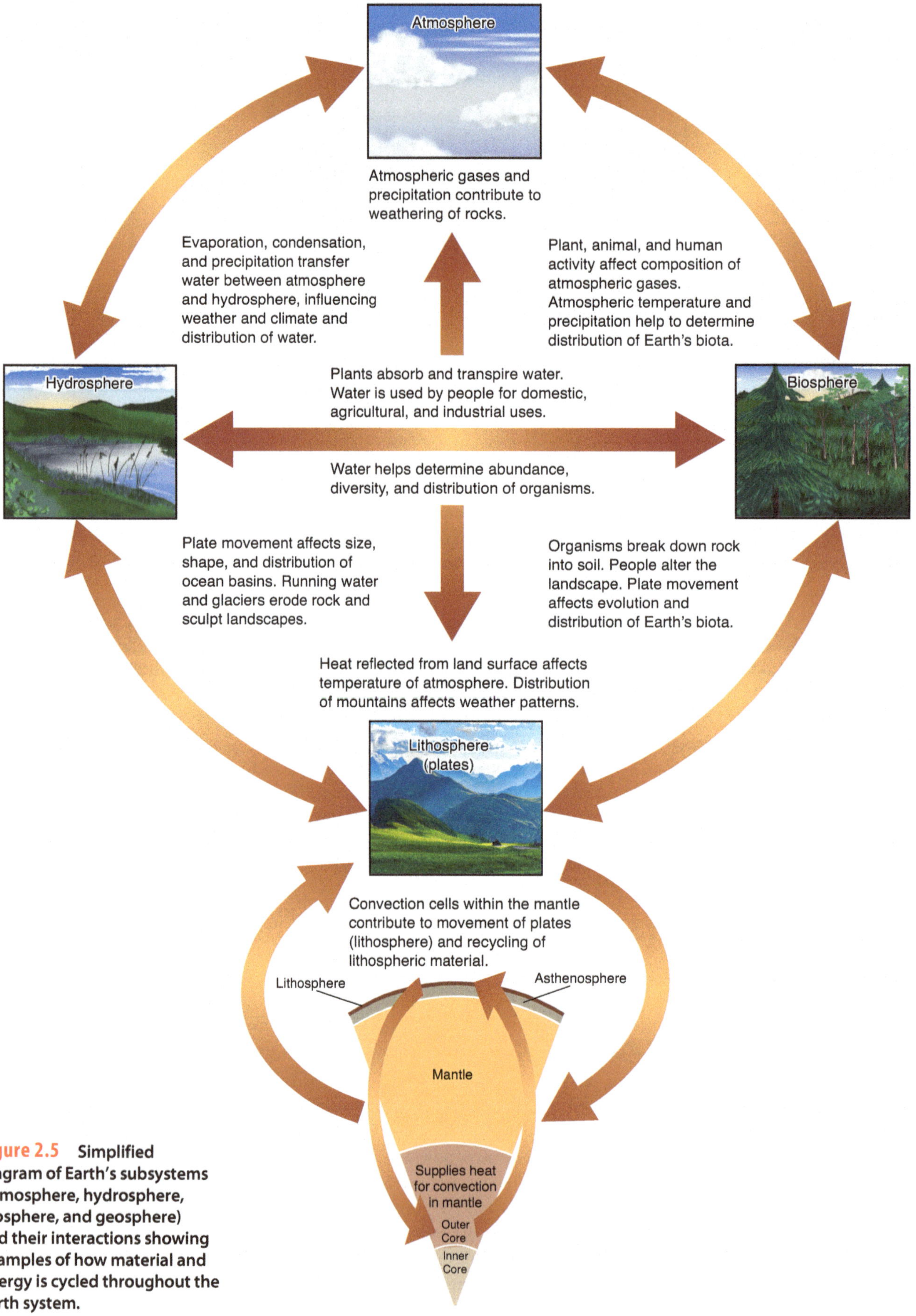

© Kendall Hunt Publishing Company.

Figure 2.5 **Simplified diagram of Earth's subsystems (atmosphere, hydrosphere, biosphere, and geosphere) and their interactions showing examples of how material and energy is cycled throughout the Earth system.**

Geosphere

Beneath our feet are the rocks and soil that make up the solid part of our planet known as the **geosphere**, or solid Earth. The solid materials of Earth are separated into layers based on chemical composition and physical properties. The compositional layers consist of the hot central core, surrounded by a thick mantle that comprises the majority of Earth's volume, and a thin cold **continental crust** at the surface. The layers separated by different physical properties change to include, from outermost to innermost, the **lithosphere**, **asthenosphere**, **outer core**, and **inner core** (Figure 2.6). The characteristics of these layers are controlled by an increase in density, temperature, and pressure with depth.

Layers of the Geosphere

The crust is a compositionally defined layer of the Earth and is the outermost and thinnest layer of the planet. It can be divided into two types based on the composition. The continental crust is thick (20–70 km, 12 to 44 mi) and consists of a wide variety of rock types but has an overall composition similar to granite, with an average density of 2.7 g/cm^3. Continental rocks are dominated by the elements silicon, oxygen, potassium, aluminum, and iron. The **oceanic crust** is thin (5–10 km, 3 to 6 mi) and consists of dark,

geosphere
The soil, sediment, and rock layers of the Earth including the crust, both continental and oceanic.

continental crust
The rocky, relatively low density, outermost layer of the Earth ranging in thickness from 30 to 50 km (19 to 31 mi).

oceanic crust
That part of the Earth's crust of the geosphere underlying the ocean basins. It is composed of basalt and has a thickness of about 5 km.

lithosphere
The outer layer of solid rock that includes the crust and uppermost mantle. This layer, up to 280 km (174 mi) thick, forms the Earth's tectonic plates and rests on the mobile, ductile asthenosphere.

asthenosphere
The uppermost layer of the mantle, located below the lithosphere. This zone of soft (plastic), easily deformed rock lies at depths between 80 and up to 800 km (50 to 500 mi).

outer core
The liquid outer layer of the core that lies directly beneath the mantle.

inner core
The solid central part of Earth's core.

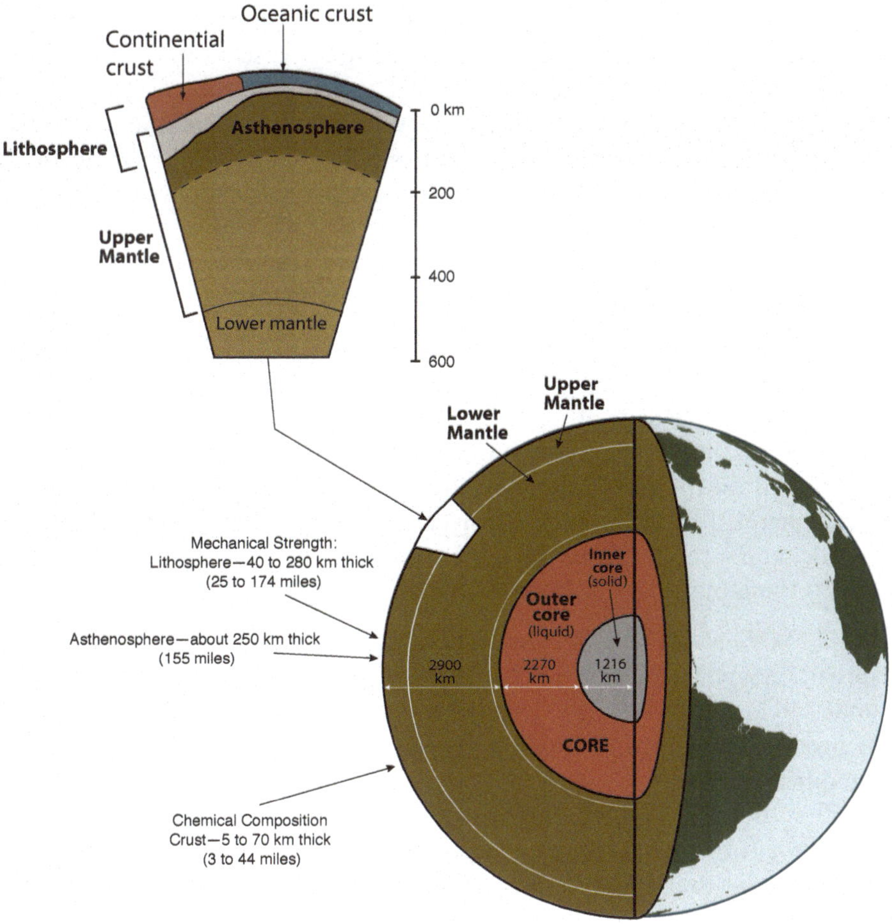

© Kendall Hunt Publishing Company.

Figure 2.6 Internal structure of the geosphere. Notice that the lithosphere includes both the crust and the uppermost layer of mantle that overlies the asthenosphere.

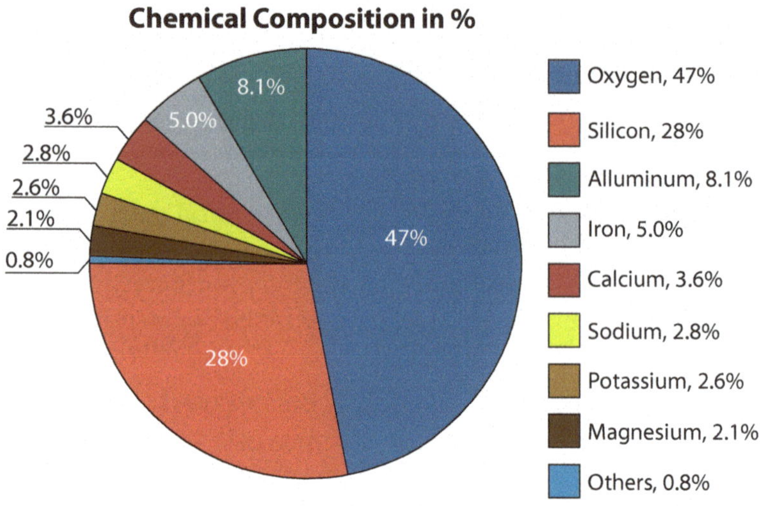

Figure 2.7 About 98 % of the Earth's crust is made up of only eight elements: oxygen, silicon, aluminum, iron, calcium, sodium, potassium, and magnesium. The remaining 2 % is made up of elements like titanium, hydrogen, phosphorous, sulfur, carbon, and others.

igneous basalt that contains more iron and magnesium than granite, thus making oceanic crust denser (3.0 g/cm^3) than continental crust. Both continental and oceanic crusts are cooler relative to the layers below, and thus consist of hard, strong rock that is rigid and behaves as a brittle material. The chemical makeup of the crust overall is dominated by eight main elements (Figure 2.7).

The compositional layer of the **mantle** is the middle layer between the crust and core and occupies about 83% of Earth's volume. It is denser than the crust with a density ranging from about 3.3 g/cm^3 in its upper part to about 5.7 g/cm^3 near the contact with the outer core. The composition is mostly of peridotite which is a dark, dense igneous rock containing **silicate** minerals (compounds of silicon and oxygen atoms) with abundant iron and magnesium like olivine and pyroxene. Although the chemical composition is uniform throughout the mantle, the temperature and pressure increase with depth. Temperatures near the top of the mantle approach 1000 °C (1800 °F), increasing to about 3300 °C (6000 °F) near the core-mantle boundary. At these temperatures the mantle should be molten rock. However, the high pressure from the thickness of overlying rocks prevents the mantle rocks from melting. This is because rocks will expand as much as 10% when they melt and the high pressures make it difficult for rock to expand and therefore inhibit melting. If the combination of temperature and pressure effects is close to the rock's melting point, the rock remains solid but loses its strength, which makes it weak and plastic. If the temperature rises, or the pressure decreases, the rock will begin to melt and form magma (molten rock). These temperature and pressure changes with depth cause the strength of the rocks to vary with depth and create layering within the mantle, as well as in the core below.

mantle

The layer of the Earth below the crust and above the core. The uppermost part of the mantle is rigid and, along with the crust, forms the 'plates' of plate tectonics. The mantle is made up of dense iron and magnesium rich (ultramafic) rock such as peridotite.

silicate

Refers to the chemical unit of the silicon tetrahedron SiO_4 which is the fundamental building block of silicate minerals. Silicate minerals represent about one third of all minerals and hence make up most rocks we see at the Earth's surface.

The uppermost mantle below the crust is relatively cool and its pressure is low, which are similar physical conditions to the crust and produce hard, strong rocks as well. The rigid physical properties of both the crust and upper mantle cause the two to combine and form an important outer layer, based on the physical properties of the rocks, called the lithosphere (rock sphere). The lithosphere, consisting of the crust (either continental or oceanic) and a thin layer of the uppermost upper mantle, averages 100 km (62 mi) thick but is as much as 280 km (174 mi) thick in some continental regions (Figure 2.8).

The asthenosphere (weak sphere) is the zone in the upper part of the mantle, about 250 km thick (155 mi), that is more ductile or plastic than the rest of the mantle. The temperatures and pressures in this zone cause the rocks to lose much of their strength and become soft and plastic, like warm putty, and they behave like a ductile layer underlying the brittle lithosphere. Partial melting within the asthenosphere generates magma (molten rock) which can rise to the surface to form volcanic eruptions. This is because the

© Kendall Hunt Publishing Company.

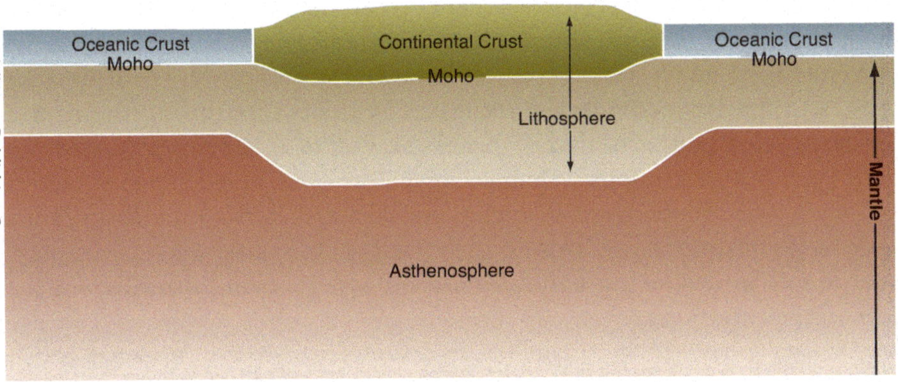

Figure 2.8 **Structure of the outer Earth based on mechanical strength. The outer portion of Earth's layers includes oceanic and continental crustal material, along with a rigid lithosphere and a hotter, plastic asthenosphere.**

magma is molten and less dense than the mantle rock from which it was derived.

Below the lithosphere and asthenosphere is the mantle which is a solid and forms most of the volume of the Earth's interior. Because the high pressure at these depths offsets the effect of high temperature, the rocks are stronger and more rigid than the overlying asthenosphere. Despite the stronger nature of the mantle, it never gets as strong as the lithosphere and remains capable of flowing slowly over geologic time.

The **core** is the innermost region of Earth and has an average calculated density of 11 g/cm^3 and comprises about 16% of the planet's total volume. The composition is considered to be predominantly metallic iron and some nickel. While the core is compositionally uniform and is a single layer of the Earth based on composition, it is subdivided into two zones based on very different physical properties: the inner core is solid and the outer core is a molten liquid. Near the center, the core's temperature is estimated to be nearly 7000 °C (12,600 °F), hotter than the Sun's surface, and pressures are 3.5 million times that of Earth's atmosphere at sea level. This extreme pressure is responsible for keeping the inner core solid despite being hotter than the outer core.

core
The innermost layer of the Earth, made up of mostly iron and nickel metal. The core is divided into a liquid outer core and a solid inner core. The core is the densest of the Earth's layers.

The combination of the rigid lithosphere lying over the non-rigid asthenosphere has extremely important geologic implications for natural hazards. The lithosphere is broken into numerous individual slabs called **plates** that slide over the asthenosphere. This motion is the reason that plate tectonics exists on Earth. Movements of the plates create earthquakes, volcanism, and mountain uplifts.

plate
Slab of rigid crust and uppermost mantle that forms the outermost brittle layer of the planet.

Earth Processes and Cycles

As a system, remember that Earth is an assemblage or combination of interacting components that form a complex whole. The system is driven by the flow of matter and energy through the components in order to try to reach equilibrium. Each of Earth's major spheres, the atmosphere, hydrosphere, biosphere, and geosphere, are usually described and studied as separate entities, each containing numerous interacting smaller systems; however, each sphere and their components interact with each other continuously, exchanging both matter and energy.

Several matter and energy cycles are fundamental to our understanding and study of natural processes and hazards connected to the dynamic movements of Earth systems. These include the **rock cycle**, the **hydrologic cycle**, and the **biogeochemical cycle**. Keep in mind that during the course of these cycles, matter is always conserved, that is, it is neither created nor destroyed, but continuously changes form. The cyclical exchange of matter and energy occurs between storage reservoirs within the spheres and on a wide range of time scales. This cyclic movement of matter and energy plays a large role in producing natural disasters when we find ourselves in the path of their motions.

rock cycle
The sequence of events in which rocks are formed, destroyed, and reformed by geological processes. Provides a way of viewing the interrelationship of internal and external processes and how the three rock groups (igneous, sedimentary, and metamorphic) relate to each other.

hydrologic cycle
The cyclic transfer of water in the hydrosphere by water movement from the oceans to the atmosphere and to the Earth and return to the atmosphere through various stages or processes such as precipitation, interception, runoff, infiltration, percolation, storage, evaporation, and transportation.

biogeochemical cycle
Natural processes that recycle nutrients in various chemical forms from the environment, to organisms, and then back to the environment. Examples are the carbon, oxygen, nitrogen, phosphorus, and hydrologic cycles.

Materials

The materials that make up the different Earth spheres, and what energy performs work on, is known as matter, which is defined as anything that has mass and occupies space. Therefore, the atmosphere, water, plants, animals, minerals, and rocks are composed of matter made up of atoms of different elements. Matter exists in three states: solids, liquids, and gases. The cycling of this matter is important in Earth's processes.

Energy

Natural processes responsible for natural disasters (such as erosion, atmospheric circulation, or plate tectonics) occur on or within the Earth and require energy for their operation of performing work on matter. Energy is defined as "the ability to do work," which can be constructive or destructive. Work itself can mean moving something, lifting something, warming something, or lighting something. Energy exists in two states: stored and waiting to do work (**potential energy**), or actively moving (**kinetic energy**). Potential energy is so named because this stored energy or work has the *potential* to change the state of other objects when released. The kinetic energy of an object is the energy it possesses because of its motion and is an expression of the fact that a moving object can do work on anything it hits and thus it quantifies the amount of work the object could do as a result of its motion. There are many different forms of energy that can be transformed from one to the other:

potential energy
The energy available in a substance because of position (e.g., water held behind a dam) or chemical composition (hydrocarbons). This form of energy can be converted to other, more useful forms (for example, hydroelectric energy from falling water).

kinetic energy
The energy inherent in a substance because of its motion, expressed as a function of its velocity and mass, or $MV^2/2$.

- **Heat energy** is exhibited by moving atoms and the more heat energy an object has, the higher its temperature, which is a measure of its kinetic energy. When gasoline is burned in a car's engine, the heat energy is converted to kinetic energy that moves the car.
- **Radiant energy** is energy carried in the form of electromagnetic waves such as visible light, ultraviolet rays, and radio waves. Most of the Sun's energy, especially light, reaches Earth in this form and is converted to heat energy.
- **Nuclear energy** is energy released from an atom's nucleus. Energy is released by three processes: radioactive decay, where an unstable radioactive nucleus spontaneously splits into two or more lighter elements; nuclear fusion, where multiple nuclei join together to form a new heavier element; and heat-producing nuclear reactions, where two nuclei merge and split to produce two different nuclei.

- **Elastic energy** is potential energy formed through deformation such as the stretching, bending, or compressing of an elastic material like a rubber band. When energy is released it is converted to kinetic energy and heat by friction. Earthquakes release energy stored in elastically deformed rocks that eventually break.
- **Electrical energy** is produced by moving electrons through material and is often converted into heat energy. Lightning storms are good examples.
- **Chemical energy** is produced by breaking or forming chemical bonds and this type of energy is usually converted to heat.
- **Gravitational energy** is produced when an object falls from higher to lower elevations. As the object falls the energy can be converted to kinetic energy or heat.

Earth's major cycles and processes are powered by energy from two sources. The Sun drives external processes that occur in the atmosphere, hydrosphere, cryosphere, biosphere, and surface of the geosphere. Climate, weather, ocean circulation, and erosional processes are the result of energy from the Sun being transformed into heat energy which powers them. The second source of energy is derived from the Earth's interior. Heat left over from the Earth's formation, along with heat produced from the constant decay of radioactive elements, powers the internal processes driving plate tectonics and forming volcanoes, earthquakes, and mountain ranges.

Energy From the Sun

The Sun is the most prominent feature in our solar system and contains approximately 98% of the total solar system mass (**Figures 2.9a–b**). The Sun is composed of hydrogen (about 74% of its mass, or 92% of its volume), helium (about 25% of its mass, 7% of its volume), and trace quantities of other elements. The Sun's outer layer is called the photosphere and has a temperature of 6000 °C (10,800 °F). The mottled appearance is due to the turbulent eruptions of energy at the surface. Solar energy is created deep within the core of the Sun, where the temperature (15,000,000 °C, 27,000,000 °F) and pressure (340 billion times Earth's air pressure at sea level) are so intense that nuclear reactions take place. Here, nuclear fusion produces energy by fusing

International Astronomical Union.

Figure 2.9a **Earth is the third planet from the Sun. When the size of Earth is compared to that of the Sun, more than one million Earths could fit inside the Sun and 109 Earths could be placed across its equator.**

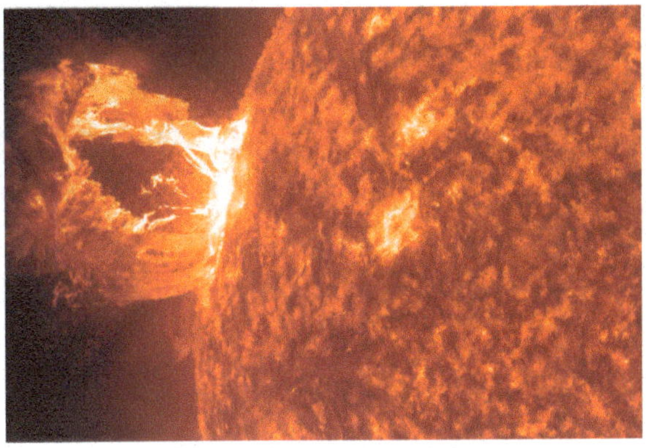

Source: NASA.

Figure 2.9b **Solar flares are sudden bursts of energy that occur at the Sun's surface. Major flares can interfere with communications on Earth.**

hydrogen atoms together to form atoms of helium. During nuclear fusion, some of the mass from the hydrogen atoms is expelled as energy and carried to the surface of the Sun by convection (see following discussion), where it is released as radiant energy. Energy generated in the Sun's core takes about a million years to reach its surface, but every second, 700 million tons of hydrogen are converted into helium and 5 million tons of pure energy are released. The Sun has been producing energy since the solar system formed about 4.6 billion years ago, and it is estimated that solar energy production will continue at this level for another 5 billion years.

The radiant energy from the Sun travels in many different wavelengths (Figure 2.10) through space, with about 43% being visible light, 49% near-infrared, and 7% ultraviolet (UV) solar radiation received on Earth. About 40% of the energy is immediately reflected back into space by the atmosphere, clouds, and the Earth's surface, mostly the oceans. Some of the energy is converted to heat and is absorbed by the atmosphere, hydrosphere, and lithosphere. Heat drives the hydrologic cycle, causing evaporation of the oceans and circulation of the atmosphere. Energy from the Sun is responsible for natural disasters such as severe weather and floods.

In the biosphere, some energy is absorbed and stored by plants during photosynthesis and used by other organisms, or is stored in fossil fuels such as coal and petroleum. This stored energy in the biosphere is released through burning to form heat energy, which also releases the stored carbon that

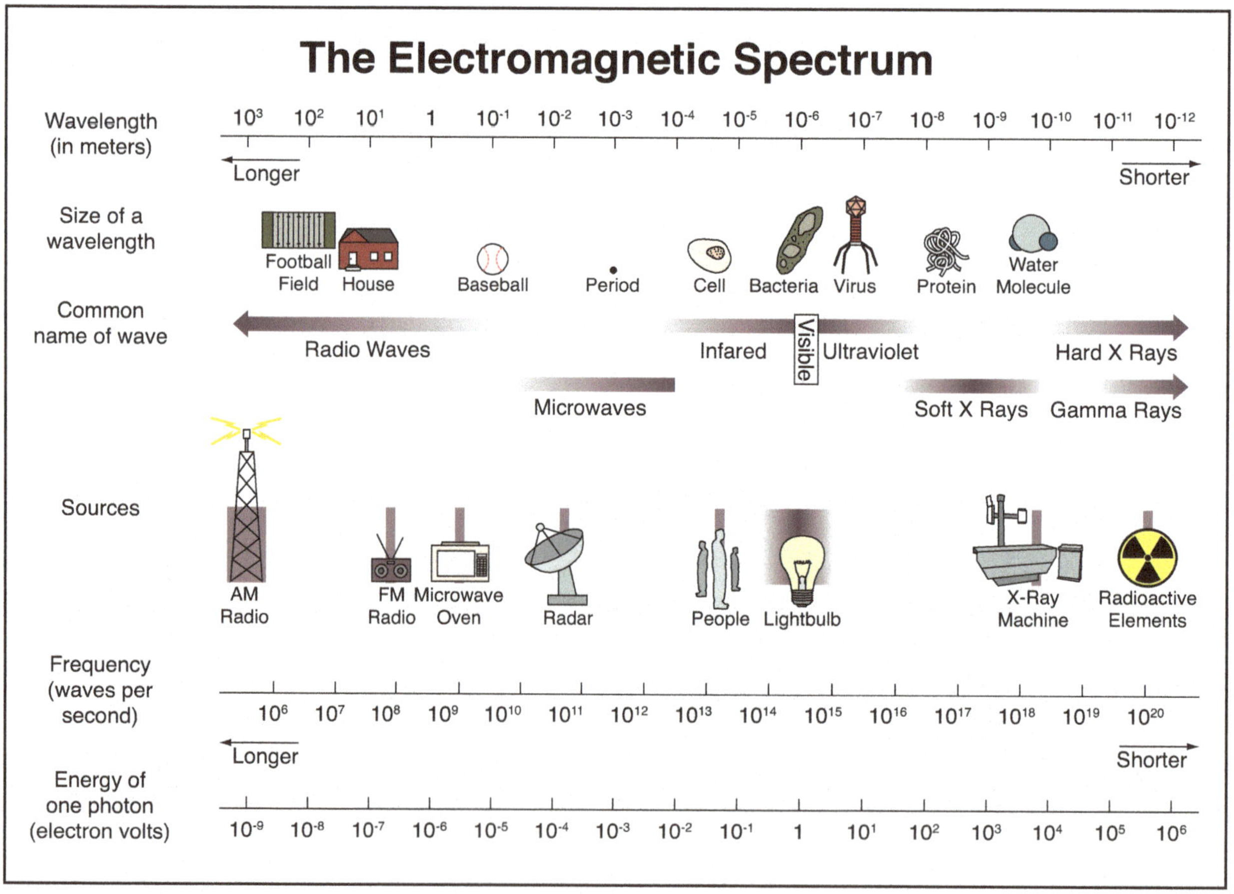

Source: NASA.

Figure 2.10 **The electromagnetic spectrum has an extremely wide range of wavelengths. Visible light is only a small portion of the spectrum.**

produces carbon dioxide, a greenhouse gas that is partially responsible for global warming.

Energy From Earth's Interior

Energy in the form of heat is generated within the Earth's interior and contributes only about 0.013% of the total energy reaching the Earth's surface, the other 99.987% is from the Sun. However, this smaller amount of energy that flows to the Earth's surface is responsible for driving plate tectonics, which creates mountains and causes earthquakes, and for melting of rocks to create magmas that result in volcanism. Heat energy is produced from three energy sources within the Earth's interior:

Radioactive decay continuously produces heat energy from radioactive isotopes such as ^{235}U (Uranium), ^{232}U, ^{232}Th (Thorium), and ^{40}K (Potassium). Radioactive isotopes have unstable nuclei that break down or decay to a more stable isotope and expel subatomic particles such as protons, neutrons, and electrons from the radioactive parent atom which can interact with surrounding rocks.

radioactive decay
Natural spontaneous decay of the nucleus of an atom where alpha or beta and/or gamma rays are released at a fixed rate.

Gravitational energy and **impact energy** produced a tremendous amount of heat during the initial formation of the Earth. Gravitational energy developed as the Earth grew in size by accretion in the original solar nebula. The deepening burial of material caused an increasing gravitational pull that continued to compact the interior and convert gravitational energy to heat. At the same time in Earth's early history, tremendous numbers of space objects such as asteroids, planetessimals, and comets were hitting the Earth. The impacts of these objects converted their energy of motion (kinetic energy) to heat. The large quantity of heat produced by the combination of these two sources did not easily escape to the surface due to the poor conductivity (see following heat transfer discussion) of rock material. The sum of heat energy generated from radioactive decay, impacts, and gravity is slowly rising to the surface of our planet through plate tectonics, and along its journey it is causing the continents to drift and ocean basins to form.

gravitational energy
The force of attraction between objects due to their mass and is produced when an object falls from higher to lower elevations.

impact energy
Cosmic impacts with a larger body convert their energy of motion (kinetic energy) to heat.

Methods of Heat Transfer

Any body, whether a solid, liquid, or gas, has a temperature. This thermal energy can be transferred to another body or its surroundings by the process of **heat transfer**. The movement occurs by heat moving from a hot body to a cold one through the processes of conduction, convection, or radiation, or any combination of these (Figure 2.11). Heat transfer never stops, it can only be slowed down. Earth's internal and external processes are powered by heat energy moving from hotter areas such as the Earth's interior to cooler ones such as the crust and ultimately into space, and so it is important to understand how heat can move through materials.

heat transfer
Heat moving from a hot body to a cold one through processes of conduction, convection, or radiation, or any combination of these.

Conduction

Conduction transfers heat directly from atom to atom. It works best in solids where the atoms are closely packed together, but also transfers heat from

conduction
Heat transfer directly from atom to atom in solids.

Figure 2.11 **Convection, conduction, and radiation often occur together. These are three ways in which heat is transferred through objects and space.**

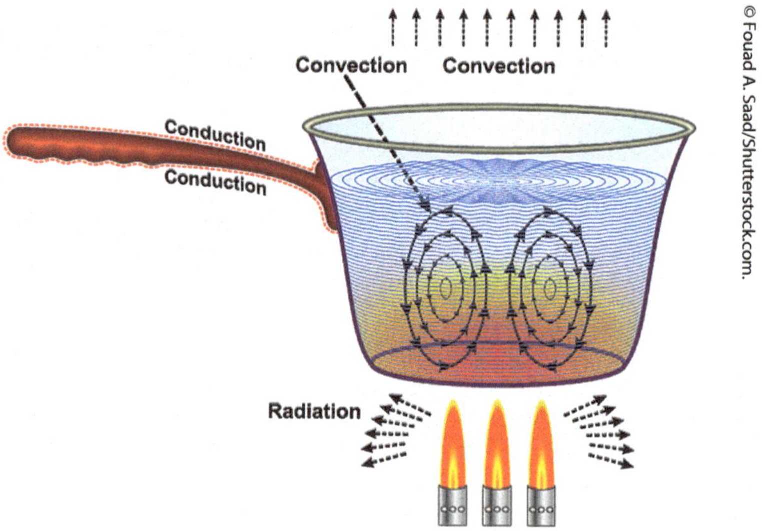

convection

(1) (Physics) Heat transfer in a gas or liquid by the circulation of currents from one region to another; also fluid motion caused by an external force such as gravity. (2) (Meteorology) The phenomenon occurring where large masses of warm air, heated by contact with a warm land surface and usually containing appreciable amounts of moisture, rise upward from the surface of the Earth.

convection cell

Within the geosphere it is the movement of the asthenosphere where heated material from close to the Earth's core becomes less dense and rises toward the solid lithosphere. At the lithosphere-asthenosphere boundary heated asthenosphere material begins to move horizontally until it cools and eventually sinks down lower into the mantle, where it is heated and rises up again, repeating the cycle.

radiation

Energy emitted in the form of electromagnetic waves. Radiation has differing characteristics depending upon the wavelength. Because the radiation from the Sun is relatively energetic, it has a short wavelength (ultra-violet, visible, and near infrared) while energy radiated from the Earth's surface and the atmosphere has a longer wavelength (infrared radiation) because the Earth is cooler than the Sun.

solids to liquids or gas. Heat energy causes atoms to vibrate against each other, and these vibrations pass from high-temperature areas of rapid vibrations to low-temperature areas of slower vibrations, causing them to heat up and vibrate faster. Heat from the Earth's interior moves through the solid crust by this mode of heat transfer, but very slowly. Each substance has a different heat conductivity based on how easily the atoms are affected by the heat energy. Metals are highly conductive, so they transfer heat energy by heating up and cooling down rapidly which is why we use metal pots for cooking, heat is transferred quickly to the contents inside. Rocks are much less conductive and so heat up and cool down more slowly, which helps heat build up inside the Earth.

Convection

Convection transfers heat by being carried along by circulation by the material that was heated. Since heat energy moves with the material, the materials that are able to move are mostly liquids and gases. We are familiar with convection of air in our houses when we heat them in the winter by a heat source, whether by a wood stove, radiator, or heater. The warmed, more bouyant air rises to the ceiling as a moving current, which then cools to a denser air mass that returns to the floor to be heated again. A complete circulation of hot air rising and cold air sinking is called a **convection cell**. Convection is an important method of moving energy through Earth materials. Inside the Earth, the mantle heats up by this method as well as through the flow of magma to the surface. On the surface, heat is transferred within and between the atmosphere and oceans by this mode, which influences climate and weather.

Radiation

Radiation transfers heat using electromagnetic waves. Radiant electromagnetic energy from the Sun (produced by nuclear reactions) is transferred by this method. On Earth radiative heat transfer is responsible for warming the oceans and atmosphere, and for reradiating heat back into space to keep our planet at a constant average temperature.

Plate Tectonics and the Cycling of the Lithosphere

Plate tectonic theory provides us with a framework for interpreting the composition, structure, and internal processes of Earth on a global scale and how they contribute to producing natural events such as earthquakes and volcanoes. **Plate tectonics** (*tekton,* "to build") is a theory developed in the late 1960s to explain how the lithospheric outer layer of the Earth moves and deforms on the ductile asthenosphere. The theory revolutionized the way we think about the Earth and has proven useful in predicting geologic hazard events and explaining many aspects of the natural processes we see on Earth.

The lithosphere is divided into segments called tectonic plates that slide continuously over the asthenosphere (Figure 2.12). They are also referred to as **lithospheric plates**, or simply plates. The plates can encompass both continental and oceanic crust, like the North American plate, but some are composed largely of oceanic crust, like the Pacific plate. The continuous movement of the plates is powered by internal heat forming convection currents within the mantle. The slow movements of different plates (usually a few centimeters per year, or about as fast as your fingernails grow) create natural hazard zones of volcanic activity and earthquakes where their margins meet.

plate tectonics

The theory that the Earth's lithosphere consists of large, rigid plates that move horizontally in response to the ductile deformation of the asthenosphere beneath them, and that interactions among the plates at their boundaries cause most major geologic activity, including the creation of oceans, continents, mountains, volcanoes, and earthquakes.

lithospheric plate

A series of rigid slabs of cold crust plus a thin layer of cold underlying mantle that make up the Earth's outer shell. These plates float on top of a softer, more plastic layer in the Earth's mantle known as the asthenosphere.

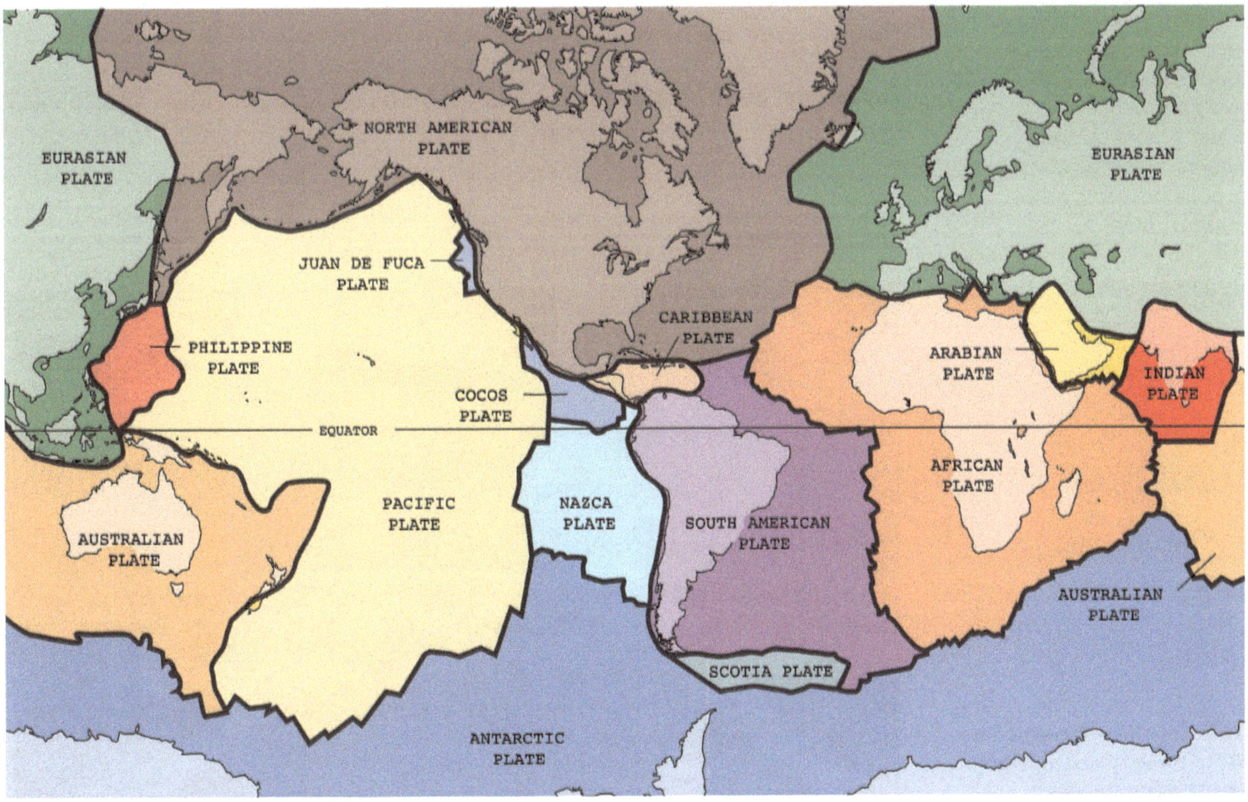

Source: USGS.

Figure 2.12 The surface of the Earth is divided into major lithospheric plates, some of which are found in the oceans, and some contain both continental and oceanic crust.

plate boundary
According to the theory of plate tectonics, the locations where the rigid plates that comprise the outer surface of the Earth meet.

divergent plate boundary
A boundary in which two lithospheric plates move apart.

mid-ocean ridge
An uplifting of the ocean floor that occurs when mantle convection currents beneath the ocean force magma up where two tectonic plates meet at a divergent boundary. The ocean ridges of the world are connected and form a global ridge system that is part of every ocean and form the longest mountain range on Earth.

convergent plate boundary
A boundary in which two plates collide. The collision can be between two continents (continental collision), a relatively dense oceanic plate and a more buoyant continental plate (subduction zone) or two oceanic plates (subduction zone).

Plate Boundaries

Along these margins, or **plate boundaries**, the plates diverge, converge, or slide horizontally past each other (Figure 2.13). Again, these plate boundaries are important because the plates interact where they meet and are zones where deformation of the Earth's lithosphere is taking place and generating potential hazard areas.

Divergent Plate Boundaries

At **divergent plate boundaries** the tectonic plates move away from each other due to tensional stresses. Divergent plate boundaries in oceanic crust generate new oceanic crust as magmas rises from the underlying asthenosphere (Figure 2.14). The magmas intrude and erupt to form basalt rock along the newly created edge of the plate. The margins of divergent plate boundaries occur mostly in the oceanic plates and they are marked by **mid-ocean ridges** such as the Mid-Atlantic Ridge, that are chains of underwater mountain ranges uplifted by the hot underlying mantle. These boundaries are also known as *oceanic spreading centers,* because oceanic lithosphere is created and spreads away on each side of the margin at the ridge as a result of underlying *convection cells* in the mantle. While most diverging plate boundaries occur at the ocean ridges, sometimes continents are split apart along zones called *rift zones,* such as at the East African Rift in eastern Africa. Volcanism and earthquakes are common along diverging plate boundaries but are of relatively low intensity.

Convergent Plate Boundaries

Convergent plate boundaries are where two tectonic plates move toward each other by compressional stresses (Figure 2.15) and where most lithosphere is

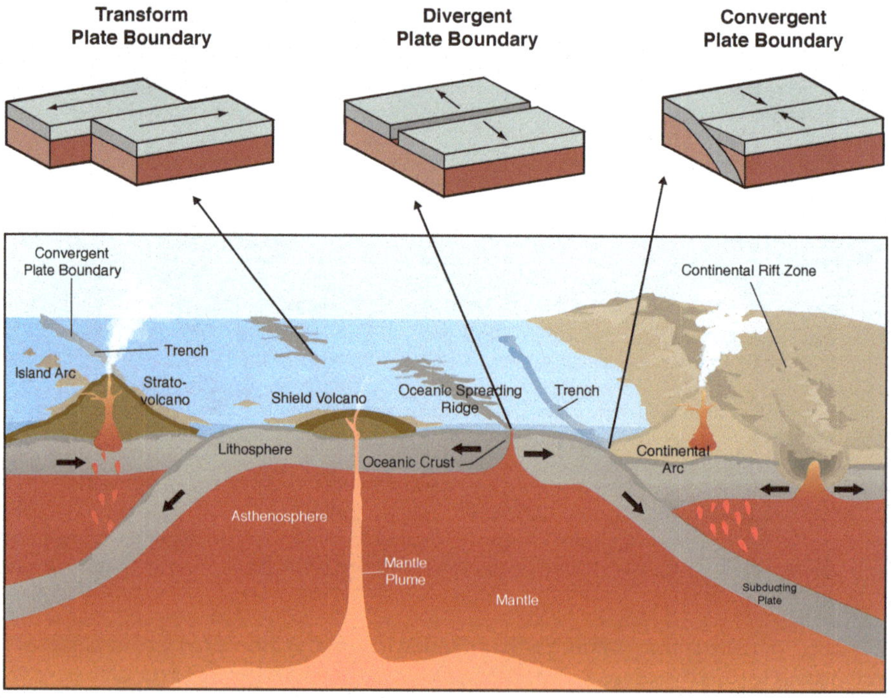

United States Geological Survey.

Figure 2.13 Plate boundaries consist of three types: divergent, where plates move apart; convergent, where plates come together; and transform, where plates slide past one another.

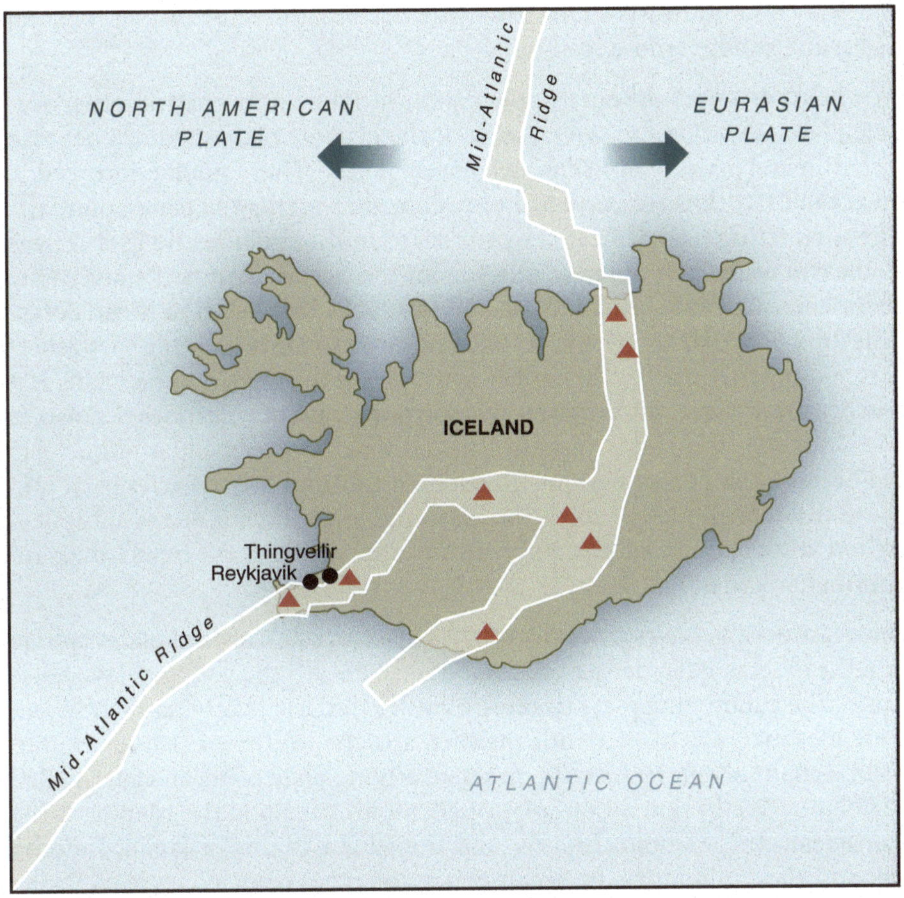

© Kendall Hunt Publishing Company.

Figure 2.14 Iceland lies along a divergent plate boundary as two plates move away from each other. The island is made of volcanic material erupted from greater depths. Active volcanoes are shown by triangles.

From Planet Earth by Renton. Copyright 2002. Reprinted by permission of Kendall Hunt Publishing Company.

Figure 2.15 Convergent plate boundaries display compressive stresses that often result in the destruction of lithospheric material.

destroyed. In this manner, the Earth maintains a global balance between the creation of new lithosphere and the destruction of old lithosphere. At these boundaries, when one of the plates is colder and denser than the other, it sinks down into the mantle in a process called **subduction**. Convergent boundaries are complex because the margins can involve convergence between two plates carrying oceanic crust, between a plate carrying oceanic crust and another carrying continental crust, or between two plates carrying continental crust. Each type of boundary behaves differently based on the density of the two different plates colliding. Continental crust is light

subduction
Process of one crustal plate sliding down and below another crustal plate as the two converge.

and buoyant and resists subduction whereas oceanic crust can be cold and dense and subduct more readily.

subduction zone
Also called a convergent plate boundary. An area where two plates meet and one is pulled beneath the other.

oceanic trench
Deep, linear, steep-sided depression on the ocean floor caused by the subduction of oceanic crustal plate beneath either other oceanic or continental crustal plates.

volcanic arc
A chain of volcanoes formed above a subducting plate. The arc forms where the downgoing plate becomes hot enough to release fluids that rise into the overlying mantle and cause it to melt.

island arc
A chain of volcanic islands produced where an oceanic plate is sinking beneath another oceanic plate.

At **subduction zones**, the oceanic plate sinks or subducts beneath another oceanic plate, *ocean-ocean convergence*, or the oceanic plate subducts beneath a continental plate, *ocean-continent convergence*. Where subduction occurs, an **oceanic trench** is formed on the seafloor that marks the plate boundary. The subducting plate is heated by the surrounding mantle and as it warms it releases water and other volatiles from the ocean crust rocks, overlying sediments, and water-rich minerals in the rocks that begin to break down. This fluid fluxes the overlying mantle and causes mantle melting that generates magma that rises to the surface through the overriding upper plate. If it reaches the surface, the erupting lava forms a chain of volcanoes (known as a **volcanic arc**). The oldest oceanic crust on Earth is about 180 million years old because the denser oceanic lithosphere continuously recycles back into the mantle at subduction zones. Rocks found in continents are as old as 3.96 billion years because subduction consumes very little of the lighter buoyant continental crust.

In ocean-ocean convergence (Figure 2.16), two oceanic crustal plates collide and the older, colder, denser crust subducts beneath the younger, less dense crust. The subducting plate releases volatiles into the overlying mantle and generates magmas from mantle melting, and the magmas rise and erupt to form a chain of volcanoes built on the overlying plate of ocean crust, called **island arcs**. Well-known examples of island arcs include the islands of the Caribbean, the Aleutian Islands, and the island chains of Japan, Indonesia, and the Philippines. In ocean-continent convergence (Figure 2.17), the magma rises through the continental plate to the surface, where it erupts to form a volcanic mountain chain along the edge of the continent. Well-known examples of this type of volcanic arc are the Cascade Volcanoes of the Northwestern United States and the Andes of South America. The hazards associated with these two boundaries include some of the largest earthquakes ever recorded, frequent volcanic eruptions, and tsunamis.

When two continental plates converge, both plates are made of light, buoyant continental material and neither one wants to be subducted. Instead,

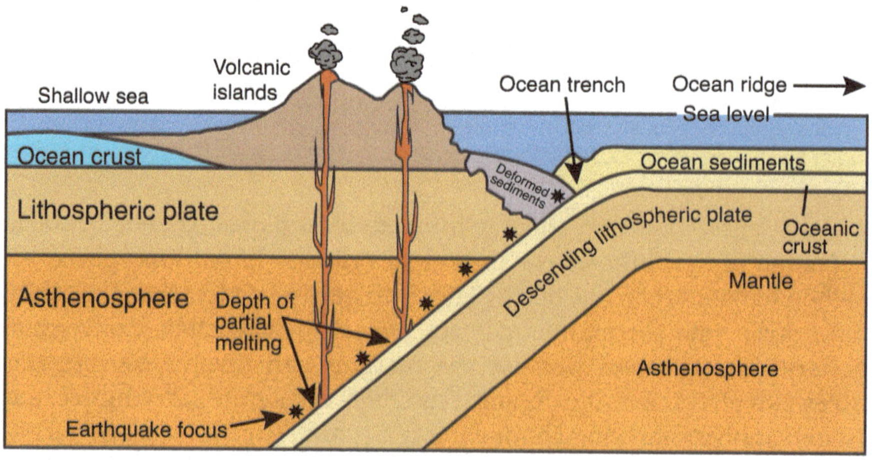

Figure 2.16 The convergence of two oceanic plates results in the subduction of one plate under the other. The molten material created by the subducting plate rises through the lithosphere and produces a string of volcanoes seated on the ocean floor.

From Physical Geology by Dallmeyer. Copyright © 2000. Reprinted by permission of Kendall Hunt Publishing Company.

they smash together to form a collision boundary (**Figure 2.18**). The force of collision forms a mountain range from the compression of the two continents at their convergent boundary. The Himalaya Mountains between India and China are being formed in this way today, as were the ancient Appalachian Mountains about 450 million years ago (**Figure 2.19**). This boundary also produces frequent and powerful earthquakes.

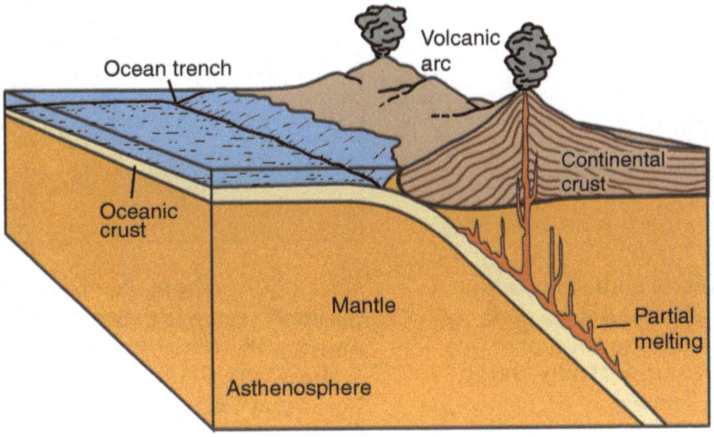

From Physical Geology by Dallmeyer. Copyright © 2000. Reprinted by permission of Kendall Hunt Publishing Company.

Figure 2.17 The convergence of an oceanic plate with a continental plate results in the subduction of the denser ocean plate under the less dense continental plate. Melting of the subducted plate produces magma that rises up through the continental lithosphere, forming a string of volcanoes on the continent.

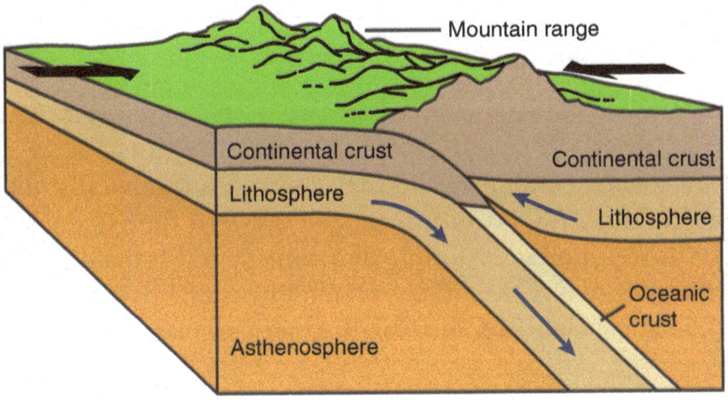

From Physical Geology by Dallmeyer. Copyright © 2000. Reprinted by permission of Kendall Hunt Publishing Company.

Figure 2.18 The collision of two continental masses does not produce subduction of the continental crust. This is because both masses are less dense than the material in the lithosphere and thus do not sink.

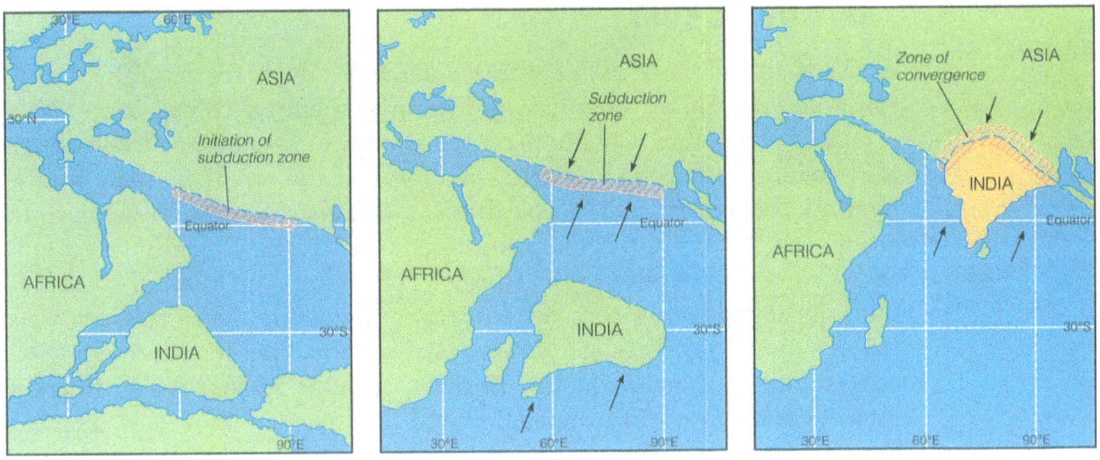

From Planet Earth by Renton. Copyright 2002. Reprinted by permission of Kendall Hunt Publishing Company.

Figure 2.19 Over the past 70 million years the Indian subcontinent has moved northward and collided with the southern edge of the Eurasian Plate. This collision has pushed up the rocks on the surface and formed the Himalaya Mountains, the highest mountain range on Earth.

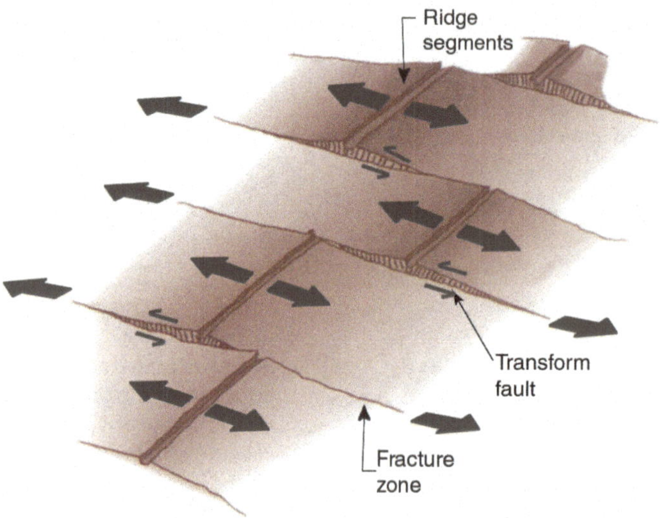

From Planet Earth by Renton. Copyright 2002. Reprinted by permission of Kendall Hunt Publishing Company.

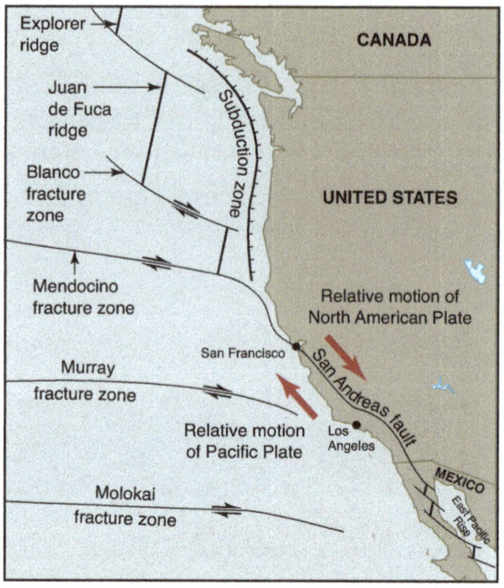

© Kendall Hunt Publishing Company.

Figure 2.20 Transform faults occur perpendicular to the oceanic ridges as the material formed at the ridge moves outward. Plates move in opposite directions between ridge segments along transform faults, and in the same direction along fracture zones.

Figure 2.21 The Pacific Plate is sliding to the north against the western edge of North America. This contact has formed the San Andreas fault, a right-lateral strike slip fault. Notice that California actually lies on two different plates.

Transform Plate Boundaries

Transform plate boundaries occur when two plates slide past one another horizontally (Figure 2.20) along faults known as **transform faults**. These are also zones of frequent and powerful earthquakes, but generally not zones of volcanism since material is not transferred between the asthenosphere and lithosphere. A well-known example of a continental transform fault is the San Andreas fault of California, US, which forms one part of the boundary between the Pacific plate and the North American plate (Figure 2.21).

The Rock Cycle

A **rock** is an aggregate of one or more **minerals** which are naturally occurring inorganic solids, each having a definite chemical composition and a crystalline structure. Chemical composition and crystalline structure are the two most important properties of a mineral and give them their unique physical properties, such as hardness. Geologists classify minerals into groups that share similar compositions and structures. The most important and common minerals are the silicates, which are composed of combinations of oxygen and silicon with (or without) metallic elements. Geologists classify rocks into three categories based on how they were formed: *igneous, sedimentary,* and *metamorphic.* Each group is made of many different rock types that vary in composition and texture, such as size, shape, and arrangement of minerals.

transform plate boundary
An area where two plates meet and are sliding side to side past each other.

transform fault
A strike-slip fault with side to side horizontal movement that offsets segments of a continental or oceanic plate.

rock
A naturally occurring aggregate of minerals. Rocks are classified by mineral and chemical composition, the texture of the constituent particles, and also by the processes that formed them. Rocks are thus separated into igneous, sedimentary, and metamorphic rocks.

mineral
Any naturally occurring inorganic substance found in the Earth's crust as a crystalline solid.

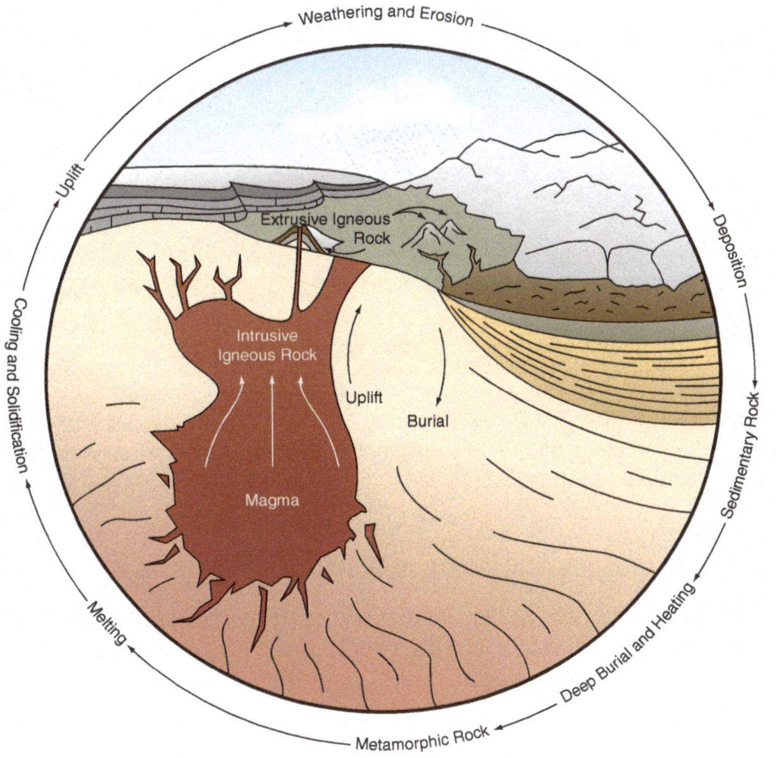

© Kendall Hunt Publishing Company.

Figure 2.22 In the rock cycle material is continually reworked to form sedimentary, igneous, and metamorphic rocks.

Rocks may seem permanent and unchanging over a human lifetime. However, over geologic time rocks are constantly exposed to different physical and chemical conditions (or environments) that change them. The processes that change rocks from one type to another are illustrated by the rock cycle (**Figure 2.22**). The interactions of energy, Earth materials, and geologic processes form and destroy rocks and minerals. The rock cycle is the slowest of Earth's cyclic processes with rocks recycled on a scale of hundreds of millions of years. This slow process is responsible for concentrating the planet's nonrenewable resources that humans depend upon. Plate tectonics is responsible for recycling rock materials and drives the rock cycle, as well as determining to a large degree how and where different rock types will form. The rock cycle provides a way of viewing the interrelationship of internal and external processes and how the three rock groups relate to each other.

Igneous Rocks

Igneous rocks result when molten rock, called magma, cools and solidifies or volcanic ejecta such as ash accumulate and consolidate (**Figure 2.23**). Magma can be formed from partial melting of the mantle or melting of continental crust. Magma that cools and solidifies (a process called crystallization) slowly beneath the surface produces *intrusive igneous rocks* whereas magma that erupts and cools on the surface produces *extrusive igneous rocks*. Igneous rocks make up approximately 95% of the Earth's crust, but their great abundance is hidden on the Earth's surface by a relatively thin but widespread layer of sedimentary and metamorphic rocks.

igneous rock
A rock formed when molten rock (magma) has cooled and solidified (crystallized). Igneous rocks can be intrusive (plutonic) or extrusive (volcanic).

© Linda_K/Shutterstock.com.

Figure 2.23 Igneous rocks can form either from the cooling of magma in the subsurface or by solidifying after being erupted onto the surface. These rocks formed following a catastrophic eruption of a volcano in the Southwestern United States more than 27 million years ago.

© Leene/Shutterstock.com.

Figure 2.24 Sedimentary rocks are typically layered, having been formed by the deposition of sediments derived from weathering of the continents.

Rocks exposed at the Earth's surface break down physically and chemically by **weathering** caused by exposure to gases and water in the atmosphere and hydrosphere. Weathering processes result in particles and dissolved ions that are reworked and redeposited as sediment and by erosional agents such as streams, glaciers, wind, or waves.

weathering

A process that includes two surface or near-surface processes that work in concert to decompose rocks. Chemical weathering involves a chemical change in at least some of the minerals within a rock. Mechanical weathering involves physically breaking rocks into fragments without changing the chemical makeup of the minerals within it.

Sedimentary Rocks

Sedimentary rocks result from the lithification (the process of turning loose sediment into stone) of unconsolidated sediment (Figure 2.24). Sediments are usually lithified into sedimentary rock when compacted by the weight of overlying sediment layers, or when cemented by minerals as subsurface water containing dissolved ions moves through the pore spaces between sediment particles and those ions are precipitated as secondary minerals filling the open spaces. Sedimentary rocks can form from the consolidation of rock fragments and mineral grains, precipitation from solution, or compaction of plant and animal remains. Since these rocks form from sediment at the Earth's surface, geologists can determine the type of environment in which sediments were deposited and the transporting agent. Sedimentary rocks can also contain evidence of past life-forms preserved as fossils and are useful in interpreting Earth's history.

sedimentary rock

A sedimentary rock is formed from pre-existing rocks or pieces of once-living organisms. They form from deposits that accumulate on the Earth's surface. Sedimentary rocks often have distinctive layering or bedding.

Metamorphic Rocks

Metamorphic rocks result from the transformation of other rocks when deeply buried and subjected to high heat and pressure as well as chemical activity with fluids (Figure 2.25). Rocks recrystallize under high heat and pressure and the new minerals align themselves in a similar orientation to become foliated; if there is no directional pressure as the new minerals crystallize, then the mineral grains form with a random orientation and are nonfoliated. Foliation is the parallel alignment of minerals due to pressure, which gives the rock a layered or banded appearance. With the addition of

metamorphic rock

A rock that has been altered physically, chemically, and mineralogically in response to strong changes in temperature, pressure, shearing stress, or by chemical action of fluids.

higher pressures or increased temperatures, the metamorphic rocks will melt, creating magma, beginning the cycle again.

The rock cycle can follow many different paths. For example, weathering may turn uplifted metamorphic rocks into sediment, which becomes lithified into a sedimentary rock. An igneous rock may become metamorphosed into a metamorphic rock. The rock cycle simply shows that rocks are not permanent but change over geologic time through internal and external processes. The rock cycle illustrates several types of interactions between Earth system's spheres. Interactions occur among rocks, the atmosphere, the hydrosphere, the cryosphere, and the biosphere. Water and air, aided by natural acids and other chemicals secreted by plants and animals, weather solid rocks to form large amounts of *clay*, an important silicate mineral, and other sediments. During these processes, water and atmospheric gases react chemically to become incorporated into the clay minerals and transfer matter from the atmosphere and hydrosphere to the solid material of the geosphere. Rain and gravity then wash the sediment into streams, which can be transported to the edge of the continents and deposited in the oceans. Energy from the Sun powers the hydrologic cycle to evaporate water to form rain, which in turn feeds flowing streams. The interaction of the hydrologic cycle is also part of the rock cycle, illustrating again that all of Earth's processes are interrelated.

© Gerry Bishop/Shutterstock.com.

Figure 2.25 **Metamorphic rocks form when rocks are heated, subjected to pressure, or have chemically active fluids introduced into their composition. They do not melt. These chemical reactions take place in a solid state and the final composition of the metamorphic rock is usually similar to the starting composition of the protolith, which is the material that was metamorphosed.**

The Hydrologic Cycle

Radiant energy from the Sun heats water at Earth's surface, causing evaporation of water molecules, which then rise into the atmosphere as vapor. The heated air and water vapor are less dense than surrounding cooler air, and so they rise and carry moisture upward into the atmosphere, where temperatures become progressively colder with altitude. As the rising air cools, the water vapor molecules combine, or condense, back into microdroplets of water through the process of condensation. Depending on the air temperature, the condensed moisture forms precipitation, where liquid rain or solid snow falls under the influence of gravity. Precipitation that falls to Earth's surface replenishes surface water and also infiltrates into the ground. Where snow falls in cold regions, it accumulates to form glacial ice. Gravity pulls on water and ice, converting their potential energy into kinetic energy as flowing streams and glaciers, which enables them to erode the surface. The rising and falling of water vapor in the atmosphere, along with heating of atmospheric gases, adds to Earth's wind systems that generate the surface currents in the oceans and create waves on the ocean surface.

From the processes described we can see that water exists in three states on Earth: liquid, vapor, and solid. Water constantly changes from one state to another, depending on temperature, and is stored in a number of dynamic reservoirs. Some examples of these dynamic reservoirs are the oceans, glaciers, lakes and streams, groundwater, clouds, plants, and animals. The

Figure 2.26 The hydrologic cycle describes the movement of water through the atmosphere, geosphere, biosphere, cryosphere, and hydrosphere.

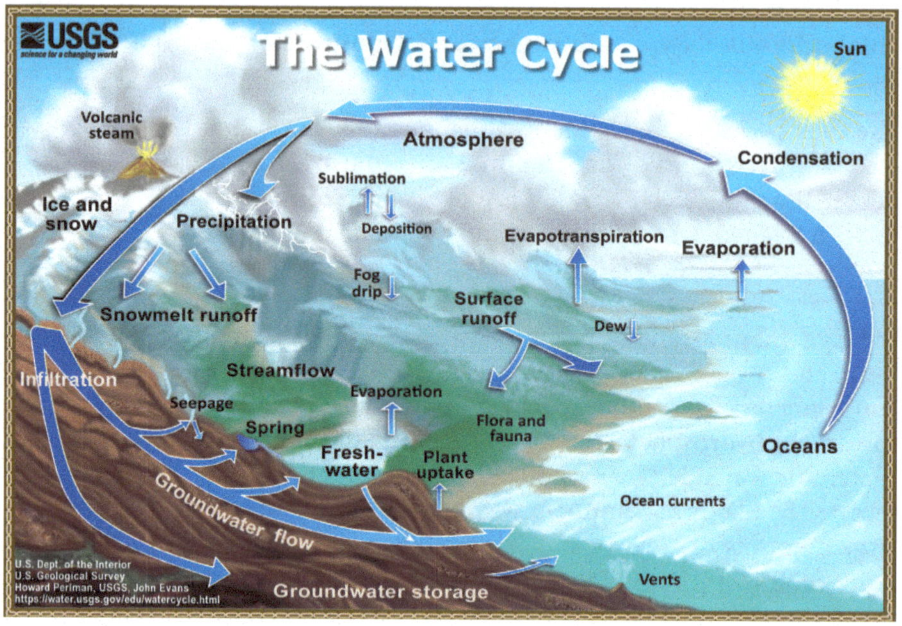

United States Geological Survey.

continual circulation of water in its three states through the atmosphere, hydrosphere, cryosphere, biosphere, and geosphere is called the hydrologic cycle, or simply, the water cycle (Figure 2.26). Overall, the hydrologic cycle operates on a much shorter time scale than the rock cycle because the exchange of water between these reservoirs can be quite rapid. This is known as residence time. Water resides in the atmosphere for only a few days, which means that the water cycle can respond to changing conditions much faster than other cycles that have a longer response time.

The Biogeochemical Cycle

Besides water cycling through the biosphere in plants and animals, other materials (e.g., carbon and nitrogen) also have a high concentration in the biosphere. Cycles that involve the interactions between the biosphere and other reservoirs utilize biological processes such as respiration, photosynthesis, and decomposition (decay), which are referred to as biogeochemical cycles. A biogeochemical cycle is a pathway by which a chemical element or molecule moves through both biotic and abiotic components of an **ecosystem** (Figure 2.27). All chemical elements occurring in organisms are part of biogeochemical cycles. These chemical elements also cycle through the other Earth spheres as well. All the chemicals, nutrients, and elements such as carbon, nitrogen, oxygen, and phosphorus that are used by living organisms are continuously recycled. An important example is the carbon cycle, which is critical in regulating our global climate by affecting levels of greenhouse gases within the atmosphere.

ecosystem

A community of plants, animals, and other organisms that interact within their given setting.

© Guenter Albers/Shutterstock.com.

Figure 2.27 A forest ecosystem contains trees, soil, moisture, and atmosphere, all of which interact to produce a dynamic, living environment.

Carbon Cycle

Carbon (C) is the basic building block of life. As the fourth most abundant element in the universe, carbon occurs in all organic substances, including DNA, bones, coal, and oil. Carbon moves through the Earth's spheres, each of which serves as a reservoir of carbon and carbon dioxide. The **carbon cycle** (Figure 2.28) involves the biogeochemical movement of carbon as it shifts between living organisms, the atmosphere, water environments, and even solid rock. The least amount of carbon is contained in the atmosphere while the largest amount is found in the lithosphere.

carbon cycle
The biogeochemical cycle by which carbon is exchanged among the biosphere, geosphere, hydrosphere, and atmosphere of the Earth.

Carbon dioxide occurs in all spheres on Earth and as a gas is readily mobile. Our increased use of fossil fuels has increased the amount of carbon dioxide in the atmosphere, with some of the gas being taken up in the hydrosphere and biosphere. Marine and freshwater organisms use carbon dioxide in their life cycles and through their incorporation into sedimentary rocks can increase the amount of carbon dioxide in the geosphere. An example is the coral reefs that are formed by corals extracting dissolved carbon dioxide from sea water.

Generally there is a dynamic equilibrium between and among the reservoirs of carbon. However, these are often perturbed by natural processes that contribute to the production and hence increased abundance of carbon dioxide. Volcanic eruptions and wildfires are two natural disasters that upset carbon equilibrium. These two examples are discussed elsewhere in the text.

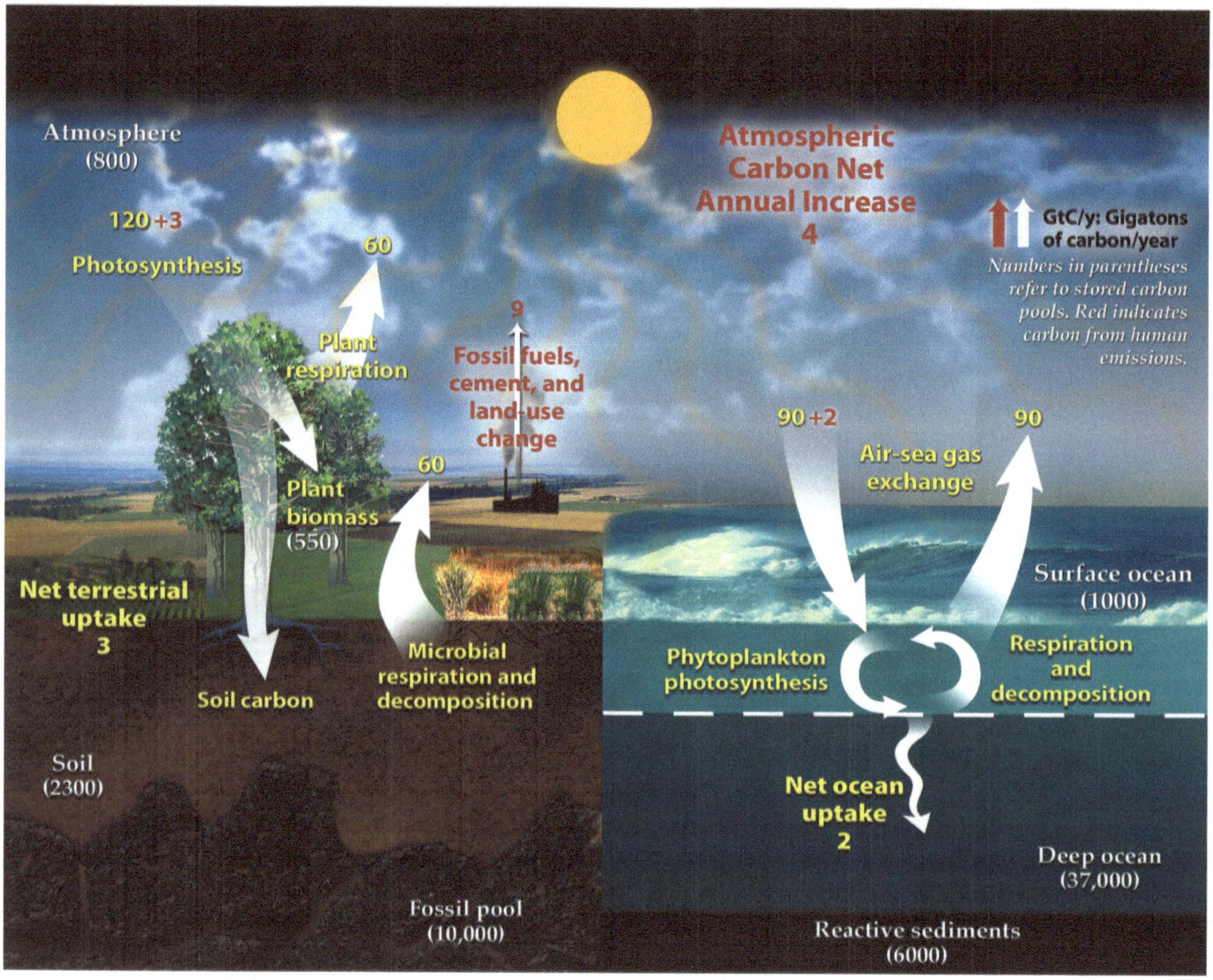

U.S. Department of Energy, Genome Management Information System, Oak Ridge National Laboratory.

Figure 2.28 The carbon cycle shows where carbon is contained in the geosphere, hydrosphere, biosphere, and atmosphere. Each year more than 3 gigatons of carbon are added to the cycle.

Key Terms

asthenosphere *(page 41)*
atmosphere *(page 36)*
biogeochemical cycle *(page 44)*
biosphere *(page 39)*
carbon cycle *(page 59)*
conduction *(page 47)*
convection *(page 48)*
convection cell *(page 48)*
convergent plate boundary *(page 50)*
core *(page 43)*
crust *(page 41)*
cryosphere *(page 39)*
divergent plate boundary *(page 50)*
Earth system *(page 36)*
ecosystem *(page 58)*
geosphere *(page 41)*
gravitational energy *(page 47)*
heat transfer *(page 47)*
hydrologic cycle *(page 44)*
hydrosphere *(page 39)*
Igneous rock *(page 55)*
impact energy *(page 47)*
inner core *(page 41)*
island arc *(page 52)*
kinetic energy *(page 44)*
lithosphere *(page 41)*
lithospheric plate *(page 49)*
mantle *(page 42)*

metamorphic rock *(page 56)*
mid-ocean ridge *(page 50)*
mineral *(page 54)*
oceanic crust *(page 41)*
oceanic trench *(page 52)*
outer core *(page 41)*
plate *(page 43)*
plate boundary *(page 50)*
plate tectonics *(page 49)*
potential energy *(page 44)*
radiation *(page 48)*
radioactive decay *(page 47)*
rock *(page 54)*
rock cycle *(page 44)*
sedimentary rock *(page 56)*
silicate *(page 42)*
stratosphere *(page 38)*
subduction *(page 52)*
subduction zone *(page 51)*
transform fault *(page 54)*
transform plate bounday *(page 54)*
troposphere *(page 37)*
volcanic arc *(page 52)*
weathering *(page 56)*

Summary

The Earth is a dynamic planet made up of chemical reservoirs that are constantly interacting: the atmosphere, hydrosphere and cryosphere, biosphere, and geosphere. The atmosphere, composed primarily of nitrogen and oxygen, is divided into five layers: troposphere, stratosphere, mesosphere, and thermosphere. The water cycle includes all of Earth's water in liquid or solid form in the hydrosphere and cryosphere. The biosphere represents all past or present life on Earth. The geosphere is made of the solid rocks of our planet, and can be classified based on chemical composition and physical properties.

Chemically, the geosphere is divided into the crust, mantle, and core. Earth's crust is the thin outermost layer of our planet and can be either continental crust or oceanic crust, which have different chemical compositions and characteristics. Continental crust is from 20 to 70 km (12 to 44 mi) thick and is made of light elements like silicon, oxygen, potassium, aluminum and iron. Oceanic crust is 5 to 10 km (3 to 6 mi) thick and is made of basaltic volcanic rock which contains more iron and magnesium and is denser than continental crust. The mantle is the thickest layer of the Earth and is made predominantly of peridotite, an igneous rock denser than basalt. The core of the Earth is made of iron-nickel metal. These layers behave in different ways depending on their physical properties. If we classify the Earth based on physical properties instead of composition, then we have a different series of layers. These are the lithosphere, asthenosphere, mantle, outer core, and inner core. The lithosphere is the brittle, cold outermost shell of Earth. It is composed of the crust and the outermost upper mantle which is cold and also behaves like a solid, brittle rock. The asthenosphere lies at depths between 80 and up to 800 km (50 to 500 mi) and is on average about 250 km thick (155 mi). Because of the increase in temperature as you move into the Earth, this upper part of the mantle behaves like a plastic, ductile layer underlying the brittle lithosphere. Below the asthenosphere is the mantle, where the extremely high pressure of overlying rocks offsets the increasingly high temperature and causes this part of the Earth to behave as a stronger and more rigid solid. The core is divided into two different layers based on physical properties. The outer core is molten liquid metal and the inner core is solid metal. The combination of the rigid lithosphere over the ductile asthenosphere allows the outer surface of Earth to break into plates that move around.

The interactions among different components of the Earth systems forms cyclical processes: the rock cycle, the hydrologic cycle, and the biogeochemical cycle. These cycles are part of the natural process of the planet trying to maintain equilibrium with respect to energy and matter. Energy can be either potential or kinetic, and comes from the Sun and the Earth's interior. Heat energy is transferred by conduction, convection, or radiation. Plate tectonic theory explains how the outer rigid layer of Earth—the lithosphere—is fragmented into plates that move and interact along plate margins. Plates can be composed of ocean crust, continental crust, or both. Plate motions can diverge, converge, or slide past each other. At divergent plate boundaries, two plates pull apart from each other. Mid-ocean ridges are divergent boundaries where new oceanic crust is created in volcanic eruptions, like the Mid-Atlantic

Ridge or Iceland. Continental rift zones are where continental crust is thinning and stretching, splitting along a rift zone, such as in the East African Rift. Convergent plate boundaries are where plates move toward each other. Ocean crust is colder and denser than continental crust so it will be subducted into the mantle, as happens along the western side of South America. If two oceanic crustal plates are converging, the older, colder plate will be subducted below the younger, less cold plate. Continental crust is buoyant and does not want to be subducted, so two continental plates collide and form large mountain ranges, as is happening today with the collision of India and the Eurasian continent forming the Himalaya Mountains. Transform boundaries are not creating or destroying crust, they are transposing it as one plate slides past another.

The rock cycle explains the complex processes of forming igneous, sedimentary, and metamorphic rocks, and how they are interrelated. The hydrologic cycle follows water through the Earth's atmosphere, cryosphere, hydrosphere, and geosphere as it moves among lakes and rivers, oceans, subsurface aquifers, and the atmosphere. The biogeochemical cycle follows all things that are or were alive through the biosphere, geosphere, and atmosphere.

References and Suggested Readings

Coch, N. K. (1995). *Geohazards: Natural and human.* Prentice Hall.

Condie, K. C. (2016). *Earth as an evolving planetary system.* Academic Press.

Duarte, J. C., & Schellart, W. P. (2016). *Plate boundaries and natural hazards.* John Wiley and Sons, Inc.

Kious, W. J., & Tilling, R. I. (1996). *This dynamic earth: The story of plate tectonics.* U.S. Geological Survey. http://geology.usgs.gov/publications/text/dynamic.html

McCall, G. J. H., Laming, D. J. C., & Scott, S. C. (1992). *Geohazards: Natural and man-made.* Chapman and Hall.

Molnar, P. H. (2015). *Plate tectonics—A very short introduction.* Oxford University Press.

Renton, J. (2011). *Physical geology across the American landscape* (3rd ed.). Kendall Hunt Publishing Company.

Stein, S., & Freymueller, J. T. (2002). *Plate boundary zones.* American Geophysical Union.

Watters, T. R., & Schultz, R. A. (Eds.). (2010). *Planetary tectonics.* Cambridge University Press.

Websites for Further Reference

https://geomaps.wr.usgs.gov/parks/pltec/

https://svs.gsfc.nasa.gov/2953

https://science.nationalgeographic.com/science/earth/the-dynamic-earth

https://www.ngdc.noaa.gov/mgg/global

https://www.platetectonics.com

https://www.usgs.gov

Questions for Thought

1. Distinguish between the processes of conduction and convection as ways to transfer heat energy within the Earth.
2. Describe the differences in formation of igneous, sedimentary, and metamorphic rocks.
3. What is the rock cycle, and what is the energy source driving it?
4. List the major internal layers of the geosphere on the basis of chemical composition and physical properties.
5. List and explain the types of plate boundaries.
6. List and briefly describe the spheres of the Earth system.
7. What are the energy sources for the Earth system?
8. How did the Earth system form?
9. Describe the effects on the Earth's surface as a result of internal and external processes.
10. Explain how the hydrologic cycle operates and its role in external processes.

Volcanic Eruptions 3

© fboudrias/Shutterstock.com.

An eruption of Volcán de Fuego (Spanish for Volcano of Fire) in Guatemala. This is one of Central America's most active volcanoes.

Living With Volcanic Risk

More than 800 million people live within 100 kilometers of a volcano, but few of the potential hazards associated with all of the active volcanoes on Earth are monitored regularly. Forecasting the hazards and impact of eruptions accurately is an ongoing issue—scientists are combining satellite measurements of ground movement and artificial intelligence to more accurately monitor volcanic activity, and may one day be able to predict volcanic eruptions. Detailed research looking at the internal 'plumbing' of volcanoes like Etna in Italy also helps us to understand eruptions and how quickly magma moves to the surface. The United States Geological Survey predicts that California's next big hazard could be a volcanic eruption rather than an earthquake. Auckland, the largest city in New Zealand, has 1.7 million people who live on a volcanic field made up of 53 different volcanoes—the youngest of which is only 600 years old. Residents may have as little as a few hours to several days of warning before the next eruption.

© Steven Bostock/Shutterstock.com.

Figure 3.0.1 **Auckland, New Zealand, city skyline with Mount Victoria volcano—now a public park and greenspace—in the foreground of the picture.**

magma
Molten rock that lies below the surface.

lava
Molten rock that flows onto the surface and cools.

The Plate Tectonic Connection

Earlier discussions of plate tectonics showed us that volcanic activity has played a major role in the development of Earth's outer crust. Volcanoes are the result of molten rock (**magma**) that has worked its way to the surface and has been extruded onto either the continents or the ocean floor as **lava**. Lava can form volcanic mountains (Figure 3.1) and flows on land (Figure 3.2).

© Ecuadorpostales/Shutterstock.com.

Figure 3.1 **Tungurahua Volcano in Ecuador was quiet for many years before it began erupting in 2000. The last activity of this current eruptive episode ended in March of 2016.**

© Wead/Shutterstock.com.

Figure 3.2 **Lava flowing down Mount Etna, Italy.**

© Kendall Hunt Publishing Company. Used with permission.

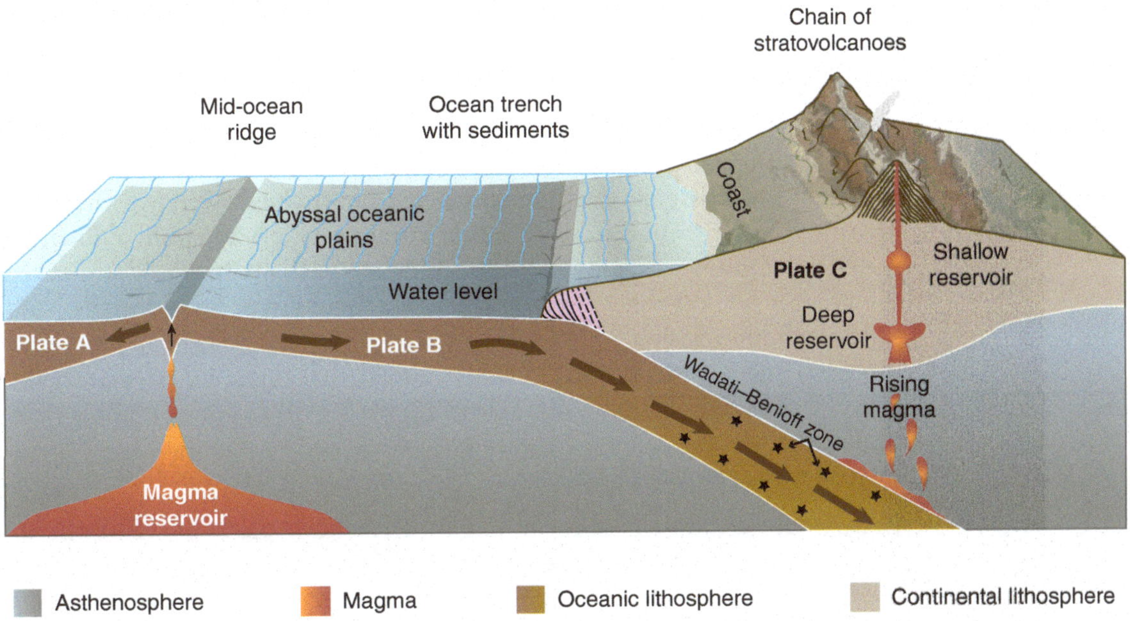

Figure 3.3 Cross section showing the divergent movement of oceanic lithospheric Plates A and B away from the mid-ocean ridge; Plate B is subducted underneath continental Plate C at a convergent plate boundary.

Basalt magma rises under the mid-oceanic ridges as plates move away from each other in a divergent manner (**Figure 3.3**). The extruded lava is cooled and quenched relatively rapidly and forms lava flows and pillow basalts, which are tubes of lava similar to toothpaste squeezed from a tube, and form because the outer shell of the lava quenches extremely fast underwater. As these tubes stack up, they drape over each other and form what looks like pillows or blobs in cross-section (**Figure 3.4**).

basalt

An extrusive, fine-grained, dark volcanic rock that contains less than 50% silica by weight and a relatively high amount of iron and magnesium.

As tectonic plates continue to move away from the ridge more magma flows out onto the ocean floor and hardens, forming the new ocean crust of basaltic oceanic lithosphere. This activity is not readily observed as it occurs under several kilometers of water in most places except Iceland, where the mid-ocean ridge is exposed above sea level.

Oceanic plates that move away from the mid-oceanic ridge in one area of the Earth eventually collide with other plates. When collision occurs between two oceanic plates, or between an oceanic plate and a continental plate, basaltic oceanic lithosphere and sediments that have accumulated on the oceanic plate are subducted and driven downward into the mantle where they heat up. The trapped water in these subducted rocks is driven out of the sediments, rocks,

Source: National Oceanic & Atmospheric Administration (NOAA).

Figure 3.4 This pillow lava from the Galapagos Islands shows a collapsed pillow lava tube with lava flowing out of it—notice the spherical shape of the pillows formed as they quench in water.

and minerals by the increased heat at depth. This dehydration process triggers melting of the overlying mantle rocks (Figure 3.5). At a depth of roughly 100 kilometers (62 mi) there is sufficient heat to begin melting and produce new magma, which rises toward the surface.

Near the continental margin the subducted oceanic lithosphere is relatively cold as it is usually old and far removed from its heat source at the mid-oceanic ridge. The descending oceanic lithosphere is being pulled down by the effect of gravity tugging on the sinking, colder slab. In Figure 3.3 we see the Wadati-Benioff zone, a region where earthquakes occur along the cold, brittle, subducting slab. As the subducted oceanic plate descends, sometimes it breaks into pieces. Seismic evidence shows that there are regions in the subsurface of subduction zones where no earthquakes are occurring. This produces a seismic gap on the surface, a situation we will look at when we discuss earthquakes in detail.

Earth's surface consists of seven major lithospheric plates as well as many smaller ones to total about 20 (Figure 3.6). Notice that continental plates such as the African plate and the North American plate also include oceanic areas as the plates extend to the mid-oceanic ridges and continues to form new crust on that boundary. The continents themselves consist of less dense, granitic material that "floats" atop the denser, underlying **asthenosphere**.

asthenosphere

The semi-plastic to molten zone beneath the rigid lithosphere.

There are about 1,500 known active volcanoes, and of these about 500 have erupted during historical times. More than 90% of these are associated with plate boundaries and approximately two-thirds of all active volcanoes are located around the Ring of Fire that surrounds the Pacific Ocean

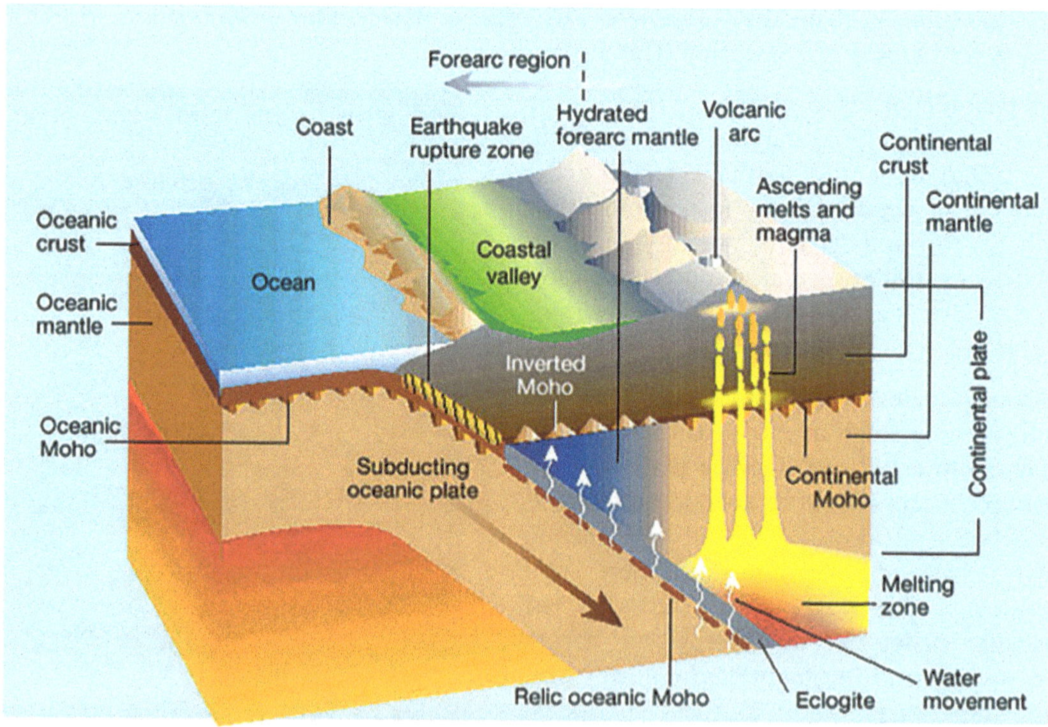

Nature (2002) v. 417, 497–498.

Figure 3.5 **Subduction of oceanic lithosphere at a continental margin. Melting of the oceanic lithosphere, subcontinental mantle, and continental lithosphere and crust intermix to produce a magma that forms a chain of volcanoes near the edge of the continent.**

From Volcanoes: An Introduction by Alwyn Scarth, 1994. Copyrighted material used with permission from Texas A & M University Press. Adapted by Kendall Hunt Publishing Company.

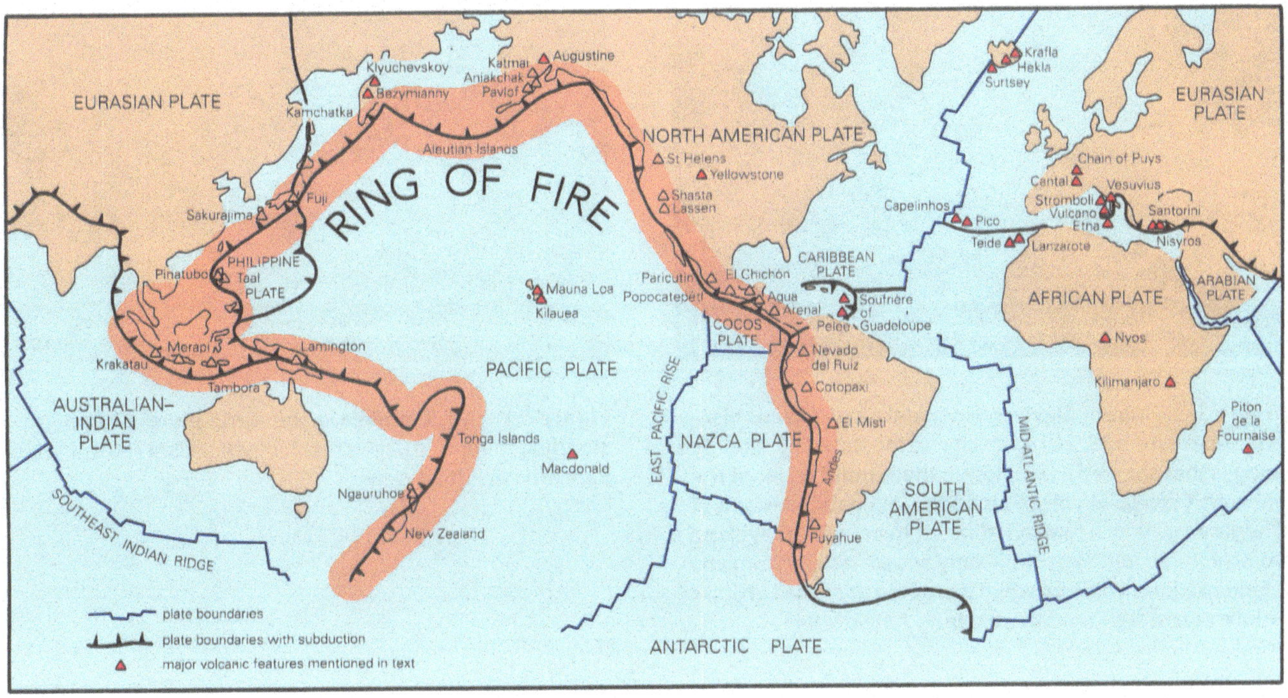

Figure 3.6 **The major oceanic and continental plates on Earth's surface. Major volcanoes are shown by the triangles. Notice that many of the volcanoes are located along convergent plate boundaries where subduction is occurring.**

(**Figure 3.7**). Those areas along the Ring of Fire that display an almost continuous line of active volcanoes, such as in the Aleutian Islands southwest of Alaska (**Figure 3.8**) or the area around Japan and the Kurile Islands, are regions where very active subduction is taking place. Both volcanic eruptions and significant earthquake activity occur in these areas as the Pacific Plate is subducted under the adjacent continental and other oceanic plates.

© Yarr65/Shutterstock.com.

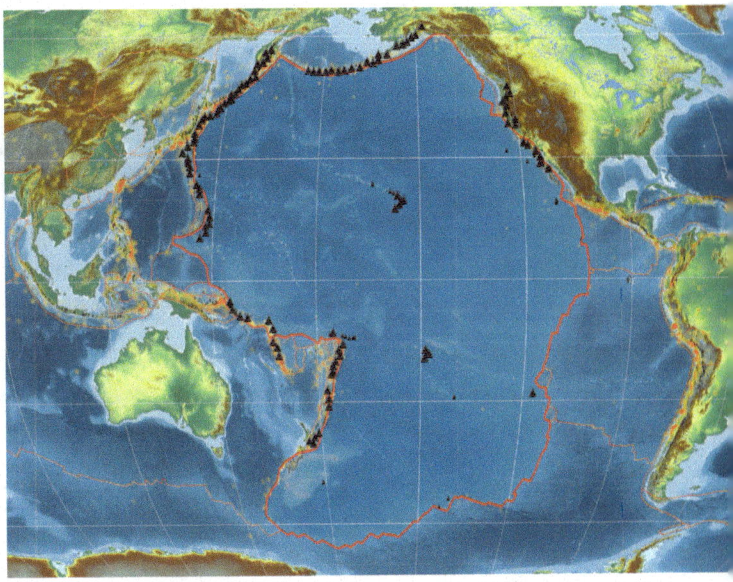

Figure 3.7 **Major volcanoes found in the circum-Pacific Ring of Fire and adjoining regions.**

Of the world's 15 most populous countries, portions or all of five of the countries (the United States, Indonesia, Japan, Mexico, and the Philippines) lie on the Ring of Fire and have areas that could experience the direct effects of major volcanic eruptions. The hazard potential for large portions of Indonesia, Japan, and the Philippines is high. Volcanic eruptions during the past 2 decades have been common in these regions.

Volcanoes also occur in **rift valleys**—regions on the Earth that are undergoing stretching due to large-scale, extensional tectonic forces. These are areas where the lithosphere is being pulled apart. The East African Rift System, which passes through Uganda, Kenya, Tanzania, and Ethiopia, is home to many volcanic peaks, including Mount Kilimanjaro (**Figure 3.9**). At an elevation of 5,895 m (19,340 ft), it is the world's highest free-standing mountain. The East African Rift results from the African plate stretching and splitting apart, which is attributed to elevated heat flow from the mantle forming thermal uplifts that stretch and fracture the outer brittle crust. This process is often associated with volcanic eruptions.

rift valley
A depression formed on the surface caused by the extension of two adjacent blocks or masses of rock.

Source: NASA.

© Volodymyr Burdiak/Shutterstock.com.

Figure 3.8 Mount Cleveland in the Aleutian Islands last erupted on June 28, 2018, which was identified by satellite images that showed a lava flow in the summit crater of the volcano. Cleveland Volcano is on the uninhabited island of Chuginadak, which makes it difficult to monitor. Cleveland volcano is responsible for the only known fatality from an Aleutian Island volcano, when it erupted as a small group of soldiers were stationed there during June of 1944.

Figure 3.9 Mt. Kilimanjaro, the world's highest free-standing mountain, is the result of volcanism related to the East African Rift System.

Magma Generation and Eruption Products

Volcanoes form molten rock, which is called magma when it is found below the surface of the Earth, and lava when it is erupted onto the surface. There are three different mechanisms to form magma. Decompression melting causes hot rock to melt when the pressure is dropped, and flux melting causes rock to melt when fluids are added. Adding heat to a solid rock increases the temperature and can cause melting. Depending on the pressure and temperature conditions and composition of the rocks, they usually melt between 750 °C and 1200 °C (1380 to 2200 °F).

Making Magma

decompression melting

Melting of hot rocks in the subsurface caused by a reduction in overlying pressure as the material rises toward the surface.

Decompression melting takes place in the earth where a solid mass of rock is kept at a constant hot temperature while the pressure is reduced as that rock moves closer to the surface (Figure 3.10). This can happen in a mantle plume and in places where the mantle rises close to the surface such as at mid-ocean ridges or continental rift zones. If the rock is hot enough to be close to its melting point and under pressure, when the pressure is released the rock will begin to melt.

flux melting

Melting of a solid rock that is caused by adding water or other volatiles to the hot rock, which lowers the melting temperature.

Flux melting takes place when fluids or volatiles are added to hot solid rock. Water is the most abundant flux melting component in subduction zones (Figure 3.11). As an oceanic plate slides down into the mantle, it heats up and releases water and other volatiles. The volatiles rise up into the overlying mantle and trigger melting of the hot mantle rocks. The down-going slab generally needs to reach depths of 80 to 160 km (50 to 100 mi) in the mantle before it heats up enough to release the trapped fluids, which is why volcanic arcs are associated with subduction zones and are offset from the subduction zone itself.

As hot magma forms and rises up to the surface, it comes in contact with cooler, solid rocks. If the magma stops or stalls, it can transfer its heat to the

© Kendall Hunt Publishing Company.

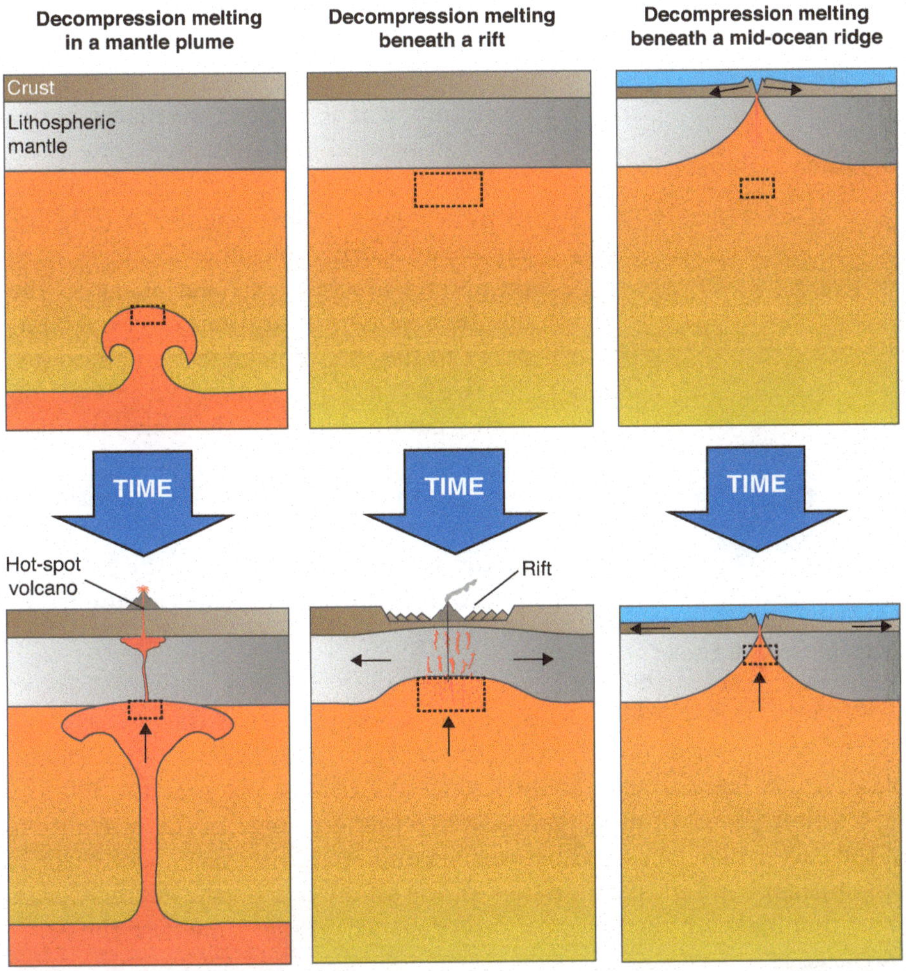

Figure 3.10 Geologic conditions associated with magma generation in different settings.

rocks around it and trigger melting of those rocks (**Figure 3.12**). This heat transfer or conductive heating works because basaltic magmas can have temperatures of from 1000 to 1250 °C (1800 to 2300 °F), and more silicate-rich rocks such as continental crust can melt at temperatures of 600 to 750 °C (1100 to 1400 °F). If there is a large enough source of heat that is maintained over a long time, then large volumes of crust can melt.

Mantle plumes are a special case of decompression melting. They represent a stationary upwelling of hot mantle from deep in the Earth, potentially from the core-mantle boundary, where the internal temperature of the Earth is estimated to be between 4000 and 5000 °C (7200 and 9000 °F). Hot mantle rises up

© Kendall Hunt Publishing Company.

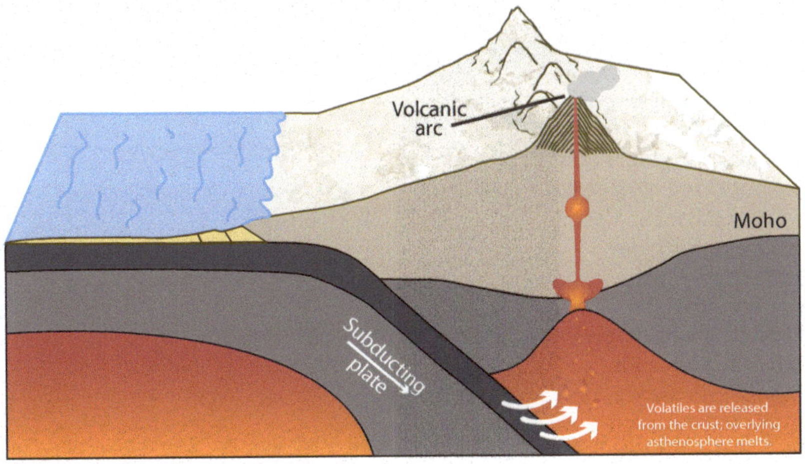

Figure 3.11 Flux melting occurs in subduction zone settings when water and other volatiles in the down-going plate are released from the wet sediments. At about 100 km (62 mi) depth hydrous mineral phases start to break down and also contribute volatiles. These volatiles lower the melting temperature of the overlying hot mantle rocks and produces magma.

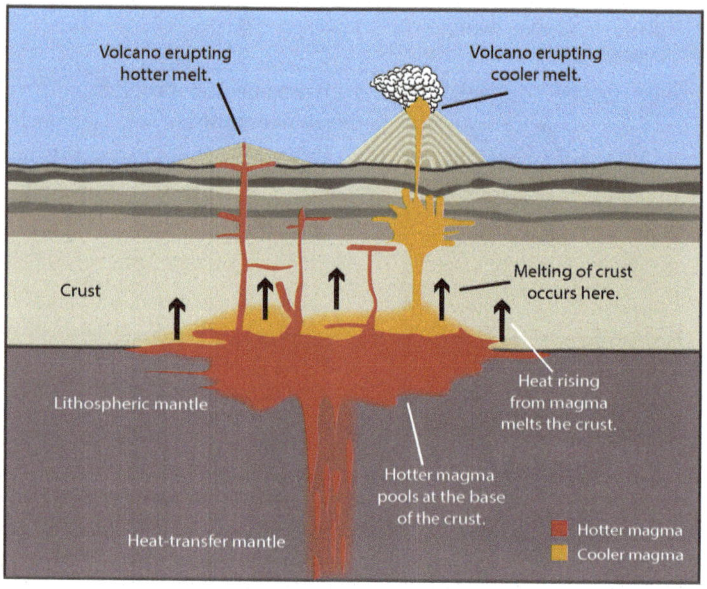

© Kendall Hunt Publishing Company.

Figure 3.12 Heat-transfer melting occurs when hot magma moves into colder crust and the heat from the magma causes the solid surrounding rocks to melt. This usually happens when hot basaltic rocks (about 1200 °C or 2200 °F) intrude into more silicic continental crust, which melts at a much lower temperature.

as a relatively stable and stationary pipe within the mantle, and as it moves toward the surface and the pressure drops it melts because it is at a temperature that is hotter than the surrounding ambient mantle by up to several hundred degrees C. The upwelling magma from a mantle plume moves to the surface and forms volcanoes. Because mantle plumes are relatively stationary in the mantle over tens of millions of years, compared to the rate of plate tectonic motions, the overlying plates will move in relation to the mantle plume. As the plate's position shifts, a new volcano is formed while the previously formed volcanoes have moved off the hot spot. The string of islands that make up the Hawaiian Islands and the Emperor Seamount chain are excellent examples of this process (Figure 3.13).

We can analyze the ages of the volcanoes and their distances from the hot spot to determine rates of motion of the Pacific plate, which has averaged about 8.6 cm (3.4 inches) per year (or 860 km [534 miles] per 10 million years). Notice that a change in the orientation of the Pacific plate occurred between 45 and 50 million years ago. There is a significant bend in the alignment of the seamounts on the Pacific Ocean crust that make up the Hawaiian hotspot track at that time near the Midway Islands (Figure 3.13).

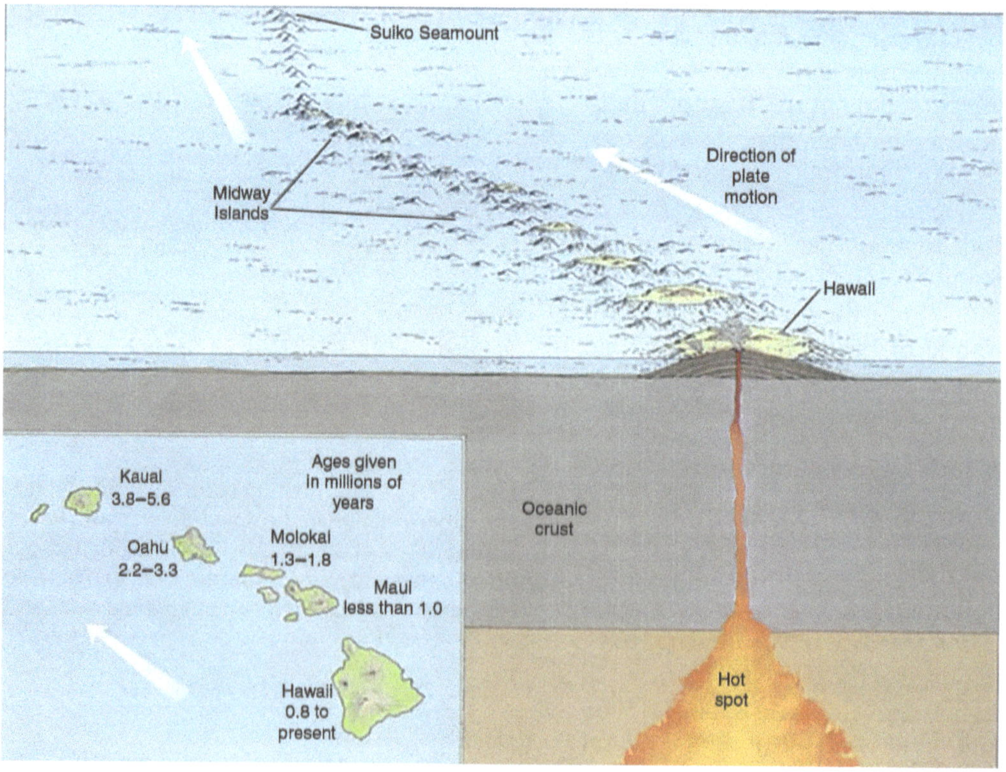

Source: NOAA. Adapted by Kendall Hunt Publishing Company.

Figure 3.13 The Hawaiian Islands and the Emperor Seamount chain extend out from the hot spot lying under the big island of Hawaii. Ages of the rocks are progressively older to the northwest—evidence that the Pacific plate has moved across the hot spot. Exposed islands lie to the southeast of Midway Island; the remaining islands to the northwest are underwater seamounts that used to be exposed above sea level but have eroded and sunk as they move off the hotspot and no longer have the heat flux from the mantle plume.

BOX 3.1 Kilauea Volcano, Hawaii: Eruption 2018

Kilauea Volcano has been erupting along its East Rift Zone since January 1983, and the lower Puna eruption is the 60-second episode of this ongoing event that began on May 3, 2018. Up until this newest phase of activity eruptions were coming from the volcanic cone Pu'u O'o, which is about 20 km (12 mi) east of Halema'uma'u.

Precursors to this new stage of eruption began in spring of 2018. In April, the lava lake at the summit of Kilauea Volcano overflowed and flooded the Halema'uma'u Crater bottom, covering more than two thirds of the crater floor with new lava (Figure 3.1.1). In early May, the lava lake rapidly and unexpectedly dropped more than 200 meters (660 ft), which resulted in explosions and clouds of ash and toxic volcanic gas as groundwater began interacting with the magma in the volcanic vent. At about the same time, the crater floor of Pu'u O'o also collapsed, as lava drained out and flowed more than 15 km (9 mi) along the East Rift Zone. Large earthquakes from magnitude 5.0 to 6.9 hit the Puna area near Hilo, and steaming cracks opened that began to erupt lava in the Leilani Estates.

© Phillip Bl. Espinasse/Shutterstock.com.

Figure 3.1.1 **The lava lake at the vent of Kilauea Volcano, in Halema'uma'u Crater rose to the rim and overflowed on April 29, 2018, a sign that the magma system was filling up.**

By May 27, 24 fissures had opened and erupted, some with lava fountains up to 90 meters high, forcing the evacuation of about 2,000 residents. By early August, lava covered 35 square kilometers of land, including Hawaii Route 132, cutting off access to residents and scientists in the area (Figure 3.1.2). At least 700 houses were destroyed (Figure 3.1.3). The Puna Geothermal Venture plant, which provided 25% of the island's electricity supply, had to shut down and remove flammable substances. The plant was later damaged by lava that flowed over a power station, a warehouse, and at least three geothermal wells.

© Fredy Thuerig/Shutterstock.com.

Figure 3.1.2 **Road closed due to lava flow.**

Figure 3.1.3 **An aerial view of Fissure 8 lava covering a housing estate in Pahoa.**

Lava has created 3.5 square kilometers (2.2 square mi) of new land as it enters the sea and builds up lava deltas (**Figure 3.1.4**), but it also covered over and buried some Hawaiian landmarks such as Kapoho Bay and Green Lake, Hawaii's largest freshwater lake. Fissure 8 became the focus of eruption activity by the end of May, creating a volcanic cone from constant lava spatter that peaked at 50 m (165 ft) high (**Figure 3.1.5**). As of July 2018, the rate of lava flowing from Fissure 8 was estimated to be 98 cubic meters per second, which is equivalent to 98,421 liters or 26,000 US gallons each minute. Enough lava was erupted to fill 320,000 Olympic-sized swimming pools.

Figure 3.1.4 **Lava from Fissure 8 entering the sea on June 6, 2018.**

Figure 3.1.5 **Fissure 8, June 5, 2018.**

In Early August, the intensity of the eruption at Fissure 8 started to decline, as did earthquake activity at Halema'uma'u Crater, although Pu'u O'o had an increase in the amount of volcanic gas venting, reaching the highest rate observed for 10 years. In December of 2018, with no active flows or volcanic activity for the previous 3 months, the Hawaiian Volcano Observatory announced that they consider this to be the end of the active eruption of Kilauea that had been continuous since 1983. Recovery costs are estimated to be more than $800 million US dollars, and Hawaii politicians are reported to be discussing a proposed ban on future home construction in areas at high risk from volcanic eruptions. In December of 2020, Kilauea began erupting again from the crater lake just moments after a magnitude 4.4 earthquake hit the south flank of the volcano.

Volcanic Eruptions

The type of material erupted from a volcano and the way in which it is erupted determine the shape and structural form a volcano takes. Magma rising from a subsurface magma chamber ascends to the surface through dikes and a central conduit. Several factors control the way in which lava and other volcanic eruptive products are spread onto the surface and thrown into the atmosphere.

© xpixel/Shutterstock.com.

Figure 3.14 Honey has a higher viscosity than water so it pours more slowly and builds up a thicker layer on the surface it is flowing over.

Chemical Composition

Volcanic eruption styles are dominantly controlled by the **viscosity** of the magma that is being erupted—low viscosity fluids flow more easily than high viscosity fluids. Water has a low viscosity and flows more easily than honey, which has a higher viscosity (Figure 3.14). Heating decreases the viscosity of a fluid, and warm honey flows more readily than cold honey. In magmas, viscosity is controlled by the amount of silica tetrahedra (Box 3.2). One silicon ion bonds with four oxygen ions to form a three-dimensional pyramid-shaped structure with a silicon in the center and an oxygen at each point. These tetrahedra can link up with each other, forming chains with the shared oxygen ions representing vertices on more than one tetrahedra. The higher the silica content of the magma, the more chains of silica tetrahedra form, resulting in higher degrees of polymerization. The more chains of tetrahedra a magma has, the higher the viscosity of the magma because those chains impede the ability of the magma to flow easily (Table 3.1).

viscosity
A measure of the internal resistance of a substance to flow; a lower viscosity means the material flows easily.

Volatiles

When magmas are generated they contain a component of gas or volatiles that are dissolved in the molten rock. The most abundant volatile is water, and other volatiles include hydrogen, carbon dioxide, fluorine, chlorine, methane, and sulfur dioxide. The amount of volatiles in magma varies with magma composition, and can range from one or two weight percent for a low-silica basaltic composition magma up to six weight percent for a high-silica rhyolite composition magma. This combination of molten rock and magmatic volatiles is under pressure at depth and the gases are dissolved in the liquid rock, but as the magma rises and the pressure drops the gases come out of solution and start to form bubbles. As the magma moves toward the surface, the bubbles expand rapidly. This is the same process that happens when we shake a bottle of carbonated soda and open it. Carbon dioxide dissolved in the soda escapes rapidly and explosively when the cap is removed and the contents are dropped from their pressurized state to atmospheric pressure.

When magmas don't contain a lot of volatiles, or the volatiles in the magma can escape easily, they tend to erupt as effusive eruptions of lava flows. This can happen in hot, fluid basalt lavas that allow the dissolved gas in the system to escape easily and the lava that erupts flows across the surface. Explosive eruptions happen in more viscous magmas where dissolved gas cannot escape easily and the expanding bubbles build up pressure. Magma can have higher viscosity because it is cooler in temperature and more sticky,

United States Geological Survey.

ash
Pyroclastic material that has an average particle size less than 2 mm (0.08 inches).

© lrosebrugh/Shutterstock.com.

lapilli
Pyroclastic material that ranges in size from 2 mm to 64 mm (0.08 to 2.5 inches).

Table 3.1 Lava Characteristics

Volcanic Rock Name	Basalt	Andesite	Dacite	Rhyolite
SiO_2 Content	48–52%	52–63%	63–68%	68–77%
Eruption Temperature	800°C			

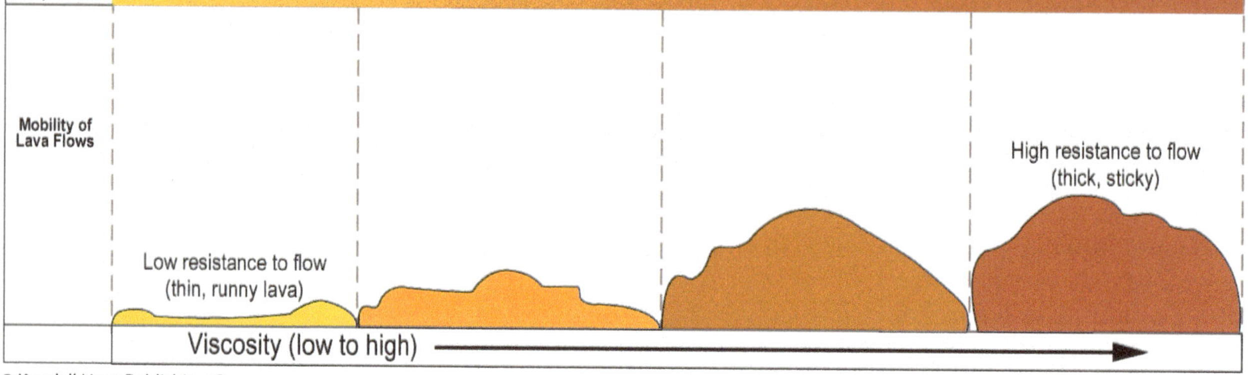

© Kendall Hunt Publishing Company.

BOX 3.2 Influence of the Silicon-Oxygen Tetrahedron on Viscosity

Many of the minerals that make up common volcanic rocks contain the silicon-oxygen tetrahedron, the basic building block for more than one third of the more than 4,000 minerals found on Earth. Oxygen and silicon are the two most common elements in Earth's oceanic and continental lithospheres. The silicon-oxygen tetrahedron consists of one silicon cation surrounded by four oxygen anions, thereby producing a negatively charged anionic complex having a charge of 24. To offset this negative charge, cations need to be present.

© Kendall Hunt Publishing Company.

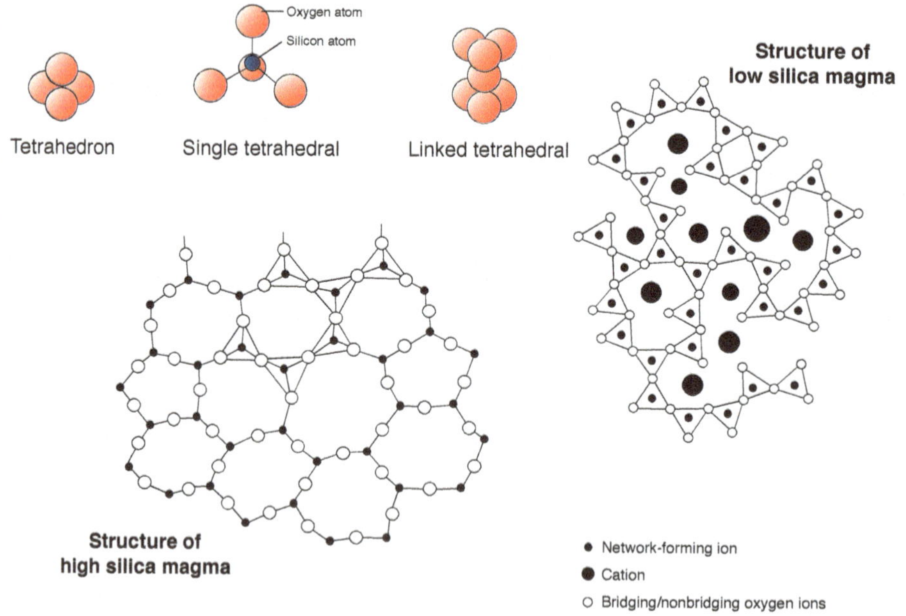

or because it is compositionally higher in silica and has more silica tetrahedra that can link together (Box 3.2). This linking process is called polymerization. Highly polymerized magmas, those that are high in silica, have high viscosities. They also tend to have higher volatiles, and those volatiles have a more difficult time escaping the magma, so the magma contains a larger volume of bubbles with higher pressure. This tends to lead to explosive eruptions.

Carbon dioxide is a common gas that is both colorless and odorless. Because its molecule contains an atom of carbon, carbon dioxide (CO_2) is heavier than oxygen (O_2). When carbon dioxide is expelled from a volcano, it can settle along low-lying areas and displace the oxygen. Numerous examples exist of animals and humans being affected by carbon dioxide. In 1986 an eruption of carbon dioxide killed many people and animals in Cameroon (see later in the chapter). In Iceland, sheep and small birds have died just before major volcanic events (Sigurdsson, 1999) and field geologists have noticed buildups of carbon dioxide in low-lying areas on the Island of Hawaii (Duffield, 2003) during eruptions (Figure 3.15).

Source: I. Ukstins.

block

A large angular fragment of lava measuring more than 64 millimeters (6.4 cm or 2.5 inches) in diameter. The author (I. Ukstins) standing on a block erupted from Vulcano Volcano, Italy.

© wanderlust86/Shutterstock.com.

Figure 3.15 **A warning sign about the dangers of carbon dioxide buildup in low-lying areas from Kilauea Volcano, Hawaii, US.**

Source: I. Ukstins.

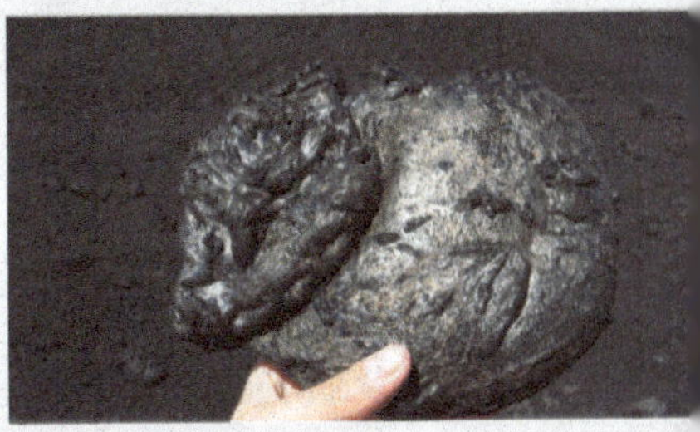

bomb

A large lava fragment larger than 64 millimeters (6.4 cm or 2.5 inches) in diameter that is erupted as a liquid and becomes rounded as it is aerodynamically shaped while traveling through the air. This bomb is from the 2014–2015 eruption of Bárðarbunga, Iceland, and is called a "cowpie" bomb for its flat patty shape that forms as the fluid bomb impacts the ground.

Measuring Volcanic Explosivity

Scientists have devised a measure of the degree of explosivity observed in volcanic eruptions, the **volcanic explosivity index (VEI)**. This index assigns a single value that describes the severity of an eruption (Table 3.2). This scale is roughly analogous to one of the magnitude scales used to measure the severity of an earthquake, a tornado, or a hurricane. Through a great deal of research, the developers of the VEI were able to assign values to ancient eruptions. Observations of the duration of the main eruption, height of the ash column, volume of material erupted, and volume of ash exploded into the upper atmosphere are used to determine the VEI of a volcanic event. Values range from 0 to 8, with the vast majority of events receiving values of 1 or 2. This range in eruptive styles generates a variety of volcanic edifice morphologies (Figure 3.16).

volcanic explosivity index (VEI)

A measure of the intensity of a volcanic eruption, with a value of 0 representing the quietest and 8 being the most explosive.

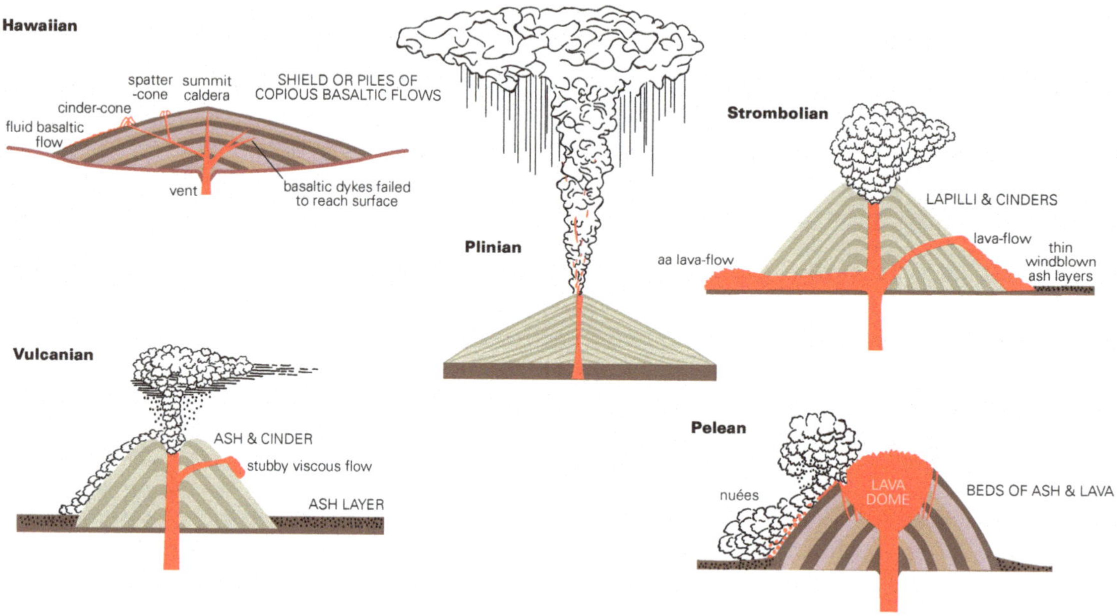

© Kendall Hunt Publishing Company.

Figure 3.16 Examples of different eruptive styles. Refer to the discussion on the VEI to compare the degree of explosiveness with the resulting eruptions.

Table 3.2 Volcanic Explosivity Index (VEI)

VEI	Volume of Ejecta (m^3)	Description	Height of Eruption Column (km)	Classification	Duration (hrs)	Troposhere	Stratosphere
0	$<10^4$	nonexplosive	<0.1	Hawaiian		negligible	none
1	10^4–10^6	small	0.1–1	Strombolian		minor	none
2	10^6–10^7	moderate	1–5	↑		moderate	none
3	10^7–10^8	moderate-large	3–15	Vulcanian	1–6	↓	possible
4	10^8–10^9	large	10–25	↓	6–12	↑	definite
5	10^9–10^{10}	very large	>25	↓		substantial	↑
6	10^{10}–10^{11}	↓	↓	Plinian	>12	↓	significant
7	10^{11}–10^{12}	↓	↓	Ultra-Plinian		↓	↓
8	$>10^{12}$	↓	↓			↓	↓

Note that 1 cu km = 10^9 cu m. Modified from Newhall & Self (1982).

Magmas that are low in silica, such as basaltic magmas, flow very easily and the lavas flow away from the vent quickly once the molten material works its way to the surface. These types of eruptions are called Hawaiian eruptions (Figure 3.16). If a large amount of gas is present in a relatively low viscosity magma, Strombolian eruptions occur. These are generated by large bubbles of gas moving up the volcanic vent and exploding intermittently. These tend to form cinder cone volcanic edifices. Intermediate composition magmas from rock types like andesite and dacite are very common at subduction zones and in volcanic arc settings. They tend to consist of combinations of flows and explosions and form composite volcanoes. A Vulcanian eruption is characterized by a short, violent explosion of relatively high viscosity magma, and usually occurs when the volcanic vent has been plugged up by viscous or cooled lava and the gas pressure in the underlying magma builds up until it explodes. Pelean eruptions result from a dome or plug of high viscosity solid magma at the volcanic vent that also allows pressure to build up, but these eruptions tend to form hot, dense mixtures of volcanic gas and rock fragments that are generated when the gas pressure in the magma below builds up enough to blast and fragment the overlying dome lava. Plinian eruptions are the most powerful eruption type, and are named after Pliny the Younger who witnessed the 79 A.D. eruption of Vesuvius Volcano in Italy. They are caused by the explosive fragmentation of glassy viscous dacite or rhyolite magma. Plinian eruptions can last for days or as long as months. They are characterized by an extremely tall eruption column of hot gas and ash that can stretch as high as 45 km (28 mi) above the surface of the Earth. Once the eruption column cools down the gas and ash begin to spread out, forming the characteristic umbrella shape that Plinian eruptions are named after (Figure 3.17).

Robert Clucas/Wikimedia Commons.

Figure 3.17 **A Plinian eruption column from Mount Redoubt, Alaska, US, from a 1989 eruption. The eruption column reached about 10 km (6 mi). The umbrella cloud, where ash is spreading out at the top of the eruption column, is clearly visible.**

Effusive Eruptions

Quiet eruptions are generally characterized by magmas that have small amounts of gases and a low silica content and thus flow easily once they are extruded onto the surface (Figure 3.18). Any gases contained in the magma are able to escape fairly quickly due to the low viscosity of the lava. Basaltic lava eruptions generally form lava flows because they have high temperatures and low silica contents which give them low viscosity, and they also have low gas contents which means they are less likely to form explosions. Most eruptions on the island of Hawaii are basaltic lava flows that slowly move along the surface. When the erupting lava is highly fluid and the molten surface solidifies into a thin glassy crust the lava flow type is called pahoehoe. These have ropy surface textures that are formed as the underlying lava drags the ductile surface crust along and folds it into convoluted shapes (Figure 3.19). As the lava cools down and begins to crystallize, its viscosity increases and it does not flow as well. It forms a chunky, rubbly lava with a molten core that moves across the landscape like a bulldozer pushing material forward. These types of lavas are called a'a (Figure 3.20). Pahoehoe flows can transform to a'a as they cool. High silica magma like rhyolite can form lava flows as well, if they erupt with very little gas, although these rhyolitic lavas are not common. These rhyolite lava flows are highly viscous and flow poorly, forming large, thick blobs of glassy high-silica rock called obsidian. Big Obsidian Flow is 1,300 years old, the youngest lava in Oregon. It is over 30 meters thick (99 ft) and covers 2.6 square km (1.6 square mi) (Figure 3.21).

© DanielFreyr/Shutterstock.com.

Figure 3.18 A basaltic lava flow from the eruption of Fagradalsfjall Volcano, in Iceland, which began on March 19, 2021. The eruption had an average lava flow discharge rate of 5.3 cubic meters per second over the first month of activity.

© Yvonne Baur/Shutterstock.com.

Figure 3.19 Black, glassy pahoehoe lava from Kilauea Volcano, Hawaii, US. Molten lava glows red before it cools to a shiny glassy crust. As the lava continues to flow forward, the upper crust is fractured or dragged to form ripples and ropes.

Figure 3.20 A'a lava flow from Kilauea Volcano, Hawaii, US. The glowing core of the a'a flow is molten but is surrounded by a rubbly, spiky mass of brecciated lava clasts.

© Ecuadorpostales/Shutterstock.com.

Figure 3.21 High-silica rhyolite can also flow as a lava if it has a low gas content. Rhyolite lavas are highly viscous and form thick glassy flows. Big Obsidian Flow in the Newberry National Volcanic Monument, Oregon, US is an example of an obsidian lava flow.

© Wead/Shutterstock.com.

Types of Volcanoes

Shield Volcanoes

Shield volcanoes are the largest volcanoes in size and are so named because they have the broad appearance of an inverted shield. They are formed when low viscosity magmas are slowly extruded onto the surface and flow as effusive lavas out over large areas. These flows build up over time and produce very large structures such as those found in the Hawaiian Islands and Iceland. Mauna Loa, the largest volcano on Earth, has an elevation of 4,175 m (13,697 ft) above sea level and rests on the ocean floor in 5,800 m (19,000 ft) of water (Figure 3.22). Generally eruptions associated with shield volcanoes are quiet and move across the surface at from 1 to 10 meters an hour (3.2 to 33 ft). Fluid lava flows travel large distances and shield volcanoes have very flat slopes of about 5 degrees.

shield volcano

A broad volcano that has gentle slopes consisting of low viscosity basaltic lava flows.

© dirkr/Shutterstock.com.

Figure 3.22 Mauna Loa, on the island of Hawaii, appears along the horizon.

Cinder Cones

Cinder cones build up through many explosions of gas-rich mafic lava that form small to medium sized **pyroclastic** particles of ash and lapilli that build up around the vent of the volcano (Figure 3.23a). As they are cast out of the volcano, they pile up and create a conical structure called a cinder cone. Because little or no lava flows out the top of the volcano, the cinders remain as loose grains that are susceptible to erosion. The unconsolidated material forms a pile based on how well the particles interlock, and the slope of the surface is called the angle of repose. Angular particles interlock well and cinder cones have steep slopes of about 30 degrees (Figure 3.23b).

Cones are generally no more than 300 m (1,000 ft) in height and pose a threat only to areas in the immediate vicinity of the eruption. Wind can transport some of the smaller ejecta over considerable distances (Figure 3.24).

cinder cone

A conical-shaped hill created by the buildup of cinders (lapilli) and other pyroclastic material around a vent.

pyroclastic

Related to material that is thrown out by a volcanic eruption.

Composite Volcanoes

Perhaps the most widely recognized volcano shape is a **composite volcano**, or stratovolcano, which has a symmetrical, generally co78nical shape. The volcanic peaks of Japan, the Philippines, Alaska, and the Pacific Northwest in

composite volcano

A volcano that forms from alternating layers of lava and pyroclastic debris; also known as a stratovolcano.

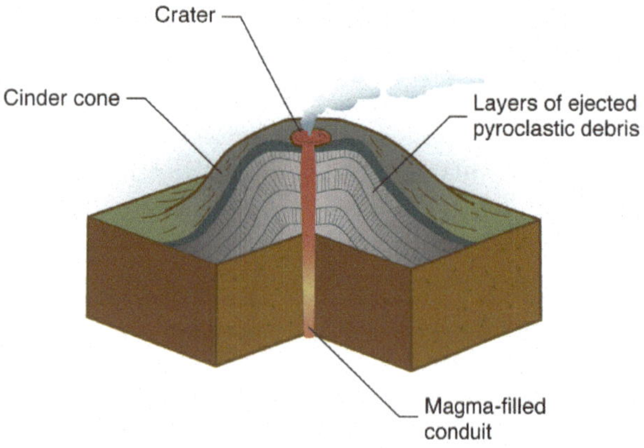

© Kendall Hunt Publishing Company.

Figure 3.23a Block diagram showing how pyroclastic debris builds up to form a cinder cone.

© Stan Jones/Shutterstock.com.

Figure 3.23b Cinder cone in Haleakala Crater, Haleaka National Park on the island of Maui, Hawaii, US.

the United States are excellent examples of composite volcanoes. They form from deposits of pyroclastic debris that build up near the vent and then are covered by occasional lava flows that move down the flanks to "glue" the looser debris in place. This repeated combination of loose material and lava produces stratified layers that create the awe-inspiring, cone-shaped mountains we associate with volcanoes (**Figure 3.25**). These are made mostly of andesite but usually include some rock types that are both higher and lower in silica. Composite volcanoes can be up to 2,500 meters (8,200 feet or 1.5 mi) from base to summit, and are second in size behind shield volcanoes. They have slopes of about 5 to 10 degrees on their flanks and as much as 30 degrees near the summit, reflecting the more fluid lavas that flow far from the vent and the pyroclastic material that builds up around the summit area. These are in many ways like a combination of shield volcanoes with lava flows and cinder cones with explosive eruptions.

© Alexander Piragis/Shutterstock.com.

Figure 3.24 An aerial view of actively erupting Tolbachik Volcano cinder cone, Kamchatka Peninsula, Russia.

© WorldStockStudio/Shutterstock.com.

Figure 3.25 Mayon volcano, rising to an elevation of 2,462 m (8,077 ft) in southeast Luzon, is the most active volcano in the Philippines. It has steep upper slopes averaging 35–40 degrees that are capped by a small summit crater. A single central conduit system has allowed the volcano to build up into a classic, symmetrical profile.

Lava Domes

During the late stage of many volcanic eruptions, as the magma in the vent cools down and begins to crystallize, the viscosity increases to the point where the lava can't be erupted but can only be pushed out as a semi-solid or solid mass, like a blob of toothpaste. Magma extruded at the vent thickens and creates a **lava dome** either just under or on the surface. Within a year or two of the major eruption of Mount St. Helens in May 1980, a dome formed in the inner crater and volcanic activity slowed down (Figure 3.26). A dome located at the top of a volcano can act like a cork in a bottle and hold in gases and new rising magma until enough pressure builds up in the system that will blow the dome out of the way. Following the main eruption of Mount St. Helens in May 1980 there have been additional eruptions that have added almost 137 million cubic meters (150 million cubic yards) of material to the dome and surrounding area (Box 3.3).

lava dome
A dome-shaped mountain formed by very viscous lava flows.

Calderas

In terms of the destructive potential of various types of volcanoes, **calderas** are by far the biggest threat. However, they are also the least likely to occur, so we are relatively safe from the mega-hazard their explosions would produce. Mega-eruptions are termed ultra-Plinian and are characterized by extreme volumes of ejected material exceeding 1 cubic km (0.6 cubic mile) of material. The largest of these result in subsidence bowls or tectono-volcanic depressions (calderas) over the magma chamber, which form toward the end of the eruption process after magma has been erupted and the overlying crust or lid to the magma chamber collapses in on the empty space created from the eruption. Yellowstone National Park, located in the northwestern corner of Wyoming, US, and stretching into portions of Idaho and Montana, is centered on a massive caldera covering almost 3,500 sq km

caldera
A large basin-like depression that is many times larger than a volcanic vent.

© Photovolcanica.com/Shutterstock.com.

Figure 3.26 Image of the glowing debris avalanche down the flanks of the lava dome in Paluweh Volcano on the island of Flores, Indonesia, during a 2012 eruption.

BOX 3.3 Whakaari/White Island Eruption, New Zealand 2019

Forty-seven people were on Whakaari/White Island when it erupted at 2:11 p.m. on December 9, 2019, killing 14 Australians, 5 Americans, 2 New Zealanders, and a German citizen. For the 26 others on the volcano during the explosion, all but three suffered severe or critical injuries and most were badly burned—some had third-degree or 'full thickness' deep tissue burns over more than 95% of their bodies. Many also ingested hot ash and gas, resulting in severe internal damage to their lungs and airways.

© eclecticworks/Shutterstock.com.

Figure 3.3.1 Whakaari/White Island after the December 9, 2019, eruption. Forty-seven people were on guided tours to the volcano, many in or near the vent, when it erupted.

Whakaari/White Island is New Zealand's most active stratovolcano which has been constructed by continuous volcanic eruptions over the last 150,000 years. It is located 48 km (30 mi) offshore of the North Island of New Zealand, and more than 70% of the 1,600 m (5,249 ft) edifice of the volcano is below sea level, leaving only the summit and active crater vent exposed. The traditional Māori name for the island is Te Puia o Whakaari, which is translated as "The Dramatic Volcano", whereas Captain Cook named it White Island—for the dense clouds of white steam emanating from it—when he first observed it on October 1, 1769.

Miners extracting sulphur intermittently worked on the island from the 1880s until 1914, when part of the western rim of the crater collapsed and generated a lahar that killed all 10 of the workers on the island at that time. The volcano has a long history of explosive activity. Near-continuous eruptions occurred from December 1975 through the 1980's, including major eruptions in 1981–1983, through the 1990's, and lasted until September 2000, making this the world's longest historic eruption episode, according to GNS Science GeoNet, New Zealand Crown Research Institute. Increased activity between July and August 2012 culminated in an eruption on August 5, 2012, and increased activity which began in January of 2013 culminated in an eruption on August 20, 2013. Additional eruptions occurred in 2016.

© NickJulia/Shutterstock.com.

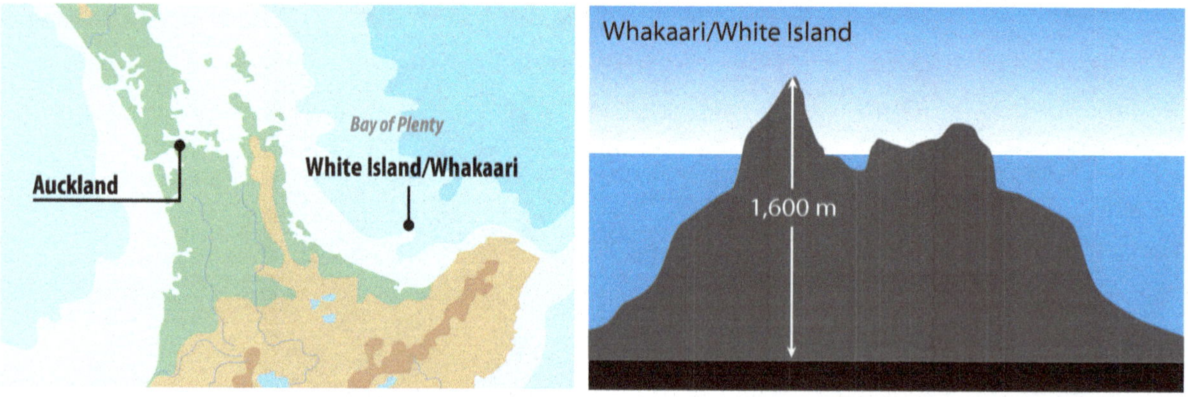

© Kendall Hunt Publishing Company.

Figure 3.3.2 Whakaari/White Island is a short 60- to 80-minute boat trip from the Whakatane coast in the Bay of Plenty, North Island of New Zealand, and was a popular tourist destination for local and international tourists and cruise ships. Most of the volcano lies below sea level.

Whakaari/White Island is a privately owned island which has been in the possession of the Buttle family for more than 100 years and the three current owners and directors oversee the Whakaari Management company. Frequent volcanic activity attracted 10,000 tourist visitors each year, with a fee of about $500 NZD ($350 US) per person to visit the volcano. Tour operators paid the family-owned business a royalty per person, bringing in around $4 million per year. Tourists arrived to the island from the mainland by boat or by helicopter, and from nearby cruise ships, like the Royal Caribbean *Ovation of the Seas,* where the majority of the 47 victims had come from that day in December. GeoNet, an agency that provides geological hazard information for New Zealand, issued multiple reports of "volcanic unrest" on the island, starting as early as late October 2019. A report from Tuesday, December 3, stated: "Moderate volcanic unrest continues at Whakaari/White Island, with substantial gas, steam and mud bursts observed at the vent located at the back of the crater lake." At the time of the eruption, the alert level was at 2, on a scale of 1 to 3, where Level 3 is an eruption.

At 2:11 p.m. on December 9, 2019, an eruption column of ash, gas, and vaporized liquid from the acid crater lake blasted 3.7 km (12,000 ft) into the air. One tour group, captured by a GNS webcam just a minute before the explosion started, walks on the path that will shortly be engulfed by the volcanic blast. Pyroclastic flows were directed down the valley with the only trail into or out of the vent, straight onto tourists waiting at the dock to be picked up after their excursion. Of the 47 people on the volcano when it erupted, the only three who were not seriously burned were at the shore, and jumped into the water as pyrocalstic flows passed over them. Boats filled with tourists who had just finished their excursions and had left the island witnessed the blast, and many turned around and went back to rescue those who were trapped onshore. Helicopter pilots from the nearby mainland risked their lives to fly in on rescue missions, hoping to find survivors. In the end, 22 people lost their lives to the eruption.

Source: GeoNet.

© Jiri Foltyn/Shutterstock.com.

Figure 3.3.3 **This webcam image, taken 1 minute before the 2019 eruption began, shows tourists walking along the trail toward the acid crater lake in the vent of Whakaari/White Island. Images from an earlier tour show what that spot looked like to those on the trail. No one in the vent area at the time of the eruption survived.**

Between December 2019 and December 2020, WorkSafe—New Zealand's government regulator for health and safety in the workplace—spent $5 million investigating the Whakaari/White Island disaster and levelled charges against 10 organizations and three individuals—several White Island tourism companies, the owners of the island, the National Emergency Management Agency, and GNS Science. All face charges relating to health and safety of tourists and staff, and if convicted, could be fined up to $1.5 million New Zealand dollars each. Two of those charged were pilots who risked their lives to save people after the eruption—but this investigation relates to the period leading up to the eruption, and not the rescue efforts afterward. The WorkSafe investigation is ongoing, and questions continue to be raised about the inherent risk of visiting an active volcano and the responsibility of tour operators to educate and alert visitors to the dangers they may encounter, as well as where the ultimate responsibility lies for deciding when and where it is safe to allow tourism.

© eclecticworks/Shutterstock.com.

Figure 3.3.4 A view of the boat dock and valley into the vent of Whakaari/White Island, with the only path into and out of the crater, and the main path of the eruption's pyroclastic surges.

Figure 3.27 Ashfall from Yellowstone eruptions 2 million years ago (in yellow) and 0.63 million years ago (in orange) compared to ash from the 1980 eruption of Mount St. Helens (in red) and Long Valley Caldera in California at 0.76 million years ago (in purple).

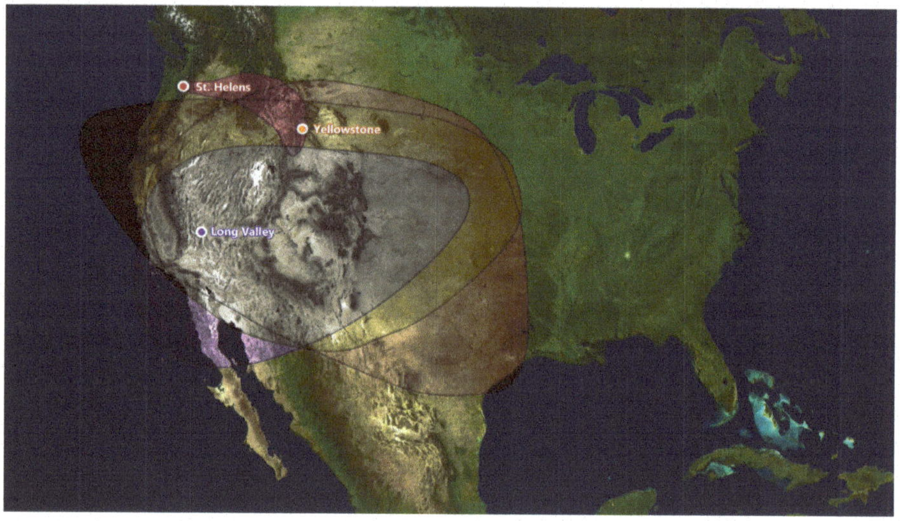

© Cromagnon/Shutterstock.com.

© Jakub Barzycki/Shutterstock.com.

Figure 3.28 Stokkur geyser in Iceland.

(1,350 sq mi). This area has experienced three major eruptions. The first occurred approximately 2.1 million years ago, followed by another about 1.3 million years ago. The most recent major eruption took place about 630,000 years ago (Figure 3.27). The result of these eruptions was the ejection of almost 3,800 cu km (900 cu mi) of volcanic ash that covered much of the continental United States. The area continues to be geologically active today with a well-developed geyser system emitting steam into the atmosphere (Figure 3.28). Earthquake activity is ongoing in the park and adjoining regions.

Volcanic Hazards

Gas Emission

In addition to gas being emitted from the volcano itself during an eruption or while it is restless, on rare occasions an eruption of only gas can occur. A documented case happened in August 1986 in Cameroon, West Africa. Lake Nyos, situated in the crater of a volcano at an elevation of 1,091 m (3,580 ft), is relatively deep given its small size (about 200 m deep, 1 to 1.5 km wide or 660 feet deep and 0.6 to 0.9 mi). Little intermixing of its cold, CO_2-saturated bottom waters with shallow, warm water normally takes place. However, the carbon dioxide built up to critical levels and the lake suddenly overturned, releasing dense suffocating carbon dioxide gas and killing more than 1,700 people along with several thousand head of cattle. The denser carbon dioxide cloud moved down several stream valleys extending almost 100 meters (330 ft) above the ground surface, killing people up to 25 km (15.5 mi) away from the lake. Some people, who were unconscious for more than 36 hours, revived once the cloud lifted, only to find all of their family, neighbors, and livestock were dead (Figure 3.29).

© Peter Turnley/Contributor/Getty.

Figure 3.29 The dense carbon dioxide from Lake Nyos stayed close to the ground, suffocating people and animals. People and animals collapsed and died in just a few breaths.

Lava Flows

Lava flows are produced when molten rock extrudes out onto Earth's surface. The ease with which they flow is controlled by the composition and temperature of the magma. Magmas that are rich in silica, such as those with a rhyolitic composition, have a high viscosity and barely flow at all. Basaltic magmas, those containing less silica, flow readily onto and across the surface. The island of Hawaii has been constructed by many episodes of eruption of basaltic lavas that continue today (Figure 3.30). As the lava cools and hardens, it forms new land on which future flows will travel and subsequently harden. Given the relatively slow movement of lava flows, it is easy to avoid them. However, infrastructure such as homes, roads, and other buildings that cannot be moved can be destroyed as it is inundated by lava.

United States Geological Survey.

Figure 3.30 Lava flowing from Pu'u 'Ō'ō, Kilauea Volcano, Hawaii, through the Royal Gardens housing subdivision in February of 2008. The lava here is about 3 meters (10 ft) across.

Lahars

lahar

A volcanic mudflow or landslide that contains unconsolidated pyroclastic material.

A **lahar** is a hot or cold mixture of water and volcanic material that forms a mudflow or debris flow. Lahars generally occur on or near composite volcanoes because of their height and steep topography. They flow down the side of the volcano, usually following river valleys. Lahars can reach speeds of 200 km/hr (120 mph) and look like a rolling, churning slurry of wet concrete which can crush, bury, or carry away almost anything in their path. Lahars can happen with or without volcanic eruptions—eruptions of hot material can melt snow and ice or mix with water from a crater lake and trigger lahars, but they can also form when a large amount or a long duration of rain occurs during or after a volcanic eruption. The fine-grained, loose volcanic material that makes up the steep slopes of composite volcanoes is easily eroded. The initial flow may be small but lahars incorporate everything in their path and accumulate additional water through addition of melting snow or ice, or river or lake water. As they move downslope, lahars increase in size and can grow to 10 times larger than their initial volume. As they move downslope and flow onto lowlands surrounding volcanoes, they slow down, spread out, and come to rest like a sheet of concrete (Figure 3.31). Lahars are one of the greatest volcanic hazards because they can occur during eruptions and also long after an eruption is over. They can occur without warning and can travel long distances beyond a volcano and can inundate areas in distal, low-lying terrain.

United States Geological Survey.

Figure 3.31 One year later, cleaning up from lahar deposits after the June 15, 1991, eruption of Mount Pinatubo, Philippines. The cataclysmic eruption was the second largest of the 20th century.

The far-reaching effects of lahars became evident during the eruption of Nevado del Ruiz in Colombia in November 1985, when the resulting lahar traveled more than 75 km (47 mi) from the volcanic vent, covering the city of Armero and surrounding villages (Figure 3.32). More than 23,000 people died. The eruption of Mt. Pinatubo in the Philippines in 1991 was followed by more than 200 lahars produced by the rainy season in the spring and summer. These continued for years after the eruption. Lahars erode and transport loose volcanic material far downstream from the volcano, and this deposition can lead to severe and chronic flooding in the river system. They can travel onto floodplains and bury entire towns and valuable agricultural land.

United States Geological Survey.

Figure 3.32 The town of Armero was destroyed and more than 20,000 people were killed by lahars from the eruption of Nevado del Ruiz, Colombia, in November 1985. This was the second-deadliest volcanic disaster of the 20th century.

Ash Fall

Erupting volcanoes eject a wide variety of material into the air, ranging in size from ash to blocks and bombs. The small particles of ash are carried up in the eruption column and fall out in a downwind plume to form ash fall deposits. These fine particles accumulate like snow, the largest pieces fall out close to the volcano and the thickest deposits occur closest to the volcano. They are glassy fragmented bubble shards and commonly have sharp edges, which makes them very abrasive. They can act as irritants to the eyes and lungs, damage airplanes, vehicles, and houses, contaminate water supplies, and damage or kill crops and livestock. If the ash layer is thick and heavy enough, it can cause roof collapse (**Figure 3.33**).

© setyo adhi pamungkas/Shutterstock.com.

Figure 3.33 Volcanic ash is highly abrasive and can be remobilized after eruptions and obstruct visibility as well as damage motor vehicles.

Pyroclastic Density Currents

A pyroclastic flow is a hot mixture of gas and volcanic particles that flows away from the volcano as a density current with speeds from 100 to 700 km/hour (60 to 430 mph) and temperatures up to 1000 °C (1800 °F). They form during explosive eruptions and usually result from collapse of an eruption column, which provides their extreme heat and speed. Pyroclastic density currents that are very dilute and have a high proportion of gas to volcanic material are called pyroclastic surges. Pyroclastic flows and surges represent two end-members on a continuum of pyroclastic density currents (**Figure 3.34**). Deposit volumes can be small, but they are also some of the largest eruptions on Earth with volumes of thousands of cubic kilometers in individual eruptions. The largest flows can travel for hundreds of kilometers and have thicknesses of 100 meters (330 ft) or more. These flows have tremendous energy and can flatten everything in their path (**Figure 3.35**). They also have enough speed that they can flow up and over topography (**Figure 3.36**). Their high temperatures and speed make them nearly impossible to escape from and they incinerate any object in their path. The residents of Pompeii and Herculaneum were killed by pyroclastic density currents from Vesuvius in 79 A.D. (**Figure 3.37**).

United States Geological Survey.

Figure 3.34 Pyroclastic density current from the 1980 eruption of Mount St. Helens, Washington State, US.

Source: J. Hughes US Forest Service.

Figure 3.35 The blast from the eruption of Mount St. Helens in 1980 flattened or damaged enough trees to build 150,000 homes.

© Photovolcanica.com/Shutterstock.com.

Figure 3.36 A pyroclastic flow from Soufriere Hills Volcano on the Caribbean island of Montserrat speeds down the Tar River Valley in January of 2010.

© Andrea De la Parra/Shutterstock.com.

Figure 3.37 Residents of Herculaneum took refuge in the boat storage sheds by the beach at the start of the 79 A.D. eruption of Vesuvius but were killed there by a pyroclastic density current that flowed through their town.

Directed Blasts

In many volcanic eruptions the vent is perpendicular to the surface and the eruption column is directed vertically, transporting a large amount of ash and volcanic particles up into the atmosphere. Large eruptions can eject massive amounts of ash that can remain in the atmosphere for months or longer causing global cooling events (**Figure 3.38a**). Evidence of this is seen in spectacular sunsets caused by this ash (**Figure 3.39**). The eruption of Tambora in Indonesia in April 1815 had a worldwide effect on weather, and in the Northeastern United States there was a significant decrease in solar radiation and temperature the following summer. The year 1816 became known as the "year without a summer." Lower temperatures had a direct effect on crops. This climatic effect was caused by dust particles and sulfur dioxide reflecting and absorbing solar radiation. Atmospheric gases, particularly sulfur derivatives such as sulfur dioxide and sulfur trioxide, are capable of absorbing large amounts of infrared radiation energy, and are capable of

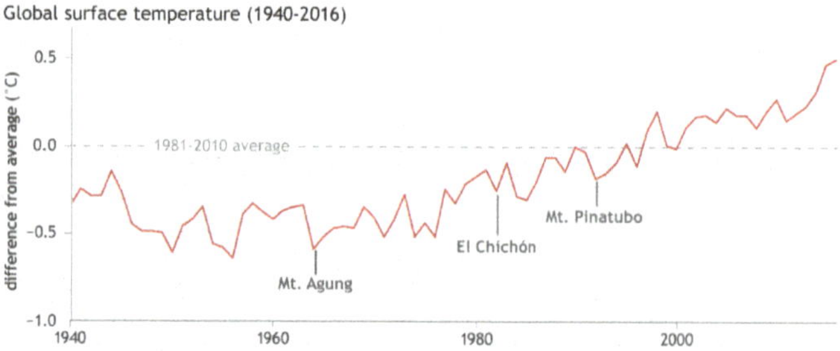

Source: NOAA.

Figure 3.38a Three volcanic eruptions in the tropics had climate-cooling effects in the second half of the 20th century—Mt. Agung (Indonesia, 1963), El Chichon (Mexico, 1982), and Mt. Pinatubo (Philippies, 1991). Volcanic gases like sulfur dioxide have to reach the stratosphere in order to cause this temporary global cooling effect.

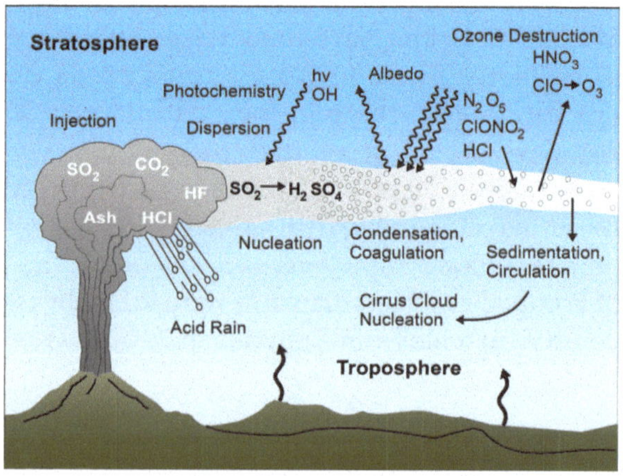

United States Geological Survey.

Figure 3.38b Volcanic gases react with the atmosphere and sulfur dioxide (SO_2) is converted to sulfuric acid (H_2SO_4). This sulfuric aerosol can cause global cooling of up to 0.5 °C (about 1 °F) that lasts for several years after an eruption, such as the 1815 eruption of Mount Tambora in Indonesia.

decreasing the solar heat warming the surface for years after an eruptive event (Figure 3.38b).

If the blast is directed laterally out the side of the volcano, the energy is oriented along the ground and can have a devastating effect on everything in its path (Figure 3.35). Such was the case with the May 1980 eruption of Mount St. Helens in southwestern Washington, US. The initial blast was directed to the northeast and thousands of acres of trees were flattened and large amounts of debris clogged the rivers and streams. The volcano sent ash and debris vertically to a height of 25 km (80,000 ft) in 15 minutes. The U.S. Geological Survey reported that ash clouds stretched across the United States in a period of 3 days and ash circled the Earth in 15 days. Ash remained aloft for several years.

Source: Bridgeman Images.

Figure 3.39 ***The Scream*** **by Edward Munch was painted in 1893, and is thought to capture the global climate effects of the 1883 eruption of Krakatoa in the red- and orange-tinged skies.**

Landslides and Tsunami

The steep slopes of a volcano are very unstable because of the interlayered lavas and clastic deposits, as well as the flux of hydrothermal fluids and sulfuric acid altering the interior of the structure. In addition to lava and other pyroclastic debris, large sectors of the volcanic edifice can collapse and race down to form large landslides with velocities up to 800 km/hr (500 mph). Volcanic eruptions may further destabilize already unstable volcanic flanks. Researchers believe that a catastrophic landslide could result if there is an eruption of Cumbre Vieja Volcano on the island of La Palma in the Canary Islands (Figure 3.40).

A review of geological evidence on the island and offshore indicates that the western edge of the volcano could slide into the Atlantic Ocean. Were this to happen, an estimated 150 to 500 cubic kilometers (93 to 310 cubic mi) of material entering the ocean will generate a tsunami that could inundate the east coast of Florida with waves exceeding 50 m high (165 ft). If this did occur, coastal residents would only have a few hours warning. Since the average elevation of the State of Florida is 17 meters (56 ft), a significant portion of the state would be directly impacted and this would be an insufficient amount of time to allow any significant evacuation of the millions of people who would be affected.

Courtesy NASA.

Figure 3.40 **Cumbre Vieja Volcano on La Palma Island in the Canary Islands.**

BOX 3.4 World Is Unprepared for Next Major Volcanic Eruption

Source: NASA.

Figure 3.4.1 This photo of Tambora was captured by an astronaut aboard the International Space Station and shows the 6 km (3.7 mile) diameter, 1,100 m (3,600 ft) deep summit caldera that formed after Tambora's 4,000 m (13,000 ft) high top was blown off in the April 10, 1815, eruption.

We haven't had a cataclysmic volcanic disaster since the VEI 7 eruption of Tambora in Indonesia in 1815—an eruption that triggered years of global cooling and resulted in tens of thousands of deaths. This event was at the low end of a magnitude 7, yet it had global consequences. The next VEI 7+ eruption will affect a very different planet. When Tambora erupted, world population was 1 billion; today it is 8 billion. In the early 1800s, communities were largely self-sufficient; today we have complex global supply chains that are vulnerable to disruption. Nearly everything we do depends on having a stable power supply; widespread, prolonged power interruptions from utilities damage will be significant. Impacts on air travel, global positioning systems (GPS), and climate will be felt by our modern, high-technology, globally interdependent world. Some volcano candidates for generating the next big eruption have up to 30 million people living within 100 km. Tens of millions of people could be in direct risk, and eruptions could destabilize financial centers, national economies, or even peace between nations, according to a study by eminent volcanologists.

Dr. Chris Newhall and Dr. Steven Self, who devised the VEI scale in 1982 (see Table 3.2), along with Prof. Alan Robock, a climate scientist, identified candidates for future VEI 7 eruptions, noting that most of these occur in subduction zone settings. Some possible sources for the next VEI 7 eruption—which could occur at any time—are: Santorini in Greece; Rotorua, Okataina, and Taupō Calderas in New Zealand; Toba in Indonesia; Lassen, Long Valley, Valles, and Yellowstone in the United States of America.

Source: NASA.

Figure 3.4.2 A photo of Santorini taken from the International Space Station shows the remnants of a large, circular volcano that exploded in a caldera-forming event, erupting 100 cubic km of material (62 cubic mi). The eruption occurred between 1600 and 1627 B.C.E. The central peak volcano formed in 1707, and there have been three eruptions in the 20th century.

There are special challenges in forecasting low-probability, high-consequence events such as these largest volcanic eruptions that occur globally only one to two times per 1,000 years, and at individual volcanic centers far less often—usually once every 10,000 to 100,000 years. VEI 7 eruptions produce at least 100 cubic km (62 cubic mi) of ash and pumice and large amounts of sulfur, carbon dioxide, and other volcanic gases. Pyroclastic flows can travel up to 100 km in any direction, potentially erasing entire cities. When they enter the sea, they can trigger local tsunami that travel several hundred kilometers, wiping out coastal communities. Ash will damage crops, buildings, and infrastructure like power and water supplies, as well as aircraft, computers, HVAC systems, machinery, and vehicles. Disruption of river systems and drainage will cause regional flooding for decades after an eruption. Coupling of oceans, sea ice, and atmosphere after large eruptions can force decade- to century-long cooling, decreased precipitation, and major impacts on agriculture and farming. The nightmare of forecasting

Source: Google Earth; modified by Kendall Hunt Publishing Company.

Figure 3.4.3 The Taupō Volcanic Zone in the North Island of New Zealand is the site of several volcanoes that could generate the next VEI 7 eruption—Taupō (which had a VEI 8 eruption 26,500 years ago), Rotorua and Okataina Calderas. Okataina Caldera has erupted six times in the last 10,000 years, including a mafic fissure eruption in 1886 at Mount Tarawera that killed an estimated 150 people and formed the world's youngest geothermal system at Waimangu Volcanic Valley. Rotorua Caldera's last major eruption was about 240,000 years ago, when it covered 4,000 square km (1,500 sq mi) in rhyolitic ash, and it hosts a highly active geothermal field.

a potential VEI 7 eruption in an urban area, and initiating the evacuation of hundreds of thousands to millions of people, potentially for weeks or months, imposes an extraordinarily heavy responsibility on scientists and officials. Experts in all fields—health, agriculture, transport, high technology—should consider how a VEI 7 eruption would impact their sector. In the words of the authors: *"A VEI 7 eruption is not unthinkable. A two in one thousand per year probability (2 x 1000) for an event with major global consequences demands our attention. Forward-thinking governments and industries will be wise to consider how a nearby VEI 7 eruption will affect their jurisdictions, and equally the effects of a VEI 7 that occurs halfway around the world. What responses will make the most economic and sociopolitical sense?"*

© N.Minton/Shutterstock.com.

Figure 3.4.4 Lake Taupō, in the middle of New Zealand's North Island, had a VEI 7 eruption in 180 A.D., with ash clouds reaching both the Roman Empire and Han Dynasty China. This could be the site of the world's next VEI 7 eruption.

Submarine and near-shore volcanic eruptions can generate tsunamis or seismic sea waves that travel at very high speeds across the open ocean. Reaching heights of several tens or even hundreds of meters, these waves wreak havoc on coastal regions. Large landslides are associated with volcanic activity near the shore on the island of Hawaii. Massive blocks measuring hundreds of cubic kilometers are moving into the sea at a relatively slow rate (several cms per year). However, the recurrence rate for large volcanic-induced tsunamis is about every 100,000 years. On the morning of November 29, 1979, a magnitude 7.7 earthquake centered below the south flank of Kilauea Volcano, Hawaii, US, triggered a landslide with ground subsidence of as much as 3.5 m (11.5 ft) in the Halape area 30 km

(18.5 mi) southwest of volcano. It also generated a tsunami with waves as high as 14.6 m (48 ft), which killed two campers in the Halape Campground in Hawaii Volcanoes National Park (**Figure 3.41**).

United States Geological Survey.

Figure 3.41 Earthquake-triggered ground subsidence of as much as 3.5 m (11.5 ft) on the flanks of Kilauea Volcano submerged a grove of coconut palms in ocean water 1.2 m (4 ft) deep and created a new shoreline 150 m (164 yards) inland from the previous location. The tsunami associated with this landslide killed two campers sleeping in the campground located in the palm grove.

Earthquakes

Magma moving toward the surface produces forces that trigger earthquakes in the vicinity of a volcano. Although relatively small in magnitude, these earthquakes can destroy buildings in the near vicinity of the volcanic eruptions before any direct effects of the volcano are felt. Earthquakes are commonly associated with volcanic activity because as the magma intrudes it adds pressure to the rock around it and heats up fluids that can also add to that pressure. This added stress may result in rocks breaking and can generate earthquakes. Mount St. Helens has had several episodes of intense earthquakes generated by magma rising up under the volcano over the last several decades—this is normal behavior for an active volcano. Signs of potential eruptions would include increased gas emissions, ground surface deformation, and shallow seismic activity within 2 km (1.2 mi) of the surface (**Figure 3.42**).

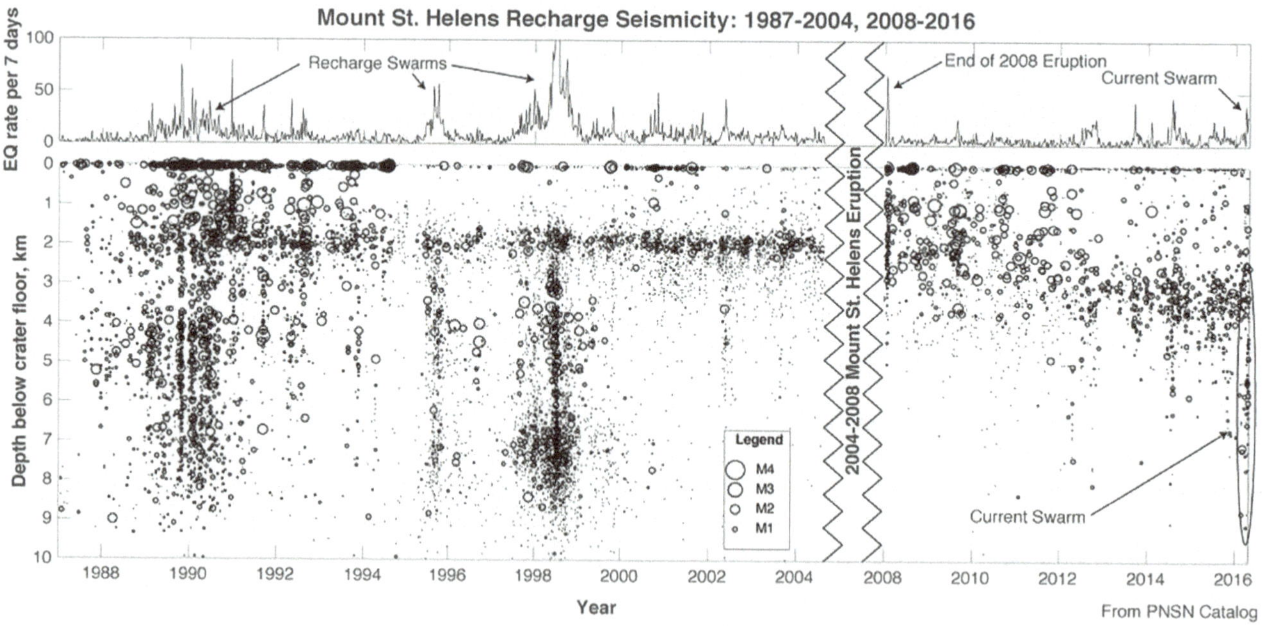

USGS Cascade Volcano Observatory.

Figure 3.42 Earthquakes occurring beneath Mount St. Helens from 1987 to 2016 show the start of a new swarm of deeper events from 2 to 7 km (1.2 to 4.3 mi) below the volcano after a period of only shallow earthquakes following the end of the 2008 eruption. USGS scientists interpret this as recharging from new magma rising up underneath the volcano.

Lessons From the Historic Record and the Human Toll

Throughout recorded history there have been many well-documented volcanic eruptions (Table 3.3). Among the earliest are those written by Pliny the Younger, who described the eruption of Mount Vesuvius in Naples, Italy, in 79 A.D. During that eruption, which destroyed the cities of Pompeii and Herculaneum, his uncle Pliny the Elder died trying to rescue people stranded near Pompeii.

Because volcanoes produce rich soils and spectacular scenery, many cities have grown up in the vicinity of these features and hence there is a large population that lives within the direct path of major eruptions. Areas that are at risk include Japan, Indonesia, the Philippine Islands, New Zealand, southern Italy, Central America, the eastern Caribbean, and the Pacific Northwest in the United States.

In the past 4 centuries, approximately two thirds of all deaths caused by volcanic eruptions have occurred in Indonesia; the Caribbean and Japan have experienced 21% of all deaths. During that time, five volcanic eruptions have accounted for more than half of all deaths: Mt. Unzen in 1792,

Table 3.3 Some Significant Volcanic Eruptions and Their Associated Loss of Life

Year	Country	Volcanic Structure	Estimated Deaths	Notes
79	Italy	Mount Vesuvius	>3,000	Pompeii and Herculaneum destroyed by pyroclastic flow
1169	Italy	Mount Etna	16,000	
1631	Italy	Mount Vesuvius	4,000–6,000	Torre del Greco destroyed
1669	Sicily	Mount Etna	20,000	Catania destroyed
1672	Indonesia	Mount Merapi	3,000	
1683	Sicily	Mount Etna	60,000	Earthquakes occurred
1782	Japan	Mount Unzen	15,000	Tsunami produced
1783–1784	Iceland	Laki fissure	9,800–10,500	Poisonous gases killed 280,000 cattle; major famine; 15 cubic km of lava (9 cubic mi)
1792	Japan	Mount Unzen	15,000	Avalanche produced tsunami
1815	East Indies	Mount Tambora	66,000–162,000	10,000 died in initial eruptions; famine killed 50,000–80,000; produced "Year without a summer"
1845	Colombia	Nevado del Ruiz	1,000	Lahars

Table 3.3 Some Significant Volcanic Eruptions and Their Associated Loss of Life (*continued*)

Year	Country	Volcanic Structure	Estimated Deaths	Notes
1883	East Indies	Krakatua	36,000	90% killed by tsunami that flooded islands; global climate change resulted
1902	West Indies	La Soufriére, St. Vincent	>1,500	Occurred 1 day before Mount Pelée eruption
1902	West Indies	Mount Pelée, Martinique	29,000	Town of St. Pierre destroyed by nuee ardente
1902	Guatemala	Santa Maria	Thousands	28 km-high column (17 mi) and ash covered 100,000 square km (62,000 square mi)
1911	Philippine Islands	Taal, near Manila	1,300	Lateral blast destroyed nearby villages
1912	Alaska	Novarupta	0	Uninhabited island saw largest eruption of 20th century
1919	Indonesia	Mount Kelud	>5,000	Boiling waters of crater lake on Mount Kelud broke through side of mountain
1931?	Indonesia	Mount Merapi	>1,300	Major ash flow
1943	Mexico	Parícutin	0	Towns buried by lava and cinders
1951	New Guinea	Mount Lamington	>3,000	Eruption similar to style of Mount Pelée; 180 sq km (112 square mi) destroyed
1963	Indonesia	Mount Agung	>1,100	Ash and mud flows
1980	United States	Mount St. Helens	57	Earthquake triggered landslide that resulted in eruption
1982	Mexico	El Chichón, Chiapas	2,000	25 km (15.5 mi) high ash cloud circled the globe; climate impact
1985	Colombia	Nevado del Ruiz	25,000	Mudslides from melting snow covered town of Armero
1986	Cameroon	Lake Nyos	1,700	Expulsion of carbon dioxide also killed >3,000 cattle
1991	Japan	Mount Unzen	43	
1991	Italy	Mount Etna	0	Lava flowed out for 473 days, longest period on record in 300 years
1991	Philippine Islands	Mount Pinatubo	754	Eruption predicted; 200,000 people evacuated; 30 km (19 mi) ash cloud circled the Earth; global climate impact

Table 3.3 Some Significant Volcanic Eruptions and Their Associated Loss of Life (*continued*)				
Year	**Country**	**Volcanic Structure**	**Estimated Deaths**	**Notes**
1993	Philippine Islands	Mayon Volcano	70	60,000 people evacuated
2002	Congo	Nyirangongo	45	Lava flows
2010	Indonesia	Mount Merapi	353	Pyroclastic flows, 350,000 people evacuated
2014	Indonesia	Mount Sinabung	16	Dormant for 400 years, pyroclastic flow and ash fall, eruptions and casualties continue through 2018
2014	Japan	Mount Ontake	63	VEI 3 explosive eruption, no warning, popular tourist attraction for hikers and hundreds of people at summit, Japan's deadliest volcanic eruption since 1902
2019	New Zealand	Whakaari/ White Island	22	Popular tourist attraction, 47 people on island when eruption started

Tambora in 1815, Krakatau in 1883, Mt. Pelee in 1902, and Nevado del Ruiz in 1985 (Figure 3.43).

Predicting Volcanic Eruptions

The terms **active**, **dormant**, and **extinct** have been used to indicate the eruptive potential of volcanoes. Different definitions exist for these terms, but generally they include the following guidelines. An active volcano is one that has shown some activity in the past several thousand years (historic time), but it could be currently inactive. A dormant volcano is one we think of as sleeping or inactive but could awaken at any time. An extinct volcano has shown no activity in historic time and is not considered likely to do so.

active
A term applied to a volcano that has erupted in recorded history or is currently erupting.

dormant
A term applied to a volcano that is not currently erupting but has the likelihood to do so in the future.

extinct
A term applied to a volcano that no longer is expected to erupt.

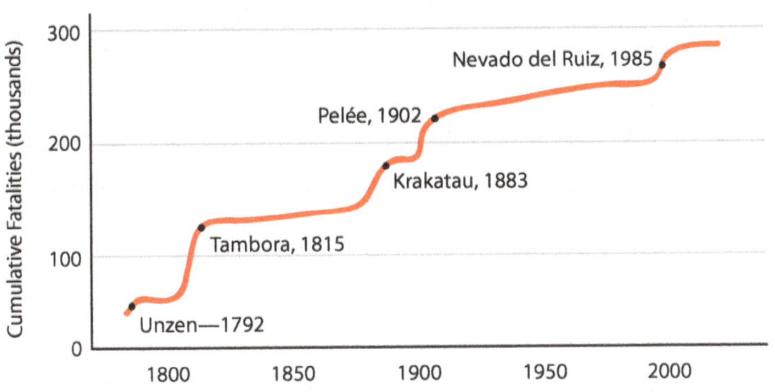

© Kendall Hunt Publishing Company.

Figure 3.43 Over the last 4 centuries there have been approximately 279,000 fatalities from more than 530 volcanic eruptions. Just a few events dominate the fatality record.

Blong (1984) points out that the terms active, dormant, and extinct are unsatisfactory because some extinct volcanoes become active. A review of past eruptions shows that many volcanoes have periods of inactivity much longer than their historic record. Also the timescale of our historic records differ for various parts of the world. The Mediterranean Sea region has a longer historic record than does the Western United States. Because of this Italy and Greece have longer periods of observed volcanic activity than in places like the Pacific Northwest of the United States, which has a much shorter historical record.

Precursors to Volcanic Eruptions

Scientists can make several observations to forecast the possible eruption of a volcano. Among these are changes in seismic activity near the volcano, an increase in measured heat on the surface, and the bulging of the volcano's surface.

© Andrey VP/Shutterstock.com.

Figure 3.44 **Seismic stations can detect, monitor, and record earthquakes locally or, if they are large enough, from anywhere in the world.**

Seismic Activity

The rise of magma is a relatively slow process that can be monitored by scientists who can detect precursors to an eruption. The convective overturn of molten material within the magma chamber produces a series of small earthquakes (usually with magnitudes <4) that have a rather specific pattern in terms of Earth movement. This is caused by the reverberation of energy within the closed magma chamber as energy bounces off the walls of the chamber. Seismic detection equipment records harmonic tremors that produce a unique pattern (**Figure 3.44**). The occurrence of these tremors is a sign that magma is rising and that an eruption could happen; an increase in seismic activity often signals that an eruption is on the verge of occurring.

Surface Heat

Ascending magma brings increased heat closer to the surface. This heat can be directly measured by heat flow instrumentation. If magma is rising beneath a volcano that is covered with snow and ice, the heat may melt the surface ice. This was observed before the eruption of Mount St. Helens in the spring of 1980. Usually the top of Mount St. Helens was covered by a solid snow pack, but scientists began to notice in March of that year the snow was becoming much thinner. This was an obvious signal to everyone that magma was moving closer to the surface. Monitoring heat flow and increased seismic activity provided evidence that magma was approaching the surface. The first eruption of Mount St. Helens in almost 125 years occurred in late March 1980 and continued with increased frequency until the main eruption on May 18.

Surface Bulge

Ascending magma also causes the Earth's surface to bulge (**Figure 3.42**). Measurements of the bulging can be taken using tiltmeters capable of measuring one unit of elevation change over a distance 1 million times greater than the rise. This is equivalent to detecting the increase in elevation produced by a US dime coin over a distance of 1 kilometer (0.6 mi) (Figure 3.45). The detected bulges show scientists where the upward forces are the greatest and give an indication of where magma could be forced out. Laser beams reflected off targets placed on a volcano will show changes in travel times of the beams and hence a change in the distance as the volcano expands or contracts.

Source: I. Ukstins.

Figure 3.45 A US dime coin is 1.2 mm (0.05 inch) thick.

Other Techniques

Other techniques such as recording gas compositions and small changes in the magnetic and electrical fields give an indication of stress and pressure variations in the ground. Changes in gas compositions appear to be effective indicators of changes in the magma chamber. Emissions of sulfur dioxide and hydrogen have been indicators of eruptions at Mount St. Helens and at several Hawaiian volcanoes.

Research into past events allows geologists to produce hazard maps for areas that are likely to experience volcanic eruptions in the future. Analysis of historical records along with detailed mapping of volcanic deposits permits scientists to generate maps that show the likelihood of volcanic deposits covering areas in the vicinity of potential eruptions. Many people living in the region south of Seattle and in the suburbs of Tacoma are located in areas that could be affected by an eruption of Mount Rainier.

Volcanic Hazard Mitigation

The term **mitigation** refers to the lessening in intensity or force. Obviously in volcanic eruptions we cannot reduce the amount of energy expended by the volcano but we can reduce the intensity of the damage caused by the event. Thus our goal would be to make the impact of an event less severe on the people and area surrounding a volcano.

mitigation
The act of making less severe or intense; measures taken to reduce adverse impacts on humans or the environment.

During the past 500 years more than 350,000 people have been killed by volcanic eruptions (Table 3.3). Usually there is some indication that eruptions may be imminent—small-scale earthquakes, in volcanic unrest such as increased gas emissions, or inflation of the edifice. Our ability to monitor in the heat flow, surface height and tilt changes, and seismic activity provide information to warn people of eruptions. A forewarning would be essential to save tens of thousands of lives if a major eruption occurred near any major metropolitan area.

Many areas that are prone to volcanic eruptions have seen significant population growth over the past several decades. The likelihood of these areas experiencing a volcanic eruption is high and the hazardousness of that potential event increases with increasing population levels. The Pacific Northwest ranks high on the list of places with volcanoes that have the potential to erupt near a major metropolitan area, and we see that areas in the vicinity of volcanoes with a historic record of activity are lying in harm's way.

Images source: United States Geological Survey.

Figure 3.46 The bulge on the north face of Mount St. Helens prior to the 1980 eruption was moving outward by more than 1.5 m (5 ft) per day. Parts of the bulge near the summit were 450 m (1,500 ft) higher than they were before the inflation began. About 10,000 earthquakes occurred from late March to mid-May within 2.5 km (1.5 mi) of the surface, centered directly beneath the bulge, generated by magma pushing into the volcano and deforming its surface. Dave Johnson (30 years old), shown here at an observation station 8 km (5 mi) from the volcano, was monitoring the bulge growth when he was caught in the blast. His body and car were never recovered.

An example of an area in such danger is the small town of Orting, Washington, which has a population of about 3,500 people. The town lies in the shadow of Mount Rainier. Built on old lahar flows that occurred several thousand years ago, the town is considered to have the greatest risk of being affected by lahars from Mount Rainier (Figure 3.47). The website for the town outlines explicit instructions for a lahar evacuation plan. The municipality is prepared in the event of an eruption, but the plan is good only if the citizens of Orting and surrounding areas are familiar with the plan and have sufficient warning to carry it out. Orting isn't the only town at risk. According to the United States Geological Survey, about 80,000 people and their homes are currently at risk in Mount Ranier's lahar-hazard zone, as well as major infrastructure such as highways, utilities, hydroelectric dams, and major seaports.

Heimaey Island

In the early morning of January 23, 1973, an area off the southwestern edge of Iceland experienced an unexpected volcanic event. On the island of Heimaey, which is part of the Vestmann Islands, there was an eruption of lava and **tephra** that came from a fissure along the northern edge of the island (Figure 3.48). The duration of the primary eruptions, which were Strombolian in nature and threw out ash, lapilli, and bombs as well as lava flows, lasted about 3 weeks. The complete episode extended for more than 5 months and ended in June 1973.

tephra
A general term for all types of pyroclastic material produced by a volcano.

The entire population of about 5,300 inhabitants was evacuated using more than 75 fishing boats that had been ported in the local harbor as a result of an earlier impending storm (Scarth, 1994, p. 112). In an orderly fashion the people, along with their cattle, cars, and even the money from the bank, were taken off the island. Because of this plan, the only items of value lost in the ensuing eruptions were more than 300 homes that were burned, inundated by lava, or crushed by the fallout. In the end, lava and other material added about 2.5 sq km (1 sq mi) of new land to the island.

Source: USGS.

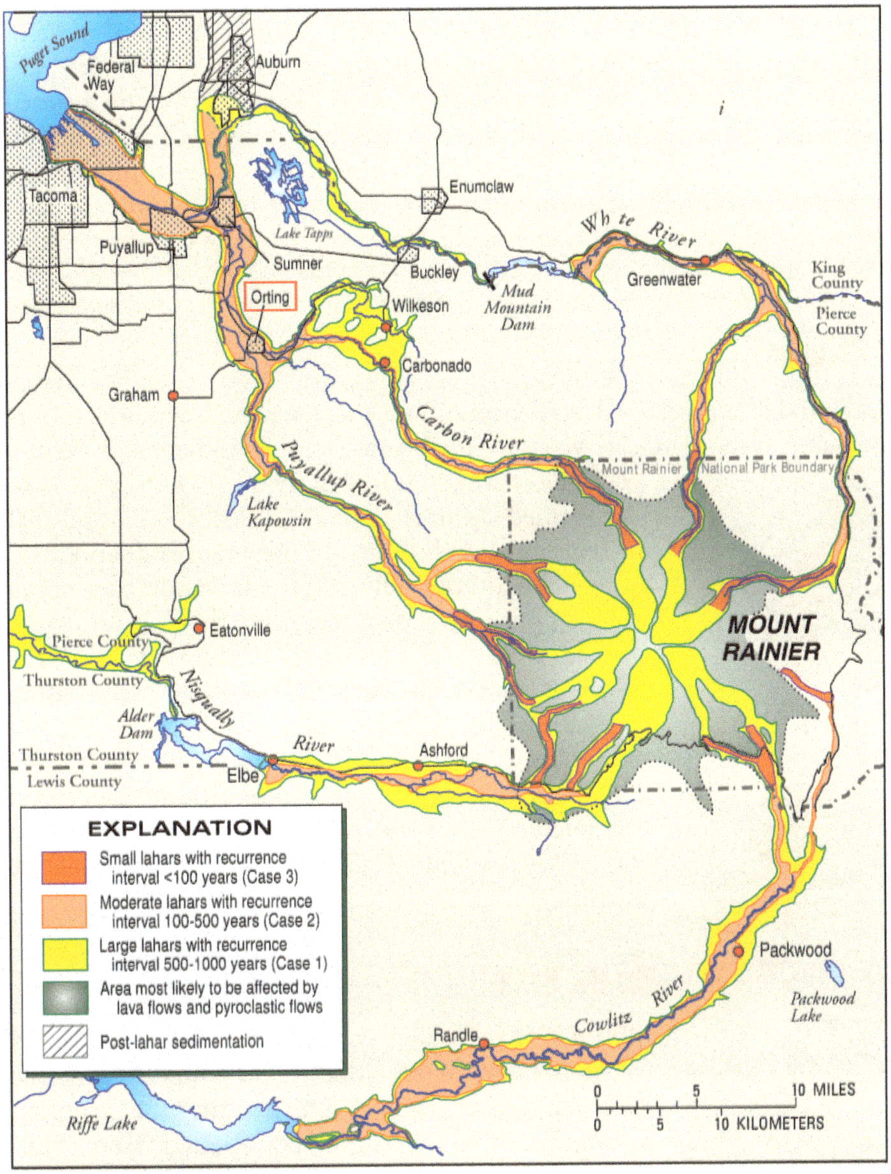

Figure 3.47 Hazard zones for lahars, lava flows, and pyroclastic flows from Mount Rainier.

© FLPA/Shutterstock.com.

Figure 3.48 Fishing boat leaving the harbor at Heimaey in the Westmann Islands, Iceland, with ash from the erupting Eldfell Volcano rising up behind the harbor.

A unique aspect of this particular event was that the U.S. Navy came in and used powerful pumps to spray seawater on the lava and lower its temperature from about 1000 °C to roughly 100 °C. This stopped the forward progress of the lava and saved the harbor that the fishing fleet depended on and much of the island needed for their livelihood. This is an instance when inhabitants rose to the challenge of nature and succeeded in saving the land and later reclaiming their homesteads. However, there are times when this is not so.

El Chichón

Fisher and others (1997) describe conditions that led up to the eruption of El Chichón, a moderate-size volcano in Chiapas, southern Mexico, in March 1982. A report submitted by two field geologists who had studied the area 18 months earlier indicated that an eruption could occur, but their report did not generate any action by authorities who could have taken steps to mitigate a disaster. After the volcano experienced a number of explosive eruptions over the course of a week, the intensity increased, terminating in a series of pyroclastic flows and surges that killed more than 2,000 people and erased more than nine villages in an 8 km (5 mile) radius around the central vent (Figure 3.49). More than 140 sq km (87 sq mi) were covered by the flows. Had local authorities heeded the warning, fewer lives would have been lost.

United States Geological Survey.

Figure 3.49 The remains of a building that was destroyed by a pyroclastic flow from a volcanic eruption in 1982.

Mount Etna

Mount Etna, located in southern Italy, is formed by subduction of the African plate underneath the Eurasian plate (Figure 3.50). During the eruption of Mount Etna that occurred nonstop from mid-December 1991 to late March 1993, numerous attempts were made to impede or direct lava flows headed toward nearby towns. Most efforts proved fruitless as the relentless inertia of the lava mass could not be stopped. In a few instances, building diversion berms with bulldozers was successful but only because of favorable circumstances and some luck. Mt. Etna continues to erupt, producing a lava flow in 2008, explosive eruptions that required shutting the airport in Catania in 2011 and 2012, and a lava fountain up to 1 km (0.6 mi) in height in 2015. In 2017 hot magma made contact with snow and exploded, injuring 10 people including a BBC News television crew.

© Alberto Masnovo/Shutterstock.com.

Figure 3.50 Mount Etna is one of several significant volcanoes in southern Italy.

Dissemination of Information

If we look back over several eruptions that occurred in modern times, we can see an unfortunate pattern in how information is disseminated to people in harm's way. Developed countries that have well-established communication and transportation systems are able to move people away from potential disaster areas. However, sometimes emergency agencies in economically stable countries experience problems, as when reports involving potential hazards contain clauses to control the sharing of findings. "This is not a problem confined to economically less developed countries [and] all contract research carries this danger. Poor countries do not have a monopoly on administrative secrecy. . . . [L]ong-term volcanic hazards are often given a lower priority than attempts to improve living standards" (Chester, 1993, pp. 210–211).

Fisher and others (1997) refer to the United Nations designated International Decade for Natural Disaster Reduction (IDNDR), which ran from 1990 to 2000. Among its themes of study was the Decade of the Volcano, which addressed detailed studies of selected volcanoes. Fifteen volcanoes were the focus of intense investigations (Table 3.4).

Table 3.4 Fifteen Volcanoes Designated by United Nations International Decade for Natural Disaster Reduction Report as Receiving Intense Study

Volcano	Country	Type or Style	Recent Eruptions
Colima	Mexico	Stratovolcano	2018
Galeras	Colombia	Stratovolcano	2014
Mauna Loa	Hawaii	Shield	1984
Merapi	Indonesia	Stratovolcano	2010
Mount Etna	Italy	Stratovolcano	2021
Mount Rainier	United States	Stratovolcano	1894
Mount Sakurajima	Japan	Stratovolcano	2021
Mount Unzen	Japan	Complex	1995
Mount Vesuvius	Italy	Complex	1944
Nyiragongo	Democratic Republic of the Congo	Stratovolcano	2016
Pico del Teide	Canary Islands, Spain	Stratovolcano	1909
Santa Maria	Guatemala	Stratovolcano	2006
Santorini (Thera)	Greece	Shield	1950
Taal	Philippine Islands	Stratovolcano	2020
Ulawun	Papua, New Guinea	Stratovolcano	2019

Source: www.volcano.si.edu

The potential for volcanic disasters is obvious. These events can occur without much warning and can have major impacts on communities, their inhabitants, and the surrounding region. Volcanoes are a threat to air travel as they throw particles into the air that can damage or destroy jet engines. Volcanoes have generated about 5% of all tsunamis over the past 250 years and have killed tens of thousands of people.

The U.S. Geological Survey reports that the United States has 170 active and dormant volcanoes. Significantly, approximately half of the 55 volcanoes that represent the biggest threat to the country are monitored with too few instruments. Volcanoes in the Cascade Range of the Pacific Northwest are the most likely to erupt in explosive forms. When this does occur, they will generate pyroclastic flows and ash falls that could threaten millions of people. More discussion of the seismic potential of this region and the Cascadia Subduction Zone of the Pacific Northwest coast is in Chapter 4, Earthquakes.

Key Terms

active *(page 97)*
ash *(page 75)*
asthenosphere *(page 68)*
basalt *(page 67)*
block *(page 77)*
bomb *(page 77)*
caldera *(page 83)*
cinder cone *(page 81)*
composite volcano *(page 81)*
decompression melting *(page 70)*
dormant *(page 97)*
extinct *(page 97)*
flux melting (page 70)
lahar *(page 88)*
lapilli *(page 75)*
lava *(page 66)*
lava dome *(page 83)*
lithosphere *(page 68)*
magma *(page 66)*
mitigation *(page 99)*
pyroclastic *(page 81)*
rift valley *(page 69)*
tephra *(page 100)*
shield volcano *(page 81)*
viscosity *(page 75)*
volcanic explosivity index (VEI) *(page 77)*

Summary

Volcanoes form from molten rock called magma moving to the surface of the earth and erupting. Plate tectonics controls the generation of magma—mid-ocean ridges and subduction zones are both settings where volcanoes are found. Of Earth's 1,500 known active volcanoes, more than 90% are associated with plate boundaries, and two thirds of all active volcanoes are found around the Ring of Fire surrounding the Pacific Ocean.

Magma forms from three mechanisms: (1) Decompression Melting; decreasing the melting point of rock by dropping the pressure; (2) Flux Melting; decreasing the melting point of rock by adding fluids; and (3) Heat-Transfer Melting: adding heat to already hot rocks. Decompression melting occurs in continental rift zones, ocean spreading centers, and mantle plume settings.

The silica content of magma controls the viscosity of the liquid, which has a strong impact on eruption style. A silicon ion forms a four-sided pyramid shape when bonded with four oxygen ions, and these tetrahedra can join together to form chains in a process called polymerization. Low-silica, gas-poor magmas like basalt are less viscous, flow easily, and tend to form lava flows like pahoehoe and a'a. Higher-silica magmas like andesite, dacite, and rhyolite are more viscous, as well as having more volatiles, and tend to generate explosive eruptions of pyroclastic particles of ash, lapilli, blocks, and bombs.

The Volcanic Explosivity Index (VEI) is a scale for quantifying the explosiveness of volcanic eruptions. Different types of volcanic eruption styles are associated with varying degrees of explosivity. Volcanic landforms are a result of the composition, explosivity, and style of eruption and can be classified as shield volcanoes, cinder cones, composite or stratovolcanoes, lava domes, and calderas.

Hazards associated with volcanoes include more than just lava flows and explosive eruptions, and can range from toxic gas emissions, lahars, ash fall deposits, pyroclastic density currents, lateral blasts, landslides, tsunamis, earthquakes, and global cooling. Volcanoes can be classified as active, dormant, or extinct, but this may be misleading as volcanoes we believe to be dormant or extinct may just have recurrence intervals of eruptions that are longer than the historical records for that volcano, and could erupt with little or no warning. Volcanoes have a wide range of associated hazards that put the people living on their flanks and nearby at risk. While disasters have happened that have killed tens of thousands of people in single events, our increasing scientific ability to predict volcanic eruptions from precursor events is helping us to more effectively live with this volcanic threat.

References and Suggested Readings

Blong, R. J. (1984). *Volcanic hazards: A sourcebook on the effects of eruptions.* Academic Press.

Chester, D. (1993). *Volcanoes and society.* Edward Arnold.

Decker, R., & Decker, B. (2006). *Volcanoes* (4th ed.). W. H. Freeman.

Duffield, W. A. (2003). *Chasing lava—A geologist's adventures at the Hawaiian Volcano Observatory.* Mountain Press.

Fisher, R. V., Heiken, G., & Hulen, J. B. (1997). *Volcanoes—Crucibles of change.* Princeton University Press.

Newhall, C., Self, S., & Robock, A. (2018). Anticipating future Volcanic Explosivity Index (VEI) 7 eruptions and their chilling impacts. *Geosphere 14*(2): 572–603.

Scarth, A. (1994). *Volcanoes.* UCL Press.

Schmincke, H.-U. (2005). *Volcanism.* Springer.

Sigurdsson, H. (1999). *Melting the Earth—The history of ideas on volcanic eruptions.* Oxford University Press.

The Encyclopedia of Volcanoes (1st ed.). (1999). Elsevier.

Web Sites for Further Reference

http://volcano.oregonstate.edu/

http://www.ivhhn/org

http://volcanoes.usgs.gov/ash

http://www.volcano.si.edu

https://volcanoes.usgs.gov/index.html

https://www.volcanodiscovery.com/

Questions for Thought

1. How does plate tectonics explain the location of active volcanoes on Earth?
2. In what settings does decompression melting to form magma take place?
3. How do flux melting and heat-transfer melting generate magma?
4. How does magma composition and changing silica content change magma viscosity?
5. What role do dissolved gases in magma play in the eruption of a volcano?
6. Explain the differences between different eruption styles and the volcanic, edifices they generate.
7. How are pahoehoe and a'a lava flows different?
8. What are the physical characteristics of shield volcanoes, composite volcanoes and cinder cones? What kinds of eruptions do they typically have to generate such different landforms?
9. Describe the different hazards associated with volcanoes and how they impact the surrounding population.
10. Potential activity of a volcano can be described by its being active, dormant, or extinct. What does each of these terms mean?
11. Describe how surface heat can be used as a precursor of a volcanic eruption.
12. Why is Mount Ranier a potential threat to residents in the area?

4

Earthquakes

FEMA by Eilis Maynard.

The Alexandria Square Building in downtown Napa, California, suffered major damage due to a magnitude 6.0 earthquake on August 29, 2014.

Cascadia Subduction Zone—What Could Possibly Happen?

The Cascadia Subduction Zone is part of the Ring of Fire, a region that encircles more than three-quarters of the Pacific Ocean (Figure. 3.7). The Ring of Fire stretches from New Zealand through Indonesia, the Philippines, Japan, and along the Aleutian Islands and the coast of Alaska. It continues southward off the coast of British Columbia, along the western edge of the United States, Mexico, and Central America, and encompasses the entire western edge of South America. This Circum-Pacific belt is extremely active seismically and has many volcanoes due to continuous collisions of oceanic plates with continental plates.

In recent years there have been major earthquakes occurring in Chile (February 2010, magnitude 8.8), New Zealand (February 2011, magnitude 6.1), and Japan (March 2011 magnitude 9.1). These events have relieved tectonic stresses in three sections of the Ring of Fire; only portions of the northeast edge, near the Shumagin Gap section of the Alaskan coast and along coastal portions of the Pacific Northwest of the United States, have been spared.

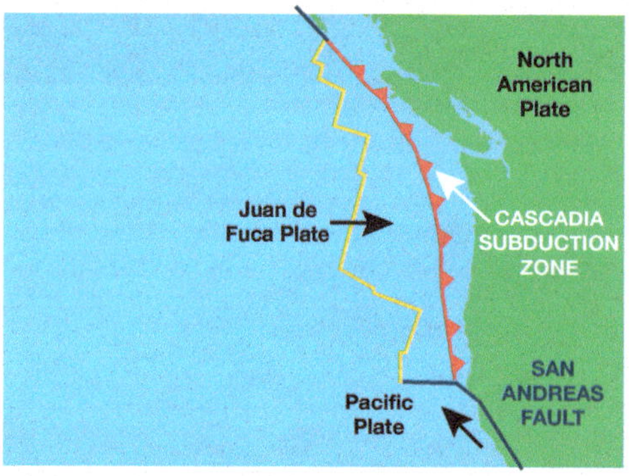

© Kendall Hunt Publishing Company.

Figure 4.0.1 The small Juan de Fuca Plate is colliding with the North American Plate. This activity generates many earthquakes in the region and has produced a string of volcanoes on the continent.

The Juan de Fuca plate lies off the coast of the Pacific Northwest (Figure 4.0.1). As it moves to the northeast, it is converging with the westward moving North American plate at a rate of approximately 3 centimeters per year and is being subducted under the North American plate. The zone of subduction is situated about 100 kilometers (60 mi) to the west of the shoreline of southwestern British Columbia, Washington, Oregon, and northern California. As the Juan de Fuca sinks into the subsurface, it generates rising magma, which has produced a line of volcanoes stretching from Lassen Peak in northern California northward to Mount Meager in British Columbia. This zone includes many well-known volcanoes, including Mount Rainier, Mount St. Helens, Crater Lake, and Mount Shasta. The subducted area is termed the Cascadia Subduction Zone, which approximately parallels the shoreline for about 1,400 kilometers (850 mi).

The North America-Juan de Fuca collision has generated many significant earthquakes in historic time, the largest of which was the magnitude 9.0 event that occurred on January 26, 1700. In the past century, numerous earthquakes of magnitude 6.0 or greater have rocked the region, including more than 10 magnitude significant events, such as the Puget Sound earthquake of 2001 (magnitude 6.8).

The absence of a great earthquake of magnitude 8.0 or larger in the region has produced growing concerns about the likelihood of such an event taking place in the foreseeable future. Several articles have brought this possibility to the attention of both disaster management personnel and the general public. Kathryn Schutz, in her 2015 Pulitzer Prize winning article, "The Really Big One," generated both recognition and concern to the people of the Pacific Northwest. Research by USGS and academic scientists has resulted in a significant amount of data that demonstrate that the region must be prepared for the potential of a catastrophic event occurring in the near future.

Unfortunately, it is virtually impossible to predict exactly when a mega earthquake will happen along the Cascadia Subduction Zone. Seismologists have pointed out that such an earthquake could be relatively soon or in several centuries. What seems certain is that the Cascadia Subduction Zone will experience many earthquakes in the future, some of which will be quite large.

Thousands of earthquakes occur every day, but fortunately very few of them are large enough to create problems. Most major earthquake activity is associated with regions where tectonic plates are colliding with one other to generate stresses that cause rocks to snap, sending seismic energy through the Earth. As we saw earlier in the discussion of global plate tectonics, earthquakes, and volcanoes—which can cause two of nature's greatest types of catastrophes—are associated with one another and will often occur together. Most earthquakes occur along plate boundaries, those areas where tectonic plates are moving apart, colliding, or sliding past one another.

Regions at Risk: The Plate Tectonic Connection

Major lithospheric plates move about on Earth's surface and eventually collide with one another, generating forces that result in earthquakes. Depending on the actual collision mechanism, which creates movement and fracturing in rocks, the earthquakes occur at varying depths, ranging from near the surface to depths approaching 700 km (420 mi). The majority of earthquakes occur at depths less than 100 km (60 mi) in the lithosphere.

Types of Plate Boundaries

There are three types of plate boundaries that produce earthquakes (Figure 4.1). **Divergent boundaries** are associated with areas where plates are moving apart—for example, along mid-oceanic ridges where plates are

divergent boundary
An area where two or more lithospheric plates move apart from each other, such as along a mid-oceanic ridge.

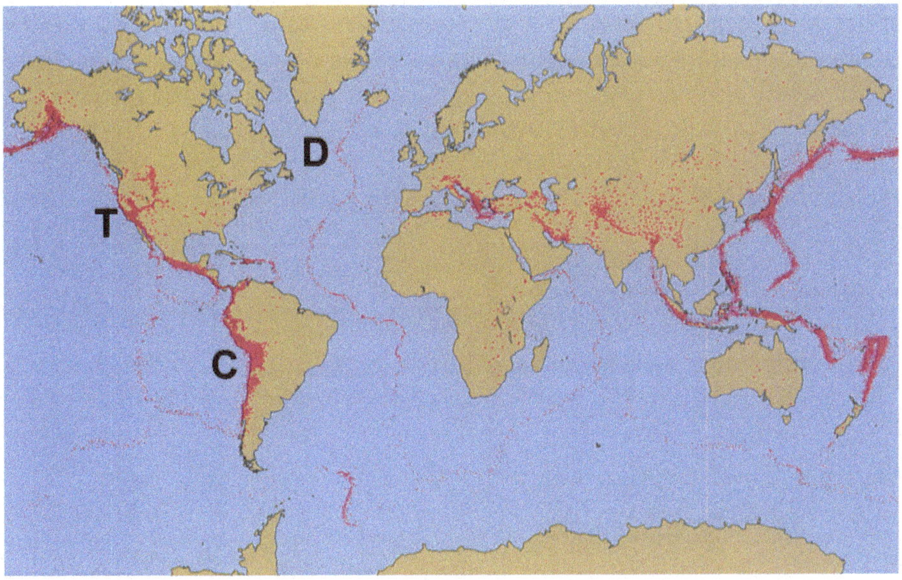

Modified from Physical Geology by Dallmeyer. Copyright © 2000. Reprinted by permission of Kendall Hunt Publishing Company.

Figure 4.1 Global map showing earthquake epicenters for events between 1961 and 1967. C, off the west coast of South America designates a convergent boundary; D, in the North Atlantic Ocean below Greenland is a divergent boundary, and T, west of California designates the transform boundary represented by the San Andreas Fault system in Southern California.

Figure 4.2 **A divergent plate boundary occurs when two plates are moving apart at the oceanic ridge. Tensional forces pull the two plates away from each other and new oceanic crust is being formed.**

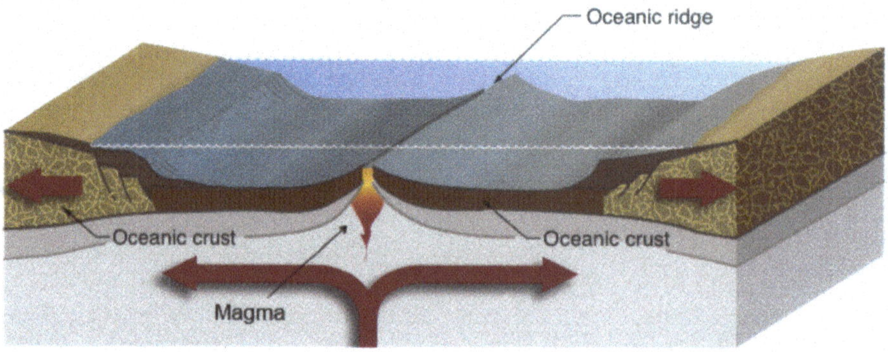

© Kendall Hunt Publishing Company.

being pushed apart. As the newly formed oceanic crust moves away from its heat source at the mid-oceanic ridge, it becomes colder. The leading edge of the oceanic plate tends to sink because it is far removed from the heat source. This sinking action creates tensional forces that pull the plate away from the ridge. Because magma is hot, stresses are unable to build up and the rocks at depth do not normally break. Seismic activity along mid-oceanic ridges is caused by rising magma as it bubbles its way to the surface. As magma ascends, the overlying material is pushed up and outward. Earthquake activity along mid-oceanic ridges is typically confined to relatively shallow depths, taking place in the upper 20 km (12 mi) of Earth's crust (Figure 4.2). Divergent motion can also occur in continental regions where the land surface is being split apart by extensional stresses. Rising heat in the mantle creates a bulge that can fracture the brittle continental crust. An example of this is in the eastern part of Africa along the East African Rift Zone.

convergent boundary

An area where two or more lithospheric plates are coming together, such as along the west coast of South America.

Convergent boundaries are regions where plates are colliding as a result of compressional forces. These conditions produce the largest earthquakes. Three types of convergent configurations exist (Figure 4.3). A collision that involves two oceanic plates is an ocean-ocean plate convergence. (Figure 4.3a). In this instance, one of the two plates is pushed under the other one, resulting in subduction of the downgoing slab. Once the slab starts its downward motion, gravitational forces pull the slab to greater depths. An example of this type of boundary is where the Caribbean Plate collides with the westward moving North American plate. The result is the string of volcanic islands that form the Lesser Antilles.

Figure 4.3 **Different types of convergent plate collisions. (a) ocean-ocean plates, (b) ocean-continent plates, (c) continent-continent plates.**

(a)

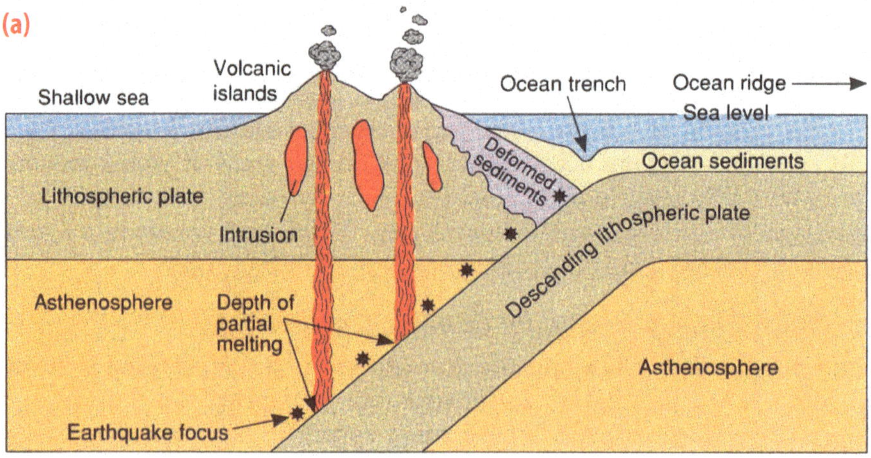

From Physical Geology by Dallmeyer. Copyright © 2000. Reprinted by permission of Kendall Hunt Publishing Company.

A convergence setting in which an oceanic plate collides with a continental plate is an ocean-continent plate boundary (**Figure 4.3b**). The oceanic plate, consisting of predominantly basalt, is denser than the continental granite so the oceanic plate is subducted. This subducted material can maintain its integrity as a solid slab to depths of several hundred kilometers.

The third example of convergent plate collisions involves the convergence of two continental plates (**Figure 4.3c**). Each plate consists of the same general rock type—predominantly granite or granodiorite. As we have seen, continents rest atop the asthenosphere and denser mantle, so this type of plate convergence will not result in any significant subduction. An example is the collision of the Indian subcontinent with the lower edge of the Eurasian plate (**Figure 4.4**). Because no subduction takes place, the two colliding land masses basically rumple themselves up, producing the Himalayas, Earth's highest mountain range, which is located in South Asia. The regions surrounding this mountain range have experienced some of the most destructive earthquakes recorded over the past century; these have been especially catastrophic because of the poor construction standards and high population densities in that region of the world. (See Box 4.1, which describes the 2005 Pakistan earthquake.)

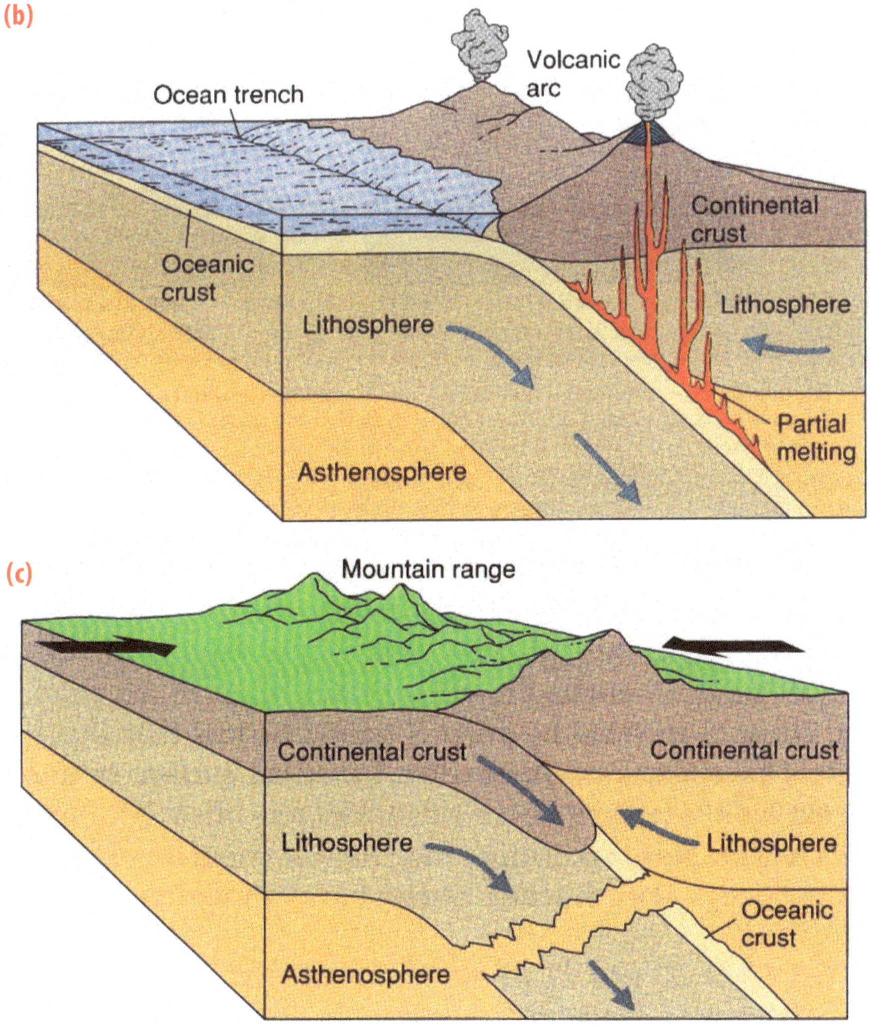

From *Physical Geology* by Dallmeyer. Copyright © 2000. Reprinted by permission of Kendall Hunt Publishing Company.

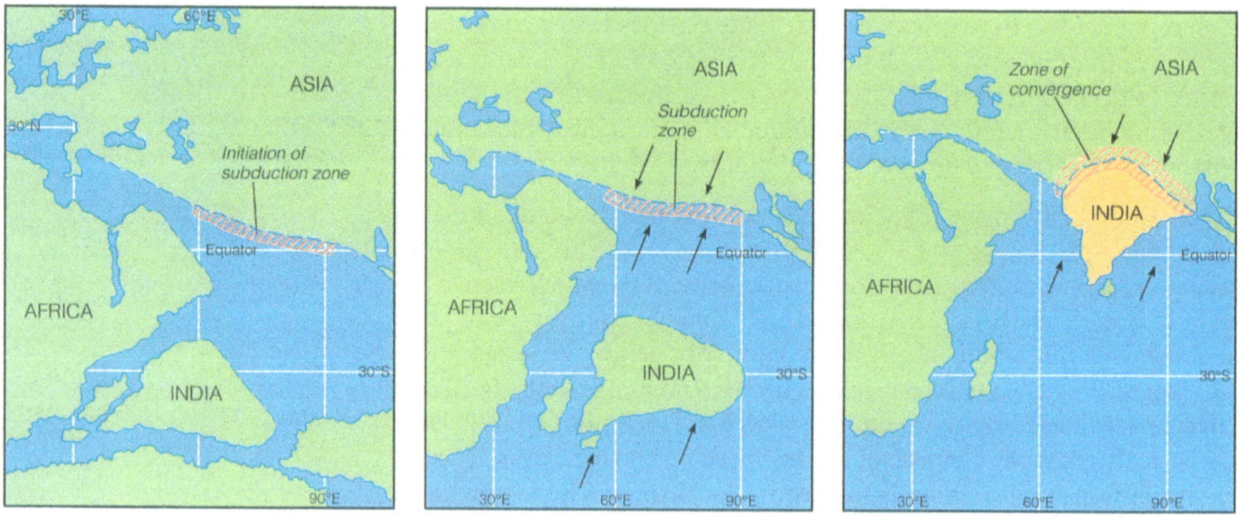

From Planet Earth by Renton. Copyright 2002. Reprinted by permission of Kendall Hunt Publishing Company.

Figure 4.4 **Indian subcontinent moving north and colliding with the Eurasian continental plate. The lack of subduction of the continental masses produced the Himalayas. This collision began approximately 50 million years ago and continues today.**

BOX 4.1 **The Pakistan Earthquake of October 2005**

The Mediterranean-Trans-Asiatic belt runs in an east-west direction above the Mediterranean Sea stretching eastward through the upper Middle East into the lower portion of Asia. Stress in this region results from the African plate moving northward and colliding with the western portion of the Eurasian plate. There are also major strike-slip faults that run through Turkey into Iran and Iraq. The entire region experiences major earthquakes on a regular basis and the results are often devastating. Buildings and infrastructure here are old and poorly constructed, so a moderate amount of ground motion destroys villages and kills many people. Major earthquakes in this region of the world kill tens of thousands of people every decade or two.

The October 8, 2005, a magnitude 7.6 earthquake in Pakistan was related to major fault systems that are part of the Mediterranean-Trans Asiatic belt. The U.S. Geological Survey reported 80,361 people killed, more than 69,000 injured, and extensive damage throughout northern Pakistan. In some areas entire villages were destroyed and more than 32,000 buildings collapsed. The maximum intensity was VIII. An estimated 4 million people in the area were left homeless. Landslides and rock falls damaged or destroyed several mountain villages, along with roads and highways, thereby cutting off access to the region for several weeks. In the western part of the country there were reports of liquefaction and sand blows. Seiches, which are waves generated in closed bodies of water that produce a sloshing motion, were observed in West Bengal, India, and many places in Bangladesh.

transform boundary

An area where two plates move past each other in a horizontal, sliding motion.

Transform boundaries are another type of large-scale features associated with plate collisions. They form when sections of two plates slide against each other. This occurs in oceanic regions when portions of oceanic plates move at different rates. Earthquakes occur along these transform faults, but the magnitudes of these events are usually small.

When transform plate boundaries occur in continental areas, the resulting motion can produce very large earthquakes. In addition to the San Andreas Fault System in California, an excellent example of this process is seen in the North Anatolian Fault, which runs through Turkey and adjoining countries (Figure 4.5). The plate motion occurs along the boundary of the Eurasian Plate which lies north of the fault and the smaller Anatolian Plate to the south.

Extending for 1,500 km, the North Anatolian Fault passes about 20 km (12 mi) south of Istanbul, a city of more than 14 million people. Since 1939, eight earthquakes of magnitude 7.0 or larger have taken place along the fault. In 1999 a magnitude 7.6 event killed more than 17,000 people near the city of Izmit. Scientists believe that the likelihood of a major earthquake occurring in the vicinity of Istanbul in the near future is high.

Seismic activity along the North Anatolian Fault has shown a migration of epicenters since 1939, except for the 1992 events. The right-lateral strike-slip motion is the result of the Anatolian Plate slipping westward relative to the Eurasian Plate.

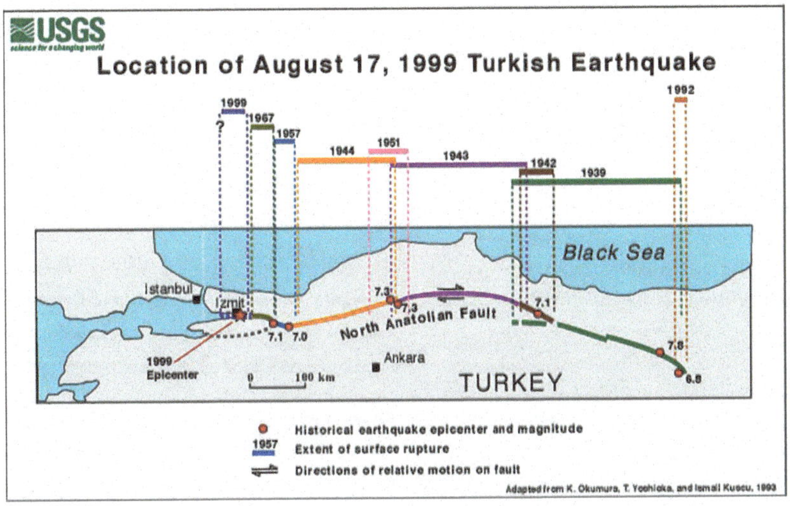

United States Geological Survey.

Figure 4.5 **The North Anatolian Fault is a very active feature. It is a major transform plate boundary displaying right-lateral strike-slip motion.**

Mechanics and Types of Earthquakes

When a force is applied to a surface, the result is termed **stress** (measured as force per unit area). Movement within the Earth occurs when the exerted stress exceeds the frictional force preventing movement from taking place. When blocks of rock move, slippage occurs along a surface or fault plane. An object that is stressed will experience **strain**, which can be elastic, thereby returning the object to its original shape. An example of this is a rubber band or piece of elastic, when pulled (extensional stress), returns to its original shape. If either the rubber band or piece of elastic is overly stressed, it breaks and the original single object is now two pieces, not the original one. If the object breaks, the strain is inelastic, such as what occurs in brittle rock.

stress
The force being applied to a surface; forces can be compressional, extensional, or shearing.

strain
The response of an object which is being stressed; strain can be elastic, in which case the object returns to its original shape or inelastic, when the object does not recover its shape, thereby being deformed.

Faults

Orientation of the fault plane can range from nearly horizontal to vertical. The fault plane separates two adjacent blocks called the hanging wall and the footwall. It is easy to assign names to each of these blocks. Picture yourself standing on the fault plane. The block underneath your feet is the footwall; the block over your head is the hanging wall.

Figure 4.6 shows three kinds of faults and the stresses related to each one. During the movement along a fault, if the hanging wall, which is the block lying above the fault plane, moves down relative to the footwall, the fault is classified as a **normal fault**. This type of fault is sometimes called a gravity fault because the downward force of gravity causes the hanging wall to move downward. Normal faults form when extensional or tensional forces are pulling on two blocks. This type of fault occurs at divergent plate boundaries, such as those found at mid-ocean ridges where oceanic plates are moving away from each other.

normal fault
A plane along which movement has occurred such that the upper block overlying the fault has moved down relative to the lower block.

reverse fault
A plane along which movement has occurred such that the upper block overlying the fault has moved up relative to the lower block.

When the hanging wall moves up relative to the footwall, a **reverse fault** is produced. Compressional forces squeezing two blocks together will generate a reverse fault. Reverse faults are commonly found at convergent plate boundaries where the plates are colliding.

A **strike-slip fault** is formed when two blocks slide one past the other with very little or no vertical displacement. In this instance a pair of parallel forces is acting one against the other, producing a shear couple. A well-known example of a strike-slip fault is the San Andreas fault in southern California (Figure 4.7a). Closer examination of movement on a strike-slip fault shows the displacement seen in Figure 4.7b.

When rocks do fracture, the initial point at which breakage occurs in the subsurface is called the **focus** (or hypocenter) of the earthquake. Energy radiates outward in a spherical pattern from this point and travels in all directions (Figure 4.8). The point on the surface directly above the focus is termed the **epicenter**, a location that can be assigned latitude and longitude coordinates indicating where the earthquake occurred.

Earthquake foci have a wide range of depths (Figure 4.9). **Shallow-focus earthquakes**, those occurring between the surface and a depth of 70 kilometers, can form as upper-plate, thrust, or intraslab events. **Intermediate-focus earthquakes** (70 to 300 km) and **deep-focus earthquakes**, those reaching depths of 700 km, take place in brittle, subducted rock. As the subducted

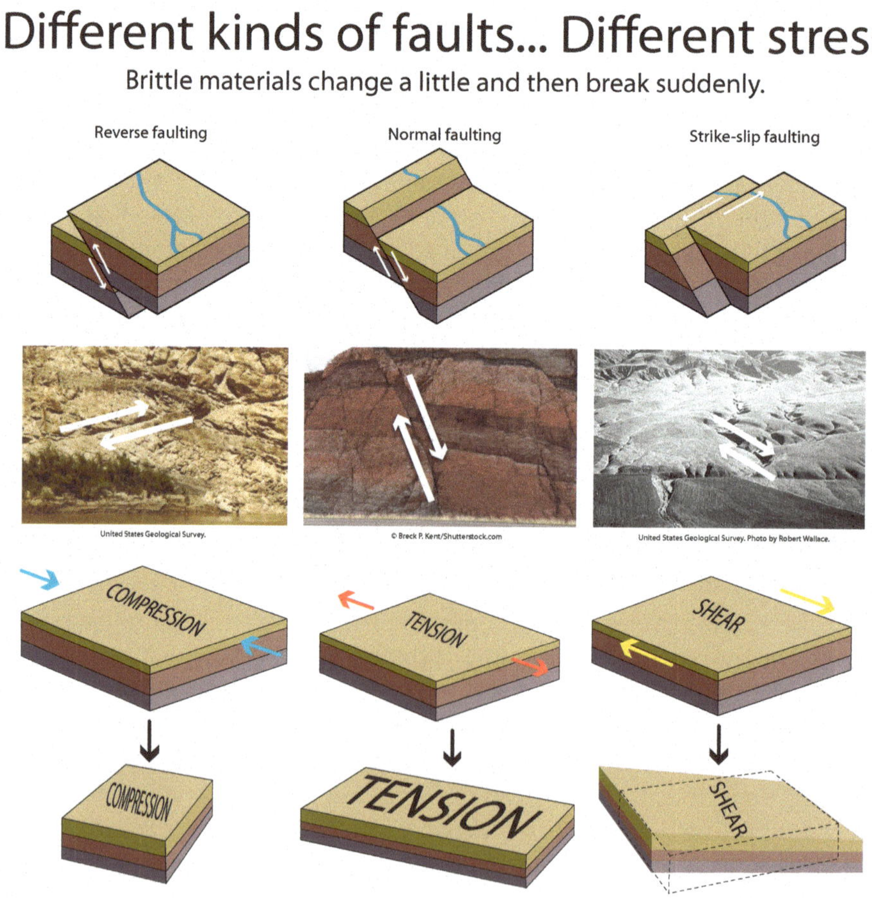

© Kendall Hunt Publishing Company.

Figure 4.6 The type of faulting depends on the type of stress applied to the rock. In the diagrams for reverse and normal faulting, the footwall and hanging wall are labeled. The terms do not apply to strike-slips faults.

strike-slip fault
A fault in which the motion of the two adjacent blocks is horizontal, with little if any vertical movement.

focus
The point in the subsurface where an earthquake first originates due to breaking and movement along a fault plane; sometimes referred to as a hypocenter.

epicenter
The point on Earth's surface directly above the focus. This is the position that is reported for the occurrence of an earthquake as a latitude and longitude value can be assigned to the point.

shallow-focus earthquake
An earthquake that has its focus located between the surface and a depth of 70 kilometers.

intermediate-focus earthquake
An earthquake that has its focus located between a depth of 70 kilometers and 300 kilometers.

deep-focus earthquake
An earthquake that has its focus located between a depth of 300 kilometers and roughly 700 kilometers.

body wave
A seismic wave that is transmitted through the Earth. P-waves and S-waves are body waves.

material sinks to depths of about 700 km, fewer events occur due to the heating of the rock, which make it more ductile and hence less likely to break.

Seismic Waves

Energy radiates out from the focus of an earthquake (**Figure 4.8**) and is transmitted as seismic waves (sound and light waves are other examples of energy being moved through space). This energy produces two types of seismic waves: **body waves** and **surface waves**.

Body Waves

Body waves are transmitted through the body or interior of the Earth (**Figure 2.5**). Body waves consist of two types. **P-waves** are the primary waves that result from a push-pull action. They are also referred to as compressional or longitudinal waves, because the back and forth motion moves in the

surface wave

A seismic wave that moves along a surface or boundary.

P-wave

The primary wave, which is a compressional wave; this is the fastest moving of all the seismic waves, and arrives first at a recording station. P-waves are a type of body wave.

United States Geological Survey.

Figure 4.7a San Andreas fault is a strike-slip fault passing through central and southern California.

United States Geological Survey.

Figure 4.7b A close view of a right-lateral strike-slip fault that displaced the fence line.

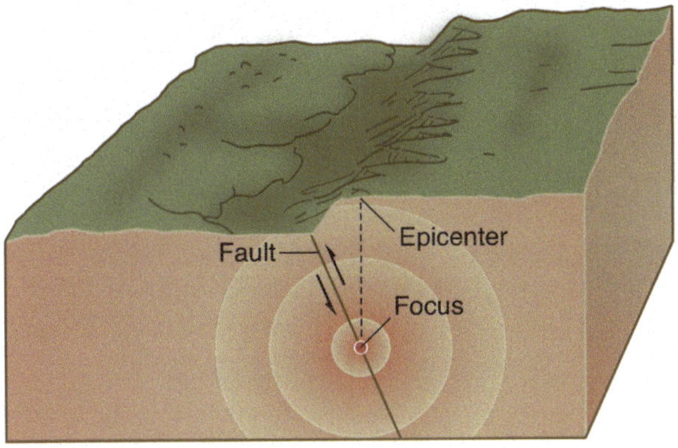

© Kendall Hunt Publishing Company.

Figure 4.8 Location of the focus, epicenter, and fault plane. Notice how seismic energy radiates out from the focus in spherical paths.

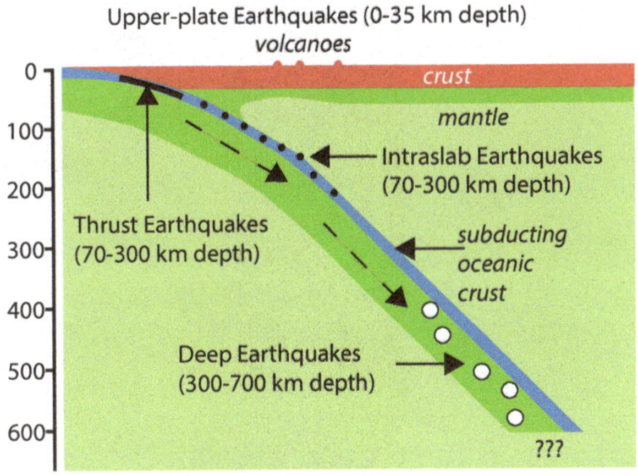

Modified from USGS.

Figure 4.9 Earthquake foci have a range of depths from the surface to approximately 700 kilometers. All of them occur in brittle rock, but fewer occur at greater depths due to heating of the rock. Depth in km.

direction of wave propagation (**Figure 4.10a**). A compressional wave moves along by squeezing the material; the compressed particles then rebound or expand, which compresses the next particles adjacent to the ones that had been compressed. P-waves are the fastest moving of all the seismic waves, and they can travel through any type of material (solid, liquid, or gas). Their velocities average approximately 6 km/sec (3.6 mi/sec), with some variations depending on rock type.

S-wave

The secondary wave, which is a shear wave that moves material back and forth in a plane perpendicular to the direction the wave is traveling. S-waves do not travel through liquids. S-waves are a type of body wave.

S-waves, also known as secondary or shear waves, are so named because they are the second ones to arrive at a detection point. They travel at about 3 km/sec and have a lower frequency and higher amplitude. S-waves are also referred to as shear or transverse waves because their actual motion causes particles to move at right angles to the direction of wave propagation (**Figure 4.10b**). An important characteristic of S-waves is that they do *not* pass through liquids. This fact has allowed seismologists to determine that certain parts of Earth's interior are liquid because S-waves are not detected at all recording stations. Seismic recording stations that are located between distances of 103° and 143° from an earthquake focus will not record S-waves because the waves hit the liquid outer core and do not get transmitted any farther.

Surface Waves

Figure 4.10c shows surface waves that travel along a surface or boundary between layers. This group of waves consists of **Rayleigh** (R) and **Love** (L) waves, two types that are named after mathematical physicists who each did research into their respective characteristics. For shallow and some intermediate-focus earthquakes, surface waves have the greatest amplitude and the most obvious signal (**Figure 4.11**). Vertically directed R-waves make the ground ripple up and down, similar to the wave motion created when you jump into a swimming pool. The lateral motion caused by faster moving L-waves is similar to that seen when a sidewinder snake moves across the desert floor. As the snake moves forward, its body moves side to side (**Figure 4.12**). The key point about each of these

Figure 4.10a–c **Seismic motion travels in different ways through material. P-waves have push-pull movement, S-waves wiggle back and forth (either vertically or horizontally) and surface waves tend to roll or slide.**

(a)

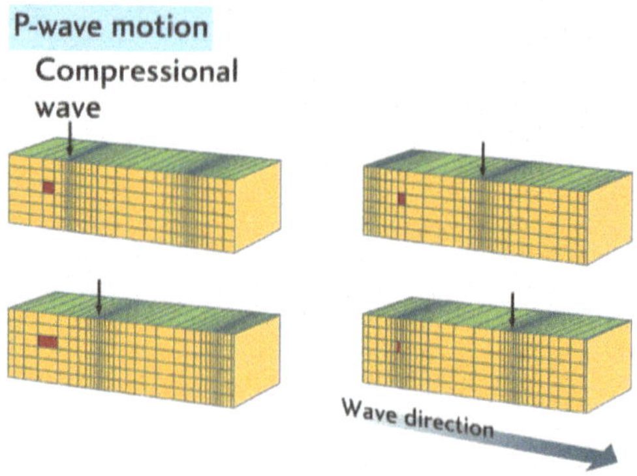

Adapted from usgs.gov.

(b)

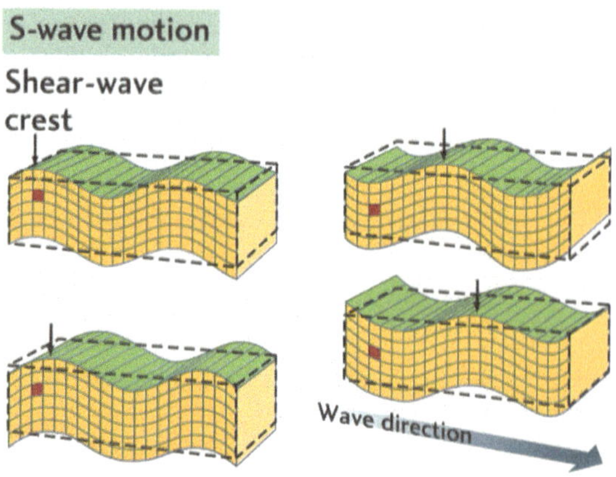

Adapted from usgs.gov.

(c)

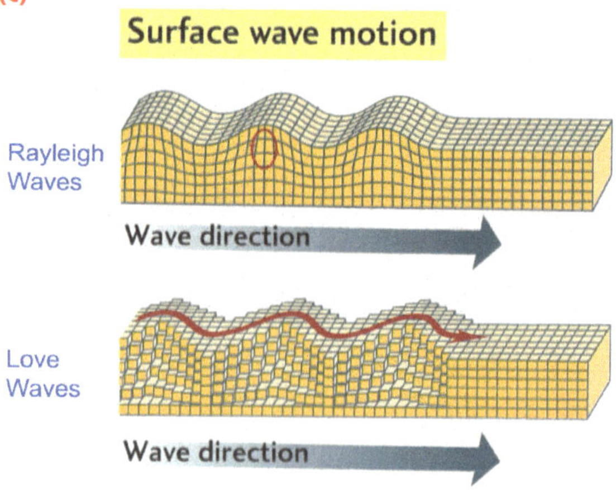

Adapted from usgs.gov.

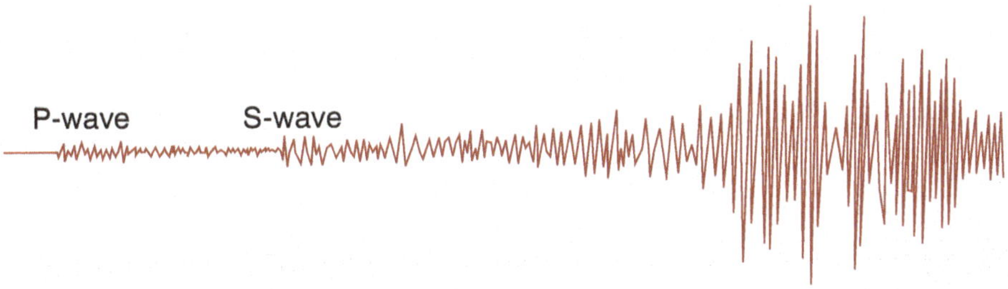

© Kendall Hunt Publishing Company.

Figure 4.11 Seismogram showing the arrivals of the P-, S-, and surface waves. Notice the changes in amplitudes of the various waves as they arrive at the detection station.

surface waves is that they involve a rolling motion of particles and they move much slower than either type of body wave. Surface waves are the most destructive because they produce the most ground motion and take longer to move through an area. Rayleigh waves have a velocity of approximately 12,500 kilometers per hour, while Love waves move at 16,000 kilometers per hour.

Courtesy Bill Caid. Used with permission.

Figure 4.12 A sidewinder rattlesnake displays a horizontal ripple motion along with forward progress similar to that of an L-wave.

When an earthquake occurs, all these types of motion (P, S, R, and L) are generated so the energy is passed through an area at different velocities, different times, and varying amplitudes. The velocity relationship is $V_P > V_S > V_L > V_R$, but the amplitudes of these waves are essentially in reverse order. The rocks and buildings that are being affected undergo a great deal of mixed stresses that produce the damage we observed after the shock waves pass. Engineers can design buildings to withstand most of these forces, but the underlying geology (bedrock, loose soil, or sediment, for example) must also be considered in the design process.

Rayleigh waves
These move particles in a rolling, up and down sense that travel in the direction of propagation of the energy.

Love waves
These move particles with a twisting, side to side motion.

Detecting, Measuring, and Locating Earthquakes

As seismic energy moves through rocks, it abruptly displaces them, sometimes on a large scale. As the ground moves, the motion is detected by a **seismograph**, a very delicate instrument that responds to small changes in the vertical and horizontal positions at a point. This movement is amplified before being sent to a recording device that produces a record of the ground motion as a **seismogram** (Figure 4.13). Careful analysis of a seismogram can provide information about the strength or size of the earthquake and its distance from the recording station. An important parameter is to know the precise time related to the recorded signal so the source region of an event can be located.

seismograph
A device that records the ground motion of an earthquake.

seismogram
The written or electronic record of ground motion detected by a seismograph.

There are numerous recording stations located throughout the world. Within the United States there are more than 500, with a significant number located in earthquake-prone regions including California and Alaska (Figure 4.14). These stations continually record all ground motion and report any unusual activity to the National Earthquake Information Center (NEIC) in Golden, Colorado, which issues reports and warnings of earthquake activity. Rapid computer analysis of earthquake data enables the NEIC to send information and warnings to all parts of the globe.

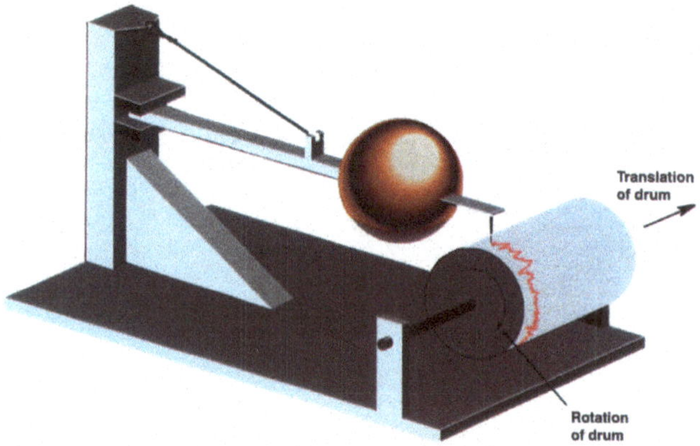

© Kendall Hunt Publishing Company.

Figure 4.13a A basic design of a seismograph shows a pendulum at the end of a horizontal bar with an attached pen that records ground motion.

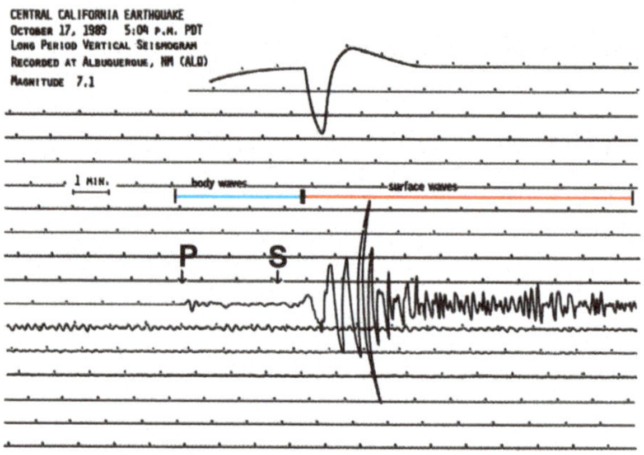

United States Geological Survey.

Figure 4.13b A recording station in Albuquerque, New Mexico, generated this seismogram of the arriving body and surface waves produced by the Loma Prieta, California, Earthquake of October 17, 1989. Given the distance of this recording station from the epicenter of the event, the P- and S-wave arrivals are small, but the surface waves are much larger.

Figure 4.14 A national network of more than 100 seismic stations throughout the continental United States provides data for detailed earthquake analysis.

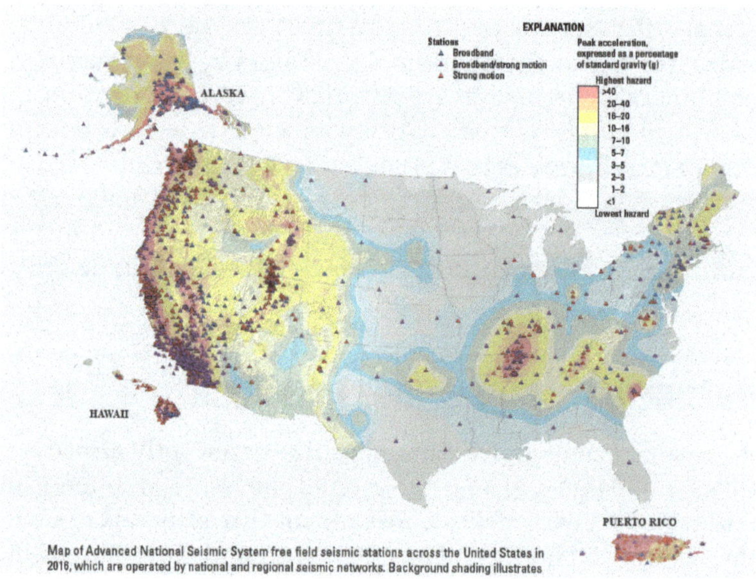

United States Geological Survey.

Measuring the Size of an Earthquake

Several factors control the recorded size of an earthquake. The amplitude of the seismogram is related to the recording station's distance from the epicenter. The greater the distance, the less pronounced is the record. Also the magnitudes of the recorded motion depend on where the seismographs are placed. Loose soil moves more rapidly than solid rock and can give a false record of the true scale of the event. Seismographs are usually located on a solid base to minimize extraneous ground movement.

Earthquakes located near a recording station will generate a sudden jolt that will be detected by the seismograph as a spiked signal. Earthquakes occurring at a greater distance tend to produce longer period signals which have a lower amplitude (**Figure 4.15**). Records of the ground motion of the Northridge, California, earthquake are shown in Figure 4.15 and the event is discussed in Box 4.2.

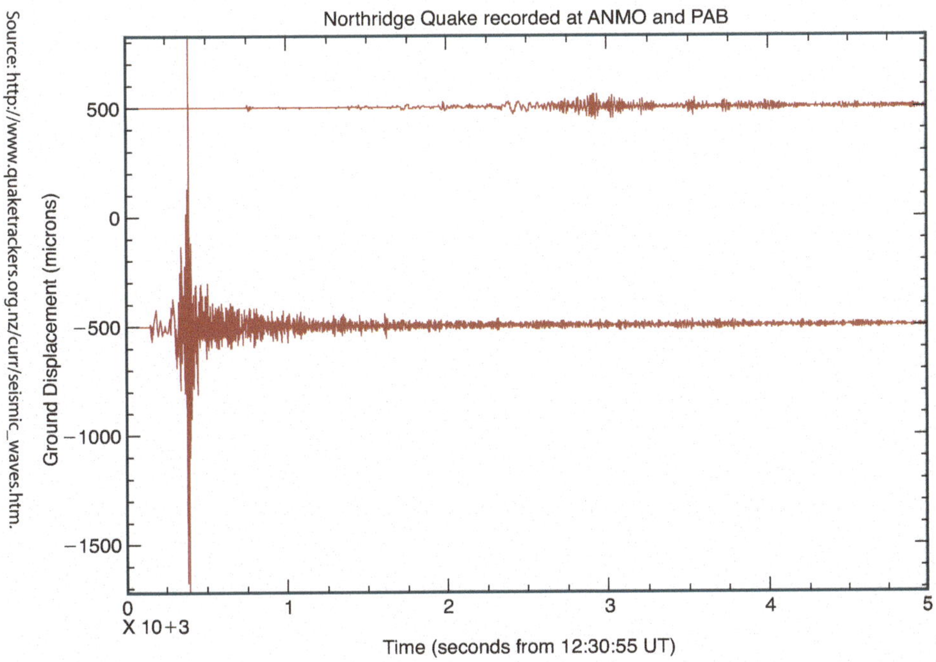

Figure 4.15 Recordings of the Northridge, California, earthquake of January 17, 1994. The upper record is from a recording station in Pablo, Spain; the lower record is from Albuquerque, New Mexico. Notice the greater amplitude for the signal recorded at the station that was much closer to the event.

BOX 4.2 Northridge Earthquake of January 1994

Many of the residents of the Los Angeles, California, area had planned to take the Martin Luther King Day holiday on Monday, January 17, 1994. Unfortunately, Mother Nature had decided it was a day of work as the city of Northridge, a suburb on the northern edge of the city of Los Angeles, was rocked by a magnitude 6.7 earthquake at 4:31 a.m. The United States Geological Survey reported that 57 people were killed and more than 8,700 injured. More than 114,000 buildings were destroyed or severely damaged; numerous gas lines ruptured and several key freeway overpasses collapsed, blocking major traffic arteries in the Los Angeles area. The Northridge earthquake still ranks among the most costly natural disasters in US history with damage estimates ranging from $13 to $44 billion. Following this earthquake major retrofitting of freeways, bridges, and overpasses in the Los Angeles area took place to minimize similar failures when the next large earthquake strikes the area.

Figure 4.2.1 Damage was very widespread in Northridge and surrounding communities. The interstate highway system was disrupted for weeks.

Figure 4.16 **Magnitude 4.0 earthquake near Cloverdale, California, on January 10, 2000. The area lies about 150 km northwest of San Francisco and faults in the area are part of the San Andreas Fault System.**

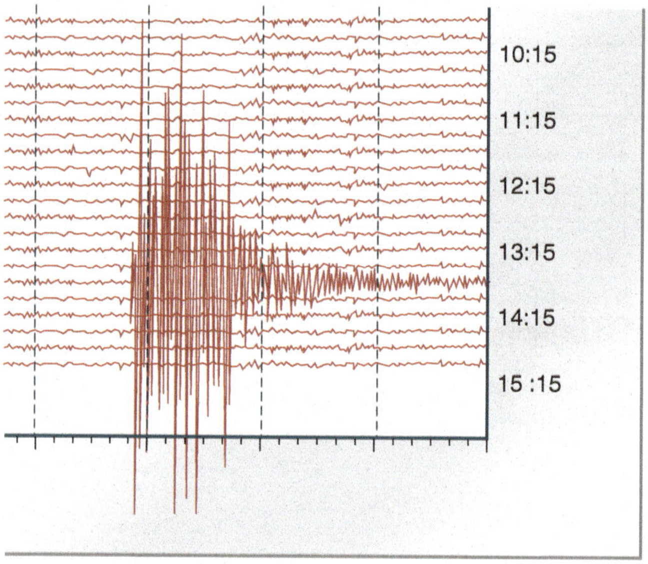

United States Geological Survey.

Figure 4.16 is a seismic record of a minor earthquake in central California. The horizontal lines are recorded by a pen that marks paper on a drum that turns once every 15 minutes, hence four lines per hour. Notice the sudden onset of this signal, which tells us this event was close to the recording seismograph. The recorded ground motion for this event lasted about 2½ minutes.

Measuring the Magnitude of an Earthquake

The magnitude of an earthquake is a number that is used to characterize the relative severity of an earthquake. Several different magnitude scales are used to describe this value. These differences are based on which of the recorded seismic waves are analyzed, but research has shown that minimal differences exist in assigned values for events of magnitude 6 or less. The majority of earthquakes that occur each year are in this smaller range. Larger events are more complex and thus require more data to determine their magnitude. We will use two scales to describe magnitudes: (1) local magnitude (M) for local or nearby earthquakes and those with values of 6.5 or less, and (2) the moment magnitude (Mw) scale for those with larger values.

local magnitude

A term that describes the size of an earthquake in an area near the epicenter; see Richter scale.

Richter scale

A measurement scale developed by Charles Richter to determine the size of earthquakes in California. This is commonly used to describe many types of earthquakes but it is more correctly used for small, localized events.

The **local magnitude**, M, is used for moderate-sized earthquakes measured close to the epicenter. This method uses the **Richter scale**, which was first developed in 1935 by Charles F. Richter of the California Institute of Technology based on data from local earthquakes in California. Richter used the maximum amplitude of the S-wave and took into account the distance from the recording station to the epicenter to establish the magnitude of an event. It is an open-ended scale, so it can include the very smallest events or the largest possible ones that occur. The U.S. Geological Survey (USGS) reports that each year more than 10,000 earthquakes occur in the southern California area. Most of them are very low magnitude with only a few hundred measuring greater than magnitude 3.0, and only about 15 to 20 being greater than magnitude 4.0.

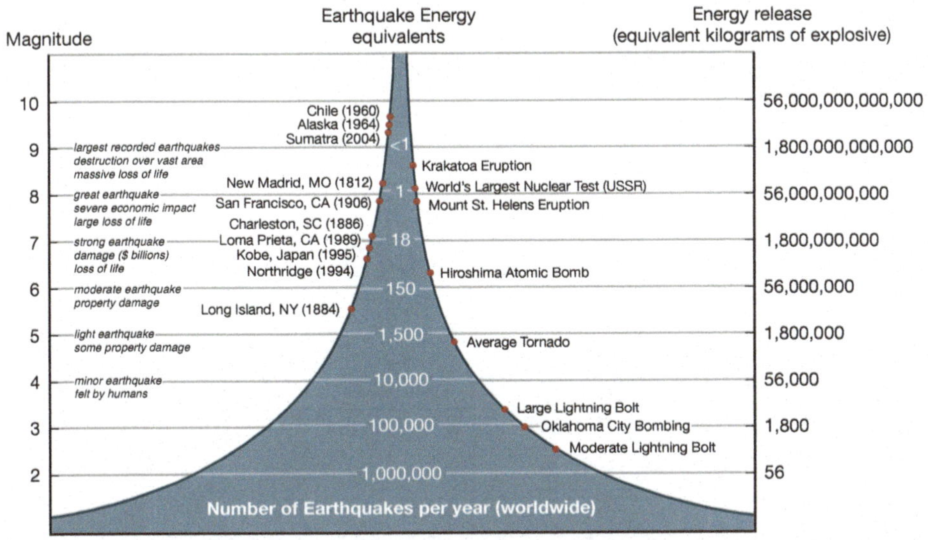

Reprinted by permission of the IRIS Consortium.

Figure 4.17 **Earthquake energy as it relates to known events. Notice the small number of larger events compared with the overall number of earthquakes. Data reflect the annual occurrence of earthquakes.**

Earthquake magnitude is a logarithmic scale signifying the size of the event. If we were the same distance from the epicenters of two different events, ground motion would be 10 times greater during an M = 6 earthquake than that of an M = 5 earthquake. This tenfold increase in ground motion translates to approximately a thirty-twofold increase in the amount of energy produced by the larger event. This larger amount of energy helps explain why higher magnitude earthquakes are generally more destructive. A magnitude 6 earthquake would have more than 1,000 times the energy of an ML = 4 event (Figure 4.17).

The **moment magnitude** (Mw) is used to measure the size of large earthquakes. It is dependent on the shear strength of the rock, the area affected by the fault rupture, and the average displacement along the fault. This scheme is used for large earthquakes having magnitudes greater than an Mw value of 8. Seismic moment is considered to be the best gauge of earthquake as it describes the amount of strain energy released by an event.

moment magnitude

A measure of an earthquake that is based on the area affected, the strength of the rocks involved, and the amount of movement along the primary fault.

A review of the 10 largest earthquakes in recorded history (Table 4.1) shows that all of them are related to active convergent plate boundaries, with nine of them occurring in the circum-Pacific Ring of Fire. The Assam, Tibet, event of August 1950 (along the India-China border) was caused by the collision of India with the Eurasian plate (Figure 4.4), which is a continent-continent plate collision.

When we examine the 10 most deadly earthquakes in history, only one of them is also listed among the 10 largest events—the December 2004 earthquake and tsunami off the coast of Sumatra. In terms of events producing the most fatalities, we notice that four of them occurred in China, a country that has always had a very large population. One other note to consider is that four of the most deadly earthquakes occurred before 1500. The fatality count for those has been determined by historians who have researched those ancient civilizations and based the estimates on historical records.

Table 4.1 Ten Largest and Ten Most Deadly Earthquakes

Ten Largest Recorded Earthquakes

Date	Location	Magnitude
May 1960	Off the coast of Chile	9.5
March 1964	Prince William Sound, Alaska	9.2
December 2004	Off the west coast of Sumatra	9.1
March 2011	Near the east coast of Honshu, Japan	9.1
November 1952	Off the east coast of Kamchatka, Russia	9.0
February 2010	Off the coast of Ecuador	8.8
January 1906	Off the coast of Ecuador	8.8
February 1965	Rat Islands, Aleutian Islands	8.7
August 1950	Tibet, near the China-India border	8.6
April 2012	Off the west coast of Northern Sumatra	8.6

Ten Most Deadly Earthquakes

Date	Location	Deaths
January 1556	Shaanxi, China	830,000
July 1976	Tangshan, China	255,000
August 1138	Aleppo, Syria	230,000
December 2004	Sumatra	228,000 The majority of deaths were due to the tsunami
December 856	Iran	200,000
December 1920	Haiyuan, China	200,000 Modified Mercalli Intensity XII
March 893	Iran	150,000
September 1923	Kanto, Japan	142,800 Extreme destruction of Tokyo and Yokohama
October 1948	Turkmenistan, Russia	110,000
September 1290	Chihli, China	100,000

Source: United States Geological Survey.

Determining the Intensity of an Earthquake

The intensity of an earthquake is a measure of the damage caused by ground shaking and how buildings and other structures respond to seismic energy. The observed damage is a subjective assessment because different observers might assign a variety of levels of severity to a given area depending on their interpretation and evaluation of the damage.

In 1902 Italian scientist Giuseppe Mercalli established a scale that assigned values to damaged areas based on how different materials responded to an earthquake. The scale was modified in 1931 by American seismologists Frank Neumann and Harry Wood, who refined the scale to better address building standards in the United States. The scale is based on 12 different levels of damage (values are assigned Roman numerals), with I being the lowest level of damage and XII being the highest (Table 4.2). The Modified Mercalli scale differs from magnitude scales because there is no mathematical basis for the values given to the damage observed in an area. The

Table 4.2 Abbreviated Descriptions for Levels of the Modified Mercalli Intensity Scale

I.	Not felt except by a very few under especially favorable conditions.
II.	Felt only by a few persons at rest, especially on upper floors of buildings.
III.	Felt quite noticeably by persons indoors, especially on upper floors of buildings. Many people do not recognize it as an earthquake. Standing motor cars may rock slightly. Vibrations similar to the passing of a truck. Duration estimated.
IV.	Felt indoors by many, outdoors by few during the day. At night, some awakened. Dishes, windows, doors disturbed; walls make cracking sound. Sensation like heavy truck striking building. Standing motor cars rocked noticeably.
V.	Felt by nearly everyone; many awakened. Some dishes, windows broken. Unstable objects overturned. Pendulum clocks may stop.
VI.	Felt by all, many frightened. Some heavy furniture moved; a few instances of fallen plaster. Damage slight.
VII.	Damage negligible in buildings of good design and construction; slight to moderate in well-built ordinary structures; considerable damage in poorly built or badly designed structures; some chimneys broken.
VIII.	Damage slight in specially designed structures; considerable damage in ordinary substantial buildings with partial collapse. Damage great in poorly built structures. Fall of chimneys, factory stacks, columns, monuments, walls. Heavy furniture overturned.
IX.	Damage considerable in specially designed structures; well-designed frame structures thrown out of plumb. Damage great in substantial buildings, with partial collapse. Buildings shifted off foundations.
X.	Some well-built wooden structures destroyed; most masonry and frame structures destroyed with foundations. Rails bent.
XI.	Few, if any (masonry) structures remain standing. Bridges destroyed. Rails bent greatly.
XII.	Damage total. Lines of sight and level are distorted. Objects thrown into the air.

Source: United States Geological Survey.

Modified Mercalli Intensity Scale is more relevant for engineers and homeowners as they can relate to the impact of the earthquake.

Figure 4.18 shows the results of the Loma Prieta earthquake of October 17, 1989 (see Box 4.3). The simplified geologic map shows those areas that consist of soft mud, sand and gravel, and bedrock. Portions of the Nimitz Freeway (I-880) were built on mud deposits, very unstable material. The roadway collapsed in those areas because the ground was severely jolted by the seismic energy. This is evident in the seismograms for the mud, sand and gravel, and bedrock outcrops. Clearly, it was faulty engineering that produced this deadly outcome.

United States Geological Survey.

United States Geological Survey.

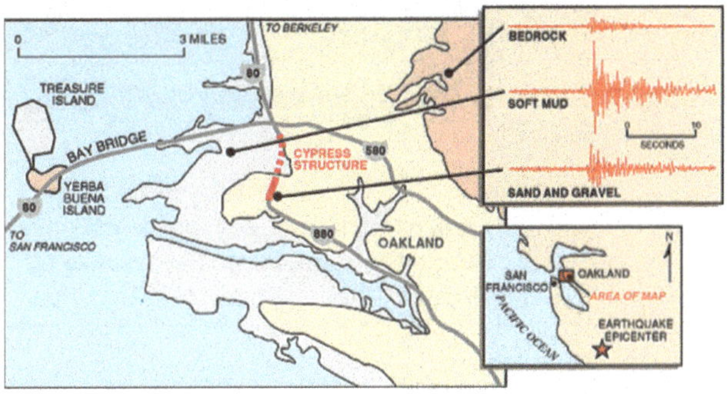

Figure 4.18 **The Loma Prieta earthquake had a profound effect on the Nimitz Freeway located on the east side of San Francisco Bay.**

BOX 4.3 The Loma Prieta Earthquake of October 1989

On October 17, 1989, at 5:04 p.m. (PDT), a magnitude 6.9 earthquake severely shook the San Francisco and Monterey Bay regions. The U.S. Geological Survey reported that the epicenter was located near Loma Prieta Peak in the Santa Cruz Mountains, approximately 14 km (9 mi) northeast of Santa Cruz and 96 km (60 mi) south-southeast of San Francisco.

As is true in many disasters, timing means a lot. The Great San Francisco Earthquake of 1906, the San Fernando quake of 1971, and the Northridge earthquake of January 1994, all struck in the early morning hours, sparing numerous people from imminent disasters. The Loma Prieta earthquake was no different in terms of timing. Baseball fans from the Bay Area were gathered in Candlestick Park in South San Francisco to watch Game 3 of the World Series between the San Francisco Giants and the Oakland Athletics.

More than 60,000 fans were in the relative safety of the stadium and off the highways. Many more were at home or in buildings watching the game so they were spared the disaster that struck the area. A portion of I-880, the Nimitz Freeway, collapsed, crushing cars and killing 47 people (this figure would have been much higher, given the rush hour timing of the earthquake, had not so many people been watching the World Series game). One section of the freeway was constructed on loose soil and the poor design of the vertical columns holding up the highest set of traffic lanes caused them to experience increased vertical resonance in the support columns. They all disintegrated under the downward force. Refer to Levy and Salvadori (1995) for details on the engineering issues.

The earthquake obviously caused the baseball game to be cancelled and, although the stadium was only slightly damaged by the seismic waves passing underneath it, the games were moved to Oakland after a 10-day postponement. The A's swept the Giants in four games.

United States Geological Survey. Photo by C. E. Meyer.

Figure 4.3.1 A section of the San Francisco-Oakland Bay Bridge collapsed during the Loma Prieta Earthquake of October 1989.

United States Geological Survey. Photo by H. G. Wilshire.

Figure 4.3.2 A section of I-880, the Cypress viaduct and Nimitz Freeway that disintegrated during the Loma Prieta earthquake. This road is usually heavily traveled at rush hour, and thousands were spared from death because of the scheduled World Series game that kept drivers off the freeways.

© Kendall Hunt Publishing Company.

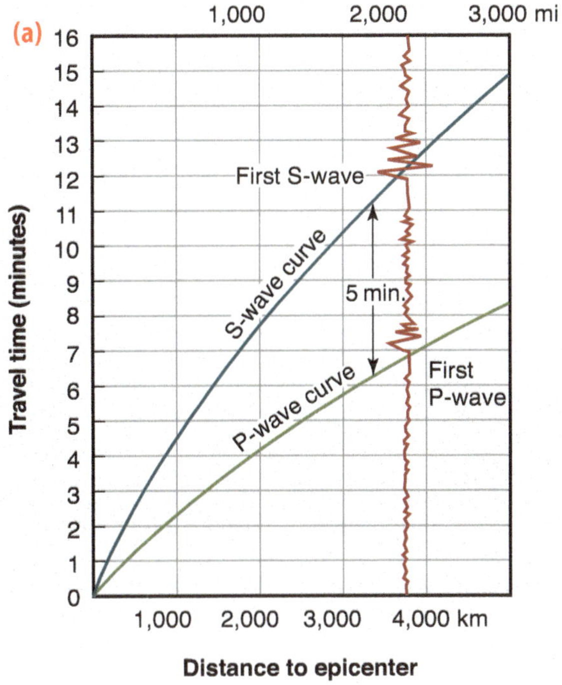

© Kendall Hunt Publishing Company.

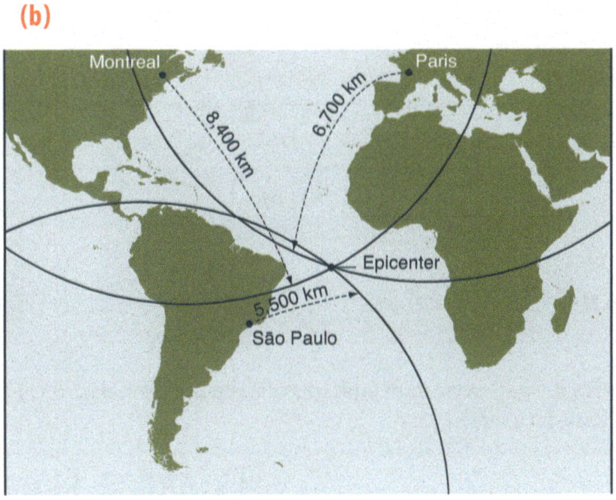

Figure 4.19a **Travel-time graph showing the P-wave and S-wave travel times as measured against the distance they have moved from the earthquake focus; the S-P time separation is 5 minutes.**

Figure 4.19b **Using arrival data from three different stations, it is possible to determine the location of the earthquake epicenter.**

Locating an Earthquake

Whenever an earthquake occurs, often several seismic stations record the event. In the case of a major earthquake, hundreds of stations produce seismic records. Today many stations are set up to digitally record earthquakes and then transmit the information to some central office where a determination is made regarding the location and magnitude of the earthquake. In the United States the primary reporting station is the National Earthquake Information Center in Golden, Colorado. If sufficient data are available at a single recording station, it is possible for scientists to make a quick determination of the distance and magnitude of a single earthquake from that station.

To get a quick fix on the location of an earthquake, arrival data are needed from a minimum of three recording stations. Each station will have a seismogram that shows the P- and S-wave arrivals of the event. The S-P time is noted and then converted to a distance using a travel-time curve (Figure 4.19a). When this distance is determined for a single station, the locus of all points located the same distance from the station is a circle. To be more precise in assigning a location for the epicenter, a similar procedure must be completed for two more stations. When this is done, the three circles should intersect at a single point, but because of minor differences in estimated rock velocities used to construct the travel-time graph, the lines usually establish a region of intersection where the earthquake occurred. As more data are analyzed from other stations, this region is eventually refined to more precisely define the epicenter of the earthquake (Figure 4.19b).

© Breck P. Kent/Shutterstock.com.

Figure 4.20a **A normal fault has offset a horizontal igneous intrusion.**

Courtesy of David M. Best.

Figure 4.20b **A reverse fault in volcanic deposits that are overlain by sedimentary material.**

Evidence From the Geologic Record

Earthquakes have occurred on Earth ever since crustal material was cold and brittle enough to break. Numerous examples exist of faults being preserved in rocks that are billions of years old. If a rock was faulted and later covered by sedimentary or volcanic material, the fault is preserved and we can reason that movement occurred prior to the later deposition of the overlying sedimentary material (Figures 4.20a–b). Undoubtedly, major earthquakes occurred throughout the world in the past, breaking up rock in many locations. This fracturing process allowed once solid material to become more easily eroded and removed, thereby erasing the record of earthquake activity. Whenever the broken material was rapidly covered by sediment and other debris, evidence of the fault was preserved.

Along the San Andreas Fault in California, geologists have dug a series of trenches across a section of the fault to detect areas where soil and sediment layers have been offset. By age-dating the layers using radiocarbon techniques, it is possible to determine when earthquakes occurred and thus generate a recurrence interval for a given area. With sufficient data they can extrapolate back in time to determine what happened and to make some educated predictions for future activity. However, because data are often not very continuous in terms of time, the recurrence intervals carry a significant amount of error.

Examples From the Historic and Recent Record and the Human Toll

Earthquakes have occurred on Earth for several billion years. Early reference to them was made by Chinese historians and by early Roman and Greek philosophers and naturalists. One of the best documented historic earthquakes was the Great Lisbon Earthquake of 1755 that devastated the capital city of Portugal. On November 1, All Saints' Day, when many people came to the city to attend church services, there was a great earthquake centered 325 km off the coast of Portugal. Estimated magnitude values fall in the range 8.3 < Mw < 8.5, making it one of the largest earthquakes ever to hit Europe. The damage suffered by the city was immense—more than 90,000 died from the earthquake, tsunami, and a major fire that raged throughout the city. Nearby coastal regions, including Lisbon and its harbor, experienced several tsunamis that ravaged seaside towns and killed thousands more. Cities in Morocco suffered also, where more than 10,000 people were killed.

Densely populated areas suffer large losses of life in natural disasters. Refer back to Table 4.1 which shows that some of the most deadly earthquakes have occurred in China, the world's most populous country. Although China is not directly affected by subduction processes related to plate tectonics, the country is laced with major fault systems that obviously experience a great deal of stress. The resulting earthquakes have killed large numbers of people over the years. However, until recently, with an opening up of news agencies to the world, we often did not know the amount of devastation and the number of lives lost in these events.

Earthquakes occur every day on Earth. They certainly disrupt the daily routine of those in the immediate area of the earthquake. Occasionally a local event can change the plans of an entire nation, such as what took place when the Loma Prieta earthquake of October 17, 1989, shook the Bay area around San Francisco, California (see Box 4.3). Millions of baseball fans were prepared to watch the third game of the World Series between the San Francisco Giants and the Oakland Athletics, only to have the game postponed and the outcome of the series delayed.

On May 12, 2008, a major earthquake of magnitude 7.9, having a shallow focal depth of 19 km (12 mi), occurred in the Sichuan province of southern China. Located about 1,600 km (1,000 mi) southwest of Beijing, the event occurred along a major fault system associated with movement of the Tibetan plateau against crustal material of southeastern China. Two weeks following the main event, an aftershock of magnitude 6 destroyed more than 70,000 buildings that had been weakened by the main event. Approximately 65,000 people were killed and an additional 23,000 people were reported missing. Major landslides dammed rivers, producing the potential for major flooding.

New Zealand is a country that is directly affected by earthquakes. From September 2010 to December 2011, several major events occurred. On September 4, 2010, a magnitude 7.0 rattled the country, followed in mid-February 2011 by a magnitude 6.3 event that was an aftershock of the earlier, larger earthquake. The February event was very destructive and killed more than 180 people (**Figure 4.21**). Ten years later another earthquake returned to the islands.

The Kermadec Trench, located in the southern Pacific Ocean, is formed by the subduction of the Pacific Plate moving under the Indo-Australian Plate. On March 4, 2021, a series of large earthquakes occurred, the first of which was a magnitude 7.3 event located 100 km (60 mi) off the east coast of the North Island of New Zealand.

© Adwo/Shutterstock.com.

Figure 4.21 Significant damage occurred in Christchurch, New Zealand, due to the February 2011 earthquake.

This event was unrelated to a series of earthquakes that struck the next day and were centered about 950 km (580 mi) to the north. The first earthquake had a magnitude of 7.4, which was considered a foreshock of the larger 8.1 magnitude event less than 2 hours later. The 8.1 earthquake was shallow, having a focus of 20 km (12 mi). Tsunami warnings were issued throughout the region, but the wave heights were quite small, only measuring 35 to 40 cm (14 to 16 in) along the northern shoreline of North Island.

Predicting and Forecasting Earthquakes

Regions that are likely to experience earthquakes on a regular basis are continually monitored for seismic activity. Scientists are aware of movement in the Earth and analyze the data for temporal and spatial patterns that might be useful in predicting future events. We do not currently have the ability to predict an earthquake with any degree of certainty. Far too many variables are involved to allow this to happen, so we view the occurrences of earthquakes as essentially random events in nature.

Short-Term Predictions

Several indicators can suggest that an earthquake could occur in a wide area. Among these are small movements in the ground or increases in recorded seismic activity in an area. Numerous sites are located along the San Andreas fault where lasers are aimed across the fault to reflectors on the opposite side. Any minute changes in movements of either block of the fault are detected and can warn of impending larger scale movements.

Monitoring seismic activity might show that a large number of small events in a short time period have been recorded, producing a swarm. Such an increase in activity could be the precursor to a larger event in the immediate future. Earthquake swarms commonly occur during phases of increased activity prior to the eruption of a volcano. They can also be present before large earthquakes, but such activity does not always precede these events. What often does follow a large earthquake is a period of smaller magnitude events. These are the result of the crust continuing to adjust to the major fracturing process.

Depending on the magnitude of the main event, the period of time following the major shock can be months. During this time there is a gradual decay of recorded activity which is typified by much smaller magnitude tremors.

A key problem with making definitive, short-term predictions is that it can send society into a panic if such an alarm were sounded. People would have to evacuate their communities and find new, temporary living conditions.

All businesses and schools would close, banks would not be available, abandoned areas would be subject to looters and criminals. Should the predicted event not happen, who would be held liable for all the problems that would follow?

There are documented cases of ill-founded predictions of major earthquakes hitting populated areas. In the mid-1970s there was an erroneous prediction of a great earthquake that was "supposed" to happen along the southeast coast of North Carolina. A similar erroneous, baseless prediction was made for the New Madrid seismic region of Missouri in late 1990. In each instance the general populace became unnecessarily alarmed about the potential for a major disaster. Needless to say the events never occurred but the experiences had been unnerving for people in those regions of the country.

Long-Term Predictions

Long-term predictions are much less precise but they do have some scientific merit. Information such as how often a certain magnitude event occurs or the location of past epicenters can be used to define regions likely to experience an increase in seismic activity. By keeping track of when large magnitude earthquakes occur, scientists can develop a time line that shows the recurrence interval of earthquakes. Collecting data over several decades allows researchers to produce a crude model to predict earthquakes. For example, if a magnitude 6 event occurred every 8 to 10 years in a region, the recurrence interval would be defined as about every 9 years. The probability is that every 9 or so years an earthquake of magnitude 6 or higher could occur. The likelihood that such an event would occur in a given year would be 1 year in 9 (1/9) or approximately 11%.

Analyses of earthquake epicenters show where events have occurred and where they have not. Plots that display gaps of epicenters indicate areas that could be likely to experience an event. A **seismic gap** is caused by a lack of crustal motion that is often explained by blocks of oceanic or continental crust being locked in place. Once the stresses are large enough to overcome the frictional forces holding the blocks or rock in place, an earthquake or series of earthquakes will most likely happen. If a long period has passed since any significant activity, a very large event could be waiting to happen because of the tremendous amount of strain accumulating in the rocks.

seismic gap
An area in a faulted region where no seismic activity has occurred in recent time. These areas are undergoing stress buildup and could be the site of future earthquakes.

High-Risk Areas in and Near the United States

In the United States and adjoining regions, several areas are considered hazardous in terms of earthquake activity. In 2006 the USGS reported that more than 75 million Americans in 39 states (25% of the country's population) are at risk from earthquakes.

The plate tectonic settings for Alaska, Washington, Oregon, and California cause those states to have a high risk factor. Over the past 300 years great earthquakes (magnitude 8 or higher) have been centered in each of these areas. In several instances tsunamis have inundated the coastlines of the states, causing significant damage. Although Hawaii rests in the middle of the Pacific plate, it has experienced great earthquakes and numerous tsunamis in the past several centuries.

In the area adjoining the Caribbean Sea, including Puerto Rico and the U.S. Virgin Islands, great earthquakes and tsunamis have killed thousands of people over the past several centuries. In addition, the entire East Coast of the United States is at risk. In 1886 a magnitude 7.3 earthquake occurred near Charleston, South Carolina, causing significant damage to the nearby communities. This event ranks among the most widely felt earthquakes in the history of the United States. The earthquake was felt in at least 23 states, ranging as far west as Arkansas and as far north as Wisconsin and New York State. Rocks in the Eastern and middle portions of the country are very old and not very highly fractured, so they transmit seismic energy more efficiently than the fractured rocks of the Western United States.

New England and eastern Canada have been subjected to significant earthquake activity over the past 3 centuries with more than 10 earthquakes exceeding magnitude 5 (**Figure 4.22**). These earthquakes are related to the St. Lawrence River and an ancient rift valley that underlies the river. A second trend is evident that connects epicenters of events through New Hampshire northwestward into Quebec. These earthquakes are thought to occur along a landward extension of the Kelvin Fracture Zone in the North Atlantic Ocean.

As we examine a map of potential earthquake activity in the United States (**Figure 4.23**), three areas stand out—Alaska, California, and an area along the east central part of the continental United States. This region, which is basically centered on the small town of New Madrid, Missouri, overlies a

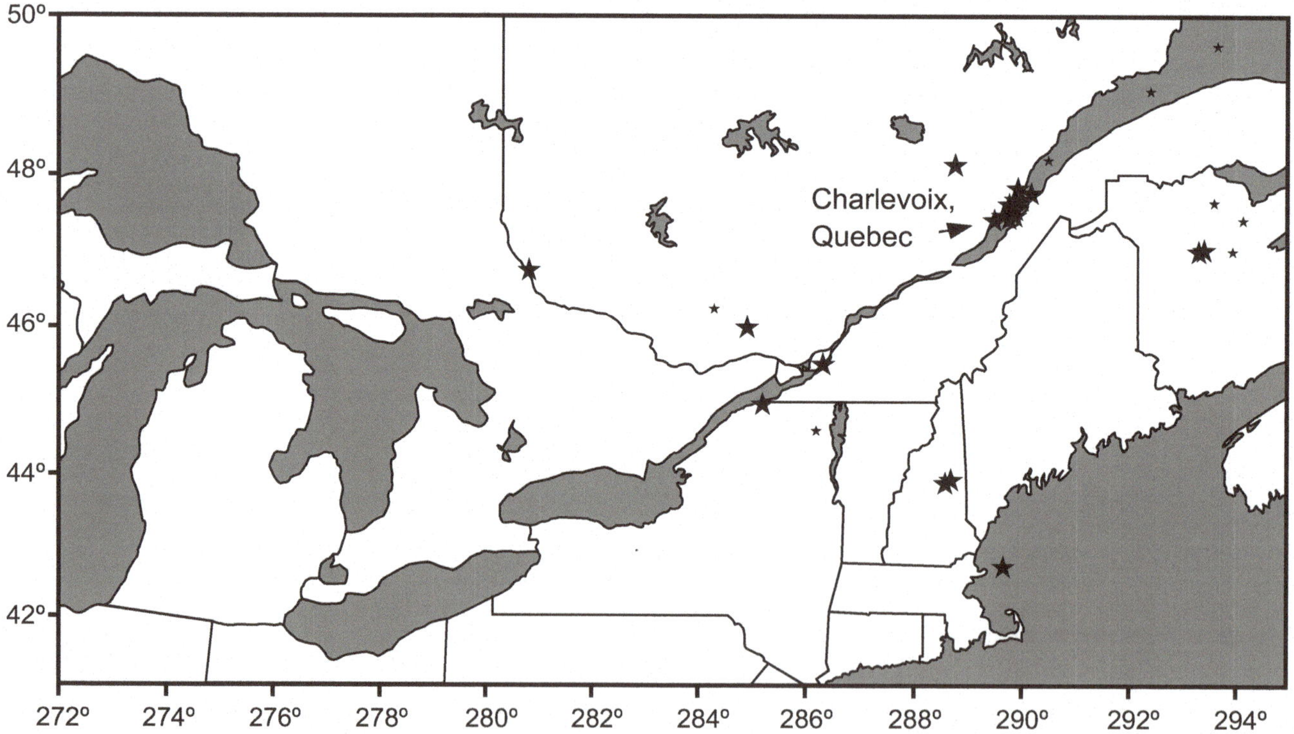

"Where earthquakes are felt," from Hough, Susan E.; Earthshaking Science, © 2002 by Princeton University Press. Reprinted by permission of Princeton University Press.

Figure 4.22 Selected earthquake activity in New England and eastern Canada, 1627–1998. Larger stars are M > 5.5 events; smaller stars are M 2 and M 3 events.

very old set of major fractures in the continent. These formed over a billion years ago and have been covered by younger material.

Over a 3 month period (December 1811 to February 1812), three major earthquakes struck the region. Estimated magnitudes were between 7.3 and 7.5, which is very unusual for mid-continent regions, however, it is obvious that the pre-existing fractures were being definitely being reactivated. More than 200 significant aftershocks were felt with 10 of them of magnitude 6 or greater.

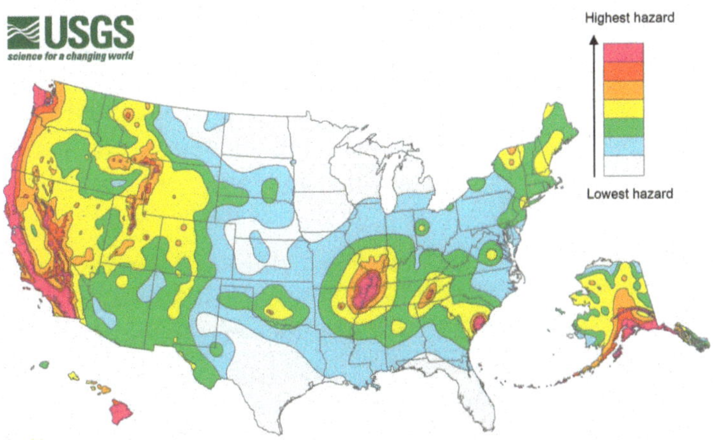

Figure 4.23 **Relative shaking hazards in the continental United States. The highest amounts of shaking occur in the Aleutian Islands of Alaska and in California.**

United States Geological Survey.

The geology of the Midwest and the East Coast consists to a large degree of sedimentary rocks, which readily transmit seismic energy. As a result, more than one third of the continent was affected by ground motion from the three earthquakes. Historic records from the East Coast and New England provided researchers with information that described the damage in those areas. Because the area near New Madrid was sparsely populated, less damage was noted, although buildings in St. Louis, Missouri, did suffer damage. One effect these events did have was that the course of the Mississippi River was permanently altered and flooding destroyed thousands of acres of farmland.

Unexpected Events in the United States

In early 2008 several earthquakes occurred in Nevada, which is the third most active state in the United States, ranking behind California and Alaska. On February 21, 2008, a magnitude 6 event shook northeastern Nevada near the town of Wells, just south of the Utah border. Within 2 months the city of Reno experienced a magnitude 5.2 earthquake, which was followed by more than 200 aftershocks, including one of magnitude 4.7 soon after the main event.

Residents of southeastern Illinois and the surrounding region were awakened on April 18, 2008, by a magnitude 5.2 earthquake. This activity is associated with the Wabash Valley Seismic Zone, which is related to the New Madrid Seismic Zone to the southwest. Evidence exists that earthquakes having estimated magnitudes between 6.5 and 7.5 have occurred in this area in the past 20,000 years.

The New England portion of the United States and eastern Canada have experienced earthquakes in the past (Figure 4.22). As reported by the USGS in their analysis of the August 23, 2011, an earthquake of magnitude 5.8 was felt by tens of millions of people. The epicenter was near the town of Mineral, Virginia, which is located at 65 km (39 mi) northwest of the capital of Richmond. This was a very shallow earthquake, having a focal depth between 6 and 8 km. Further study showed that it occurred on a newly discovered fault.

Due to the dense population along the East Coast, almost one third of the population of the United States felt the tremors. More than 148,000 people reported their experience on the USGS *Did You Feel It*? website link. Economic losses throughout the region were between $200 and $300 million. Among the many buildings affected by the ground motion was the Washington Monument, which was closed until May 2014 to undergo examination and repairs.

These three episodes demonstrate that seismic activity can recur in regions that are usually thought of as having little or no movement in the crust, although examination of a longer time frame shows that such events have indeed happened there.

Earthquake Hazard Mitigation

Many countries lie in regions that are affected by earthquakes, especially those on the edges of the circum-Pacific belt (**Figure 4.1**). Numerous densely populated communities are in danger of being affected by seismic activity that has historically devastated these same cities and towns, resulting in massive loss of property and lives. Communities tend to rebuild on the established sites. This is true following all types of natural disasters. Any attempt to lessen or mitigate future damage would be welcomed.

Of the world's 12 most populous countries in 2020 (Table 4.3), only Brazil, Bangladesh, and Nigeria are exempt from suffering significant seismic activity. The countries with the highest likelihood of major earthquakes have 4.05 billion people or roughly 55% of the world's population. The majority of these countries are developing regions that do not have rigid building standards in place to lessen the effects of earthquake damage. Some countries are almost totally within seismic zones, such as the Philippine Islands, Indonesia, and Japan, while others would be affected in only relatively isolated regions, such as the West Coast of the United States or the eastern islands of Russia.

Table 4.3 **Twelve Most Populous Countries as of 2020. Only Brazil, Nigeria, and Bangladesh Are Spared Major Earthquake Activity**

China	1.398 billion
India	1.366 billion
United States	328 million
Indonesia	270 million
Pakistan	217 million
Brazil	211 million
Nigeria	201 million
Bangladesh	163 million
Russia	144 million
Mexico	128 million
Japan	126 million
Ethiopia	112 million

Source: U.S. Census Bureau; World Atlas.

The loss of life and amount of damage inflicted on buildings are a function of several variables. The major factor controlling damage is the distance a populated area is from the epicenter of a quake. The focal depth of the source is important in determining how much motion is felt on the surface. Building construction style and engineering integrity respond directly to the energy that travels through an area. Buildings constructed on bedrock are more stable than those situated on loose or unconsolidated material.

However, problems can be reduced if buildings are designed to dampen seismic energy, much like shock absorbers are designed to lessen the bumps an automobile encounters on the road. This concept utilizes Newton's law of inertia, which states that a body (for example, a building) remains at rest unless a force (for example, seismic energy) is applied to it. To prevent a building from being destroyed, it should be "disconnected" or decoupled from the ground. Levy and Salvadori (1995) outline several good examples of ways to do this in earthquake-prone areas.

Mexico City experienced a major earthquake in 1985 that literally flattened scores of buildings. The city is constructed on ancient lake beds that provide a very weak geologic foundation. When the seismic energy passed through the area, it shook the ground as if the city rested on gelatin. Numerous buildings collapsed in a pancake fashion and many others crumbled (Figure 4.24). More than 9,500 people were killed and an estimated 30,000 were injured.

The central region of Mexico experienced a magnitude Mw 7.1 earthquake at 1:14 p.m. on September 19, 2017. Occurring at a depth of 51 km (32 mi), the epicenter was located approximately 120 km (75 mi) southeast of Mexico City. Severe ground shaking lasted for more than 20 seconds and produced

United States Geological Survey.

Figure 4.24 Damage was extensive as the General Hospital in Mexico City was destroyed following the September 19, 1985, earthquake that was centered off the west coast near the subduction zone, where the Cocos plate dives under the North American plate.

major damage in the epicentral area. The surrounding area, which included Mexico City, experienced significant damage, as many structurally weak buildings collapsed. More than 360 people died throughout the region, with more than 240 deaths occurring in Mexico City. The devastation and loss of life in Mexico City is attributable to the city having been built on an ancient lake bed. When seismic energy moves through the underlying sand and clay, ground motion is greatly amplified. The USGS reported that there have been 19 earthquakes of magnitude 6.5 or greater within 250 km (150 mi) of the September 19, 2017, event.

Two interesting facts are related to this event. The date, September 19, is the anniversary of the 1985 Mexico City earthquake which killed approximately 10,000 people. As a reminder of the event, annual earthquake drills are held throughout the country on that day. September 19, 2017, was no exception as the drill began at 11 a.m., a little more than 2 hours before the earthquake struck. The city has a warning system in place but many people failed to heed the true warning just prior to the onset of the earthquake because they thought it was a continuation of the scheduled drill.

Regions that experience earthquakes must strengthen their building codes and attempt to retrofit existing structures that tend not to withstand seismic events well. Following the 1994 Northridge, California, earthquake, the state has spent billions of dollars to reinforce bridges and other structures to try to preserve the critical infrastructures that are often devastated by major earthquakes.

Earthquakes of equal or very similar magnitudes can produce extremely different effects, depending on where they occur. In December 2003 an earthquake of M 5 6.6 killed 26,200 people in southeastern Iran, while a magnitude 6.7 event in Hawaii in October 2006 did not kill anyone. The primary cause of the large death toll in Iran was the poor construction standards used in many of the older buildings. The lack of reinforced concrete or other building material allows buildings to collapse very easily. Careful planning and the enforcement of strict building codes has reduced the effects of earthquakes in many developed countries.

BOX 4.4

Haiti Earthquake of January 12, 2010

A major earthquake (M 7.0) struck the country of Haiti on January 12, 2010, at 4:53 p.m. local time. The region lies south of the boundary of the westward moving North American plate and the eastward moving Caribbean plate. This left-lateral strike slip motion has produced numerous earthquakes to the east in the Dominican Republic, the country that lies on the eastern side of the island of Hispaniola.

Two major east-west trending, strike-slip fault systems occur in the area. The Septentrional fault system is found in northern Haiti and the Enriquillo-Plantain Garden fault system traverses the southern part of the country in a similar manner. It was the latter fault system that ruptured, producing the earthquake. The last major earthquake in the vicinity of Port-au-Prince occurred in 1770. Thus stresses in the rock had been accumulating for 240 years, even though the average movement along these faults is about 21 mm per year.

More than four dozen significant aftershocks rocked the area, creating very unsafe conditions for rescue workers struggling to free trapped people from the rubble. Eight days after the main shock an aftershock of M 5.9 occurred within 10 km of the main shock.

The capital city of Port-au-Prince, with a population of more than two million people, suffered severe damage. Thousands of buildings were destroyed and an estimated 250,000 to 300,000 people died with projections reaching as high as 230,000 dead. More than 300,000 people were injured and more than 1.5 million people are homeless. The shallow depth of the event, about 13 km (8 mi) below the surface, added to the large amount of destruction.

Haiti, which is among the poorest on Earth, has experienced other major disasters in recent years including four hurricanes in 2008. The country lacks the resources and infrastructure to cope with such events and has been the recipient of worldwide aid to begin rebuilding.

The geological setting for this area in Haiti is very similar to that in southern California, where the right-lateral strike slip motion of the San Andreas fault system passes under heavily populated areas. In the next few decades possible movement along the southern San Andreas fault could possibly generate a similar catastrophe.

Source: Department of Defense.

Figure 4.4.1 **Thousands of buildings were destroyed or severely damaged, including many poorly constructed homes.**

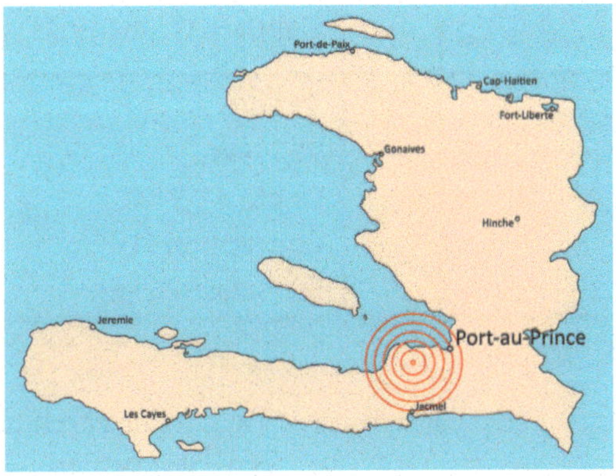

© 21th Design/Shutterstock.com.

Figure 4.4.2 **Epicenter of the Haiti earthquake of January 12, 2010, was located at a distance about 25 km (15 mi) west-southwest of Port-au-Prince.**

BOX 4.5 The Great Chilean Earthquake of February 2010

On February 27, 2010, at 3:34 a.m. local time, the earth shook in the vicinity of Maule, Chile, located about 115 km (70 mi) north-northeast of the city of Conception. This event measured M 8.8, making it the fifth largest earthquake in recorded history. The region lies at the boundary between the eastward-moving Nazca plate and the westward-moving South American plate. Over the past 40 years, the region has experienced 30 earthquakes of magnitude 7 or greater. The epicenter for this event is located about 275 km (165 mi) north of the location of the May 1960 magnitude 9.5 earthquake, the largest earthquake in recorded history.

© Yai/Shutterstock.com.

Figure 4.5.1 **Buildings and homes were severely damaged or destroyed by the February 2010 earthquake.**

Given the proximity of the earthquake to the Pacific Ocean, the Pacific Tsunami Warning Center in Hawaii issued several tsunami warnings to alert more than 50 nations and territories bordering the Pacific Ocean. However, the warnings were canceled after it was determined that these areas would be spared from the destructive waves. Countries as far away as Japan and Russia were expected to receive tsunamis about 21 hours following the main shock, but only a slight wave arrived in those areas. One contributing factor to the warning was the relatively shallow depth of the focus—only 35 km below the surface. More than 150 significant aftershocks were generated in the first 2 days following the main shock. These, of course, can be destructive from the standpoint of destroyed buildings that were weakened by the main shock. The aftershock region extended for more than 700 km (420 mi) along the Chilean coastline. By the end of the third day the number of fatalities reached more than 700 people.

Two Largest Earthquakes in the History of Chile

Date	Mw	Location and Depth	Death toll	Notes
May 22, 1960	9.5	38.29S 73.05W Depth = 33 km (est.)	1,655	Largest earthquake in recorded history; epicenter located in the southern portion of the Nazca-South American plate boundary
February 27, 2010	8.8	35.846S 72.719W Depth = 35 km Conception 115 km Santiago 325 km	525	Fifth largest earthquake in recorded history; epicenter located in the south-central portion of the collision of the Nazca and South American plates

Two major earthquakes in early 2010 produced major damage in the Western Hemisphere. The earthquakes in Haiti (Box 4.4) and Chile (Box 4.5) are among the most deadly and strongest in decades. Notice that these two earthquakes were unrelated as they occurred on different tectonic plates. In 2011 Japan was hit with a great earthquake that produced an extremely deadly tsunami (refer to the effects of this event in Chapter 5). The tsunami killed far more people than did the actual earthquake.

On August 14, 2021, Haiti was struck by a magnitude 7.2 earthquake in the early morning. The epicenter was located 100 km west of the capital of Port-au-Prince, which had experienced a magnitude 7 event in January 2010. Initial reports were that more than 2,000 were killed and 12,000 injured. Destruction in the region included more than 130,000 homes. Two days later, Tropical Storm Grace dumped more than 25 cm on rain of the region, generating more than 150 landslides.

Mitigation involves reducing the risk that earthquakes create. Revising building codes to strengthen existing buildings, freeway bridges, and other structures lessens the damage they incur when they undergo shaking by seismic energy. Buildings experience damage because of the direct waves that pass through an area; if these do not destroy buildings, they can weaken them to the point that energy from aftershocks can finish the job of leveling the edifice.

Earthquakes will always occur because of the dynamics associated with plate tectonics, whether the events are the result of direct collision of plates or the result of volcanic activity that is also attributable to plate motion. Being aware of regions on Earth that are most likely to generate earthquakes will make people understand why they occur, although the actual timing of the events is not predictable.

Key Terms

body wave *(page 114)*
convergent boundary *(page 110)*
deep-focus earthquake (page *117)*
divergent boundary *(page 109)*
epicenter *(page 114)*
focus *(page 114)*
intermediate-focus earthquake *(page 114)*
local magnitude *(page 117)*
Love waves *(page 107)*
moment magnitude *(page 121)*
normal fault *(page 113)*
P-wave *(page 115)*
Rayleigh waves *(page 117)*
recurrence interval *(page 109)*
reverse fault *(page 113)*
Richter scale *(page 120)*
S-wave *(page 116)*
seismic gap *(page 129)*
seismograph *(page 117)*
seismogram *(page 117)*
shallow-focus earthquake *(page 114)*
strain *(page 113)*
stress *(page 113)*
strike-slip fault *(page 114)*
surface wave *(page 115)*
transform boundary *(page 112)*

Summary

Earthquakes are a daily occurrence, happening somewhere almost at any given moment. Fortunately, most of them are rather small in magnitude and do not affect people's lives. Seismically active regions are spread around the world and are most often associated with plate boundaries. Seismic activity is associated with the three types of plate boundaries: convergent, divergent, and transform. The most active boundary is the convergent type, which produces the majority of large magnitude earthquakes. These earthquakes are very common in the circum-Pacific belt as well as the Mediterranean-Trans-Asiatic belt. The Circum-Pacific regions are typified by subduction zone tectonics, whereas the Trans-Asiatic and San Andreas fault zones experience transform, or strike-slip, motion. Divergent plate boundaries, such as those along the midoceanic ridges, usually generate fewer and smaller events; these are the result of magma rising under the oceanic crust.

When stresses build up in rocks, they fracture along fault planes. Normal faults are due to extensional forces, while reverse faults are created by compressional forces. Strike-slip and transform faulting is caused by a pair of forces moving side-by-side in opposite directions. These forces can produce seismic energy that radiates out from a central point in the subsurface that is termed the focus. Along subduction zones, focal depths can reach as deep as 700 kilometers. The point on the surface directly above the focus is the epicenter. Seismic waves move through the Earth in several forms. P-waves are compressional and travel the fastest, while S-waves (shear) move more slowly. S-waves and surface waves are the most destructive.

Earthquakes are described in terms of their magnitude and intensity. Magnitudes are determined by examining the seismic records of numerous recording stations and have a numerical value. The Richter scale measures magnitude for local events and the moment magnitude measures large earthquakes. Intensity is a subjective value that is a function of the amount of damage in a given area. The Mercalli Scale uses Roman numerals to assign the degree of destruction.

Historically millions of people have been killed by earthquakes, but in the United States, the numbers are quite small due to better building codes and a lower population density. Within the past 40 years, California and Alaska have been hit with several large earthquakes, but the loss of life has been relatively low. During that same time span, several hundred thousand people have died in other parts of the world; these areas have a higher population density and do not have the engineering standards and the response resources used in the United States. In addition to California and Alaska, there are highly populated areas of the United States that have a history of seismic activity that can be reactivated without warning, thus putting tens of millions of people in harm's way.

The ability to predict earthquakes does not exist. Earthquakes are random events and have numerous variables at work that cause them to occur where they do, when they do. Scientists can provide some idea as to where future events will occur, but they cannot forecast when they will strike. The best way to lessen the impact and damage in any area is to plan for what in many areas is the inevitable and employ good construction practices, especially in areas where the population density is high.

References and Suggested Readings

Bolt, B. A. (2006). *Earthquakes* (5th ed.). W. H. Freeman.

Brumbaugh, D. S. (2010). *Earthquakes—Science and society* (2nd ed.). Prentice-Hall.

Doughton, S. (2013). *Full-Rip 9.0 the next big earthquake in the Pacific Northwest.* Sasquatch Books.

Fountain, H. (2017). *The great quake.* Crown Publishing Company.

Gates, A. E., & Ritchie, D. (2007). *Encyclopedia of earthquakes and volcanoes* (3rd ed.). Facts on File.

Hough, S. E. (2010). *Predicting the unpredictable: The tumultuous science of earthquake prediction.* Princeton University Press.

Levy, M., & Salvadori, M. (1995). *Why the earth quakes.* W. W. Norton.

Musson, R. (2012). *The million death quake—The science of predicting Earth's deadliest natural disaster.* Palgrave MacMillian.

Petrone, C. M., Scandone, R., & Whittaker, A. (2019). *Volcanoes and earthquakes—A guide to our unquiet earth.* Smithsonian Books.

Prager, E. J. (2000). *Furious earth—The science and nature of earthquakes, volcanoes, and tsunamis.* McGraw-Hill.

Prager, E. J. (2020). *Dangerous earth—What we wish we knew about volcanoes, hurricanes, climate change, earthquakes and more.* The University of Chicago Press.

Renton, J. J. (2011). *Physical geology across the American landscape* (3rd ed.). Kendall Hunt Publishing Company.

Ulin, D. L. (2004). *The myth of solid ground—Earthquakes, prediction, and the fault line between reason and faith.* Penguin Group.

Yeats, R. S. (2012). *Active faults of the world.* Cambridge University Press.

Yeats, R. S. (2015). *Earthquake time bombs.* Cambridge University Press.

Yeats, R. S., Sieh, K., & Allen, C. R. (1997). *The geology of earthquakes.* Oxford University Press.

Websites for Further Reference

https://earthquaketrack.com/recent
https://earthquake.usgs.gov
https://scedc.caltech.edu
https://www.epa.gov/natural-disasters
https://www.iris.edu/hq
https://www.ngdc.noaa.gov/hazards.shtml
https://www.scec.org
https://www.volcanodiscovery.com
https://www2.usgs.gov/natural_hazards

Questions for Thought

1. What are the three types of plate boundaries? Give one example of each type.
2. Which of the plate boundaries has the largest number of earthquakes and why?
3. Explain why the San Andreas fault system in California has earthquakes but not volcanic activity.
4. What are the differences between P-waves and S-waves?
5. How does a seismograph detect seismic activity?
6. Compare the Richter scale and the Mercalli scale. What does each measure?
7. How are earthquake epicenters located?
8. What reasons can you give for larger losses of life from earthquakes in China than in the United States?
9. How successful are scientists in predicting earthquakes?
10. How likely is an earthquake in your hometown?
11. Why are the locations of many volcanoes and earthquake epicenters often found in the same locations?
12. Compare the potential damage caused by earthquakes of the same magnitude if one occurred in California and one occurred in Oklahoma.
13. Explain why Brazil and Nigeria do not lie in areas likely to have earthquakes.
14. Explain why there are very few earthquakes occurring in the middle of the United States.

5 Tsunami: The Utmost in Destruction From the Sea

© Frans Delian/Superstock.com.

The December 26, 2004, tsunami was extremely destructive in South Asia and adjoining regions, as more than 230,000 people were killed and several million injured, as the wave trains traveled across the Indian Ocean and nearby seas.

So, you think we are safe?

Lurking off the coast of the Pacific Northwest, the Juan de Fuca Plate is inching its way to the east. In doing so it is colliding with a portion of the North American Plate which is moving westward. As we described in the previous chapter, there is a real potential for a mega-earthquake occurring in the region, triggering a catastrophic tsunami.

This event will cause utter destruction and loss of life, as did the March 2011 earthquake and tsunami that occurred in Japan. In that event 98% of the more than one million destroyed or damaged homes were the result of the tsunami, not ground shaking caused by the earthquake. Ninety-two percent of the more than 20,000 fatalities were caused by the tsunami, not the earthquake itself. The low-lying coastal regions of northern California, Oregon, and Washington are heavily developed and densely populated. Given the close proximity to the potential epicenter, there will be very little time to alert communities that are in the path of the tsunami. In addition, the highway systems will be virtually impassable due to debris from the event and countless vehicles on the roadways attempting to flee the area.

Tsunami have existed on Earth for millions of years, but it took the events of December 26, 2004, to remind the world of their existence and destructive power. The earthquake and resulting tsunami stunned the world when the destructive power of a mega-tsunami struck the coastal regions of Southeast Asia. Off the west coast of the island of Sumatra in Indonesia, two oceanic plates converged. The resulting earthquake was a M 9.1 event, the fourth largest in recorded history. The vertical movement of the ocean floor was more than 16 meters (50 ft) along a length of 1,200 kilometers (720 mi), the distance between Chicago and New York. This displaced thousands of cubic kilometers of water, generating a tsunami that raced across the adjoining oceans, killing an estimated 230,000 people. Those people who lived near the source had very little time to react to the event, as there is no tsunami warning system in place in the Indian Ocean, unlike the Pacific Ocean. Once officials learned what had happened, countries farther away were notified and thus were able to warn residents of the impending disaster. Coastal regions in East Africa did not experience the large loss of life that occurred in Sumatra, Sri Lanka, India, and Thailand because of the warnings. Obviously if a warning network had been in place throughout the region around the Indian Ocean, more inhabitants would have been informed in time to save many lives.

tsunami

A series of giant, long wavelength waves produced by the displacement of large amounts of ocean water by an earthquake, volcanic eruption, landslide, or meteorite impact.

seismic sea wave

A large ocean wave that is produced by a major disturbance in the ocean, such as an earthquake, volcanic eruption, or a landslide.

What Is a Tsunami?

The word **tsunami** (soo-NAH-mee) is derived from two Japanese characters: *tsu* meaning harbor, and *nami* which means wave (the word is both singular and plural in Japanese, as is the word *deer* in English). A tsunami (also termed **seismic sea wave**) is generated by a large displacement of seawater. A tsunami can be generated by an earthquake, a volcanic eruption, a landslide, or a meteorite impact striking the ocean. The energy moves out from the source, producing a series of waves that have long wavelengths. The time between the wave pulses can be as much as an hour or more.

Throughout the long documented history of Japan, these waves have occurred rather frequently. Almost 200 tsunami have been recorded hitting the islands. The location of Japan on the northwest edge of the Pacific Ocean is a focal point for these destructive waves as they travel across the Pacific. The shoreline of the Japanese islands consists of many harbors and inlets. When a seismic sea wave hits these indentations, the energy is rapidly focused into a small area. Depending on slope and elevation of the ground along the coast, water can move inland up to several kilometers. Other areas bordering the Pacific Ocean have also been struck by these waves, including Alaska, Hawaii, Guam, and California.

© Michael Ledray/Shutterstock.com.

Figure 5.1 Evacuation routes are set up in areas that can experience tsunami. These routes direct people to higher ground and a safer environment.

Tsunami commonly occur several times a year throughout the world as active tectonic plates produce large magnitude earthquakes that displace massive amounts of seawater. Although the majority of tsunami result from the vertical displacement of the sea floor or adjoining areas, strike-slip motion is thought to produce about 15% of all tsunami. Energy from in collisional subduction zones strike-slip mechanisms tends to remain close to the source, thus not producing large-scale disasters. The lateral slip motion produces little, if any, vertical offset that could displace the seawater. As populations have increased in the areas surrounding the Pacific, more people have settled in low-lying, coastal areas that are in harm's way when these waves strike the shoreline. Coastal inhabitants are aware of the potential disasters and have developed escape routes to allow them to evacuate low areas (Figure 5.1).

Characteristics of Tsunami

Tsunami are most often generated in deep water when vertical movement occurs along a fault or water that is displaced by an earthquake, a volcanic eruption, or a landslide. Plate boundaries are areas where movement is often very slow. The interface along the boundary edges can be thought of as being stuck in place. However, the buildup of force can cause the rocks to suddenly break. When this occurs, there is often some degree of vertical displacement. The initial vertical motion projects water upward, which then collapses, creating a splash that generates waves in all directions. This upward motion is not very large, perhaps on the order of several meters. However, a massive volume of water is moved. Successive waves move out from this point, separated by a distance measured in tens or hundreds of kilometers. A similar condition is produced when you drop a stone into a pond. The waves radiate from the point where the water was displaced. If you were on a ship in the open ocean when a tsunami passed underneath, it would be difficult to detect it. In addition to the long **wavelengths** (the distances between the crests of adjacent waves), the amplitude of the wave is very small, on the order of a meter or two. However, ocean buoys can be set in deep water to detect these waves and transmit the information to warning centers.

wavelength
The horizontal distance between any two similar points of the wave, such as two crests or two troughs.

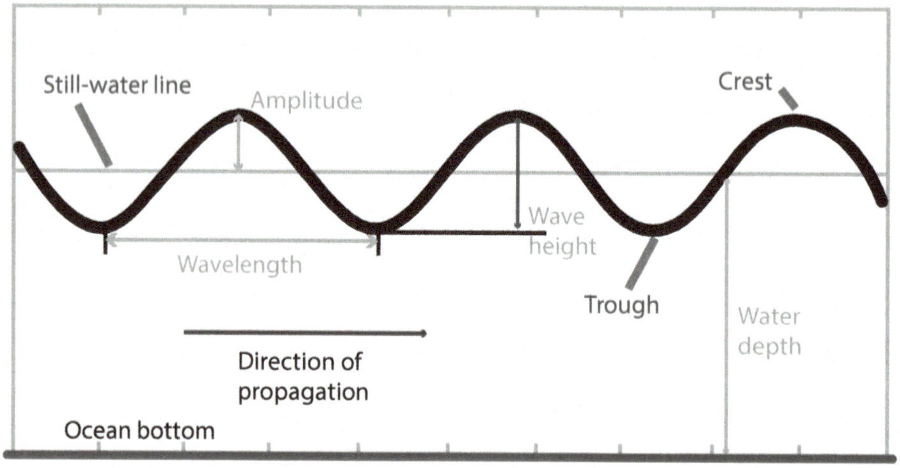

© Kendall Hunt Publishing Company.

Figure 5.2 Example of a general wave form. The period is the time it takes for two adjacent crests to pass a fixed point.

Energy in the oceans is transmitted by waves that have the characteristics shown in Figure 5.2. Tsunami differ from wind-generated waves in that tsunami have a much longer period, the time separating adjacent crests of the waves, and they also have much larger wavelengths. The periods of tsunami range from about 5 minutes to as many as 60 or more minutes. Normal, wind-generated waves typically have periods of between 5 and 20 seconds. Wavelengths for tsunami range between tens and hundreds of kilometers, while wind-generated waves average around 100 to 200 meters (330 to 660 ft).

Tsunami have been referred to in the past as tidal waves, although their genesis is not related to tides at all. Perhaps the misnomer came about because, as tsunami approach the shore, they appear similar to breaking waves. Tsunami come ashore and continue moving inland; they do not recede in an ebb-and-flow sense that is typical of normal tides.

Life Cycle of a Tsunami

Tsunami are relatively short-lived events. The United States Geological Survey (USGS) recognizes four stages of a tsunami: initiation or creation, splitting, amplification, and runup.

The first stage, initiation, is when some energy source displaces water. Vertical movement on the sea floor caused by an earthquake pushes water upward, and an accompanying drop can pull water down. A volcanic explosion could displace water both vertically and horizontally, depending on how the volcano erupted. Vertical displacement in deep water generates on a relatively small uplift of water. This can occur over several tens or hundreds of kilometers of distance, displacing enormous volumes of water.

The uplift and subsequent collapse of water at the surface causes it to move radially outward (Figure 5.3). If the point of origin for the source of the tsunami is near shore, two distinct tsunami are created. The one that travels out into deep water is termed the distant tsunami, while the one closer to the coast is the local tsunami. The local tsunami reaches land faster because of the shorter travel distance and does the initial damage. The velocity with which the tsunami travels is related to the square root of the water depth

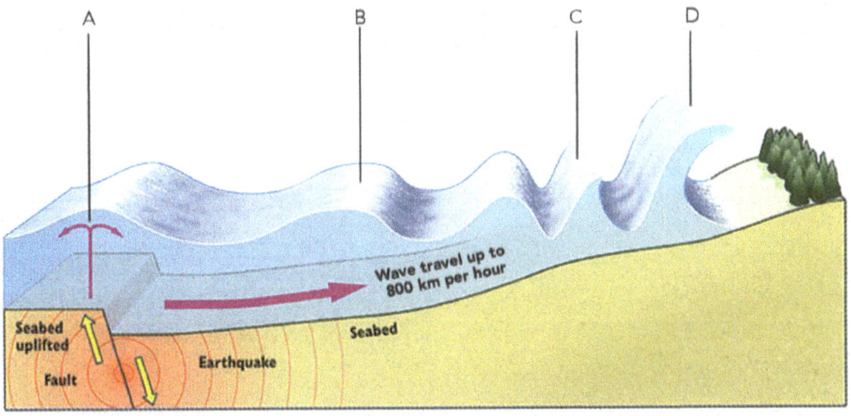

Source: Ingrid Ukstins.

Figure 5.3 (A) Energy is transfered to the water, generating a tsunami. (B) Waves in open ocean are usually 1 meter or less in height in deeper water. (C) Waves slow down in shallow water and stack up. (D) At the seashore wave heights can reach 30 meters or more.

(refer to the equation below). For this reason, the distant tsunami travel much faster than the local tsunami.

As the tsunami approaches more shallow water near the continent, the energy is now focused in a much smaller volume of water. The amplitude of the waves increases dramatically as the velocity and wavelength decrease. As the wave approaches the shore, if the trough portion arrives first, this gives an appearance of the tide and water going out to sea. This is termed **drawdown**. Shortly after the water recedes, the next crest in the wave train arrives, producing **runup**.

drawdown

A lowering of water level at the beach when the trough of a tsunami comes ashore.

runup

The height to which water comes up onto the shoreline, either from natural wave action or the occurrence of a tsunami.

Water arrives as a surge as the sea level constantly rises (Figure 5.4). Runup consists of seawater that rushes ashore as the tsunami hits the beach. If the wave moves into a closed inlet or the mouth of a river, it is sometimes classified as a tidal bore, although in this instance it is the tsunami and not the tides creating the feature. The amount of runup can be variable, because it is controlled by offshore seafloor topography and the outline of the coast. Areas with a shallow offshore topography are overrun by water, while those beaches with a steep offshore seafloor are not inundated as severely. The steeper sea floor deflects the energy and thus lessens the incoming force of the water. Strong currents develop as massive volumes of water move in and out of confined areas. Tsunami generally have several major runup events separated by periods of drawdown. In the December 2004 tsunami, beaches that are close to each other experienced extremely different effects due to difference in their sea floor geometries.

© Kirsz Marcin/Shutterstock.com.

Figure 5.4 Runup has an appearance similar to this wave, but it continues inland for a kilometer or more, covering the flat coastal areas.

Propagation of Tsunami Through the Oceans

The origin of most tsunami is in deep water, although some can occur along the continental shelf. Because the wavelength of tsunami is much greater than the depth of the water through which they travel, the velocity of the wave is only dependent on the water depth (D). So the velocity V is determined by

$$V = \sqrt{gD}$$

where g is the gravitational constant (9.8 meters per second) and D is the water depth, measured in meters.

The average depth of the world's oceans is 3,700 meters. Therefore the velocity of a tsunami traveling in water that deep is

$V = \sqrt{9.8 \times 3700} = 190$ meters per second or 684 kilometers per hour (410 miles per hour).

© JupiterImages Corporation.

Figure 5.5 Wind-generated waves tend to break near the shore, whereas tsunami produce a rapid rise in water level.

This is about the average speed of most commercial airplane flights. If you were on the ocean in a small boat, the wave would move so fast you could not detect it. In addition in deep water the vertical movement of the water might be one or two meters, no different that most waves in the open ocean.

At sea tsunami are basically harmless waves. Very little energy is lost as they travel through the open ocean. When the energy reaches shallow water near the shoreline, the waves rise up and become very destructive. They may have an appearance of normal wind-generated waves (Figure 5.5), but when tsunami hit the shoreline, they continue to move inland as a relentless flow of water.

Frequency of Tsunami

The National Centers for Environmental Information (NCEI), a division of the National Oceanic and Atmospheric Administration (NOAA), maintains a database of tsunami activity since 2000 B.C.E. to the present. A key part of the destruction of tsunami is the runup of water on land adjacent to the ocean. Numerous sources of information have been examined to develop a better understanding of where tsunami and runup events occur (Table 5.1). Eighty-three percent of all tsunami have occurred in the Pacific Ocean and the Mediterranean Sea. Major tsunami that affect a large area occur about once per decade.

Extensive documentation exists, especially in Japan and the Mediterranean areas, where inhabitants have recorded the occurrence of natural disasters for centuries. The first recorded tsunami occurred off the coast of Syria in 2000 B.C.E. Since 1900 (the beginning of instrumentally located earthquakes), most tsunami have been generated in Japan, Peru, Chile, New Guinea, and the Solomon Islands. However, the only regions that have generated remote-source tsunami affecting the entire Pacific Basin are the Kamchatka Peninsula, the Aleutian Islands, the Gulf of Alaska, and the coast of South America. All of these are very active areas of tectonic activity.

Table 5.1 Comparative Amounts of Tsunami and Runup Events, by Region

Region	Percentage of Tsunami	Percentage of Runup Events
Pacific Ocean	61	82
Mediterranean Sea	22	4
Indian Ocean	6	9
Caribbean Sea	4	2
Atlantic Ocean	7	3
Black Sea	1	0
Red Sea	< 1	0

Source: National Centers for Environmental Information, NOAA.

Hawaii, because of its location in the center of the Pacific Basin, has experienced tsunami generated in all parts of the Pacific.

Regions around the Mediterranean and the Caribbean have experienced numerous, locally destructive tsunami. A small subduction zone exists where the African Plate is colliding with the Eurasian Plate. The Caribbean Plate is surrounded by the North American and South American Plates on its eastern edge and the North American and Cocos Plates on its western edge.

Only a few tsunami have been generated in the Atlantic and Indian Oceans. In the Atlantic Ocean, there are no subduction zones at the edges of plate boundaries to generate such waves, except small subduction zones under the Caribbean and Scotia Arc, which is located at the southern end of South America. Subduction is active along the eastern border of the Indian Ocean, where the Indo-Australian plate is being driven beneath the Eurasian plate at its eastern margin. Thus, most tsunami generated in this region are propagated toward the southwest shores of Java and Sumatra, rather than into the Indian Ocean. The December 2004 Indonesian tsunami was anomalous in that it affected all the countries in the immediate and distant areas because of the size of the earthquake that generated the seismic sea wave (Figure 5.6).

© capturefoto/Shutterstock.com.

Figure 5.6 Damage caused by the December 2004 tsunami was far-reaching. Areas such as this one near Banda Aceh, Indonesia, were overrun by water that stripped the landscape clean.

Not all major earthquakes in the oceans produce tsunami. In a 2 week period in January 2018 two such events did not generate any significant wave action. A magnitude 7.6 earthquake off the coast of Honduras was determined to be related to strike-slip motion along a fault bordering the contact of the Caribbean Plate and the North American Plate.

On January 23, 2018, a magnitude 7.9 event, located 220 kilometers southeast of Kodiak Island, Alaska, occurred along a strike-slip portion of the Pacific Plate–North American Plate boundary. Although a warning was issued, no tsunami materialized. Two previous large events (magnitudes 7.8 and 7.9) occurred in the same region in the late 1980s, producing no known damage.

Historical Occurrences of Tsunami

Throughout recorded history tsunami have affected millions of people, especially in countries surrounding the Pacific Ocean. Table 5.2 lists selected tsunami that have been generated by several different sources. The major events are all associated with sites that border the oceans. Also significant tsunami were generated by the four largest earthquakes in recorded history—Chile 1960 M 9.5, Alaska 1964 M 9.2, Indonesia 2004 M 9.1, and Japan 2011 M 9.1. These events were each caused by active subduction of oceanic plates under adjoining continental or oceanic plates.

Unimak, Alaska, April 1, 1946

Unimak Island, located 1,100 kilometers (660 mi) southwest of Anchorage, Alaska, lies in the Aleutian Island chain at the western end of the Alaskan Peninsula. In the early morning of April 1, 1946, an earthquake of magnitude 8.6 occurred approximately 150 kilometers (90 mi) south of the island in a section of the Aleutian Subduction Zone. The resulting energy lifted the seafloor, thereby generating a significant tsunami with a height of approximately 35 meters (115 ft). On Unimak Island the Scotch Gap lighthouse, which was built in 1940 and situated about 15 meters (50 ft) above the mean tides of the area, was totally destroyed (Figure 5.7). Five people who were on duty in the station at the time the tsunami came ashore were killed.

Figure 5.7 **(a) Scotch Gap lighthouse was a reinforced concrete structure (b) that was demolished by a large tsunami produced by a magnitude 8.1 earthquake in Alaska in 1946.**

Table 5.2 Significant Tsunami That Have Occurred Since 1755

Date	Location	Source Mechanism	Death Toll	Interesting Facts
November 1, 1755	Lisbon, Portugal	Earthquake estimate M 9	40,000 to 50,000	Portugal, Spain, and Morocco hit by tsunami; tsunami crossed the Atlantic Ocean and struck West Indies.
August 27, 1883	Krakatoa, Indonesia	Volcanic eruption	36,000	Collapse of top of volcano generated large tsunami that reached the English Channel.
June 15, 1896	Japan	Earthquake	26,000	Largest wave was 30 m (100 ft).
November 18, 1929	Grand Banks, Newfoundland	Submarine landslide	29	Rate of motion calculated by determining breaks in transatlantic transmission cables.
April 1, 1946	Hawaii	Earthquake M 8.1	164	Pacific Tsunami Warning System was set up after this; 159 lives lost in Hawaii; 5 in Alaska.
July 9, 1958	Lituya Bay, Alaska	Landslide	Fewer than 10	Waves reached 525 m (17,230 ft) up the slope opposite the slide, the largest recorded runup in history.
May 22, 1960	Chile	Earthquake M 9.5	5,700	Largest earthquake in recorded history generated a tsunami that traveled across the Pacific Ocean; 61 died in Hawaii, 142 in Japan.
March 27, 1964	Alaska	Earthquake M 9.2	132	Second largest earthquake in recorded history generated a major tsunami that killed 132, including 12 in Crescent City, CA, who went to watch the incoming waves.
August 16, 1976	Philippine Islands	Earthquake M 7.9	7,000	Moro Gulf region hit by tsunami; 700 km (420 mi) of coastline hit.
September 1, 1992	Nicaragua	Earthquake M 7.6	170	Slow moving fault generated weak seismic waves that were not felt, so people were unprepared for the tsunami.
July 12, 1993	Hokkaido, Japan	Earthquake M 7.6	202	Hundreds missing; repeated warning went out but some ignored; can be seen on YouTube.
July 17, 1998	Papua New Guinea	Earthquake M 7.1 triggered submarine landslide	3,000	Epicenter of M 7.1 event just off northwest coast. No time to generate a warning. In less than 10 minutes the tsunami hit the shoreline.
December 26, 2004	Off west coast of Sumatra, Indonesia	Earthquake M 9.1	230,000	Third largest recorded earthquake in history generated a tsunami that traveled around the globe.
March 11, 2011	Off east coast of Japan	Earthquake M 9.1	18,000 to 20,000	Highest tsunami was 40.5 m (135 ft) that produced nuclear plant meltdown.
September 28, 2018	Sulawesi, Indonesia	Earthquake M 7.5	4,340	Strike-slip faulting produced tsunami that traveled at 15,400 kph (9,600 mph).
December 22, 2018	Java, Sumatra	Volcanic eruption landslide	426	Anak Krakatoa erupted, related to 1883 event.

Within 5 hours the initial tsunami had raced across the Pacific Ocean where it struck the Hawaiian Islands. Because no warning system was in place at the time, the inhabitants of the islands did not know what was headed their way. Waves ranging in heights between 10 and 18 meters (33 to 60 ft) hit the shorelines, destroying many of the coastal buildings. The waterfront of the city of Hilo was totally obliterated and the final death toll for the islands was 159 people. To the south, coastal sections of Washington, Oregon, and California were also affected by the tsunami waves.

Because of the widespread destruction in the region, in 1949 the United States government established the Pacific Tsunami Warning Center (PTWC), which is located near Honolulu, Hawaii. The PTWC is the informational source for tsunami warnings for almost all countries lying on the Pacific Ocean rim, as well as islands within the ocean.

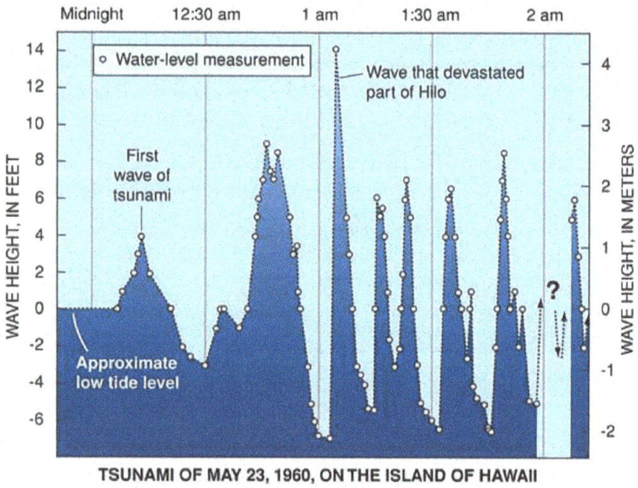

USGS Circular 1187.

Figure 5.8 A series of tsunami waves struck the island of Hawaii over a period of 2 hours following the Chilean earthquake of March 22, 1960. Notice the fluctuation in the high and low levels of the water.

Chile, May 22, 1960

Off the west coast of South America the Nazca Plate is moving eastward at the rate of about 15 cm (6 in) per year. It is being subducted under the South American Plate, forming the Peru-Chile Trench, and creating numerous large earthquakes.

In the early evening of May 22, 1960, the sea floor gave way, producing the largest recorded earthquake in history (M 9.5). Numerous strong foreshocks had occurred in the days prior to this event. The epicenter of the quake was located approximately 150 kilometers (90 mi) west of Valdivia in west-central Chile. Although many buildings in the city were heavily damaged or destroyed, the 25-meter (80 ft) high tsunami that arrived about 15 minutes later was the major cause of the estimated 2,000 deaths in the coastal communities of Chile with another 3,000 injured. Damage was reported throughout the Pacific Ocean basin, with the Philippines, Japan, and Russia experiencing large tsunami waves. These waves killed another 3,500 people as they struck flat-lying shorelines including many islands of the Pacific.

Within 12 hours of the initial shock, tsunami reached the Hawaiian Islands. Some waves were as high as 12 meters (40 ft). The city of Hilo on the island of Hawaii experienced at least eight waves (Figure 5.8). Note that the highest wave to hit Hilo was not the first one. The series of waves had a range of periods between 10 and 30 minutes, but luckily they all struck during low tide, lessening the damage.

Alaska, March 27, 1964

On Good Friday 1964, the second largest recorded earthquake in history and the largest to ever occur in the United States, struck near Prince

BOX 5.1 An Experience of Living Through the Good Friday Great Alaskan Earthquake of March 27, 1964

The week was coming to a close for Darrell Boomgaarden and everyone in the area was preparing for the upcoming Easter weekend. As he was making his way across the U.S. Navy base in Kodiak, delivering newspapers along his route, he remembers the events that would be history making. "My strongest memory was the sound, which I heard about three to five minutes before the ground motion hit. I could hear it bouncing off the nearby hills as I said to myself, 'What is that?'"

- And then the motion hit. As he looked down the street, all the lights were moving and cars were bouncing up and down. "I couldn't believe concrete could move like it did, and it seemed to keep going on forever. There were several minutes of motion."
- To Darrell the houses looked fine and there were no poles that appeared to have fallen. The residential area didn't seem too different. After completing his paper route, he headed home for dinner. About 1 hour later the siren went off. The base was evacuated, which was challenging as it was dark by this time. Everyone went up the mountain to safer ground, to a location about 20 feet above the water. They all spent the night there.
- Following the M 9.2 earthquake that occurred at 5:36 p.m. local time (the epicenter was about 455 km (275 mi) to the northeast in Prince William Sound), several tsunami struck the town of Kodiak and did significant damage (Figure 5.1.1). The townspeople consisted mainly of fishermen and cannery workers. The oil-fired power plant on the base was destroyed, which meant there was no electric power. The naval base was inundated with oil-filled sea water. All food in the commissary was lost. Darrell's father was the officer-in-charge of ground control at the naval air station, so he had to ensure the runway and its radar were not going to be flooded. Thousands of sandbags were used to protect the runway from incoming water.

United States Geological Survey.

Figure 5.1.1 Damage along the shoreline in Kodiak following the earthquake of March 27, 1964.

- Slowly, basic services were restored. A seaplane tender ship was sent to the naval base to generate power. There was no television in the region, so news was slow to reach everyone.
- It was several days before Darrell made it into town. "I could not believe that boats could get moved that far inland—several hundred yards from their docks onto land." School was canceled for 2 weeks, and teenagers joined in the effort to clean up the area. Everything was covered in oil.
- Darrell later learned that several of his friends died when they were driving along the shoreline after dark. Their car was hit by a tsunami, which they were unable to see. Recovery in the region took many months and the events of Good Friday, 1964, still live in the minds of those who survived.

Source: NOAA.

Figure 5.9 An overturned ship and a crumpled chemical truck, along with other debris, show the force exerted by the tsunami associated with the March 27, 1964, earthquake in Alaska.

Source: NOAA.

Figure 5.10 Damaged caused by a tsunami and fires that followed destroyed a significant part of Okushiri Island following a nearby major earthquake in July 1993.

William Sound, Alaska. This event was caused by the northward movement of the Pacific Plate as it dives under the North American Plate. The M 9.2 quake generated a series of tsunami waves that traveled throughout the Pacific Ocean, affecting numerous coastal localities. The harbor in Valdez was hit by several tsunami that were created when the harbor seafloor slid downslope. The port area of Seward, Alaska, was destroyed, with ships, vehicles, and buildings strewn across the landscape (Figure 5.9). More details, as well as a firsthand account, are provided in Box 5.1.

Hokkaido, Japan, July 12, 1993

On July 12, 1993, the island of Hokkaido, the northernmost island in Japan, experienced the effects of a magnitude 7.7 earthquake. The epicenter was located in the Sea of Japan 25 kilometers (15 mi) from the western edge of the island. This location was unusual as most earthquakes occur in the Pacific Ocean lying to the east of Hokkaido. Debate existed as to the cause, but later research showed that the Pacific Plate moved under the eastern edge of the Eurasian Plate. The relatively shallow focal depth of 15 kilometers (9 mi) contributed to the formation of the destructive tsunami.

Surface movement produced by the event resulted in the nearby small island of Okushuri dropping as much as 80 centimeters in elevation. In less than 5 minutes, insufficient time to warn people, the island was hit with one of the largest tsunami to ever occur in Japan (Figure 5.10). The highest wave reached more than 30 meters (100 ft), killing 230 people.

Papua, New Guinea, July 17, 1998

submarine landslide

The collapse of land material either underwater or from the land that slides into water, producing a massive wave.

The southwestern Pacific Ocean is an area that has a great deal of tectonic activity, as colliding plates produce numerous large earthquakes. Such was the case in mid-July 1998 when a magnitude 7.1 earthquake occurred just north of the coast of Papua, New Guinea. Analysis after the event showed that a major **submarine landslide** was triggered by the earthquake.

A tsunami formed very quickly following the main event, sending several waves onto the island. Wave heights ranged between 10 and 15 meters (33 to 50 ft) and pushed landward, destroying several villages. More than 2,000 people died.

Sumatra, Indonesia, December 26, 2004

December 26, 2004, began very quietly for many vacationers who were visiting the beaches of Sumatra, Thailand, India, and nearby countries. Unknown to many people, one of the largest earthquakes in history occurred, displacing trillions of gallons of water in a tsunami wave that ended up going all the way around the globe.

© Frans Delian/Shutterstock.com.

Figure 5.11 **Banda Aceh, Indonesia, was one of the most devastated cities along the Sumatra coast, following the December 26, 2004, tsunami.**

At 7:58 in the morning local time, a tsunami generated by a huge (M 9.1) earthquake began its journey across the Indian Ocean. Within about 15 minutes, it struck Sumatra, killing over 130,000 people. Damage in Sumatra was complete—trees were ripped out of the ground by their roots, entire towns were leveled, and water flooded the land to a depth of up to 25 meters (82 ft) (Figure 5.11). Within two hours, the tsunami had struck Sri Lanka and Thailand, claiming another 40,000 lives.

In all, at least 230,000 people died in one of the worst natural disasters in human history. The waves dramatically pointed out the vulnerability of coastlines to tsunamis, and the devastation caused when little warning is available.

Palu, Sulawesi, Indonesia, September 28, 2018

Palu, a city of 367,000 people, is the provincial capital of the island of Sulawesi. It lies at the southern end of a 30 km (19 mi) long bay. On September 28, 2018, a M 7.5 earthquake occurred 80 km (50 mi) north of the city. Numerous tsunami were generated, the largest of which was 10.7 m (35 ft) high. Significant sections of the city were destroyed (Figure 5.12) and an estimated 4,340 people died, 1,252 by the largest tsunami which struck at high tide. An unusual observation was that scientists believed the major fault movement that generated the tsunami was a strike-slip fault, a type not known to generate vertical water movement in the ocean. Another oddity was that the tsunami traveled at a rate of 9,600 mph (15,400 kph), which is approximately 25% faster than the typical tsunami.

© Kurniawan Rizqi/Shutterstock.com.

Figure 5.12 **Devastation was widespread in the city of Palu, Sulawesi, when a tsunami struck the coast.**

BOX 5.2 Tohoku, Japan, Earthquake and Tsunami, March 11, 2011

As the Pacific Plate moves to the northwest and collides with a portion of the North American Plate, the resulting subduction has produced the volcanic islands that form Japan. The region has historically experienced many earthquakes including a major historic event in 1923.

On March 11, 2011, the earth ruptured 130 kilometers (80 mi) east of Sendai, a city of one million people on the eastern coast of Honshu. The shallow event had a focal depth of 24 kilometers and produced a magnitude 9.1 earthquake, which ranks as the worst in Japan's long history of seismicity and the world's fourth largest recorded since 1900. As often occurs with great earthquakes (magnitude 8.0 or larger), there were several major foreshocks, including a magnitude 7.2 event 2 days earlier. More than 11,000 aftershocks were recorded. The vast majority of the 20,000 or more deaths and missing persons resulted from the tsunami that was formed by the sudden fault movement offshore. Damage estimates were more than $424 billion US dollars.

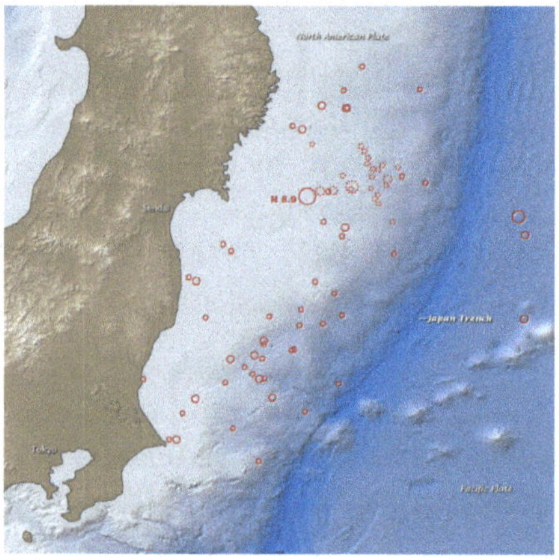

Source: NASA.

Figure 5.2.1 Location of the epicenter of the March 11, 2011, magnitude 9.1 earthquake. The Pacific Plate is being subducted under the North American Plate.

The resulting tsunami raced across the Pacific Ocean at the speed of a jet airplane. Within an hour of the earthquake, a wave 2.8 meters (9 ft) high struck the city of Hanasaki on the island of Hokkaido. Numerous other reporting stations throughout the Pacific reported waves, including Midway Island (5 hours later, wave height 1.3 meters), Maui, Hawaii (8 hours later, wave height 1.7 meters or 5 ft), and Crescent City, California (11 hours later, wave height 2 meters or 6 ft). Areas that have arcuate coasts and harbors experienced large surges due to the focusing of the wave energy. The Japanese government estimated that more than 1.5 million tons of debris floated out into the Pacific Ocean, some of which made its way to the West Coast of the United States.

Seawater driven ashore by the tsunami damaged the cooling systems of the Fukushima Daiichi Nuclear Power Station. Three nuclear reactors overheated, producing a meltdown of one reactor and releasing radioactive material into the atmosphere and surrounding water bodies. Nuclear power plants throughout the country were shut until inspections and maintenance were completed.

© Smallcreative/Shutterstock.com.

Figure 5.2.2 Extensive damage was caused by the March 11, 2011, tsunami as it came ashore at Sendai, Japan.

Tsunami Prediction, Warning, and Possible Mitigation

Tsunami are often associated with large-magnitude earthquakes that occur in or near oceans. When a large earthquake occurs in a region near the ocean, instruments alert scientists about the potential for tsunami to be formed and warnings can be issued.

In the Pacific Ocean the Pacific Tsunami Warning Center located at Ewa Beach, near Honolulu, Hawaii, becomes the focal point of activity when an earthquake could potentially generate a tsunami. This facility is operated by the National Weather Service (NWS) and serves as the headquarters for the Operational Tsunami Warning System, a group that coordinates the monitoring and reporting of seismic activity around the Pacific Ocean, the Indian Ocean, and the Caribbean Sea. The West Coast/Alaska Tsunami Warning Center, located in Palmer, Alaska, monitors the northern Pacific Ocean, including Alaska and the Aleutian Islands. It also oversees tsunami activity along all coastal regions of the United States (except Hawaii), the coastal provinces of Canada, Puerto Rico, and the Virgin Islands.

Warnings are only as good as the people who heed them. If people decide to rush to the shore to watch waves come in, they will obviously be placing their lives in danger; such was the case in Crescent City, California. In March 1964 people went to the shore to watch tsunami generated by the great earthquake in Alaska, resulting in 12 deaths.

In order to lessen the loss of life and property damage, fewer people must live near the shoreline, especially in areas where there is very little relief above sea level. Oncoming waves that are as much as 15 meters (50 ft) or higher will move rapidly onshore and flood flat-lying areas that are only slightly above sea level. All property will be inundated and likely totally destroyed from the onslaught of water.

Tsunami Detection Devices

The National Oceanic and Atmospheric Administration (NOAA) operates a network of stationary tsunami detection buoys in several oceans throughout the world. The system that employs the recording technology is the Deep-Ocean Assessment and Reporting of Tsunamis (DART). These buoys are primarily deployed in the Pacific Ocean, but there are several in the western Atlantic Ocean and in the Gulf of Mexico (**Figure 5.13**). Only six buoys were operational in the Pacific prior to the 2004 Indonesian event, and none were located in the Indian Ocean. Therefore there was no means to notify nations around the Indian Ocean of the impending disaster coming their way.

The buoys float on the water surface (**Figure 5.14**) and receive a signal from a source located on the seafloor. As a tsunami passes the sensor, there is a rapid change in pressure which is sent to the hydrophone underneath the buoy. The instruments are programmed to recognize this change and then transmit a signal to a warning center, which then sends the appropriate information to areas in the possible path of the tsunami.

National Data Buoy Center, NOAA.

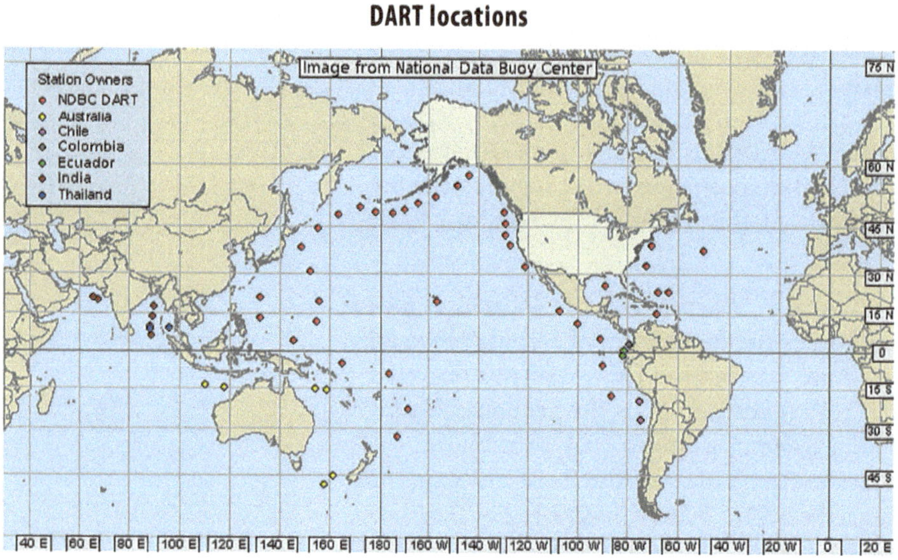

Figure 5.13 Detection buoys provide early warnings around the Pacific Ocean, the Caribbean and the East Coast of the United States for areas prone to tsunami.

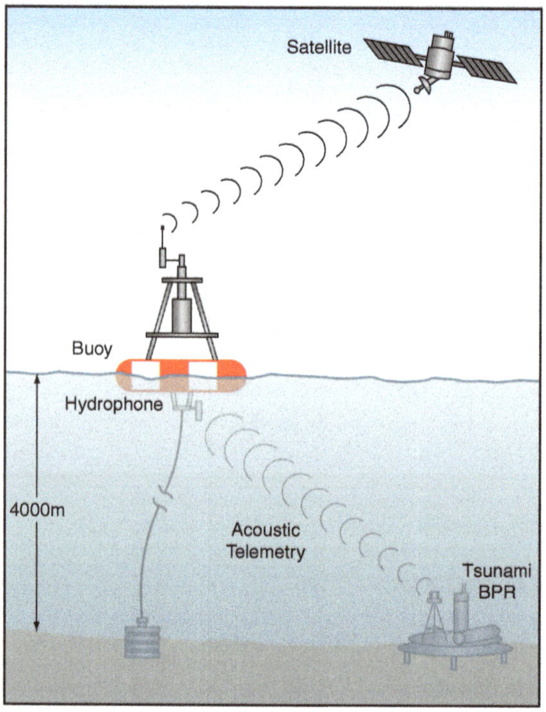

Source: NOAA.

Source: NOAA.

Figure 5.14 The real-time tsunami system operated by NOAA is capable of detecting a tsunami as it passes through the deep waters of the ocean. Information is transmitted via satellite to the warning centers. Buoys float on the surface of the ocean.

If a tsunami is created, the first information available is related to the earthquake and its source area. This information gives scientists a general idea of the area that could be affected. As the tsunami travels across the ocean, its movement is tracked by data furnished from the recording buoys. As more data become available, the path of the tsunami can be projected more accurately, thus allowing steps to be taken to reduce the destruction ahead of the waves.

What Happens During a Tsunami Warning?

Within seconds of an earthquake's occurrence, seismographs located around the Pacific Basin send data on the earthquake to the Pacific Tsunami Warning Center in Hawaii. Computer programs determine a preliminary magnitude and location for the earthquake. If the magnitude is large enough, the system alerts the warning center staff that a large earthquake has occurred. Because of this, all warning center staff must live on the grounds of the warning center, with one staff member on the grounds at all times. Based on the magnitude of the earthquake, the staff may choose to issue an information bulletin, an advisory, a watch, or a warning.

- An **information bulletin**. These are informational statements to scientists and disaster specialists that an earthquake has occurred. They generally mean that a tsunami has not occurred, and are intended to prevent needless evacuation.
- An **advisory**. These are issued when it is believed that a tsunami may have been generated. They are issued to areas outside the current warning or watch area to inform disaster managers to remain alert and wait for further expansion of the warning or watch areas.
- A **watch**. Tsunami watches are issued based on seismic information, without confirmation that a tsunami has been generated. The size of the watch area depends on the magnitude of the earthquake. NOAA sets guidelines which allow tsunami warning centers to use the following rules of thumb for issuing a watch:

  For earthquakes over magnitude 7.0, the watch area is 1 hour tsunami travel time outside the warning zone.

  For all earthquakes over magnitude 7.5, the watch area is 3 hours tsunami travel time outside the warning zone.

  The watch will either be upgraded to a warning in subsequent bulletins or will be cancelled, depending on the severity of the tsunami. Watches are issued to provide disaster managers with the earliest possible indication that a tsunami may be on the way.
- A **warning**. This is the highest level of tsunami alert. Initially, a warning center may issue a tsunami warning for areas close to an earthquake before confirming that a tsunami has been generated, in order to give maximum time for disaster managers to evacuate the coastline. In successive statements, however, warning systems issue warnings only when tsunami generation has been confirmed (either through DART buoys or shoreline observation). Tsunami warning centers use these NOAA rules for issuing a warning:

  Earthquakes over magnitude 7.0 trigger a warning covering the coastal regions within 2 hours tsunami travel time from the epicenter.

  When the magnitude is over 7.5, the warned area is increased to 3 hours tsunami travel time.

information bulletin
Notification to scientists and researchers that an earthquake has occurred, but not necessarily a tsunami was generated.

advisory
An announcement that is issued when a tsunami could result from a large earthquake that occurs near a coastal region.

watch
An alert that a tsunami could potentially occur in an area.

warning
The highest alert level that is given when a tsunami has been formed and its arrival is highly possible.

As water level data showing the tsunami are recorded, the warning will either be cancelled, restricted, expanded incrementally, or expanded in the event of a major tsunami.

Initial warnings are issued based on the magnitude of the earthquake. Changes are made to previous warnings as size and location of the earthquake, along with tide gauge and DART information, are updated.

Tsunami Safety Rules

In case you are ever in an area where there is a threat of tsunami, heed the following tsunami safety rules from the West Coast/Alaska Tsunami Warning Center:

- A strong earthquake felt in a low-lying coastal area is a natural warning of possible, immediate danger. Keep calm and quickly move to higher ground away from the coast.
- All large earthquakes do not cause tsunami, but many do. If the quake is located near or directly under the ocean, the probability of a tsunami increases. When you hear that an earthquake has occurred in the ocean or coastline regions, prepare for a tsunami emergency.
- A tsunami is not a single wave, but a series of waves. Stay out of danger until an "all clear" is issued by a competent authority.
- Approaching tsunami are sometimes heralded by a noticeable rise or fall of coastal waters. This is nature's tsunami warning and should be heeded.
- A small tsunami at one beach can be a giant a few miles away. Do not let the modest size of one make you lose respect for all.
- Never go down to the beach to watch for a tsunami! **If you can see the wave, you are too close to escape.** Tsunami can move faster than a person can run!
- Homes and other buildings located in low-lying coastal areas are not safe. Do **not** stay in such buildings if there is a tsunami warning.
- The upper floors of high, multi-story, reinforced concrete hotels can provide refuge if there is no time to quickly move inland or to higher ground.

Key Terms

advisory *(page 157)*
drawdown *(page 145)*
information bulletin *(page 157)*
runup *(page 145)*
seismic sea wave *(page 142)*
submarine landslide *(page 152)*
tsunami *(page 142)*
warning *(page 157)*
watch *(page 157)*
wavelength *(page 143)*

Summary

Tsunami are perhaps the most destructive of all natural disasters. They are produced by large magnitude events that occur in or near oceans: earthquakes, volcanic eruptions, submarine landslides, or meteorite impacts. Tsunami (or seismic sea waves) travel very rapidly across open oceans because their velocities are related directly to the depth of the water through which the energy is traveling. Velocities of 800 kph (500 mph) are commonly recorded in deep oceans. Throughout history there have been numerous disastrous tsunami. The deadliest recorded one occurred in December 2004 off the coast of Indonesia, killing more than 230,000 people. In March 2011 a major tsunami destroyed a nuclear power plant on the east coast of Japan, spreading radioactive contamination across the Pacific Ocean.

Tsunami typically involve a series of waves that hit the coastline. The largest wave is generally not the first in the series. The devastation produced by these waves is far-reaching when tsunami hit the flat shorelines. Countries around the Pacific Ocean are very likely to experience tsunami, because of the number of large earthquakes associated with active plate subduction around the Circum-Pacific region. Almost every year there are tsunami warnings issued in the region, oftentimes affecting many coastlines. The DART detection systems are in place in several oceans but there still needs to be better coverage in the world's oceans, particularly in the Indian Ocean. Whenever people are in areas that are prone to experience tsunami, it is critical to heed warnings and be aware of actions they should take to survive such an event.

References and Suggested Readings

Atwater, B. F., Cisternas V. M., Bourgeois, J., Dudley, W. C., Hendley II, J. W., & Stauffer, P. H. (1999). *Surviving a tsunami-lessons from Chile, Hawaii and Japan.* USGS Circular 1187.

Birmingham, L. (2012). *Strong in the rain-surviving Japan's earthquake, tsunami and Fukushima nuclear disaster.* Palgrave MacMillian.

Bryant, E. (2008). *Tsunami: The underrated hazard.* Springer.

Duchamp, L. T. (2007). *Tsunami.* Aqueduct Press.

Dudley, W. C., & Lee, M. (1998). *Tsunami!* (2nd ed.). University of Hawaii Press.

Ehrlich, G. (2013). *Facing the wave—A journey in the wake of the tsunami.* Pantheon Press.

Kling, A. A. (2003). *Tsunami.* Lucent Books.

Lace, W. W. (2008). *The Indian Ocean tsunami of 2004.* Chelsea House Press.

Samuels, R. J. (2013). *3.11-Disaster and change in Japan.* Cornell University Press.

Svarney, T. E., & Barnes-Svarney, P. (2000). *The handy ocean answer book.* Visible Ink Press.

U.S. Geological Survey. *Tsunami hazards—A national threat.* USGS Fact Sheet 2006-3023.

Yamakawa, M., & Yamamoto, D. (Eds.) (2017). *Unraveling the Fukushima disaster.* Routledge, Taylor & Francis Group.

Websites for Further Reference

https://nctr.pmel.noaa.gov
https://walrus.wr.usgs.gov/tsunami/
https://www.tsunami.noaa.gov
https://woodshole.er.usgs.gov
https://www.usgs.gov/

Questions for Thought

1. What is the origin of the word *tsunami*?
2. Why does Japan experience so many tsunami?
3. How do tsunami and wind-generated waves differ?
4. Explain the difference between runup and drawdown.
5. How does vertical movement associated with earthquakes on the ocean floor generate tsunami?
6. How does a change in water depth affect the velocity of a tsunami?
7. Why is Hawaii prone to having a large number of tsunami hit it?
8. Explain how landslides on the continents or in the oceans create tsunami.
9. Explain how a tsunami detection buoy provides information on possible tsunami.
10. Give an example of how a tsunami warning system saves lives.

Unstable Ground: Mass Movements and Subsidence

6

United States Geological Survey. Photo by R. L. Schuster.

La Conchita, California, has experienced two massive landslides in recent years. This one occurred in 1995. In 2005 another one affected the same slope after a series of rain storms saturated the slopes. Ten people died in the latter event.

It Won't Happen to Me—We'll See

The small town of Union Gap, Washington (population 6,110), is located about 6 kilometers south of Yakima. In early October 2017 townspeople noticed a crack forming on the surface of Rattlesnake Hills, located just east of town. At first the hill was moving to the south at the rate of approximately 30 centimeters per week. Geologists and engineers set up monitoring equipment to better observe the activity on the 20-acre tract of land. The hill consists of basalt overlying a weaker sedimentary layer. Within a few months the movement had increased to almost 70 centimeters a week. The potential is for an estimated three million cubic meters of material to slide downslope. Steps have been taken to slow the movement with large barricades at the base of the hill, which is slowing moving toward Interstate 82. A quarry at the foot of the hill has closed, and nearby homeowners have been evacuated. Although water does not appear to be a factor, the potential certainly exists for a heavy rainfall to work its way into the surface crack and flow downhill along the interface between the basalt and the sedimentary rock. Only time will tell if this disaster will occur.

Although the ground we live on may appear to be stable and unchanging, it is in fact a dynamic place where the force of gravity is constantly acting on Earth's materials, over time causing the downward movement of rock and loose surface material. **Mass movement** (also called *mass wasting*) is the downward movement of Earth's surficial material from one place to another under the direct influence of gravity. Earth's surface consists mainly of slopes ranging from gentle to steep and often covered with several components of the surface. **Regolith** consists of all material lying above unaltered bedrock, including unconsolidated and fragmental material which breaks down to form soil. Although gravity acts constantly on all slopes, the strength and resistance of materials usually hold the slope in place. However, natural processes or human activity may destabilize a slope, causing failure and mass movement of material. This movement may be so slow that it is almost undetectable (known as *creep*), or be sudden and swift, as in devastating **landslides**. Mass movements do not always have to involve failure on a slope, however. **Subsidence** is the vertical (downward) motion of earth materials, and includes vertical movements such as the gradual sinking of the land surface when fluids are withdrawn, or the sudden collapse into subterranean voids creating **sinkholes**.

mass movement (mass wasting)
A term used more by geologists to describe the downslope movement of material.

regolith
The layer of varied material that overlies unaltered bedrock and includes unconsolidated and fragmental particles.

landslide
A general term used to describe the downslope movement of material under the force of gravity.

subsidence
A relatively slow drop in the ground surface caused by the removal of rock or water located under the surface.

sinkhole
A circular depression on the surface that forms from the collapse of material into an underlying void.

Mass movement is an important natural erosional process but can be a serious hazard where expanding populations are building homes on or near steep hillsides or over abandoned mine areas. Rapid downslope movements mostly involve landslides, which every year cause loss of life, property damage, or economic hardship to people around the world.

Landslides have occurred in all 50 states (Figure 6.1), and damage from these landslides exceeds $1 billion every year, while claiming 25 to 50 lives. In 1985, a massive landslide in Puerto Rico killed 129 people, making it the greatest loss of life from a single landslide in US history. The Thistle, Utah, landslide of 1983 caused an estimated $400 million in losses, making it one of the most expensive single landslides in US history, while the 1997–1998 El Niño rainstorms in the San Francisco Bay area produced thousands of landslides, causing over $150 million in losses. Landslides can also be

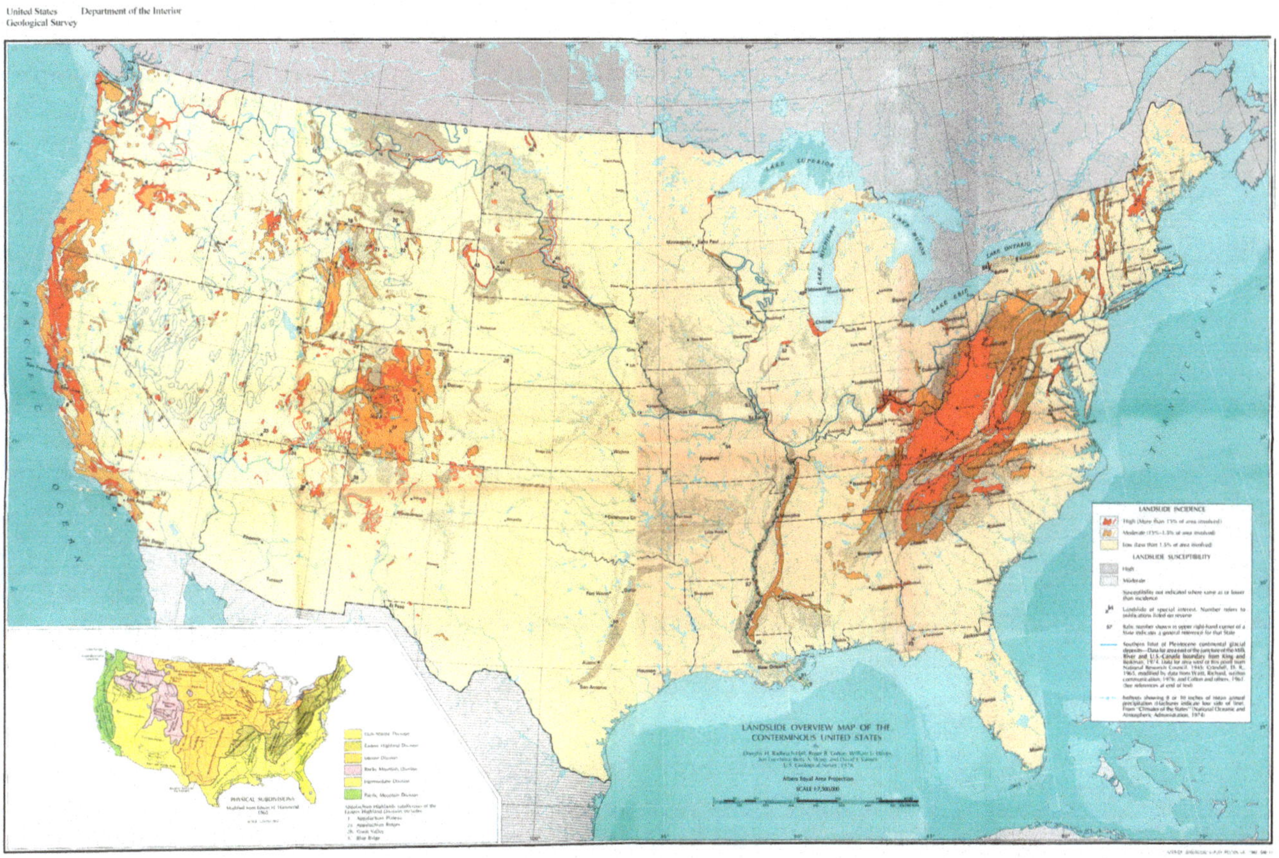

United States Geological Survey.

Figure 6.1 **Landslides occur in all the lower 48 states. Moist, unstable terrain in the Eastern United States is especially affected by landslides. Darker shading shows areas that experience the highest likelihood of landslides.**

associated with other hazards, such as earthquakes, floods, storm surges, severe storms, and volcanic activity. **Slope failures** are a secondary effect of wildfires that remove the vegetation holding soil and surface materials in place. Often they are more damaging and deadly than the associated hazard event.

slope failure
The downslope movement of an unstable area.

What makes many mass movements, such as landslides and collapse sinkholes, so hazardous is that they can occur at any time and almost any place with little or no warning, and the event is over very quickly. Knowledge about the relationships between the local geology and mass movement processes can aid in better development planning and a reduction in vulnerability to such hazards. Therefore, we will look at the various types of mass movements, their underlying causes, factors that affect ground and slope stability, and what we can do to reduce the vulnerability and risk of living on unstable ground.

Landslides: Mass Movements on Slopes

Earth's surface materials that make up slopes comprise a complex physical, chemical, and biological system. These materials are a mixture of rock, regolith, and soil, with a variable amount of water and organic material. The amount of water and organic material (mostly vegetation) may vary on a slope from season to season and year to year. This variability depends upon such factors as the nature and extent of precipitation, rates of materials added or removed to the slope by deposition or erosion, and activities that affect the vegetation, such as human impact or fire. These complex interactions involving the strength and composition of earth materials are responsible for the different shapes of slopes, and how slopes may fail under the influence of gravity. The resulting mass movement plays an important part in the erosional process, moving material downslope from higher to lower elevations.

The term *landslide* is used in the general sense to describe a wide variety of mass movement landforms and processes involving the downslope transport of earth material under gravitational influence. Landslides have a great variety of shapes (morphologies), rates of motion, and types of movement, and can range in size from a small area to a region of many square kilometers. Landslides play a significant role in the evolution of the landscape, and are among the most widespread natural hazards on Earth.

Slope Processes and Stability: The Balance Between Driving Forces and Resisting Forces

driving force
A force that produces downslope movement caused by gravity.

resisting force
A force that tends to prevent downhill movement.

normal force
A force that is acting perpendicular to a surface.

shear force
A force that acts parallel to a surface.

Two types of forces are involved in mass movements on slopes: **driving forces** and **resisting forces**. Driving forces are those that promote movement, while resisting forces are those that tend to prevent movement. The main *driving force* responsible for mass movement is gravity. Gravity acts everywhere on the Earth's surface, pulling everything in a vertical direction toward the center of the Earth. On a flat surface the force of gravity acts directly downward, so, as long as the material remains on the flat surface, it will not move. On a slope, the force of gravity can be divided into two components: one acting perpendicular to the slope (the **normal force**) and one acting parallel to the slope (the **shear force**) (Figure 6.2). The normal force holds the material in place on the slope, whereas the shear force moves the material down the slope.

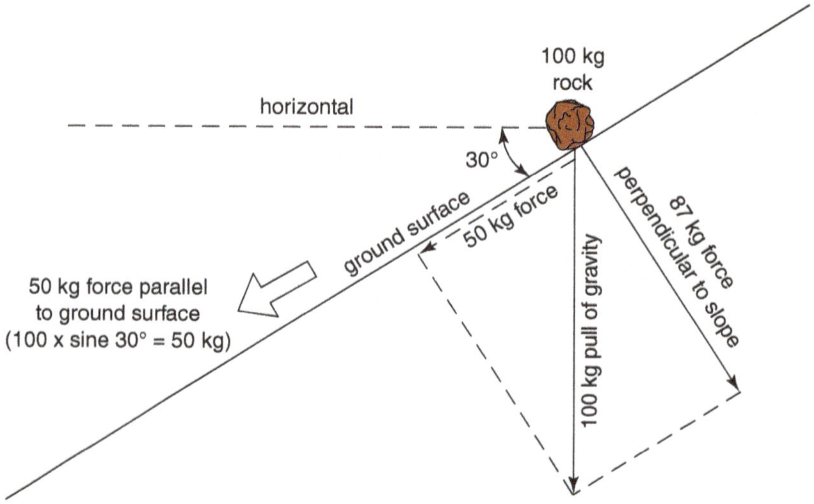

© Kendall Hunt Publishing Company.

Figure 6.2 Forces acting on a 100 kilogram boulder resting on a 30° hillslope.

On and below the ground surface, resisting forces, such as friction and particle cohesion, act in opposition to the driving forces. When the driving forces acting downhill exceed the resisting forces, mass movement results.

The **Factor of Safety (FoS)** can be defined for a slope as:

$$\text{FoS} = \text{resisting forces}/\text{driving forces}.$$

Factor of Safety (FoS)

The ratio of resisting force to driving force.

If the resisting forces are greater than the driving forces, the FoS is greater than 1 and the slope is stable. If the driving forces are greater than the resisting forces, the FoS is less than 1 and slope failure occurs. The magnitudes of the resisting and driving forces can change as a result of weather conditions or slope configuration.

Slopes are in a continuous state of dynamic equilibrium by constantly adjusting to new conditions caused by changes in the driving and resisting forces. Although we view mass wasting events as disruptive and often destructive processes, it is simply a way for a slope to adjust to new and changing conditions. For example, when a building or a road is constructed on a hillslope, the equilibrium of the slope is upset and the slope may become unstable. This could lead to slope failure as the slope adjusts to the new set of conditions (additional mass) placed upon it. The role of gravity is to level out all slopes and create a horizontal landscape.

Causes and Triggers That Influence Slope Stability

Rock, regolith, and soil remain on a slope only when the driving forces are unable to overcome the resisting forces keeping the material in place. Long before a landslide occurs, slopes undergo various processes that weaken rock material and gradually make the surface more susceptible to the pull of gravity. Eventually, the strength of the slope is weakened to the point that some event (known as a *trigger*) allows it to suddenly cross from a state of stability to one of instability.

Following a landslide, geologists and engineers often look for clues (the contributing factors) to explain why the slope failed. The reasons for slope failure can be broken down into two categories: *causes* and *triggers* (Table 6.1a–b). The difference between causes and triggers of landslides is subtle, but important to understand. **Causes** are considered to be the controlling factors involved in making the slope vulnerable to failure in the first place; in other words, they contribute to the slope becoming weak and unstable. A **trigger** is the single event that temporarily disturbs the equilibrium of the slope and initiates mass movement, resulting in slope failure. Triggers that initiate movement include earthquakes, heavy rainfall, and volcanic activity. The causes are the explanation for why a landslide occurred in a certain location, and include geological factors, morphological factors, and factors associated with human activity. Landslides most often have multiple causes that combine to make a slope vulnerable to failure, and then a trigger finally initiates the movement.

cause

A controlling factor that makes a slope vulnerable to failure.

trigger

An event that disturbs the equilibrium of a slope, causing movement to occur.

In most cases it is relatively easy to determine the trigger after the landslide has occurred, but the causes leading up to slope failure can be much more difficult to identify, because once a landslide occurs, any pre-existing evidence is usually destroyed.

Table 6.1a Common Landslide Causes

1. Natural Geological Causes
a. Weak or sensitive materials
b. Weathered materials
c. Sheared, jointed, or fissured materials
d. Adversely oriented discontinuity (bedding, schistosity, fault, unconformity, contact, and so forth)
e. Contrast in permeability and/or stiffness of materials
2. Natural Morphological Causes
a. Tectonic or volcanic uplift
b. Glacial rebound
c. Fluvial, wave, or glacial erosion of slope toe or lateral margins
d. Subterranean erosion (solution, piping)
e. Deposition loading slope or its crest
f. Vegetation removal (by fire, drought)
g. Thawing
h. Freeze-and-thaw weathering
i. Shrink-and-swell weathering
3. Human Causes
a. Excavation of slope or its toe
b. Loading of slope or its crest
c. Drawdown (of reservoirs)
d. Deforestation
e. Irrigation
f. Mining
g. Artificial vibration
h. Water leakage from utilities

Table 6.1b Common Landslide Triggers

1. Physical Triggers
a. Intense rainfall
b. Rapid snowmelt
c. Prolonged intense precipitation
d. Rapid drawdown (of floods and tides) or filling
e. Earthquake
f. Volcanic eruption
g. Thawing
h. Freeze-and-thaw weathering
i. Shrink-and-swell weathering
j. Flooding

Source: USGS Fact Sheet 2004-3072 and Circular 1325-508.

Common Controlling Factors (Causes) in Slope Stability

The following is a discussion of the more common controlling factors that influence slope stabilities, and the role they play in making them vulnerable to failure. Although they are discussed separately, most of them are interrelated and can collectively affect a slope's stability.

The Role of Slope Materials

Some rocks are inherently more stable than others. Massive, uniformly-textured rocks such as granite, basalt, and quartzite, have interlocking crystals and grains that give them nearly equal strength in all directions. Such rocks will hold their position against gravity, and form steep slopes only when they become fractured and jointed. Over time, chemical and physical weathering processes produce unconsolidated regolith, and change its resistance to gravity. Freezing and thawing of water in cracks, or shrinking and swelling of clay minerals decrease friction and cohesiveness and promote slope failure.

Unconsolidated earth materials such as soil, clay, sand, and gravel are more susceptible to movement. Clay becomes plastic and weak when it absorbs water, causing it to act as a lubricant for mass movements. For sediment coarser than clay, stability on slopes is influenced by grain shape (whether angular or round) and grain roughness. The presence of plant root networks and mineral are very effective in holding material in place, thus promoting stability.

The Role of Water

The water content in slope material strongly influences its stability in several ways, and is a critical factor in many causes of mass movements, as well as being a trigger. As a driving force, slope material can become saturated with water (after heavy rains, melting snow, or rising groundwater) by filling pores and fractures. This increases the weight (load) of material on the slope and increases the stress acting parallel to the surface (water weighs 1 kg per liter or 8.35 lbs per gallon). However, this extra weight of water is probably less important than the reduction of the resisting forces of the material. Water percolating through slope material helps decrease friction between grains and contributes to a loss of cohesion among particles.

Another aspect of water that affects slope stability is fluid pressure, which in some cases can build in such a way that water in the pores can actually support the weight of the overlying material. When this occurs, friction is greatly reduced, and thus the resisting forces holding the material on the slope is also reduced, causing slope failure. Water can also reduce rock strength by circulating through the pores of some rocks and dissolving soluble cementing materials, such as calcium carbonate. This process reduces cohesion as well. Another effect of water is that it can soften layers of shale, and even cause some types of clay minerals to expand, reducing friction between rock layers. Clay consists of platy particles that easily slide over each other when wet. This lubricating effect of water with clay is the reason why clay beds are frequently the slippery layer along which overlying material slide downslope.

angle of repose

The natural angle of a slope that forms in a pile of unconsolidated material.

undercutting

A process whereby a slope or hillside has supporting material removed by erosion.

The Role of Slope Angle

Slope angle is a major factor that influences mass movement. Commonly, the steeper the slope, the less stable it is, and therefore, steep slopes are more likely to experience mass movement than gentle slopes. The steepest angle that a slope can maintain without collapsing is called its **angle of repose** (Figure 6.3). For dry, unconsolidated materials, the angle of repose increases with increasing grain size, but usually lies between about 33° and 37° on naturally formed slopes. At this angle, the resisting forces of the slopes material counterbalance the force of gravity. Stronger material such as massive bedrock can maintain a steeper slope. If the slope angle is increased, the rock and debris will adjust by moving downslope.

© Djordje Zoric/Shutterstock.com.

Figure 6.3 **The angle of repose is the natural angle of a slope formed by a pile of material.**

A number of processes can make a slope too steep and cause it to become unstable. **Undercutting** by stream and wave action is a common process that removes the base of a slope, increasing the slope angle, and increasing the driving force acting parallel to the slope. Waves pounding against the base of a cliff, especially during storms, often result in mass movements along the shores of oceans and lakes. Human activities also often create unstable slopes that become prime sites for mass movement. Mass movement occurs when grading and cutting into the slope too steeply increases the downhill forces and the slope is then no longer strong enough to maintain the steep angle. Such actions by humans is analogous to undercutting by streams and has the same result, explaining why so many mountain roads and building sites are plagued by frequent mass movements.

The Role of Vegetation

Plants protect regolith and soil against erosion, and contribute to the stability of slopes (Figure 6.4). Vegetation helps absorb rainfall and leads to a decrease in water saturation of the slope material that would otherwise lead to a loss of shear strength. The root systems of plants also help stabilize the slope by binding soil particles together and holding the soil to bedrock. Where vegetation is lacking, mass movement is enhanced. The removal of vegetation by natural events (fires, drought) or human processes (timber, farming, or development) frequently results in unconsolidated materials moving downslope.

© Paul Marcus/Shutterstock.com.

Figure 6.4 **Vegetation on slopes is effective in holding the soil and regolith in place as root systems help stabilize the material. Whenever grasses and trees are removed, soil and regolith can readily move downslope.**

The Role of Overloading

Overloading occurs when additional weight is added to a slope. As mentioned earlier, water can become a driving force by increasing the weight of slope material after prolonged periods of rain. However, overloading is most often the result of human activity, and typically results from the dumping, filling, or piling up of material on a slope (Figure 6.5). The additional weight created by the overloading increases the water pressure within the material, which in turn decreases the stability of the slope. If enough material is added to the slope, it will eventually fail, sometimes with tragic consequences. Water can become a driving force by increasing the weight of slope material after prolonged periods of rain by completely filling pore spaces.

The Role of Geologic Structures

Plate tectonic forces can reorient rocks after their formation, causing them to be rotated and tilted. If the rocks underlying a slope dip in the same direction as the slope, mass movement is more likely to occur than if the rocks are horizontal or dip in the opposite direction of the slope. When rocks dip in the same direction as the slope, water can percolate along the bedding planes and decrease the cohesiveness and friction between adjacent rock layers (**Figure 6.6**). This is particularly important when clay layers are present, because they become slippery when wet.

Even if the rocks are horizontal or dip into the opposite direction of the slope, other structural weaknesses (such as joints, foliations, and faults) may dip in the same direction as the slope. The water percolating through the rock structures weathers and expands the openings further until the weight of the overlying rock causes it to fail.

The Role of Human Activity

There are many ways in which human activities can affect slope stability and promote landslides. One is by the clearing of stabilizing vegetation during logging operations, exposing sloping soil to rain. Many types of construction projects can lead to over-steepening of slopes. Highway roadcuts, quarrying, or open-pit mining, and construction of homes on benches cut into hillsides are among the most common activities that can cause problems (**Figure 6.7**). Where dipping layers of rock are present, removal of material at the lower end of the layers may leave large masses of rock unsupported.

In the 1870s throughout Europe there was a large demand for slate, which was used to make blackboards. To meet this demand, slate quarry miners near Elm, Switzerland, dug slate quarry at the base of a steep cliff containing excellent planar foliation that dipped toward the quarry. By September 1881, the quarry was 180 m long and 60 m into the hill below the cliff, and a fissure above the cliff had opened to 30 m wide. Falling rocks were frequent in the quarry and almost continuous noises were heard coming from the overhang above the quarry. On September 11, 1881, a 10 million cubic meter mass of rock above the quarry suddenly fell to the quarry floor, and produced an avalanche moving at 180 km/hr that traveled over 2 km, burying the village of Elm and killing 115 people.

© Harvey Hessler/Shutterstock.com.

Figure 6.5 Unstable slopes are created when tailings and other material from mining operations are piled up. These slopes are unable to develop soil layers that can support vegetation, so in addition to being unsightly they are likely to move if a triggering mechanism occurs nearby.

© Maksimilian/Shutterstock.com.

Figure 6.6 Bedding planes dipping in slope direction lead to slope failure.

© Baxtar/Shutterstock.com.

Figure 6.7 Terraces are cut in an open pit mine in order to help stabilize the steep walls. Notice the trucks on the roadway.

© Alexander Chaikin/Shutterstock.com.

Figure 6.8 This home in Valparaiso, Chile, is situated on solid rock. However, earthquakes are a common occurrence in Chile, and this home could easily collapse down the hill.

Slopes cut into unconsolidated material at angles higher than the angle of repose without planting stabilizing vegetation becomes extremely unstable. The very act of building a house above a naturally unstable or artificially steepened slope adds weight to the slope, thereby increasing the stress acting parallel to the slope. Other activities connected with the presence of housing developments on hillsides include watering the lawn, use of septic tanks, or an in-ground swimming pool from which water can seep slowly. These are all activities that increase the water content of the soil and can render the slope more susceptible to movement. Even homes built on apparently solid rock could be prone to destruction (Figure 6.8).

Common Triggers of Slope Failure

The factors discussed thus far all contribute to slope instability and are considered causes that make the slope vulnerable to failure. A sudden triggering event may then happen that initiates movement on the unstable hillside. Movement could eventually occur without a trigger if slope conditions became more unstable over time. However, sudden, rapidly moving landslides usually involve a triggering mechanism. Water, volcanic activity, and seismic activity are three common landslide trigger mechanisms that can occur either singly or in combination (refer to Table 6.1a–b).

Water as a Trigger

Rainfall. When a slope becomes saturated with water, the increased mass of the water, and its lubricating effects, create instability. Intense or prolonged rainstorms act as a trigger of the mass movement. Porous soil and layers of underlying rock allow the pore pressure to increase, thereby decreasing the adhesion of the grains. Failure results. Heavy rainfall occurs during tropical storms and hurricanes, short-lived, severe thunderstorms, and prolonged rainfall related to slow-moving weather systems. Flooding and landslides are commonly the result (Figure 6.9).

© PHOTO999/Shutterstock.com.

Figure 6.9 These houses were destroyed by landslides associated with Typhoon Morakat, which hit Taiwan in August 2009.

In early December 2020, areas along the southeast coast of Alaska received massive rainfall in a short period of time. During the event, dubbed a 200-year rainfall, more than 10 inches of rain fell, generating a large landslide in the community of Haines. Several dozen homes were destroyed or severely damaged. Tragically, two people were killed.

An increase in landslides and mudslides has occurred throughout the Western United States as the result of extensive wildfires. These fires have removed all vegetation that once adsorbed or impeded the flow of rainwater across the surface, particularly on hillsides. Heavy rains in late January 2021 caused a large landslide to erode 50 meters (165 ft) of California Highway

1, the primary coastal route in the state. This resulted in a 23-mile (37 km) closure that was estimated to last several months while the roadway was rebuilt. A larger section of the same road was closed for more than a year in 2017, due to an even larger event. Repairs are very costly and often only provide temporary solutions.

Snowmelt. In regions that experience prolonged cold weather that allows snow packs to increase in size, seasonal melting will produce large amounts of water in relatively short time periods. Melting can be accelerated by short periods of unseasonably warm weather. In addition to direct runoff of melt waters, water can percolate into the subsurface and saturate the soil above the zone of permanently frozen ground (permafrost). If there is any slope to the saturated layer, the fluids and solid material will rapidly move downhill as increased pore pressure reduces the shear strength of the soil. Depending on the amount of water present and the angle of the slope, rapid or slow mass movement can occur.

River and Wave Undercutting. Erosional processes caused by moving water in streams, or wave action along a shoreline, can remove material at the base of a slope. If sufficient removal occurs, the slope is no longer supported and slope failure occurs. A large-scale example of this is seen at Niagara Falls along the border of the United States and Canada (Figure 6.10). Water cascading over the falls erodes the relatively soft underlying shale layers. The upper rocks are no longer supported and collapse. Erosion there is occurring at the rate of about 1 meter per year, causing the falls to retreat upstream. The falls are retreating about 1 meter per year.

Source: David M. Best.

Figure 6.10 Undercutting occurs as water flows over Niagara Falls. The softer, underlying rock (shale) is eroded, thereby removing support for the more resistant overlying dolomite, which subsequently falls into the area below.

Volcanic Activity as a Trigger

Some of the largest and most destructive landslides known have been associated with large volcanic eruptions. These occur either in association with a violent eruption of the volcano itself, or as a result of mobilization of its weak deposits that are formed as a consequence of volcanic activity. Volcanic eruptions produce seismic shaking similar to high-energy explosions and earthquakes, and can trigger flank collapses that create fast-moving debris avalanches. Hot volcanic lava or ash eruptions can melt snow and ice on a volcano summit at a rapid rate, causing a flow of rock, soil, ash, and water that accelerates rapidly on the steep slopes of volcanoes. These volcanic debris flows (known as *lahars*) can reach great distances, once they leave the flanks of the volcano, destroying anything in its path.

The 1980 eruption of Mount St. Helens triggered a massive debris avalanche on the north flank of the volcano (Figure 6.11). The 1985 Nevado del Ruiz eruption in Colombia created pyroclastic flows that melted ice and snow at the summit, forming lahars up to 50 meters thick. These massive flows traveled more than 100 kilometers down several river valleys, destroying many houses and towns. The town of Armero was completely covered by the lahar, killing approximately

United States Geological Survey.

Figure 6.11 A massive debris avalanche in the Coldwater Lake area near Mount St. Helens, Washington.

23,000 people, making it the second-deadliest volcanic disaster in the 20th century (after the 1902 eruption of Mount Pelée, which killed 29,000 people).

Seismic Activity as a Trigger

Earthquake activity can cause potentially unstable slopes to fail as seismic energy passing through rock shakes the material. Consolidated, solid rock, such as granite is not affected by the energy passing through it. However, loose, unconsolidated material or mud can be turned into a liquid if sufficient water is present. A combination of vertical and horizontal stresses creates failure in slopes when the earthquake trigger occurs. In regions where water is common, pore pressure is increased by the seismic waves, and liquefaction occurs.

Figure 6.12 **The ground in Anchorage, Alaska, was affected greatly by the M 9.2 earthquake of March 1964. Up to 3 m of subsidence occurred on the left side of the photo.**

Most of the monetary losses due to the 1964 Great Alaska Earthquake were caused by widespread slope failures and other ground movement (Figure 6.12). In other areas of the United States, such as California, Oregon, and the Puget Sound region of Washington, slides, lateral spreading, and other types of ground failure due to moderate to large earthquakes have been experienced in recent years. Worldwide, landslides caused by earthquakes kill more people and damage more structures than in the United States.

Classification and Types of Mass Movements

Often, the term *landslide* is used in the general sense to describe many types of slow to rapidly descending masses of material down a slope. However, the term does not tell us anything about the processes involved in the movement. Most geologists prefer the term *mass movement,* because slope movements can occur on the ocean floor and not just on land. Also, many slope movements occur by mechanisms such as falling, flowing, and creeping rather than by sliding and more descriptive terms such as *rockfall, debris flow,* and *rockslide* are used. Geologists and civil engineers distinguish different types of mass movement based on three characteristics:

- **Type of Material.** The type of material involved in a mass movement depends upon whether the descending mass began as unconsolidated material or as bedrock. If regolith dominates, terms such as *debris, mud,* or *earth* are used in the description. The term *earth* refers to material that is composed mainly of sand-sized or finer particles, and debris is composed of coarser fragments. If the massive bedrock breaks loose and moves downslope, the term *rock* is used as part of the description.
- **Type of Motion.** The type of motion describes the way material moves down the slope. The most common motions described are *fall, slide,* or *flow.* Falls move unimpeded through the air and land at the base of a slope. Slides move in contact with the underlying surface. Flows are plastic or liquid movements in which the mass breaks up during movement. Other movements include topples and lateral spreads.

BOX 6.1 Landslides in Washington State

Each year the Pacific Northwest is subjected to a large amount of moisture that comes off the Gulf of Alaska and the Pacific Ocean. The result of these weather systems is the formation of numerous landslides, usually located at road cuts and naturally occurring slopes along highways. Early in the morning of October 11, 2009, a massive landslide cascaded down a hill in southwest Washington, about 35 km northwest of Yakima. Covering approximately 80 acres, the slide moved more than one million cubic meters of material, damming the Naches River. State Highway 410 was covered in debris that was more than 12 meters deep, cutting off more than 600 people from nearby towns.

Less than 5 years later, a much larger landslide occurred on March 22, 2014. After almost 6 weeks of heavy rainfall (about 200% of normal), a hillside gave way near the small community of Oso, 80 km northeast of Seattle. The slide was 460 m long, 1,300 m wide and ranged in depth from 7 to 21 m. The Oso mudslide, which killed 43 people, ranks as the deadliest mass movement event in the United States, excluding those associated with earthquakes or volcanic eruptions.

Source: Washington State Department of Natural Resources.

Figure 6.1.1 In October 2009, more than 1 million cubic meters of material slid down this hillside, destroying about 800 m of highway and temporarily damming the Naches River.

© Real Windows Creative/Shutterstock.com.

Figure 6.1.2 The scar of the Oso landslide shows the large amount of material that destroyed the forest and blocked the North Fork of the Stillquamish River.

- **Rate of Movement.** The movement of slope material occurs at a wide range of speeds. Some mass movement occurs so slowly that it may take years before the movement is noticeable from the downslope displacement of trees, fences, and walls. Other mass movements occur very rapidly and reach speeds in excess of 200 kilometers per hour. Water also affects the rate of movement. Commonly, the higher the water content, the faster the rate of movement.

Landslide Hazards

Since the hazards associated with landslides are highly variable and depend on the material involved and mechanism of movement, the common types

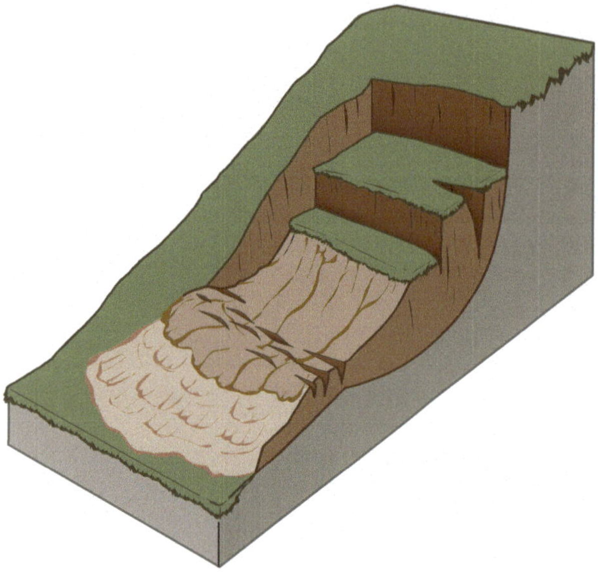

Source: United States Geological Survey.

Figure 6.13a A rotational slide has a curved plane along its base, giving the sense of rotation of the material.

Source: NOAA.

Figure 6.13b Slumping along the Pacific Palisades of Southern California.

of landslides are discussed below, along with their associated hazards. These definitions are based mainly on the terminology discussed in USGS Fact Sheet 2004-3072, which explains landslide types and processes. The type of movement describes the actual internal mechanics of how the landslide mass is displaced: fall, topple, slide, spread, or flow. Thus, landslides are generally described using two terms that refer respectively to material and movement; for example, rockfall, debris flow, rock avalanche, and earthflow.

Slides

slide

The movement of material along a curved or flat plane.

Slides refer to mass movements where there is a distinct zone or surface of weakness that separates the overlying slide material from more stable underlying material. The two major types of slides are rotational slides and translational slides, which are characterized by the surface of rupture.

Rotational Slide. This is a slide in which the surface of rupture is curved concavely upward and movement is roughly rotational about an axis that is parallel to the ground surface and perpendicular across the slide (**Figure 6.13a**). Rotational slides occur most frequently in homogeneous materials and are the most common landslides occurring in loose or unconsolidated material. They are usually associated with slopes ranging from 20 to 40 degrees, and travel extremely slowly (less than 0.3 meter every 5 years) to moderately fast (1.5 meters per month) to rapid. Most are triggered by intense and/or sustained rainfall or rapid snowmelt that can lead to the saturation of material resting on a slope.

slump

A mass movement in which generally unconsolidated material moves down a hillside along a curve, rotational subsurface plane.

A **slump** is an example of a small rotational landslide and commonly occurs in unconsolidated sediments and in some weaker rocks (**Figure 6.13b**). Slumping is commonly caused by erosion at the base of a slope, which removes support for the slope material. The erosion may be natural, such as cutting away of the base of a coastal cliff by waves, undermining of a river bank by stream flow, or the result of human activity, such as road construction. When slope failure occurs, the slump block rotates downward, a *scarp*

United States Geological Survey.

Figure 6.14 Severe slumping occurred in the Turnagain Heights region of Anchorage due to the March 27, 1964, earthquake. A combination of clay and water created unstable slopes that collapsed.

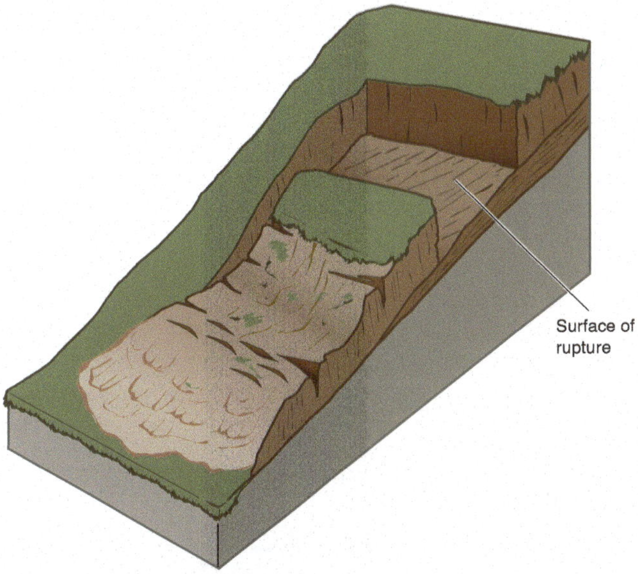

Source: United States Geological Survey.

Figure 6.15 A translational slide often has a solid block of rock sliding down a surface, which has been translated to a new, lower position on the hillside.

(cliff) is formed at the head of the slope and the toe moves outward over the slope below. The leading edge or toe provides the primary resisting force, so when it is upset by erosion or removal, the material generally moves downhill.

Slumping is an especially serious hazard where structures are built on cliffs above a shoreline when wave energy erodes the base of the cliffs, causing slope failure (especially during storms). In many cases, homes built along the cliffs are destroyed and the scarp progresses inland, posing a hazard for the next line of houses on the cliff.

The Great Alaska Earthquake of March 27, 1964 (M 9.2, see Chapter 4) caused serious land displacement in the city of Anchorage, which is mostly underlain by the Bootlegger Cove Clay. This former marine clay layer originally held saltwater in its pores that was later replaced by freshwater, making this area an unstable *quickclay.* The ground shaking that accompanied the earthquake caused the clay to liquefy and the ground fail, producing a series of slump scars. Some of the worst damage was in the subdivision of Turnagain Heights built on a flat-topped bluff some 30 m above the level of Cook Inlet (Figure 6.14). Slumping began in the overlying glacial deposits as soon as the underlying clay lost its cohesion. Within minutes, the flat bluff area was changed into a mass of rotated slump blocks, covered with twisted trees and destroyed homes.

Translational Slide. In this type of slide, the movement occurs along a roughly planar surface with little rotation or backward tilting (Figure 6.15). A *block slide* is a translational slide in which the moving mass consists of a single unit or a few closely-related units that move downslope as a relatively coherent mass. Often called *rockslides,* the sliding surface is commonly a bedding plane, but rockslides can develop on other surfaces such as fractures that cut across layered rocks. One of the most common types of landslides, translational slides are found throughout the world. They generally occur close to the surface, and can range from small (residential-lot size) failures to very large, regional landslides that are kilometers wide. Movement may initially be slow (1.5 meters per month) but many are moderate in velocity

Source: David M. Best.

Source: United States Geological Survey.

Figure 6.16 Material at the base of the cliff forms a talus slope. Weathered rock falls downhill and builds up a slope which is usually unstable.

Figure 6.17 A rockfall is often a free-fall of rock to the surface, where it can break into smaller pieces or roll along the surface.

(1.5 meters per day) and some extremely rapid. With increased velocity, the landslide mass of translational failures may disintegrate and develop into a debris flow. They are triggered primarily by intense rainfall, rise in ground water within the slide due to rainfall, snowmelt, flooding, or other inundation of water resulting from irrigation, or leakage from pipes or human-related disturbances, such as undercutting. These types of landslides can also be earthquake-induced. Occasionally the translating block will ride on a cushion of air, which greatly reduces the frictional forces, thereby causing material to move at high speeds. These situations are preceded downslope by an air blast that often flattens trees.

Falls

Falls are abrupt movements of masses of geologic materials, such as rocks and boulders, that become detached from steep slopes or cliffs. Rocks can separate along potential planes of weakness such as joints, fractures, or bedding planes. When material falls under the force of gravity, it comes to rest and accumulates at the base of the hillside or slope. The material at the base is termed **talus** (Figure 6.16). If the buildup is significant, the result is a talus slope.

talus

The natural pile of material that builds up at the base of a cliff or hillside by material falling from above.

Rockfalls. Rockfalls are abrupt, downward movements of rock, earth, or both, that detach from steep slopes or cliffs (Figure 6.17). The falling material usually strikes the lower slope at angles less than the angle of fall, causing bouncing. The falling mass may break on impact, or may begin rolling on steeper slopes, and may continue until the terrain flattens. They are common worldwide on steep or vertical slopes, in coastal areas, and along rocky banks of rivers and streams. The volume of material in a fall can vary substantially, from individual rocks or clumps of soil to massive blocks thousands of cubic meters in size. They are rapid to extremely rapid, free-fall, bouncing and rolling of detached soil, rock, and boulders. The rolling velocity depends on slope steepness. Rockfalls are commonly triggered by undercutting of slope by natural processes such as streams and rivers or differential weathering (such as the freeze/thaw cycle), human activities such as excavation during road building and (or) maintenance, and earthquake shaking or other intense vibration.

Topples. Toppling failures are distinguished by the forward rotation of a unit or units about some pivotal point below or low in the unit (Figure 6.18). Toppling is driven by gravity exerted on the weight of the displaced mass. Toppling can be caused by water or ice expanding in cracks or voids in the mass. Topples can consist of rock, debris (coarse material), or earth materials (fine-grained material). They are known to occur globally, often prevalent in columnar-jointed volcanic terrain, as well as along stream and river courses where the banks are steep. Velocities can range from extremely slow to extremely rapid, sometimes accelerating throughout the movement, depending on distance of travel.

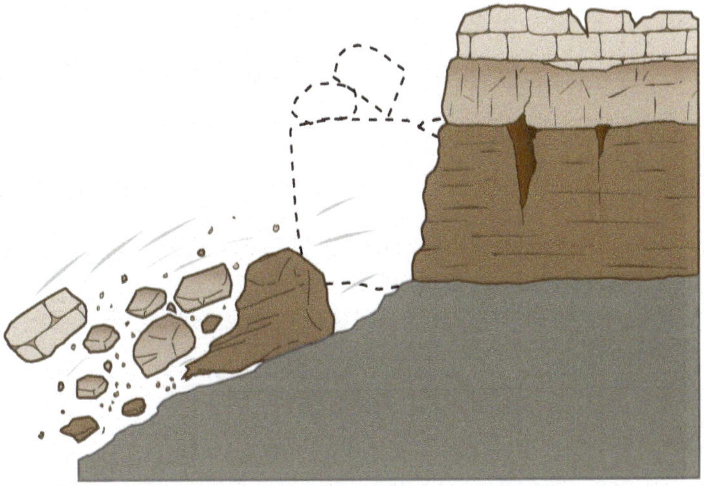

Source: United States Geological Survey.

Figure 6.18 Topples result from rocks rotating about a pivot point at the base of a cliff and the rocks basically tilt over. The dashed blocks designate the position of the rocks prior to collapsing.

Flows

Flows are mass movements in which the material behaves like a viscous fluid. In many cases, mass movements that start as falls, slides, or slumps are transformed into flows farther downslope. Flows have broad characteristics and encompass wettest to driest, and fastest to slowest types of mass movement.

Debris Flow. A debris flow is a form of rapid mass movement in which a combination of loose soil, rock, organic matter, air, and water mobilize as a slurry that flows downslope (Figure 6.19). Debris flows typically consist of less than 50% fine grained material that has been carried along by extreme fluid flow generated by heavy rainfall or snowmelt. Debris flows are

Source: United States Geological Survey.

Figure 6.19 A debris flow is characterized by a low viscosity combination of soil, rock, and water that easily spreads out across the surface.

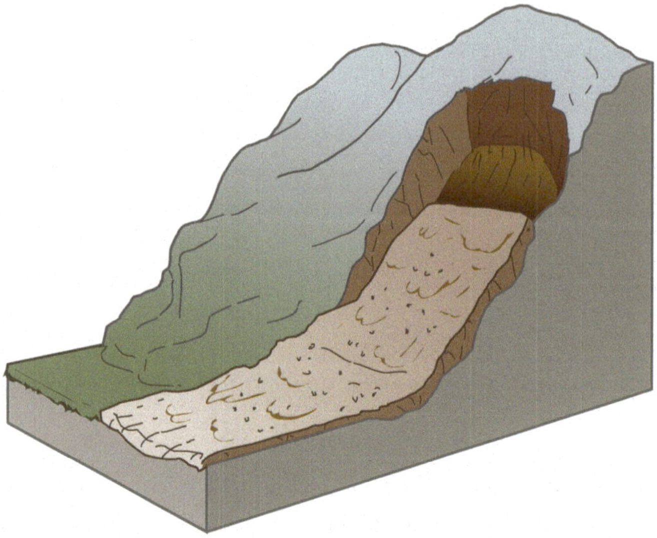

Source: United States Geological Survey.

Figure 6.20 **A debris avalanche consists of a viscous mixture of rock, soil debris, and a small amount of fluid that travel a short distance before becoming its own impedance.**

Source: David M. Best.

Figure 6.21 **Rock avalanches consist of large pieces of angular rock that have been broken off and moved downslope.**

common in areas that have experienced wildfires, and are also associated with volcanic deposits.

These types of flows can exhibit a range of viscosities, being thin and watery or thick with sediment and debris, and are usually confined to the dimensions of the steep gullies that facilitate their downward movement. Movement across the surface is relatively shallow, and the runout is both long and narrow, sometimes extending for kilometers in steep terrain. The debris and mud usually terminate at the base of the slopes and create fanlike, triangular deposits called debris fans, which may also be unstable. They can be rapid to extremely rapid, depending on consistency and slope angle.

Debris and Rock Avalanche. Debris avalanches are large, extremely rapid flows formed when an unstable slope collapses and the resulting fragmented debris is rapidly transported away from the slope (Figure 6.20). In some cases, snow and ice will contribute to the movement if sufficient water is present, and the flow may become a debris flow and/or a lahar if volcanic conditions exist. They occur worldwide in steep terrain environments and on very steep volcanoes, where they may follow drainage courses. Rock avalanches can occur whenever shear rock walls collapse and can travel several kilometers from their source. They are rapid to extremely rapid and can move up to 100 meters per second (Figure 6.21).

Figure 6.22 **Earthflows are relatively fluid masses that include a mixture of soils and weathered bedrock.**

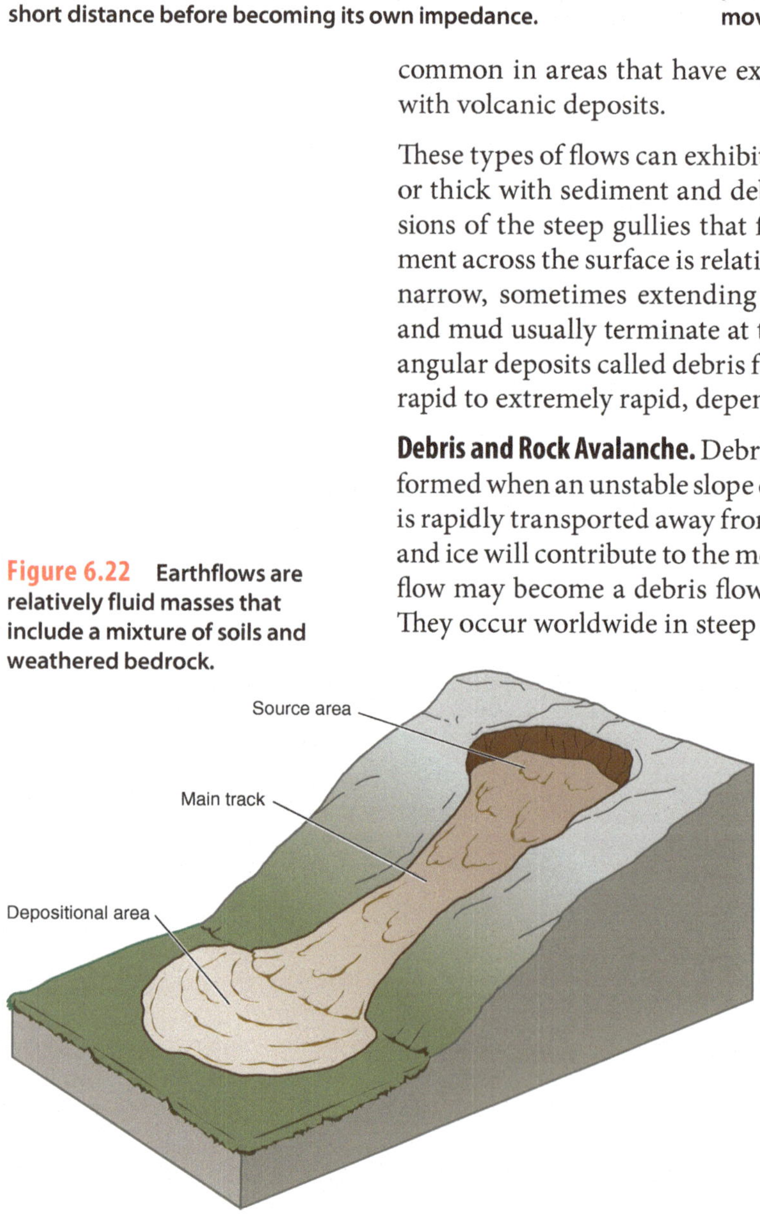

Source: United States Geological Survey.

Earthflow. Spring and early summer are prime times for earthflows, when the upper layers of the regolith trap water. This reduces friction and grain cohesion, which allows the mass to make a slow downhill movement. Earthflows can occur on gentle to moderate slopes, generally in fine-grained soil, commonly clay or silt, but also in very weathered, clay-bearing bedrock (Figure 6.22). The mass in an earthflow moves as a plastic or viscous flow with strong internal deformation. Slides or lateral spreads may also evolve downslope into earthflows. Earthflows range from very slow (creep) to rapid and catastrophic. A *mudflow* is an earthflow consisting of material that is wet enough to flow rapidly and that contains at least 50% sand, silt, and clay-sized particles. These are often caused by earthquakes, volcanic eruptions and excessive rainfall. Newspaper reports often refer to mudflows and debris flows as "mudslides."

The most disastrous mudslides in recent history have been related to either volcanic eruptions or torrential rainfalls (Table 6.2). Four of the five deadliest events were caused by massive rainfall; only the 1985 Armero event was the result of the eruption of Nevado del Ruiz volcano in Colombia.

Creep. Creep is the imperceptibly slow, steady, downward movement of slope-forming soil or rock. Evidence of creep is seen in curved tree trunks, tilted walls or fence posts, or small ridges or ripples on the surface of the soil. The driving forces acting on the slope are sufficient enough to permanently deform the material but not large enough to cause shear failure. There are three types of creep:

1. Seasonal, in which changes in soil moisture and temperature allow movement by expansion of the slope surface (Figure 6.23),
2. Continuous, where the driving force always exceeds the strength of the material (resisting force), and
3. Progressive, where slopes are reaching the point of failure as other types of mass movements affect the slope.

Table 6.2 Five Deadliest Mudslide Disasters

Rank	Mudslide Name	Location	Estimated Fatalities
1	1999 Vargas Tragedy	Vargas, Venezuela	30,000
2	1985 Armero Tragedy	Tolima, Colombia	20,000
3	2013 India Monsoons	Uttarakhand, India	5,700
4	2010 Gansu Mudslide	Zhouqu County, China	1,287
5	2006 Leyte Mudslide	Leyte, Philippines	1,126

Source: World Atlas

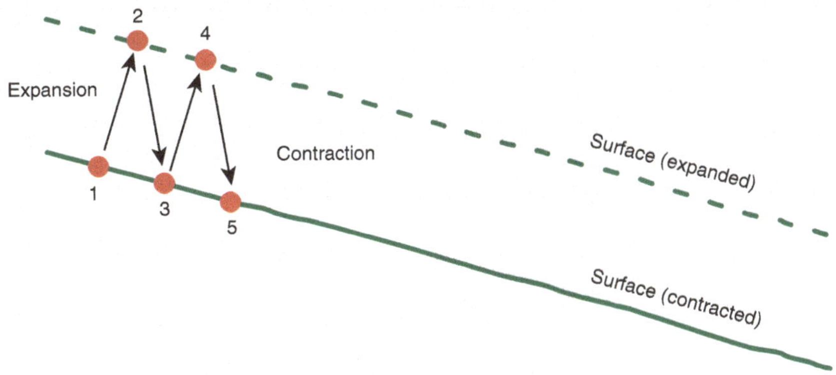

© Kendall Hunt Publishing Company.

Figure 6.23 Creep occurs when a surface is expanded by the addition of moisture or frost action. Dessication or thawing causes the expanded surface to fall back under the force of gravity, with the result that a particle beginning at point 1 is repositioned at point 2 and so forth.

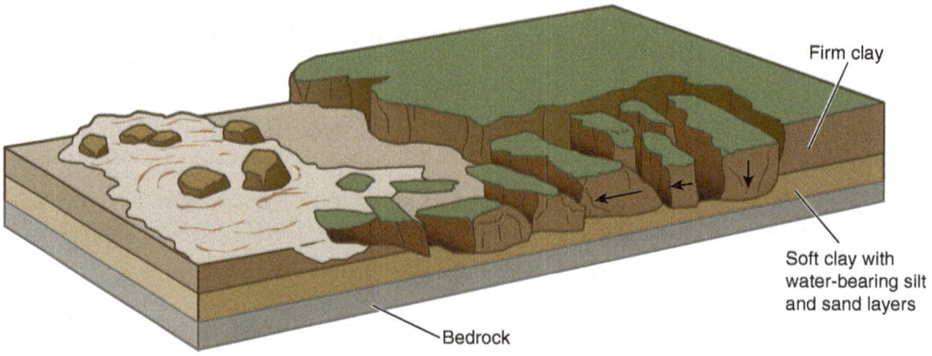

Figure 6.24 Lateral spreads result from the horizontal movement of a relatively firm layer of material moving over a deeper layer of water-rich silt and sand. Very little slope is required for this to occur.

Creep is widespread around the world and is probably the most common type of mass movement, often preceding more rapid and damaging types of landslides. Creep can occur over large regions (tens of square kilometers) or simply be confined to small areas. It is difficult to define the boundaries of creep since the event itself is so slow, and surface features representing perceptible deformation may be lacking. The rates of movement range from very slow to extremely slow, usually less than 1 meter per decade. Wherever seasonal creep is occurring, rainfall and snowmelt are the typical triggers, whereas other types of creep could have numerous causes, including chemical or physical weathering, leaking pipes, poor drainage, or destabilizing types of construction.

Lateral Spreads

Lateral spreads are easily identified because they usually occur on very gentle slopes or flat terrain (**Figure 6.24**). Movement is caused by horizontal tension that pulls the surface material and produces shear or tensile fractures. **Liquefaction** is the primary cause of lateral spreads, developing when loose sediments such as sand and silt become saturated. Water holds the particles in suspension so any movement will cause the liquid and suspended solids to move rapidly across the surface. Usually two layers are present; the overlying one is a more cohesive layer lying above material that can undergo liquefaction. If liquefaction occurs, the upper layer then responds by subsiding, rotating, or disintegrating, and begins to either flow or liquefy. The initial failure occurs in a small area that spreads rapidly as liquefaction intensifies.

liquefaction

A condition that exists when an overabundance of liquid, usually water, is present.

Lateral spreads are known to occur where there are liquefiable soils. The area affected may start small in size and have a few cracks that spread quickly, affecting areas hundreds of meters in width. Rate of movement may be slow to moderate; however, if the event is triggered by an earthquake or volcanic eruption, movement will be very rapid. The ground may then slowly spread from a few millimeters per day to tens of meters per day. Other causes could be:

- Natural or anthropogenic overloading of the ground above an unstable slope;
- Saturation of underlying weaker layer due to precipitation;
- Snowmelt, and (or) ground-water changes;

- Liquefaction of underlying sensitive marine clay;
- Disturbance at the base of a river bank or slope; and
- Plastic deformation of unstable material at depth such as thick deposits of buried salt.

Submarine and Subaqueous Landslides

Submarine and subaqueous landslides include rotational and translational landslides, debris flows and mudflows, and sand and silt liquefaction flows that occur underwater in coastal and offshore marine areas or in lakes and reservoirs. The failure of underwater slopes can result from overloading by rapid sedimentation, release of methane gas in sediments, storm waves, current scour, or earthquake seismic shaking. Subaqueous landslides pose problems for offshore and river engineering, jetties, piers, levees, offshore platforms and facilities, and pipelines and telecommunications cables, as well as producing tsunami hazards.

United States Geological Survey.

Figure 6.25 The harbor at Kodiak, Alaska, was totally overrun by a tsunami that was generated by the Great Alaskan Earthquake of March 27, 1964. The city is located near the major fault that moved during the event.

On November 18, 1929, a magnitude 7.2 earthquake occurred about 250 km south of Newfoundland near the Great Bank. The entire region felt the event, with reports of ground motion as far away as New York and Montreal. The earthquake triggered a large submarine slump estimated at 200 cubic kilometers of material. At the time, major undersea transatlantic cables stretched from eastern Canada to Great Britain. An analysis of numerous cable breakages was used to determine that the slump moved across the ocean floor at 95 kilometers per hour as it rolled down the continental slope. The event also produced a tsunami that killed 28 people along the Canadian coast.

The much larger 1964 Alaska earthquake (M 9.2) resulted in almost instantaneous catastrophic failure of the steep submerged shore in the harbor of Kodiak, Alaska (Figure 6.25). The submarine slide retrogressed beyond the shoreline, submerging areas of coastal land and harbor facilities, and almost 75 million cubic meters of land of Valdez harbor disappeared into the sea.

Lessons From the Geologic Record

Slopes have always existed on Earth, and are constantly changing as a result of plate tectonic and erosional processes. Since mass movement is an erosional process that removes surface material, ancient landslide deposits are not as frequently found in the ancient rock record. However, the world's largest landslides discovered are prehistoric and are important to geologists in recognizing danger zones for future development. The following are examples of some of the largest mass movements recorded in the geologic record. Landslides of this magnitude have never been witnessed in historic times, and if they were to occur in today's populated areas, the destruction would be devastating.

United States Geological Survey. Photo by Harry Glicken.

Figure 6.26 Hummocky hills in the foreground of Mount Shasta are thought to have formed from a massive landslide that occurred about 300,000 years ago.

Heart Mountain Detachment

Heart Mountain is located in northwestern Wyoming and represents the largest known rockslide on Earth. It covers about 1,300 sq km of surface area and has a thickness of several kilometers. This volume of rock rests on a gentle slope of 1° to 2°. The leading edge of the land mass moved a minimum of 40 km while the entire mass covered more than 3,400 sq km, slightly larger than the state of Rhode Island. Erosion has removed most of the oldest part of the slide sheet, which completed its move about 48 million years ago. How could such a massive amount of material move? Although many hypothetical models have been put forward, geologists who have studied the mountain and surrounding area in detail concluded that volcanism held the key to the source of motion. An eruption possibly shook loose the rock which then began its downslope movement. Frictional heating of underlying limestone generated free CO_2 along with water vapor that served to reduce friction along the major fault plane.

Mount Shasta

One of the largest landslides triggered by volcanic activity is a debris avalanche formed about 300,000 years ago in northern California. The most likely cause of the hummocky terrane was a major eruption of 4,317-m-high Mount Shasta in the Cascade Range of northern California (**Figure 6.26**). The debris avalanche today extends 43 km westward from the base of the volcano and has an estimated volume of 26 km^3. This huge deposit and its moon-like surface was not recognized as a landslide until after the 1980 eruption of Mount St. Helens, which resulted in a similar, but smaller, disruption of Earth's surface.

Examples From the Recent Past

Landslides can cause flooding by forming landslide dams that block valleys and stream channels, allowing large amounts of water to back up. This causes backwater flooding and, if the dam fails, subsequent downstream flooding. Also, solid landslide debris can upset any stream equilibrium by adding volume and density to otherwise normal stream flow. New debris in a channel can produce blockages and diversions, creating flood conditions or localized erosion. Landslides can also cause overtopping of reservoirs and/or reduced capacity of reservoirs to store water.

Usoi Landslide Dam and Lake Sarez, Pamir Mountains, Tajikistan

The world's largest documented landslide occurred in February 1911 near the Pamir Mountains of southeastern Tajikistan, the land-locked country lying to the north of Afghanistan and west of China. A rock slide and avalanche resulted from a magnitude 7.4 earthquake in the region which created a dam that is 567 m high. This is the highest dam, either natural or man-made, on Earth. Material dammed the Murgab River, producing Sarez Lake, a body of water 75 km long that averages 200 m in depth. The structure remains relatively stable, but some seepage out of the dam in the form of several outlet springs allows water to continue flowing downstream through the Murgab River. Although the system appears to be in equilibrium, the entire structure could fail suddenly in the event of a major earthquake. This would put the more than five million people living downstream at risk.

Gros Ventre, Wyoming

United States Geological Survey. Photo by Dickson Cunningham.

Figure 6.27 **The Gros Ventre slide in June 1925 moved 40 million cubic meters of material that dammed the Gros Ventre River until it was breeched 2 years later by heavy rainfall and snow melt.**

One of the largest landslides to occur in recent times, the Gros Ventre (pronounced GROW vaunt) slide, took place on June 23, 1925, in the area east of Jackson Hole valley in Wyoming. The landslide was caused by a geologic setting in which permeable sandstone layers rested on top of impermeable shale layers. Following several weeks of heavy rainfall, the friction holding up the sandstone was reduced by the water and caused almost 40 million cubic meters of rock to slide down the north face of Sheep Mountain (Figure 6.27). The slide crossed the Gros Ventre River and moved up the opposing side of the valley. The landslide formed a dam over 60 m high and 400 m wide, creating Lower Slide Lake. This remained in place for only almost 2 years, when on May 18, 1927, part of the dam failed following heavy rains that melted snowpack in the region. Flood waters reached more than 5 m in depth and destroyed the town of Kelly, located 10 km downstream, killing several people. Water continued to travel for almost 30 more kilometers.

Nevados de Huascarán, Peru

Nevados de Huascarán (elevation 6,786 m) is the highest peak situated in the central Andes Mountains of Peru. The summit is covered with glacial ice throughout the year. Between 1962 and 1970 two catastrophic rock avalanches occurred, killing more than 22,000 people (Figure 6.28). In January 1962, a massive slab of glacial ice and rock fell without warning and without any apparent triggering mechanism. The resulting avalanche, consisting of 10 million cubic meters of material, rapidly covered the valley at the base of the mountain, killing 4,000 people in a town at the base. The larger city of Yungay was spared any disaster, as the flow was stopped by a hill on the valley floor.

This section of South America experiences many major earthquakes. On May 31, 1970, an M 7.7 earthquake occurred on the subduction zone about 135 km away from the Nevados de Huascarán (Figure 6.29). The duration of ground motion was 45 seconds, long enough to shake loose a large block of material from the Nevados de Huascarán between 5,400 and 6,600 m elevation. The ensuing rock and ice avalanche (estimated at a minimum of 50 million cubic meters) slid across the glacier at the mountain base, reaching velocities exceeding 300 km per hour. The avalanche gained speed as it dropped 4 km in elevation over a distance of 16 km.

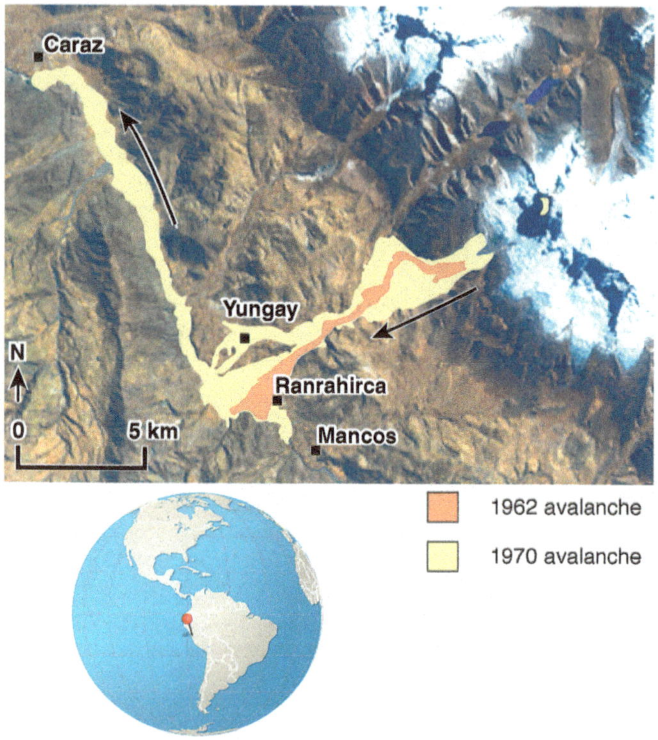

Figure 6.28 The 1970 avalanche was much farther reaching than that of 1962, although the two had a parallel course in the upper portions of the valley. Source: Data from the World Glacier Monitoring Service, Zurich, Switzerland, and based on figure by UNEP's DEWA/GRID-Europe, Geneva, Switzerland, and *Hugo Ahlenius, UNEP/GRID-Arendal.*

United States Geological Survey. Photo by George Pflaker.

Figure 6.29 Peru Earthquake May 31, 1970. Lower part of the Huascaran debris avalanche of May 31,1970. The combined Yungay and Ranrahirea debris lobes cover an area of about 8 kilometers and probably contain close to 50 million cubic meters of material.

Almost 18,000 residents of the city of Yungay died as the material covered the valley floor and moved up the opposite side of the valley, where an additional 600 people died. The city has been reoccupied but, given its tenuous location, it will undoubtedly be in the path of another catastrophic mass movement in the future.

Utah Geological Survey.

Figure 6.30 Heavy rains and a melting snow pack caused the most costly landslide in the history of the United States in spring 1983. The town of Thistle, Utah, was flooded by a lake that formed when the Spanish Fork River was dammed by the landslide.

Thistle, Utah

On April 14, 1983, a massive landslide in Utah County, Utah moved part of the mountain and blocked two creeks, forming an earthen dam (Figure 6.30). The Thistle landslide began moving in the spring in response to groundwater buildup from heavy rains the previous September and melting snowpack from the winter of 1983. Abnormally high precipitation in 1982–1984, related to a strong El Niño Southern Oscillation, caused thousands of landslides in mountain areas of the Western United States. Within a few weeks, the landslide dammed the Spanish Fork River, consequently obliterating U.S. Highway 6 and the main line of the Denver and Rio Grande Western Railroad.

The town of Thistle was inundated by the floodwaters rising behind the landslide dam. Eventually a drain system was engineered to drain the lake and avert a potential disaster. The landslide reached a state of equilibrium across the valley, but fears of reactivation caused the railway to construct a tunnel through bedrock around the slide zone, at a cost of millions of dollars. The highway likewise was realigned around the landslide. When the lake was drained, residual muck partially buried the town, and virtually no one returned to Thistle. It is now a ghost town. Total costs (direct and indirect) incurred by this landslide exceeded $400 million, making this one of the most costly landslide events in American history.

Madison County, Virginia

The Appalachian Mountains of eastern North America have experienced major rainfall events over the past several decades. Some of the storms have been related to tropical cyclones, such as Tropical Storm Agnes in 1972, which affected West Virginia and Pennsylvania (see Chapter 11). Other storms such as a very intense rain storm in June 1995 are isolated events in time. However, both types of storms generate large amounts of rain that produce landslides.

Madison County, Virginia, located in the north central part of the state, received 30 inches of rain in a 16-hour period in late June 1995. Although local hillsides are covered in heavy vegetation, the amount of rain was sufficient to saturate the ground and produce hundreds of debris flows that devastated the region. This event was only the most recent of several that have hit the area. In 1969 the remnants of Hurricane Camille (a tropical storm when it struck Virginia) saturated an area about 150 km south of Madison County with the result being 150 people killed and more than $100 million (in 1969 dollars) damage. In November 1985 3 days of storms struck Virginia and West Virginia, producing flooding and debris flows that killed 70 people and generated $1.3 billion in property damages.

Every year, numerous landslides occur worldwide and cause thousands of casualties and billions of dollars in monetary losses, making landslides a serious natural hazard. These annual costs are approximately equal to the damage caused by earthquakes over a 20-year interval. Yet losses are increasing in the United States and worldwide as development expands under pressures of increasing populations. Improper construction techniques lead to many of the problems. In most instances, losses are due to the public's lack of recognition or concern of the dangers. The resulting encroachment of developments into hazardous areas, expansion of transportation infrastructure, deforestation of landslide-prone areas, and changing climate patterns may lead to continually increasing losses.

Northern California Coastline

The coastal region of California north of the San Francisco Bay area experiences landslides every year. Shallow landslides, those less than 3 to 5 meters in depth, are usually caused by rains that saturate the ground. Debris flows result as rapid movements of soil, trees, and boulders destroy property, causing injuries and even deaths. Intense rainfall produced by early winter storms coming off the Pacific Ocean is the trigger for these events.

Source: United States Geological Survey.

Figure 6.31a A major landslide along California Highway 1 disrupted traffic for months.

Source: United States Geological Survey.

Figure 6.31b Heavy rainfall on a burned area in northern California eroded California highway 1 in January 2021.

Deep-seated landslides are those ranging in depths greater than 3 to 5 meters. Generated by prolonged rainfall, these commonly occur near the end of the winter season. The slides move more slowly than shallow landslides but can produce more widespread damage across the affected areas.

In recent years portions of California have been affected by large wildfires. These events have removed ground vegetation that impeded surface water flow, thereby causing extensive debris flows and erosion in steep terrains. One area that has been particularly hard hit is along California Highway 1, the Pacific Coast Highway. This route is a popular scenic one along the ocean, as well as a means of connecting numerous coastal communities.

The Mud Creek landslide, which occurred on May 20, 2017, was a massive event located 230 km south of San Francisco (Figure 6.31a). Almost 2 million cubic meters of material buried California Highway 1 to depths exceeding 20 meters. More than 50,000 square meters (13 acres) of new land was added to the coast.

Less than 4 years later, in late January 2021, another slide closed the highway about 30 km south of Big Sur. Over a 2-day period, more than 20 cm of rain fell around Rat Creek. Most of the watershed was in an area devastated by the Dolan Fire which occurred in the fall of 2020. Debris clogged drainage pipes, sending water and material over California Highway 1. This caused erosion on the cliff below, producing a 60-meter-wide gap in the roadway (Figure 6.31b).

Landslide Prediction and Mitigation

The potential for landslides to occur in an area can be assessed by examining the geology and geologic setting to determine if conditions could result in some type of mass movement. Several factors need to be considered including the actual location of an area and what types of human influence might add to the instability of the landscape. Past landslide history must be determined to develop a possible set of recurrence conditions.

Mass movements are inevitable where the terrain has a slope and where weather and other conditions can contribute to downhill movement. The effects of landslides can be reduced by preventing construction on or near hazardous zones. Governments can reduce the effects of landslides by establishing regulations and policies that minimize or reduce commercial and residential development. Governmental agencies that have control over the development of areas that could experience landslides should carefully monitor requests to use those areas. Prior to any construction, the services and knowledge of engineering geologists, civil engineers, and geotechnical professionals should be sought in order to properly assess the potential for future land failures.

The potential hazards of landslides can be mitigated if construction does not occur on steep slopes or on areas that have experienced landslides. Stabilization of slopes can be increased by constructing walls to reinforce the hillside (Figure 6.32). The removal of surface water from the top of a slide area increases stability, as the water cannot percolate into the subsurface and contribute to slope failure. Channels at the tops of slopes and drainage pipes leading from the lower portions of retaining walls direct water out of the soil and away from unstable areas.

Source: David M. Best.

Figure 6.32 **The retaining wall was built following the 1995 slope failure at La Conchita, California.**

Subsidence and Collapse: Vertical Mass Movement

Under the steady pull of gravity, unsupported material at the Earth's surface can subside or collapse due to subsurface vertical movement. Land subsidence is a gradual settling of the Earth's surface, while **collapse** is the sudden sinking into voids. Subsidence and collapse can be natural processes, but they are often caused by human actions. For hundreds of years, humans have removed large amounts of Earth's underground resources: groundwater, oil, coal, metal ores, and rock. Removal of these materials creates many potential problems underground. When fluids are removed (water, oil, or natural gas), the pores in which they are found are emptied. When solids are removed (coal, salt, or ores), large void spaces are created. In both of these cases, support is decreased for the overlying surface materials.

collapse
The sudden sinking of the surface into a subsurface void.

We continue to withdraw increasing volumes of these fluids and solids from the subsurface to meet the demands of our growing population and industrial economy. Subsidence and collapse are serious global hazards and growing more common due to land development expanding into areas that are more vulnerable. In the United States, an area roughly the size of New Hampshire and Vermont combined, has been directly affected. More than 80% of the identified subsidence and collapse in the United States is a consequence of our removal of groundwater resources.

The growing demand for groundwater in many communities lowers the water table and reduces pore pressure, thus causing more subsidence. In some areas, underground mines or caves have collapsed, causing damage to new developments on the surface. The increasing exploitation of land and water resources threatens to accelerate existing subsidence and collapse problems and initiate new ones.

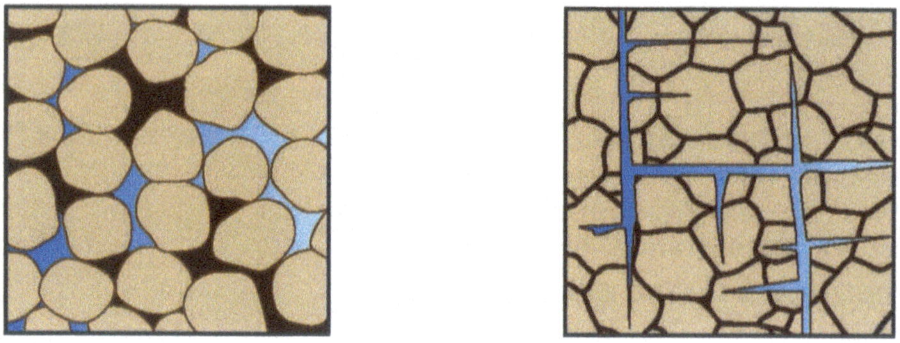

Figure 6.33 Pore space between minerals grains (left) or fractures (right) can be occupied by fluids.

© Kendall Hunt Publishing Company.

The Effect of Removal of Fluids From Under the Surface

In the subsurface, pore space exists between grains of soil, rock, and other buried materials. Void space also exists in fractures within the rock. These voids can collect fluids such as water, oil, or gas (Figure 6.33). The fluids exert a force (pore pressure) that helps keep the grains apart. Removal of the fluid reduces this pore pressure, and compaction and collapse can occur.

The single greatest cause of subsidence is the overpumping of groundwater. If the level of the water table in the subsurface is near the ground surface, collapse is relatively rapid and nonreversible. The largest user of groundwater is agriculture, especially in the arid regions of the Western and Southwestern United States.

Areas surrounding Las Vegas, Nevada, a large city located in a desert, have experienced subsidence as more people occupy a region that has little surface water. Water is drawn from both Lake Mead and from groundwater. Similar conditions have developed in southern Arizona as expanding agriculture has used increasing amounts of groundwater. Extensive fissuring and subsidence has lowered the landscape between Phoenix and Tucson and has caused large dessication cracks in the ground.

Two major agricultural valleys in California, the San Joaquin and the Santa Clara Valleys, have subsided as much as 15 m since widespread pumping of groundwater began in the early 1900s. The conditions only worsen as more people migrate to these areas.

Once the ground surface has sunk, the process cannot be reversed by pumping water into the subsurface.

Removal of oil and natural gas began in a field between Galveston and Houston, Texas, in 1917. For decades these fluids along with water were removed from the subsurface, causing the ground to subside as much as 2.5 m. The sinking of the land increased as several million people in the area began using groundwater as their water source. Readjustment of the land has reactivated old faults and created new ones in the region. Coastal seawater has moved landward into low-lying areas that dropped during the subsidence process, thereby causing pollution of the groundwater.

The Natural Removal of "Solid" Rock

Rock is generally considered to be solid—"hard as rock" is the saying. However, certain types of rock, especially sedimentary rock consisting of soft mineral grains or grains that can be chemically eroded, are prone to removal by natural means. Water can erode soft sediments, causing collapse of any overlying material. The **dissolution** or chemical solutioning of limestone produces a landscape that forms by the collapse of surface material into large subsurface voids.

dissolution
The natural dissolving of a substance by chemical reactions.

Limestone is a common sedimentary rock whose primary mineral is calcite ($CaCO_3$). The atmosphere naturally contains carbon dioxide (CO_2) which combines with rainwater or surface water and forms carbonic acid (H_2CO_3). The naturally occurring reaction between these two compounds, calcite and carbonic acid, is

$$\underset{\text{Calcite}}{CaCO_3} + \underset{\text{carbonic acid}}{H_2CO_3} \rightarrow \underset{\text{soluble compound}}{Ca^{++} + 2HCO_3^-}$$

Measuring the pH of a liquid determines its acidity. Normal rainwater has a pH of 5.6 (neutral pure water has a pH of 7). Therefore, falling rain has sufficient acidity to dissolve limestone. In addition, surface water from lakes and streams can flow through cracks in limestone and react with the rock below the surface.

The soluble compound in the equation above is now moved around by any excess water present and taken away from the source area. When the water leaves an area, such as by a period of drought or perhaps a drop in the water table level, the dissolved rock produces a void that forms a cave or cavern (Figure 6.34). This cave can expand in volume through time as more dissolving of the rock takes place. Eventually the roof of the cave will not have sufficient support to hold it in place and it will collapse. The ground surface now lies within a depression on the surface. The collapse of subsurface voids can have a negative effect on the landscape, particularly if it occurs in developed areas.

Subsurface mining of minerals such as halite (common salt, NaCl), and the excavation of coal, have produced numerous examples of ground failure through subsidence. Once the surface has collapsed the area is more susceptible to erosion.

About 25% of the conterminous United States is underlain by limestone and other dissolvable rocks, including salt. The landscape that forms from the processes described earlier, or by the dissolving of salt in water, are termed *karst topography*. The name karst comes from the Karst region along the coast of the Adriatic Sea along Croatia and Slovenia in southern Europe.

Source: David M. Best.

Figure 6.34 The solutioning of limestone can result in intricate cave formations. Natural chemical reactions have dissolved limestone and produced these subsurface voids.

In addition to the removal of dissolvable rock, excessive excavation of salts or extraction of groundwater obviously increase the likelihood of collapse. Such excavations must be done thoughtfully in order to minimize the possible failure of the surface.

In such areas it is important to map the rock units along with the flow of water, and identify fracture patterns in the rock. Once the potential hazard zones are recognized, these areas should be placed off-limits for future development. Continuous monitoring is necessary to forecast any sudden changes that might occur in the near-surface geologic conditions. The withdrawal of groundwater and other fluids should be closely controlled in order to minimize upsetting the balance between the forces of the fluids and those of the rocks. Numerous examples of significant land subsidence are present in the United States. One well-documented case is in the northern Santa Clara Valley, northwest of San Jose, California, thousands of acres of land have subsided as much as 2.5 m in a 50-year period. This was caused by overdrafting groundwater for agricultural and human uses. The northwest end of the valley has experienced the landward movement of seawater. Efforts have been made to slow the process but it is not fully reversible.

Key Terms

angle of repose *(page 168)*
cause *(page 165)*
collapse *(page 187)*
dissolution *(page 189)*
driving force *(page 164)*
Factor of Safety (FoS) *(page 168)*
landslide *(page 162)*
liquefaction *(page 180)*
mass movement (mass wasting) *(page 162)*
normal force *(page 164)*
regolith *(page 162)*
resisting force *(page 164)*
shear force *(page 164)*
sinkhole *(page 162)*
slide *(page 174)*
slope failure *(page 163)*
slump *(page 174)*
subsidence *(page 162)*
talus *(page 176)*
trigger *(page 165)*
undercutting *(page 168)*

Summary

Mass movements result when driving forces exceed resisting forces. The major driving force is gravity, assisted primarily by water. Resisting forces include friction and the shear strength of the material. Rock structures, such as bedding and joints, also play an important role in facilitating mass movement, especially when these features are inclined toward a valley or excavation. The removal of material on slopes, either by human activity or natural events such as wildfires, increases the likelihood of slope failure. The causes of mass movement can be natural (geological or morphological) or human-caused. Triggering events upset the equilibrium and lead to slope failure. Small-scale landslides usually affect single hills or slopes. Large landslides cause extensive damage and often permanently alter the landscape. Examples of such events can be generated by earthquakes and volcanic eruptions.

Mass movements are mainly characterized as slides, falls, or flows, depending on the nature of the movement. Translational slides occur when solid rock slides along a flat plane. Rotational slides (or slumps) occur when sediment or poorly consolidated rock moves down along a curved surface. In falls, material moves through the air to land as a pile on the ground. Slides are movements along a well-defined surface. Flows can be mudflows, debris flows, or avalanches, depending on the amount of water versus solid material being moved.

Oftentimes mass movements occur slowly and do not create obvious hazards to people. However, many disastrous examples gave no warning. Several of the deadliest occurrences have been associated with torrential rainfall in areas at the base of steep mountainous terrane. Rapid mass movements can occur when earthquakes and volcanic eruptions occur.

Subsidence is the vertical lowering of the land surface, while collapse is the relatively fast or sudden opening of the land surface and movement of surface material into cavities below. Removal of rock in the subsurface through chemical action or anthropogenic mining creates unstable situations in which the overlying rock collapses, either slowly in large-scale subsidence or by rapid collapse. Regions underlain by limestone are affected by the dissolving of the rock which then forms subsurface voids. Changes in groundwater levels allow collapse of the surface, forming sinkholes that devour homes and other property.

References and Suggested Readings

Browning, J. M. (1973). *Catastrophic rock slides. Mount Huascarán, north-central Peru, May 31, 1970. American Association of Petroleum Geologists Bulletin, 57,* 1335–1341.

Coch, N. K. (1995). *Geohazards, natural and human.* Prentice Hall.

De Graff, J. V., & Evans, S. G. (2002). *Catastrophic landslides: Effects, occurrences, and mechanisms.* Reviews in Engineering Geology.

Galloway, D., Jones, D. R., & Ingebritsen, S. E. (1999). *Land subsidence in the United States.* U. S. Geological Survey Circular 1182.

Highland, L. M., & Bobrowsky, P. (2008). *The landslide handbook—A guide to understanding landslides.* U.S. Geological Survey Circular 1325.

Jibson, R. W. (2005). *Landslide hazards at La Conchita, California.* U.S. Geological Survey Open-File Report 2005-1067.

Risley, J., Walder, J., & Denlinger, R. (2006). *Usoi dam wave overtopping and flood routing in the Bartang and Panj Rivers, Tajikistan.* U.S. Geological Survey Water-Resources Investigations Report 03-4004.

U.S. Geological Survey. (1997). *Debris-Flow hazards in the United States.* USGS Fact Sheet 176-97.

U.S. Geological Survey. (1999). *Real-time monitoring of active landslides—Reducing landslide hazards in the United States.* USGS Fact Sheet 091-99.

U.S. Geological Survey. (2004). *Landslide types and processes.* USGS Fact Sheet 2004-3072.

U.S. Geological Survey. (2005). *Landslide hazards—A national threat.* USGS Fact Sheet 2005-3156.

Websites for Further Reference

https://geochange.er.usgs.gov/sw/changes/anthropogenic/subside/
https://pubs.usgs.gov/of/2001/ofr-01-0276/
https://usgs.gov/mission-areas/natural-hazards
https://usgs.gov/natural-hazards/landslide-hazards
https://www.cdc.gov/disasters/

Questions for Thought

1. How does the orientation of rock features such as bedding and joints affect the development of mass movement?
2. Distinguish among the types of mass movement called falls, slides, and flows.
3. Give several examples of how human activities can cause mass movement.
4. Many mass movement events occur after rainfalls. Why?
5. Describe the conditions that result in a debris avalanche.
6. How do rockslides and slumps differ?
7. Describe five different problems resulting from surface subsidence.
8. Explain why subsidence is an especially serious problem in coastal areas.
9. Describe ways in which human activities have resulted in sinkhole development, subsidence, and collapse in limestone areas such as Florida.
10. Explain the basic chemistry involved with the dissolving of limestone by natural processes.

7 Threats From Space

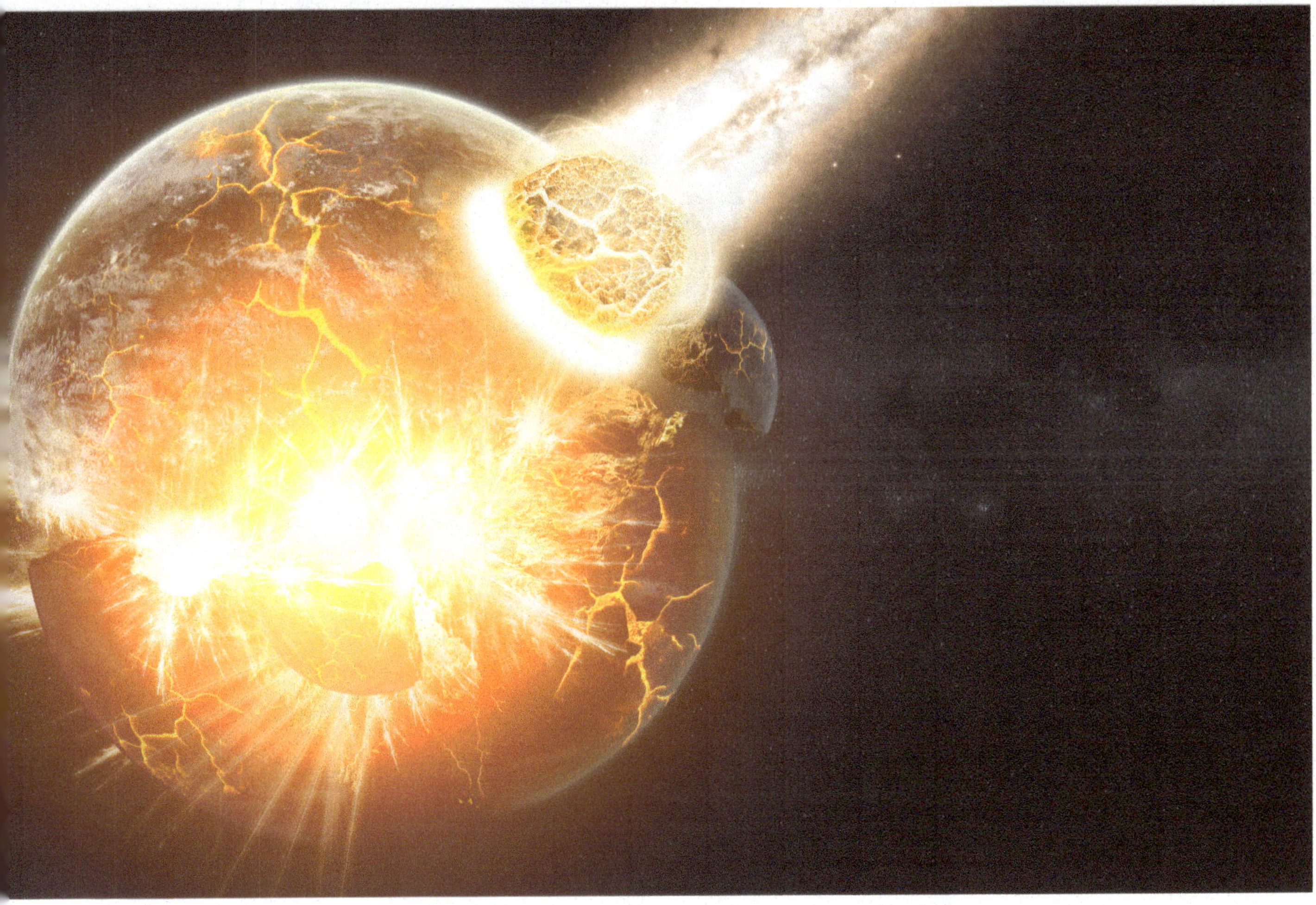

© sdecoret/Shutterstock.com.

Meteorite impacts are one of the most destructive forces in the solar system. Meteorite impacts are thought to have triggered mass extinction events, but also may have brought some of the building blocks of life to Earth in its early history. The largest known impact crater on Earth is more than 300 km (186 mi) in diameter, but Mars has visible craters up to 2,300 km (1,430 mi) in diameter.

Threats From Space

A 1.1 km (0.7 mile) diameter asteroid named 1950DA is approaching Earth. There is currently a 1 in 8,300 chance that this will hit our planet on March 16 in the year 2880—a 0.012 % probability seems small right now. If it happens, massive tsunami, global firestorms and nuclear winter caused by sunlight-blocking dust and debris injected into the stratosphere would devastate most life on Earth. Stephen Hawking considered asteroid impact to be the biggest threat to our planet, and we are unprepared for such an event. The B612 Foundation, dedicated to planetary defense, believes that we are 100% certain to be hit by a devastating asteroid, but we are not 100% sure when. Predicting an asteroid impact is extremely difficult and only a handful of actual impacts have been detected just hours in advance—these were small, hit wilderness or ocean, and didn't hurt anyone. There is a low rate of success and so far has not allowed any time for an appropriate response—especially considering that it takes years to launch a space mission. Plus, NASA scientists now believe that asteroids are stronger and harder to destroy than previously thought. It is only a matter of time before these problems go from being just thoughts to defining our response to a major planetary threat.

Big Bang Theory

The idea that the Universe formed from an initial point mass that exploded about 13.7 billion years ago and moved outward in an ever-expanding fashion.

In 2013, the European Space Agency's Planck mission measured light first emitted from a very young Universe, and determined that the Universe is 13.82 billion years old (Figure 7.1). The **Big Bang Theory** postulates that the Universe first existed as a single, extremely dense point of matter. It exploded, spreading material throughout space, and has been continually expanding. Astronomical observations also show that other galaxies are moving away from this central point at a rate of 67.3 kilometers per second (42 miles per second) per megaparsec, where megaparsec is an astronomical unit equal to 3.26 million light years, based on Planck's measurements. Astronomers estimate that there are billions and billions of objects in the observable universe—25 billion galaxy groups, 350 billion galaxies, 7 trillion dwarf galaxies, and 30 billion trillion stars—that's 3×10^{22}. Astronomers estimate that there are 300 billion stars in the Milky Way Galaxy, and even if only a small number of them have planets, there may be tens to hundreds of billions of solar systems in our own galaxy. There could be as many as 40 billion Earth-sized planets orbiting in the habitable zones of Sun-like stars and red dwarfs in the Milky Way, and there are a hundred billion galaxies in the Universe. One estimate suggests that for every grain of sand on Earth, there are as many as 10 Earth-like planets in the Universe. Our search for life on other planets is largely focused on telescopes that can study the atmosphere of distant planets and look for signs of biological processes, and this desire to answer the question "Are we alone?" drives space exploration.

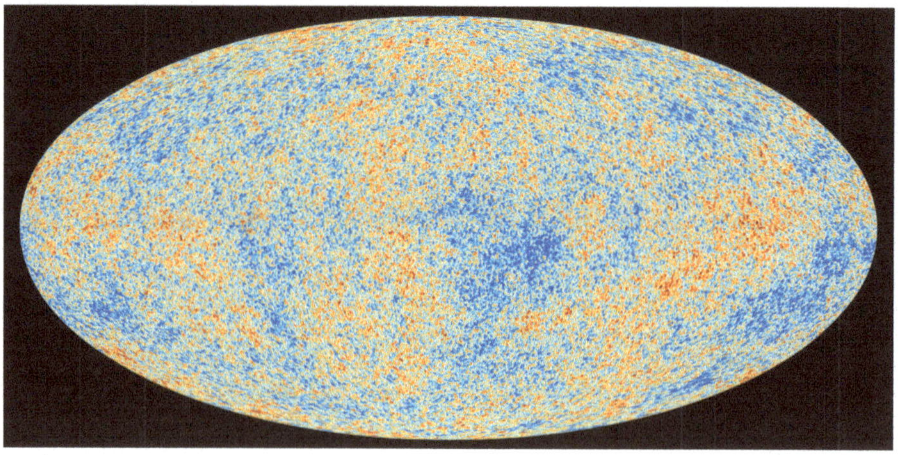

Source: NASA/ESA and the Planck Collaboration.

Figure 7.1 This map shows the oldest light in the universe—cosmic microwave background—that was imprinted in the sky when the universe was 370,000 years old. Tiny temperature fluctuations correspond to variations in density that would go on to become all of the stars and galaxies that we see today. This data comes from NASA's Planck mission.

Earth as Part of Our Solar System

© Kendall Hunt Publishing Company.

Figure 7.2 A schematic view of the solar system from afar.

Our Sun and the Solar System formed about 4.6 billion years ago (**Figure 7.2**). The oldest inclusions found in meteorites are thought to be the first solid material formed in the pre-solar nebula, and date to 4,568.2 million years old. The Solar System is divided into two groups of planets: the inner or terrestrial planets that are Earth-like, and the gas giant or Jovian planets (**Figure 7.2**). The four inner planets, those closest to the Sun, are Mercury, Venus, Earth, and Mars. These four have several similar characteristics including their average densities and their general compositions. The giant planets—Jupiter, Saturn, Uranus, and Neptune—are much larger than the inner planets and have characteristics that are indicative of their frozen, gaseous compositions.

Earth is the third planet from the Sun and the only one in our Solar System that has an environment capable of sustaining life as we know it. Discussions later in the text provide information about the biosphere, atmosphere, hydrosphere, and geosphere in which life exists. The composition of Earth is discussed in Chapter 2.

As Earth travels around the Sun and moves through space, the planet encounters extraterrestrial objects that are moving through the Solar System. Throughout its existence Earth has been bombarded by particles ranging in size from microscopic dust to large masses weighing billions of tons. A combination of metallic, stony, and icy material has contributed to the growth of the planet and the addition of water and chemical elements that have been incorporated into its composition.

The Solar System

The solar system consists of a group of objects that interact with one another. It is composed of eight planets, five dwarf planets, 182 planetary moons, millions of asteroids, and perhaps a trillion comets, all orbiting a star, the Sun, in an elliptical path on roughly the same plane. The Sun comprises over 99.8% of the mass in the solar system, and it is the Sun's gravity that causes the orbit of objects in the Solar System.

Asteroids

An **asteroid**, from the Greek word for "star-like," is a rocky or metallic object orbiting the sun that has a diameter from 1 meter or larger. Objects between 10 microns and 1 meter in size are called meteoroids (0.0004 inches to 3.3 ft). The largest currently known asteroid is Ceres, also listed as a dwarf planet, with a diameter of 975 km (605 mi), followed by 4 Vesta, 525 km (326 mi) in diameter. Most of the known asteroids lie in the **Main Asteroid belt**, a region

asteroid

A rocky or metallic body ranging in size from a meter to almost 1,000 km (621 mi) in size.

Main Asteroid belt

A region around the Sun lying between Mars and Jupiter that is the most common source of asteroids.

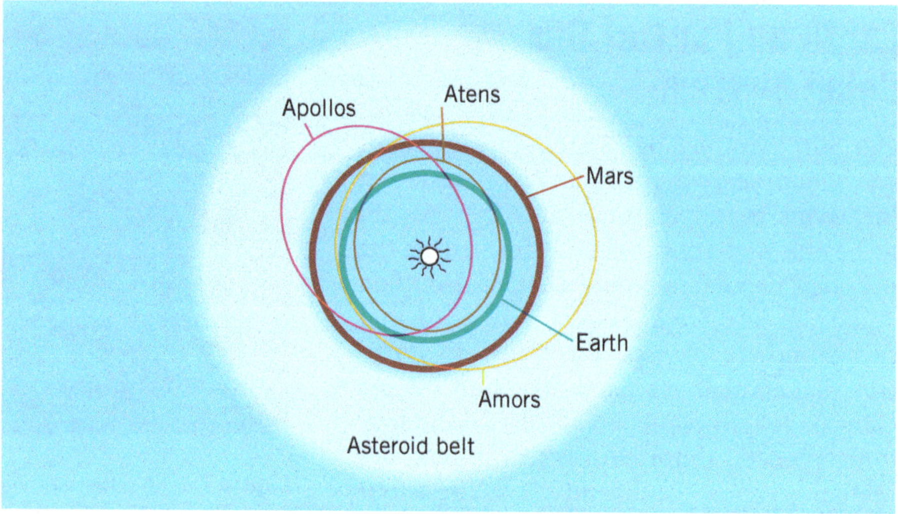

© Kendall Hunt Publishing Company.

Figure 7.3 **Different families of asteroids follow similar orbits—Amor asteroids cross Mars's orbit but don't cross Earth, whereas Apollo asteroids have an Earth-crossing path. Atens asteroids orbit in a path that is mostly inside that of Earth, but can also cross our planet's orbit.**

located between the orbits of Mars and Jupiter (Figure 7.3). In the main asteroid belt there are tens of millions of asteroids smaller than 1 km, and between 1.1 and 1.9 million that are larger than 1 km (0.6 mile). The largest ones, like Ceres, are roughly spherical and are thought to be surviving protoplanets from the early history of the solar system. Smaller asteroids are irregularly shaped, and are either fragments of larger bodies or planetesimals made of unmodified (not changed in any way) material from the beginning of the solar system. The mass of all the objects in the asteroid belt is about 4% of the mass of the Moon, and Ceres is about 30% of the total.

astronomical unit (AU)

The distance of the Earth from the Sun, averaging 150 million km or 93 million miles; used to describe large distances between bodies in space.

When astronomers determined the distances of the planets from the Sun, the position of the asteroid belt corresponded to the orbital distance that a possible planet could have had. The ring of material extends from about 300 million km to about 600 million km (186 to 373 million mi), or 2 AUs to 4 AUs from the Sun (1 AU equals 1 **astronomical unit**, equivalent to the distance separating the Sun and Earth, 150 million km or 93 million mi). About half of these asteroids lie near a distance of 2.8 AUs. Depending on the distance from the Sun, material takes between 3 and 6 years to complete a revolution around the Sun. The material is spread out in a manner that prevents it from accreting and forming a planet because of the strong gravitational forces of Jupiter. Trojan asteroids may be as numerous as those in the Main Asteroid Belt. They share the same orbital path as Jupiter, and lie in two groups about 60 degrees ahead and behind the planet. The largest known Trojan is about 200 km in diameter (124 mi). These asteroids are thought to have been captured by Jupiter's gravity about 500 million years after the Solar System formed.

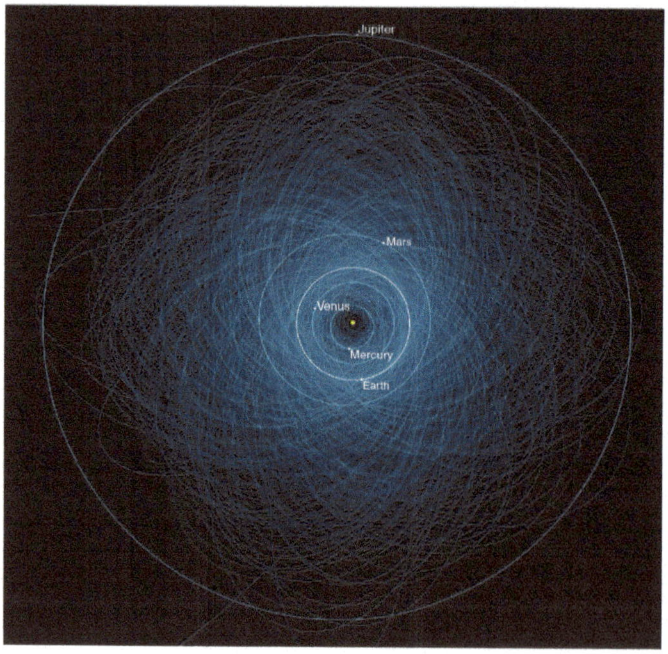

Source: NASA.

Figure 7.4 **The orbits of all of the Potentially Hazardous Asteroids (PHAs) are shown on this diagram with the orbits of Mercury, Venus, Earth, and Jupiter also shown. These are considered hazardous because they are large—at least 140 meters long—and because their orbits are close to that of Earth's.**

A near-Earth object (NEO) is any small object in the Solar System that has an orbital path that

© Kendall Hunt Publishing Company.

Amors
Earth-approaching Near Earth Asteroids with orbits outside Earth but inside Mars, named after asteroid 1221 Amor.

Atens
Earth-crossing Near Earth Asteroids with orbits slightly larger than Earth, named after asteroid 1892 Apollo.)

Apollos
Earth-crossing Near Earth Asteroids with orbits slightly smaller than Earth, named after asteroid 2062 Aten.

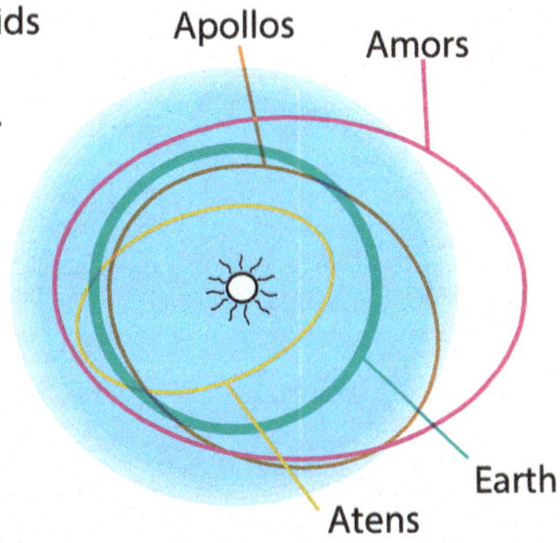

Figure 7.5 Three main orbital groups.

comes close to Earth (Figure 7.4). The critical definition of close, in Solar System terms, is a distance of less than 1.3 AU. Over 25,000 asteroids have been discovered that fall into the NEO category, and these are termed near-Earth asteroids (NEA). Near-Earth asteroids are divided into groups based on their orbit in relation to Earth's orbit (Figure 7.3). The three main orbital groups are (Figure 7.5):

1. **Amor asteroids**—these have an orbit that is larger than, and entirely outside of, Earth's orbital path. They do not cross Earth's orbit and are not an immediate threat, but their orbit may change to become earth-crossing in the future.
2. **Aten asteroids**—these have orbital axes that are close to, but less than, 1 AU, or about the same distance from the Sun as Earth's orbit. These objects cross Earth's orbit.
3. **Apollo asteroids**—these have an orbital axis that is greater than 1 AU and also cross Earth's orbital path.

If a NEA has an orbit that crosses Earth's orbit, and it has a size greater than 150 m (500 ft) meters in diameter—which means it is large enough to cause significant regional damage in the event of an impact—it is considered to be a potentially hazardous asteroid (PHA) (Figure 7.4). There are several thousand known PHAs, about 11% of all known near-Earth asteroids, and 157 of these PHAs are estimated to be larger than 1 km (0.6 mi) in diameter. Most of these are Apollo asteroids (85%) and the remainder are Aten asteroids. Potentially hazardous asteroids are usually only a risk on the timescale of 100 years as their orbits become more irregular due to interactions with Earth's or other planets' gravity or by collisions with other objects that shift their path.

The Psyche mission is a NASA-funded project to visit the metallic asteroid 16 Psyche (Figure 7.6). This asteroid is about 222 km (138 mi) in diameter and is made out of about 95% metal. It is thought to be the exposed iron core of an early proto planet that may have been as big as Mars which lost its outer surface in a series of violent collisions. Dr. Lindy Elkins-Tanton of Arizona State University is the Principal Investigator, and proposed this project as a way to study metallic planetary cores like the one Earth has—everything we know about or own planet's core has been learned from indirect study. It's

Figure 7.6 NASA's Psyche mission is our first mission to a unique world made of metal, an asteroid orbiting the Sun between Mars and Jupiter. This allows us to explore a completely new type of world, one not made of rock or ice.

too deep and too hot to drill a hole to our core, but this metal asteroid gives us a way to study a planet's metal core directly. The Psyche space mission departs Earth in 2022 and arrives at the asteroid Psyche in early 2026, where data the mission collects will be beamed to the public and scientists at the same time, allowing everyone to share in the discoveries.

Meteoroids, Meteors, and Meteorites

meteoroid

A relatively small object that moves through space; they range in size from dust to 1 m in size; composed of rock, metal, or ice.

meteorite

A meteor that hits Earth's surface.

meteor

A rapidly moving body that passes into the atmosphere and begins to burn up, producing what is sometimes referred to as a "shooting star."

The term **meteoroid** is used to describe small chunks of rock and debris in space. Meteoroids are significantly smaller than asteroids, ranging in size from small grains to 1-meter-wide objects, and most are fragments of asteroids or comets. These objects travel through space at speeds of 20 km per second or more (72,000 km/hour or 45,000 mph), and when they enter Earth's atmosphere aerodynamic heating of the object from friction causes it to heat up and glow as well as emit a trail of particles as it disintegrates, both of which give it the appearance of a streak of light. If the object survives passing through the atmosphere and hits the surface of Earth, it is called a **meteorite** (Figure 7.7).

Micrometeoroids as small as a few microns (millionths of a meter or 0.00004 inches) continually strike the atmosphere and the resulting dust falls to Earth and adds an estimated 100,000 tons of mass to the planet each day.

Several times each year, Earth's orbit places it in the path of a comet that crossed the orbit hundreds or thousands of years ago. Like bugs hitting the windshield of a fast-moving car, **meteors** hit Earth's atmosphere when the planet passes through a particularly dense part of the stream of debris left behind by a comet (Figure 7.8). The sky is filled with many streaks of light as clouds of material from the comet generate numerous showers of light. The Perseid meteor shower each August and the Leonid meteor shower each November are examples of such events. These are so named because the meteor showers appear to come from each of these respective constellations.

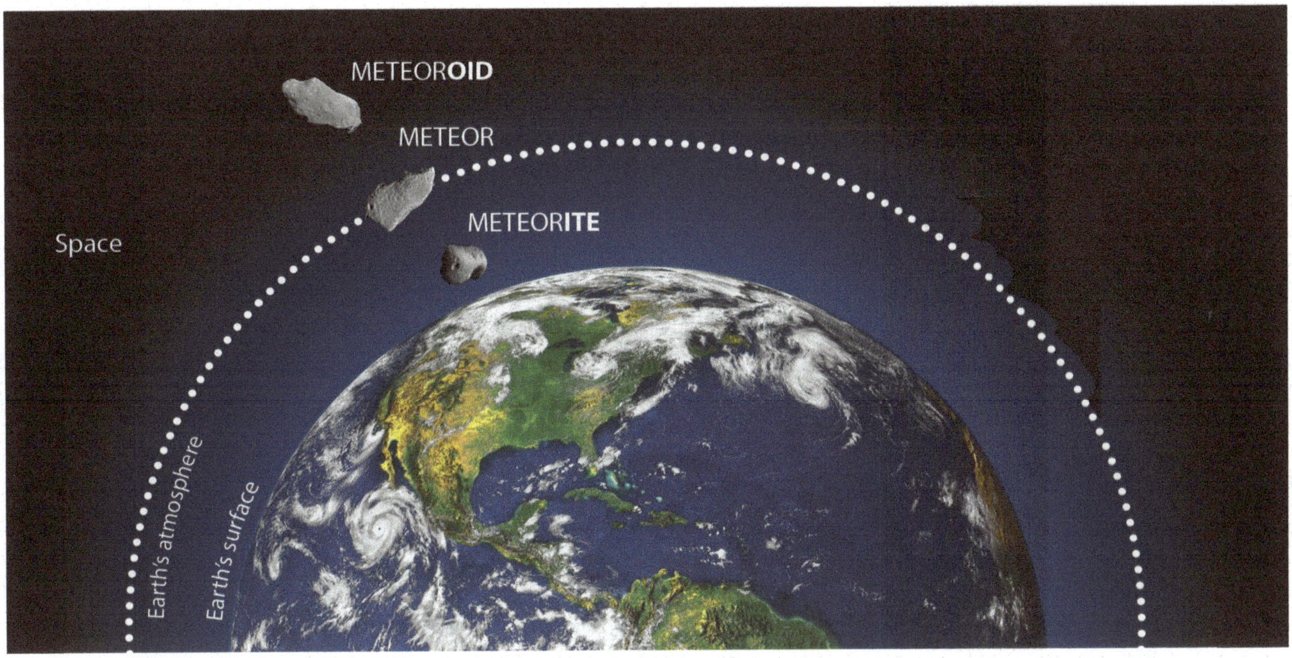

Source: NASA. Adapted by Kendall Hunt Publishing Company.

Figure 7.7 Meteorite, meteor, and meteoroid are all similar words that mean similar things, but reflect the different positions of an asteroid approaching Earth. Meteoroids are objects—asteroids or comets—that are outside Earth's atmosphere. As soon as they enter Earth's atmosphere, we call them meteors—these are also known as shooting stars. If the space object hits the surface of the Earth, it is called a meteorite.

© Allexxandar/Shutterstock.com. Adapted by © Kendall Hunt Publishing Company.

Figure 7.8 The bright shooting star streaks we see in a meteor shower are from glowing hot air as meteors blast through Earth's atmosphere. These objects are usually small, from dust to basketball to small car-sized, and almost always burn up in the atmosphere. In a meteor shower, the tails always point back to the same spot in the sky, which is where we are passing through the dust trail of a comet.

Meteorite Classification

Meteorites are divided into three general categories based on composition and whether they are dominantly composed of rocky material (stony meteorites), metallic material (iron meteorites), or a mixture of rocky and metallic material (stony-iron meteorites) (**Figure 7.9**). Detailed chemical and microscopic analyses are required to uniquely identify and classify a meteorite; these analyses can differentiate two different meteorites that fell in the same area at separate times, or two or more pieces of a single meteorite that were found in separate places or at different times.

meteorite fall

This is also called an observed fall, and is a meteorite collected after its fall from space was observed by people or automated devices.

meteorite find

Any meteorite found by people, but whose fall was not observed.

Meteorites are naturally transported from the celestial body on which they formed to a region beyond that body's gravitational field, and later collide with the surface of Earth. A **fall** is a meteorite that was observed to fall to Earth and was subsequently collected, whereas a **find** is a meteorite that was not observed to fall but was recognized by distinct features and collected at some point after reaching the surface of Earth. Most meteorites that are observed to fall are not found, and very few are observed hitting the ground.

Meteorites range in age from some of the oldest materials in our solar system, with calcium and aluminum-rich particles dated at 4.65 billion years old, to fragments from the Moon ranging in age from 4.5 to 2.9 billion years, and fragments from Mars ranging in age from 4.5 billion to 200 million years old. Most meteorites are thought to originate in the asteroid belt between Mars and Jupiter and were formed early in the history of the Solar System. Meteorites from asteroids are all about 4.56 billion years old. No Earth rocks are this old because the rocks on our planet have repeatedly been modified by terrestrial geologic processes like erosion, volcanism, and plate tectonics. These extraterrestrial objects were either knocked out of their orbit into Earth-crossing orbits through collisions with other objects, or through interactions with gravitational forces exerted by the Sun and other planets like Jupiter.

© Kendall Hunt Publishing Company.

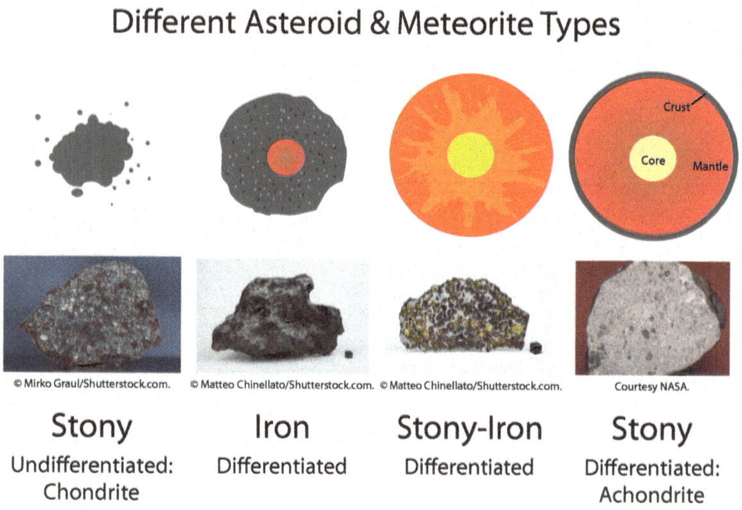

Figure 7.9 **Meteorites are grouped into three overall categories based on the most dominant component of their composition—rocky material in stony meteorites, metallic material in iron meteorites, and a mixture of rock and metal in stony-iron meteorites. Stony meteorites are divided into two groups based on texture and formation mechanism: chondrites have undergone very little change since they originally formed, and achondrites have been significantly modified since their original formation by planetary differentiation processes.**

Meteorite Locations

It is estimated that each year one meteorite per square kilometer (0.6 mi) of surface falls to Earth. Meteorites land randomly across Earth, and with water covering 70% of the planet's surface, most fall into the oceans and are never recovered. For those that fall on land, geologically stable desert areas are the places where meteorites are most easily preserved and identified. Hot and cold deserts with low precipitation, like the Nullabor Plain in Australia or ice fields in Antarctica, help preserve the meteorites from weathering, allow for accumulation over time, and make

Source: NASA.

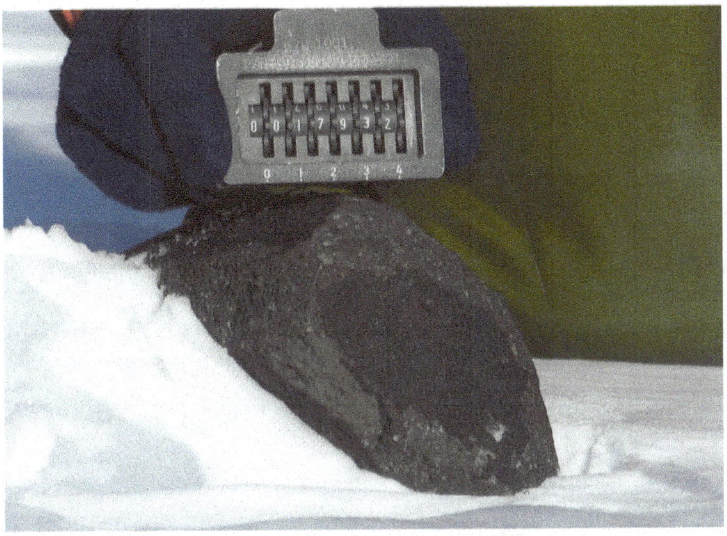

Figure 7.10 Scientists search the ice of Antarctica for meteorites that have fallen there. The contrast between the stony or metallic meteorites and the ice and snow makes them easier to identify.

© Dmitry Pichugin/Shutterstock.com.

Figure 7.11 The Hoba iron meteorite in Namibia is the largest-known single piece of meteorite and weighs more than 60 tons.

them easier to find because of the lack of vegetation and different lithology of terrestrial material, like ice in Antarctica (Figure 7.10). Meteorite collecting teams can spend several weeks on the ice fields in Antarctica and may recover 400 or more meteorites during their expedition. Some desert regions have dozens of different meteorites per square kilometer, although they can sometimes be difficult to distinguish from normal terrestrial rocks.

Meteorite Characteristics

Size. Meteorites can range in size from a few millimeters to several meters in diameter. The largest known individual meteorite is the Hoba meteorite from Namibia, which measures 2.7 meters (8.9 ft) wide (Figure 7.11).

Shape. Meteorites are typically irregular in shape with rounded edges (Figure 7.12).

Color. Fresh meteorite surfaces are black and shiny. Frictional heating and abrasion melts the outer surface of the meteorite and forms a black, glossy crust. Exposure to water and weathering on Earth's surface can erode the fusion crust, chemically weather materials in the meteorite, rust the metallic components, and mechanically weather the meteorite and fragment it into smaller pieces. The longer the meteorite has been on Earth, the more weathered it will become (Figures 13a–b).

Surface. Most meteorites have a smooth surface with no holes, but some show surface textures called regmaglypts that are thin flow lines of once molten fusion crust surface material or thumbprint-like scoops that are likely caused by the extreme abrasion and melting of the meteorite surface as it passes through the atmosphere (Figure 7.14).

© DenVDen/Shutterstock.com.

Figure 7.12 This fragment of the iron meteorite from Sikhote-Alin in Southern Russia fell in 1947 and is part of 100,000 kilograms (220,462 lbs) of material that was strewn across 1.3 square kilometers (0.8 square mi).

© Bjoern Wylezich/Shutterstock.com.

Figure 7.13a Fragments of the Chelyabinsk meteorite that fell over the southern Ural region of Russia on February 15, 2013, show a fresh and glassy fusion crust.

© Bjoern Wylezich/Shutterstock.com.

Figure 7.13b This piece of the Canyon Diablo meteorite has a rusty weathered surface from exposure to the elements on Earth's surface.

© Galyna Andrushko/Shutterstock.com.

Figure 7.14 These flow structures form as the meteorite is hurling through the atmosphere and the frictional heating melts the outer surface.

© Mirko Graul/Shutterstock.com.

Figure 7.15 Chondrules are round grains that form as molten or partly molten droplets in space that are gathered together in chondrites, and represent one of the oldest solid materials in our solar system.

Weight. Meteorites are generally heavier than terrestrial rocks of the same size because of their higher iron-nickel metal content. Naturally occurring terrestrial rocks are usually poor in metal, especially nickel. An apple-sized iron meteorite (9 cm or 3.4 inches in diameter) can weigh as much as 2.5 kilograms (5.5 pounds).

Interior. Most meteorites that otherwise look like terrestrial rocks or stones will contain small metallic flecks on broken, cut, or polished surfaces. Many stony meteorites will also contain small round chondrules, which range in size from a few micrometers to over a centimeter, which are partly molten droplets that formed in space before being accreted to their parent asteroid. These chondrules are some of the oldest solid materials within the solar system and represent the building blocks of planetary systems (Figure 7.15).

Metallic meteorites are almost entirely made of metal and have distinctive nickel compositions and interlocking crystal structure called Widmanstätten patterns that are described below. Solid metal objects are relatively uncommon on the surface of Earth, so these meteorites tend to stand out from surrounding terrestrial rocks on the surface. Stony-iron meteorites are made of about half silicate minerals and half metal and also are distinctive because of their unique physical appearance.

Magnetism. Because of their high iron and nickel content, almost all meteorites are magnetic.

Meteorite Types

Stony Meteorites

Stony meteorites are the most common type of meteorite and represent about 94% of observed meteorite falls. However, of all the meteorite types they can be most easily misidentified as terrestrial rocks, and therefore are more difficult to identify when already on the surface, especially after weathering. Because of this, they represent only about half of meteorite finds. If we want to infer the proportion of different types of meteorites as they exist in space, we need to use the proportion of falls rather than finds.

Stony meteorites are made of about 75 to 90% silicate minerals, 10 to 25% iron-nickel alloy, and trace amounts of iron sulfide. There are two different types of stony meteorites: chondrites and achondrites. They have different formation mechanisms.

Chondrites

Chondrites are the most abundant type of meteorite, making up about 86% of all meteorites that fall to Earth's surface. There are over 27,000 recognized chondrite meteorites weighing up to 1,770 kg (3,900 pounds), and some chondrite falls are made up of thousands of stones. In 1912 northern Arizona was showered in an estimated 14,000 meteorites that rained down in the Holbrook fall.

Chondrite meteorites represent early solar system materials that have not been modified by melting or differentiation of the parent asteroids after they formed. They are made of dust and small grains that accreted together. They are characterized by, and named after, features called chondrules, which are round grains that form as molten or partly molten droplets in space before being accreted to larger bodies (Figure 7.16). Chondrules represent some of the oldest solid materials in the solar system, and make up from 20 to 80% of chondrites by volume. Most chondrules are rich in the silicate minerals olivine and pyroxene, and some contain calcium and aluminum rich inclusions and pre-solar grains, which are the oldest substances in our solar system.

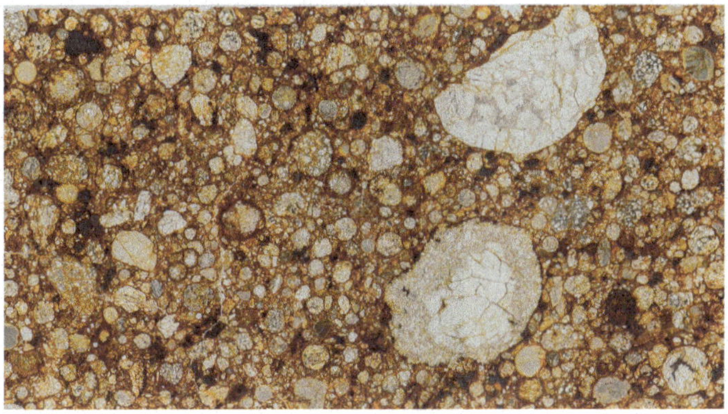

© Mirko Graul/Shutterstock.com.

Figure 7.16 This piece of meteorite NWA 7859 shows small glassy chondrules and fragments of larger chondrules that have accreted together.

Chondrules are thought to form from flash heating to about 1000 Kelvin (727 °C or 1,340 °F) in under a minute, which resulted in melting of solid dust aggregates. The environmental setting and source material are not known, but the starting material had approximately Solar composition and high levels of radiation and shock waves—potentially up to 10 kilometers per second (6 mi), such as those generated in a supernova—and could have melted the chondrules. The fine-grained matrix the chondrules were embedded in after their accretion in a chondrite parent body was probably condensed directly from the solar nebula. There are about 15 different groups of chondrites based on their distinct mineralogy and chemical composition. These chondrite groups probably represent separate asteroids or groups of asteroids that are the parent bodies for the different chondrite meteorite types.

Because chondrites have not undergone significant modification after they formed, and they haven't melted or had any major chemical interactions with any other objects, they represent very primitive materials and provide important information about early solar system formation. One particular type of chondrite meteorite, called carbonaceous chondrites, contains more than 600 organic compounds, including amino acids identical to those used by life on Earth. Amino acids on Earth have a "left-handed" molecular structure (twisting to the left), while **amino acids** in meteorites contain both "left-handed" and "right-handed" structures.

amino acids

Organic compounds composed of nitrogen, carbon, hydrogen, and oxygen, along with an additional alkyl side chain group.

Achondrites

Achondrites are stony meteorites made of material that is similar to the basaltic or plutonic volcanic rocks that are found on Earth, material that has melted and recrystallized on or within meteorite parent bodies. They do not contain chondrules and have a distinct texture and mineralogy that indicates igneous processes formed them. This means that they originated on differentiated parent bodies—asteroids, planets, or moons with a distinct core and crust—and were formed from molten or brecciated fragments that were ejected into space as a result of another collision (Figure 7.17). Only about 8% of all meteorites are achondrites, and about 60% of achondrites are related and thought to originate from the crust of the asteroid 4 Vesta. Meteorites from Martian and lunar crust are also part of this group. Because achondrites very closely resemble terrestrial rocks on sight, it is much more difficult to identify them as finds.

Source: NASA.

Figure 7.17 This brecciated meteorite is thought to come from the asteroid Vesta and formed from grinding and fusion during meteorite impacts on that asteroid's surface.

Stony-Iron Meteorites

Stony-iron meteorites contain both silicate and metal and are the least common of the three types of meteorites. They represent about 1% of observed falls and 5% of finds. There are two types of stony-irons: pallasites and mesosiderites. Pallasites are made of about half iron-nickel metal and half olivine crystals (Figure 7.18). Olivine crystals, which also form the gemstone peridot, are found in Earth's mantle.

Pallasites are believed to form between the silicate mantle, or outer shell, and molten metal core of a differentiated asteroid. They represent the layer of contact between the core and mantle of the asteroid. Of the approximately 60,000 officially recognized meteorites, only about 300 are pallasites, making them extremely rare.

© R G Meier/Shutterstock.com.

Figure 7.18 This pallasite meteorite has been sliced and polished to show the large, clear olivine crystals surrounded by metal matrix.

Mesosiderites are also stony-iron meteorites that contain nearly equal amounts of metal and silicates, but these are **brecciated**. They have an irregular texture with silicates and metal occurring as lumps or clasts, as well as in fine-grained intergrowths. They are thought to form from violent collisions between metal-rich and silicate-rich asteroids, and the silicate portion is composed mainly of igneous rock fragments representing the crust of asteroids. Only about 200 mesosiderites have been identified, about 50 of which came from Antarctica, and only seven are observed falls that were then collected. While they represent a small number of meteorites, some mesosiderites are among the largest meteorites known based on volume. Vaca Muerta was found in the Atacama Desert of Chile in 1861, when a prospector mistook the shiny metal in meteorite fragments for silver ore. Many fragments were found over a large area, and the total known mass is about 3,800 kilograms, or 8,400 pounds. A mesosiderite fell over Estherville, Iowa, USA, on May 10, 1879 (Figure 7.19). This is the most massive of the seven observed falls of mesosiderite. Eyewitness accounts describe a terrible, loud explosion late in the afternoon, followed by several additional thunderous blasts. After a brilliant fireball, a shower of several large fragments and many small ones fell, totaling 320 kilograms (710 pounds). Witnesses found a crater that was 4 meters wide and 2 meters (6.5 ft) deep, with a 195 kg (430 lb) mass inside. A second large mass weighed 65 kg (143 lb). The asteroid 16 Psyche is thought to be a candidate for the parent body of mesosiderites such as these.

brecciated
Mesosiderites broken into angular fragments.

© Matteo Chinellato/Shutterstock.com.

Figure 7.19 This piece of mesosiderite is from the Estherville, Iowa, meteorite that fell in 1879. Portions of the meteorite are displayed around the world: in the Smithsonian Museum of Natural History, the Museum Reich der Kristalle in Munich, Germany, the Naturhistorisches Museum in Vienna, Austria, and the Estherville Public Library in Estherville, Iowa, US.

Iron Meteorites

Iron meteorites are made entirely of metal, and are thought to originate in the cores of large asteroids that have melted and differentiated. When metal asteroid cores are exposed to the cold vacuum of space after large collisions fragment them, the molten metal cools and crystalizes, forming intergrown solid metallic masses. Earth's metal core is made of similar materials to these meteorites. They represent about 5% of observed falls, but because of their very distinctive appearance, they are more easily recognized on the surface of the Earth when compared to terrestrial rocks, and represent about 40% of surface finds.

© Mirko Graul/Shutterstock.com.

Figure 7.20 This fragment of iron meteorite is from the Henbury crater field in the Northern Territory of Australia. Over a dozen craters were formed during this impact event, ranging in size from 7 to 180 meters (23 to 590 ft), and this is one of the few impacts to have occurred in an area that was populated. This event happened about 4,000 years ago and was likely witnessed by the native Aboriginal groups who lived in the area at the time.

lamellae
Thin layers, sheets, or plates.

Iron meteorites are composed predominantly of iron, with the remainder consisting of nickel and elements such as cobalt, copper, gallium, platinum, iridium, and other trace metals. Pure iron almost never occurs naturally on Earth; it is usually found as an oxide chemically combined with oxygen or in other mineral ores. Meteorite iron and nickel alloys are classified based on their composition. Kamacite has 90 to 95% iron and 5 to 10% nickel, and Taenite has 20 to 65% nickel and the remainder iron. Both of these alloys are found on Earth almost exclusively in meteorites and nowhere else. These alloys form **lamellae** of long crystals interwoven in an octahedral pattern called a Widmanstätten pattern (Figure 7.20). This pattern becomes visible on metal meteorite surfaces that have been cut and polished and then etched with acid to highlight the crystal pattern.

Formation of Impact Craters

Entering Earth's Atmosphere

Between 100 million and 10 billion kg (220 million to 22 billion lb) of space debris hit Earth's atmosphere every year. There is a constant rain of dust-sized material, potentially up to 300 tons per day, which enters the atmosphere at speeds of from 38,000 to 248,000 kilometers per hour (24,000 to 154,000 mph), depending on whether they are orbiting in the same direction or the opposite direction to the Earth's motion around the Sun. These particles collide with molecules in the air as they enter the atmosphere, and this causes very rapid heating to temperatures greater than 1600 °C (about 2900 °F). Most of the dust that enters the atmosphere is too small to be detected without specialized meteor radars, but if the particle has a diameter greater than about 2 millimeters—about the size of a tomato seed—it can produce a "shooting star." One of these happens about every 30 seconds. This fine-grained cosmic dust provides a nucleation surface for ice crystals in high-level clouds in the atmosphere, and iron from the dust fertilizes the oceans and supports the growth of marine phytoplankton.

BOX 7.1 Case Study: Chelyabinsk Meteor, 2013.

On February 15, 2013, people around the globe were waiting anxiously for the flyby of a near-Earth asteroid called 2012 DA14, which was supposed to pass 27,700 kilometers (17,200 mi) above Earth's surface, a distance of only 4.3 Earth radii. However, the world was shocked and amazed when a different, completely unexpected and totally unrelated meteor exploded in the atmosphere over Russia at about 9:20 a.m. local time (Figure 7.1.1). By the time 2012 DA14 (now called 367943 Duende) passed by 16 hours later, it was old news.

Alex Alishevskikh/Flickr.com.

Figure 7.1.1 The Chelyabinsk meteor vapor cloud trail left in the atmosphere as the meteor entered and exploded on February 15, 2013. The meteor entered on the right of the image and the glow in that portion of the cloud is where the greatest amount of frictional heating and ablation took place. The large mushroom in the cloud is where the first explosion happened as the meteor fragmented.

The Chelyabinsk meteor was about 20 meters (66 ft) in diameter and hit Earth's atmosphere at a speed of 19.16 kilometers per second, or about 65,000 kilometers per hour (40,000 mph). It exploded 29.7 kilometers above Earth's surface (18.5 mi), centered over the southern Urals region of Chelyabinsk, with the energy equivalent to 400–500 kilotons of TNT, or about 30 times as much energy as the Hiroshima atomic bomb. No one was killed, but 1,491 people were seriously injured and required medical treatment. People saw the flash of light from the explosion and went to their windows to see what it was, and when the shock wave arrived minutes later, the windows were blown in, blasting them with broken glass (Figure 7.1.2). More than 7,000 buildings in six cities across an area about 100 kilometers (62 mi) wide and a few tens of kilometers long were damaged, most with broken windows, but some suffered collapsed roofs. The repairs are estimated to cost $33 million US dollars and had to be completed in sub-freezing −15 °C temperatures.

© The Asahi Shimbun/Contributor/Getty Images.

Figure 7.1.2 Shattered glass from exploding windows was the main cause of injury during the meteorite explosion.

The meteorite was about 12,000 metric tons, heavier than the Eiffel Tower, and is the largest object known to enter the Earth's atmosphere since 1908, when there was an airburst over Tunguska, Siberia (Figure 7.1.3). As the meteor exploded it formed a fireball 30 times brighter than the sun and fragmented into 253 known pieces, the largest of which hit the surface of Lake Chebarkul, punching a hole in the ice that was 6 meters (20 ft) wide (Figure 7.1.4). Only 0.03 to 0.05% of the initial meteorite survived entry into the atmosphere; 76% of the meteorite evaporated and the remainder was fragmented to dust. The explosion as the meteorite fragmented registered on seismographs in the area as a magnitude 2.7 earthquake, and was detected

© Bjoern Wylezich/Shutterstock.com.

Figure 7.1.3 This fragment of the meteorite shows a dark fusion crust and an interior that has the texture of an ordinary chondrite. It contains about 10% iron, olivine, and sulfides, and has shock melted black veins running through it that show it has had a complex history.

© Kyodo News/Contributor/Getty Images.

Figure 7.1.4 This 6 m (20 ft) hole in the ice of Chebarkul Lake was made by the largest surviving fragment of the Chelyabinsk meteor after it exploded in the atmosphere.

by monitoring stations run by the Comprehensive Test Ban Treaty Organization which were designed to pick up nuclear weapons testing, some as far away as Antarctica. Those waves reverberated around the world several times and took over a day to dissipate. In June of that year, scientists recovered a 60 cm, 654-kilogram (2 ft, 1,442 lb) fragment from the lake bottom mud (Figure 7.1.5).

© Migel/Shutterstock.com.

Figure 7.1.5 The largest fragment of the Chelyabinsk meteorite recovered from Lake Chebarkul is now on display in the State History Museum of South Ural, Russia.

This meteorite hit Earth without warning, and it served as a wake-up call to show us that the entire planet is vulnerable to potentially hazardous objects. Airbursts from objects of this size (about 20 m) occur about every 60 years. On that day, in the face of 2012 DA14 approaching Earth, the United Nations had proposed creating an "Action team in Near-Earth Objects" and a global asteroid warning network system. The UN now sanctions June 30 as Asteroid Day, a global awareness campaign co-founded by Dr. Brian May—astrophysicist and lead guitarist of QUEEN; Danica Remy—president of the B612 Foundation; Rusty Schweickart—Apollo 9 Astronaut; and filmmaker Grig Richters. Asteroid day currently has over 11,000,000 online participants and 2,500 events in 193 countries (https://asteroidday.org/).

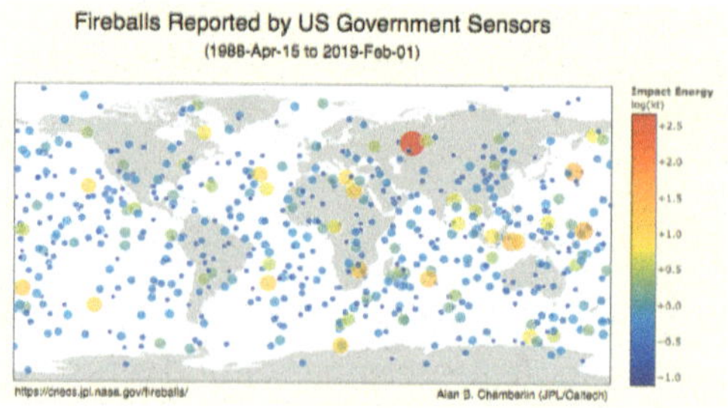

Source: NASA, Alan B. Chamberlin (JPL/Caltech).

Figure 7.1.6 Reported fireball events over a 30-year time period show the global distribution of events and highlight the magnitude of the Chelyabinsk airburst.

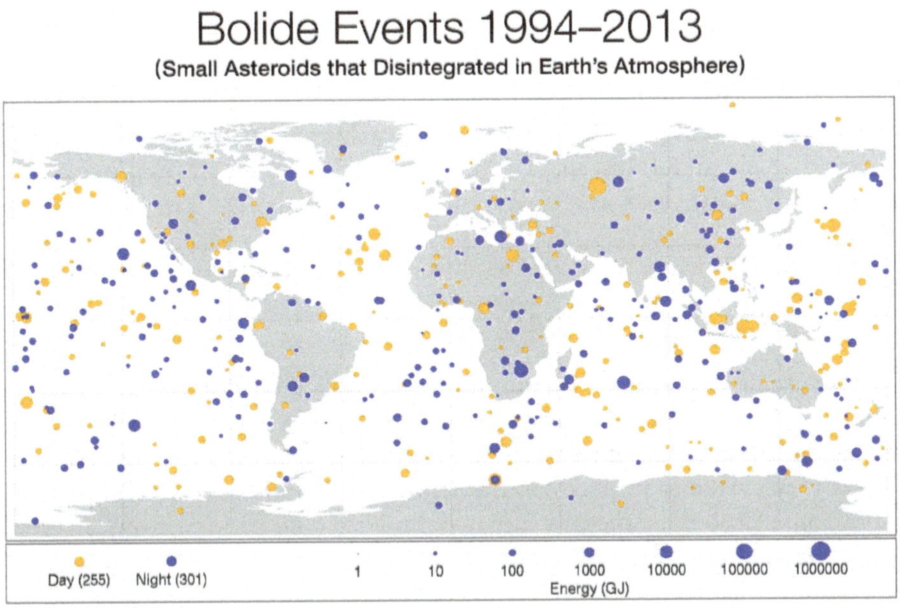

Figure 7.21 This shows all of the meteors that impacted Earth's atmosphere and disintegrated from 1994 through 2013. Orange dots represent daytime events, blue dots represent nighttime events, and the size of the symbol represents the energy released. These objects ranged from about 1 to 20 m (3 to 66 ft) in size.

Statistically, we can also expect larger objects to hit Earth on a regular basis (Figure 7.21, Table 7.1). About once each year a stony asteroid the size of an automobile (4 meters or 13 ft) hits Earth's atmosphere, creating a fireball as it burns up before reaching the surface. If the asteroid is about 7 meters in diameter, it has as much energy as the atomic bomb dropped on Hiroshima—about 14 million kg or 32 million pounds of TNT—and objects of this size enter the atmosphere once every 5 years or so. Small asteroids like these, less than 20 m (66 ft) in diameter, usually explode in the upper atmosphere, and most or all of the solid material is vaporized. As asteroids increase in size, they generate more powerful airbursts. An object

kiloton

1 kiloton or metric kiloton (unit of mass) is equal to 1,000 metric tons. A metric ton is exactly 1,000 kilograms (SI base unit) making a kiloton equal to 1,000,000 kilograms. 1 kt = 1,000,000 kg.

megaton

1 megaton or metric megaton (unit of mass) is equal to 1,000,000 metric tons. A metric ton is exactly 1,000 kilograms (SI base unit) making a megaton equal to 1,000,000,000 kilograms. 1 mt = 1,000,000,000 kg.

Table 7.1 Stony Asteroids Entering Earth's Atmosphere

Diameter (m)	Diameter (ft)	Frequency (years)	Kinetic Energy at atmospheric entry*	Size of crater (km)	Size of crater (miles)
4	13	1.3	3 kt	-	-
10	33	10	47 kt	-	-
20	66	60	376 kt	-	-
50	160	764	5.9 mt	-	-
100	330	5,200	47 mt	1.2	0.75
250	820	59,000	734 mt	3.8	2.4
400	1,300	100,000	3,010 mt	6	3.7
1,000	3,300	440,000	47,000 mt	13.6	8.5
5,000	15,840	20,000,000	26,000,000 mt	100	62

*kt = kiloton; mt = megaton

like Chelyabinsk, which was estimated to be about 20 m (66 ft) in diameter, strikes Earth about twice each century and creates an explosion as large as 30 Hiroshima bombs. Larger objects may survive the trip through the atmosphere and impact Earth's surface, creating a crater (Figure 7.22). Stony asteroids over about 100 m (330 ft) in diameter hit Earth approximately once every 5,000 years and form craters that are about 1 km (0.6 mile) in diameter.

Formation of Impact Craters

Impact crater formation is a dynamic and complicated process that involves tremendous amounts of energy transferred from an asteroid or comet traveling faster than the speed of sound as it contacts Earth's surface. These hypervelocity impact conditions occur for objects larger than about 7,000 kg (15,400 lb), generally larger than 50 m (164 ft) for a stony asteroid and greater than 20 m (66 ft) for a denser iron-nickel asteroid. Those that are smaller experience enough drag in the atmosphere to slow them down from their cosmic travel velocities of from 11 to 72 km per second (6.8 to 45 miles per second), depending on the orbital path of the object compared to that of Earth. The median impact velocity is about 20 kilometers per second (12.4 miles per second). The larger the object, the easier it can pass through Earth's atmosphere without experiencing any slowing. An object of about 9,000 kg (20,000 lb) retains about 6% of its original velocity, one of 900,000 kg (2,000,000 lb) retains about 70%, and extremely large objects are not slowed down by the atmosphere at all. This is important because the kinetic energy of a body is equal to half the product of the mass and the square of the speed, so while the size and mass of a meteoroid contributes to the energy produced in an impact, the speed or velocity is even more important. An object doubling its speed has four times as much kinetic energy.

The impact process can be divided into three different stages: (1) initial contact and compression, (2) excavation, and (3) modification and collapse (Figures 7.23a–b).

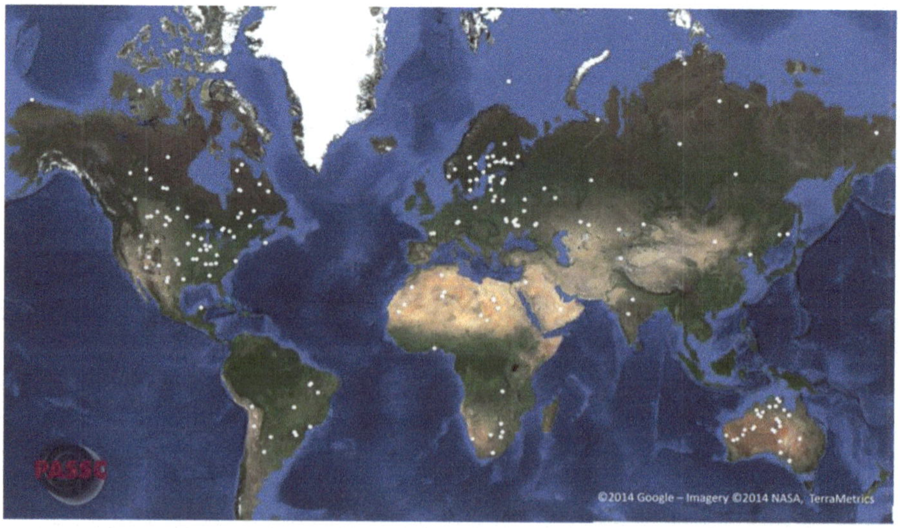

Source: NASA.

Figure 7.22 **Locations of all known meteorite impact craters on Earth.**

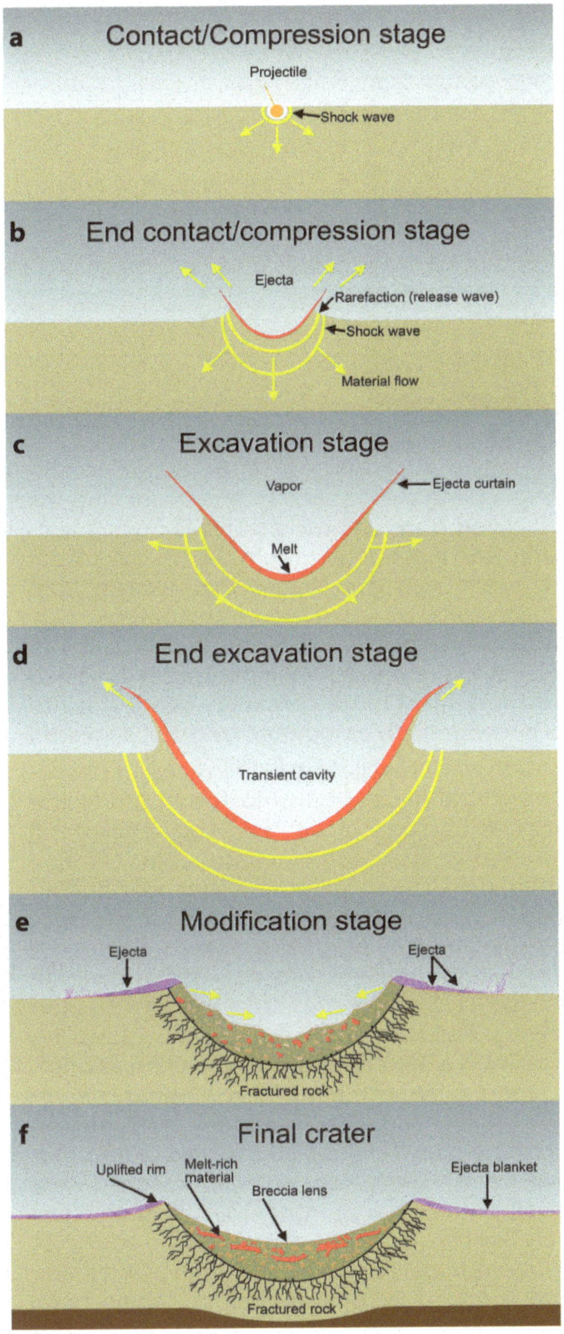

Figure 7.23a Small meteorite impact craters have a simple, bowl-like shape. They form when a meteorite hits Earth's surface, generating a shock wave that radiates into the target rock (panels a and b). The target rock is pushed downwards and outwards, forming a large cavity and blasting material out in an ejecta curtain (panels c and d). Unstable rocks on the crater walls fall or slump into the newly formed crater (panel e). The final crater is bowl-shaped and has a floor of brecciated rock (panel f).

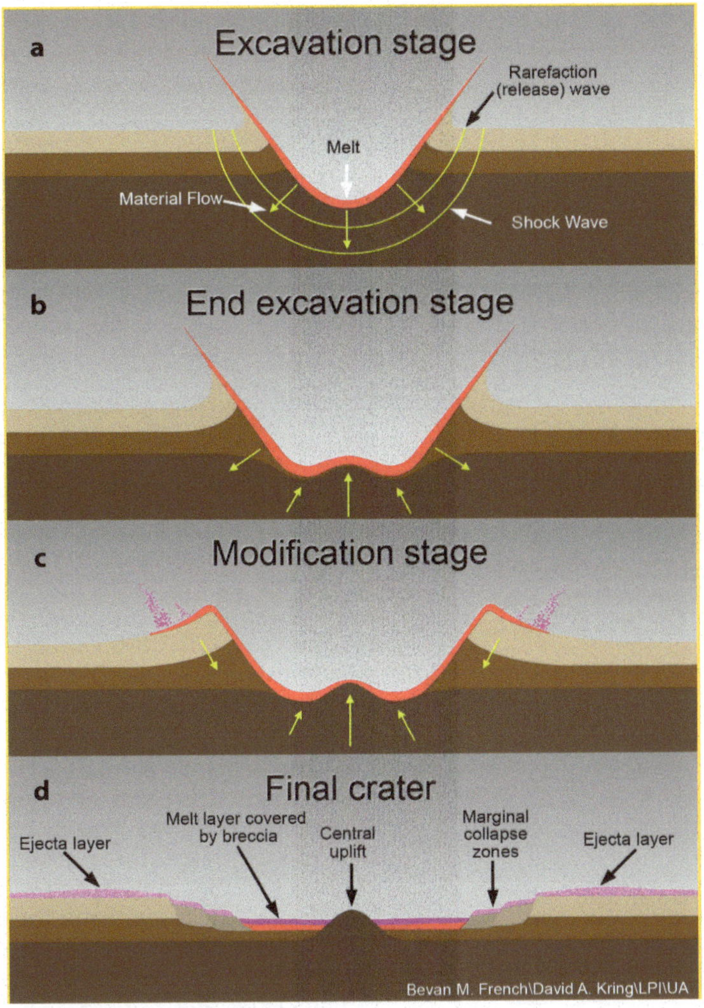

Source: NASA.

Figure 7.23b Large impacts form complex craters with walls that are so steep that they collapse down into the crater as coherent blocks, forming stepped rings, and the rebound of rocks in the center of the crater forms an uplifted central peak. The initial stage of crater formation is similar to a simple crater, in that a bowl-shaped cavity forms (panel a). After the shock wave passes through the crust, the material at the base of the crater rebounds and rises back up, forming a central peak (panel b). The crater walls are extremely unstable because they are so high, and they collapse in under the effect of gravity (panel c). These big impacts release so much energy that a large part of the target rocks are melted, forming a melt sheet in the crater (panel d).

BOX 7.2 Case Study: Man-Made Space Hazards Threaten Earth.

© zhangjin_net/Shutterstock.com.

Figure 7.2.1 **China's Long March 5 rockets are designed for delivering large, heavy cargo to a low-Earth orbit.**

In late April 2021, China began building a permanent space station, launching the first of 11 components into space on a Long March B5 rocket. This is China's largest rocket, measuring 57 m (187 ft) long with a diameter of 5 m (16 ft) and a mass of 854,500 kg (1,883,900 lb). Rockets are launched in stages that contain from two to five phases of propulsion which are fueled by separate core and booster rockets. Shortly after liftoff, the used core and booster rockets disengage and fall back into the atmosphere, usually landing harmlessly in the ocean. While this rocket launched successfully, something went wrong with the 30 m (100 ft) core stage, sending it into low-Earth orbit, where it began to spin uncontrollably as it drifted back toward Earth.

The Long March 5B core stage is the fourth most massive object to make an uncontrolled descent to Earth, and is equivalent to a 10-story building. It took a rare and unexpected path over heavily populated areas, passing over both Los Angeles and New York City just before it entered the atmosphere over the Atlantic. China's foreign ministry said that most debris would burn on re-entry, but other agencies including the US Space Command and the European Space Agency were tracking the rocket's trajectory and predicted that between 20% and 40% of the dry mass could survive, equivalent to several tons of debris impacting the Earth. Any debris that survived re-entry could land between 41° North and 41° South, a strip of Earth that runs from southern Spain, Portugal, and Italy down to Australia. The rocket core was travelling at speeds of 7.7 km per second (4.8 mps), with a relatively flat trajectory. It was orbiting the Earth once every 90 minutes, and a difference of just a few minutes in re-entry time could result in a landing point that changed by thousands of kilometers. The US Air Force predicted its re-entry time plus or minus 30 minutes, but in that time the rocket could travel three quarters of the Earth. As re-entry approached, New Zealand, Australia, and Costa Rica were all on the list of countries that could be hit by debris. Twitter was awash with #ChineseRocket memes. Re-entry took place at 2:24 a.m. GMT, Sunday May 9, according to China's Manned Space Engineering Office, and debris fell with a blazing trail of light over the Arabian Peninsula before landing in the Indian Ocean near the Maldives.

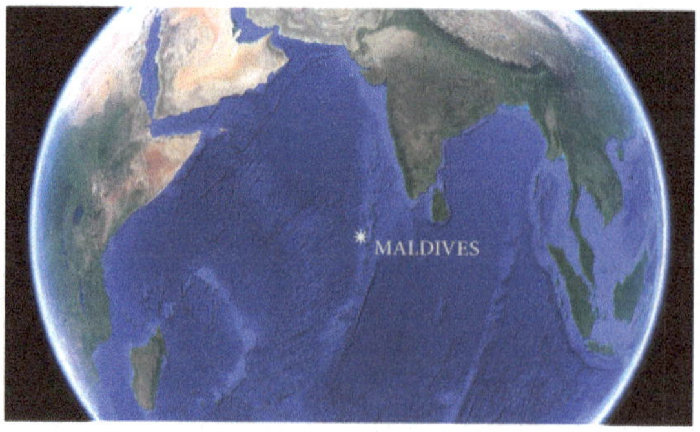

Source: Google Earth

Figure 7.2.2 **Rocket debris crashed into the Indian Ocean after a long and tense wait, with several countries such as Costa Rica and New Zealand anticipating possible impact.**

1. Initial Contact and Compression Stage. The crater formation process begins when the impacting body (referred to as the projectile) first touches the surface (referred to as the target). If the target is solid rock, the projectile penetrates below the surface about one to two times its own diameter in a fraction of a second. Once its kinetic energy is transferred to the target rocks by the shock waves generated at the contact between the projectile and the target rock, the projectile stops. These shock waves are much stronger than those produced by normal terrestrial processes, and generate pressures of several hundred gigaPascals to greater than one teraPascal in large impacts. This is equivalent to pressures found deep in the interiors of planets. Temperatures can be tens of thousands of degrees at the point of impact, where the projectile and a large part of the target rock may be vaporized or melted. Crystalline materials can be transformed into higher-density phases by shock wave compression; for example, quartz can be transformed to coesite or stishovite. As the shock waves move further into the target material, they spread out over an increasingly larger hemispherical area with radial distance, which dilutes the energy across a larger and larger volume. Additional energy is used in the target rocks through heating, deformation, and motion of the material. The entire contact and compression stage lasts only a second or two in even the largest of impacts and is equal to the time it takes the projectile to travel the distance of about one diameter below the point of contact at its original velocity. This is about 2 seconds for a 50 km or 31 mile diameter projectile travelling at 25 kilometers per second (16 miles per second) and less than 0.01 seconds for a 100-meter diameter (330 ft) projectile at the same speed. At the end of this stage the impactor has transferred all of its energy and does not have anything to do with the formation of the impact crater itself. The actual excavation of the crater is due to shock waves expanding through the target rocks. The impactor material may expand out of the crater as part of a vapor plume if it was vaporized, and the remainder, which is virtually all melted, may be mixed into the melted and brecciated target rocks.

2. Excavation Stage. The extremely brief impact stage, which occurs within a few tenths of a second for all but the largest impact events, immediately transitions into the excavation stage once the impactor has transferred all of its energy to the target material. The excavation stage is when the actual impact crater forms from the interaction of the shock wave with the original ground surface. During excavation, the crater grows as the accelerated target material moves away from the point of impact. Shock pressures remain high at significant distances away from the point of impact, and this is why the final crater is many times larger than the diameter of the projectile itself, typically 20 to 30 times greater.

At the end of the contact and compression stage, the projectile has penetrated approximately a diameter length deep into the target and is surrounded by a hemispherical-shaped envelope of shock waves that expand rapidly. The center of this hemisphere lies within the target rock at some distance below the original surface. Shock waves that move up hit the original ground surface and fracture and shatter the target rocks, which are blasted outwards at high velocities of up to several kilometers per second.

This may be enough to uplift near-surface rocks to form a rim, as well as excavate material and blast it out beyond the rim of the crater. At lower levels, below the impactor, target material is pushed down and out, and this produces a bowl-shaped depression in the target rocks. The initial crater that is formed is called the **transient crater** at this stage because the shape of the crater continues to change with time.

transient crater

The initial approximately bowl-shaped meteorite impact crater with a topographically elevated crater rim that has reached its maximum size.

As long as the expanding shock waves have enough energy to eject or compress material, the transient crater continues to expand and near-surface rocks are lifted up into the rim or blasted out of the crater. These waves continuously lose energy as they deform and eject the target rocks they pass through as they spread out. Eventually, the shock waves don't have enough energy left to excavate or displace target rocks, and at that point the growth of the transient crater stops. Now the transient crater is at its largest extent, the excavation stage stops, and the modification stage begins immediately.

The excavation stage lasts longer than the contact and compression stage but is still extremely fast; a 200-kilometer diameter (124 mile) transient crater can be excavated in about 90 seconds. A 1 km (0.62 mile) diameter crater similar to Barringer Meteor Crater in Arizona, US, takes about 6 seconds to excavate (Figure 7.24). At the end of this stage, the transient crater that has formed is a bowl-shaped depression with an uplifted rim. It has a maximum depth that is about one third its diameter. All impact structures go through this transient bowl-shaped crater stage.

3. Modification Stage. The modification stage begins immediately after the transient crater reaches its maximum size, when the expanding shock waves no longer have enough energy to deform the target material and gravity begins to modify the transient crater shape. In most cases, the transient crater is oversteepened and unstable, and begins to collapse immediately. The initial modification stage also lasts only a brief period of time—less than a minute for small craters and only a few minutes for larger ones—and can be thought of as "when things stop falling." Continuing modification

© Frank Bach/Shutterstock.com.

Figure 7.24 Barringer Crater in northern Arizona, US, is the best preserved impact feature on Earth. Its estimated age is 49,000 years. This crater is 1,300 m (0.8 mi) in diameter and 174 m (570 ft) deep.

of the crater does not have a clear end as the immediate crater modification processes merge into normal geological processes like mass movement, erosion, and sedimentation.

The extent that the transient crater shape changes depends on the size of the crater. On Earth, in craters that are smaller than about 2 to 4 kilometers (1.3 to 2.5 mi), the walls of the transient crater are unstable and material from near the rim, along with brecciated debris, collapses back down into the crater. The final structure forms a bowl-shaped depression that is similar in size and shape to the transient crater and is called a **simple crater** (Figure 7.23a). Larger craters have a more complicated modification process, and the resulting final crater shape is called a **complex crater** (Figure 7.23b). Complex craters have an uplifted central region, a flat crater floor, and extensive inward collapse around the rim. The diameter of the final crater may be one and a half to two times greater than the diameter of the original transient crater. In these large craters, the target rock rebounds after impact like the ripples around a drop of water. These processes are taking place in solid rock and can occur over hundreds of kilometers. The amount of uplift is equal to about one tenth of the final diameter of the structure. For large impact events with crater diameters of 100 to 200 km (62 to 124 mi), the crust beneath the crater is uplifted by 10 to 20 km (6 to 13 mi) during the early modification stage. This central uplift happens over only a few minutes, and in the largest craters may take less than 15 minutes. The central uplift is preserved as a **central peak structure** in craters up to about 20 to 25 (13 to 15.5 mi) in diameter, and larger craters are even more complicated as the central peak is replaced by a series of concentric rings and basins called **peak-ring structures** (Figure 7.25).

The largest known meteorite impact craters are several hundred kilometers to more than 1,000 km (620 mi) in diameter, and they look like huge bullseyes with multiple rings. These have been identified on the Moon, Mercury, Mars, and some moons of Jupiter. These are produced by impacts with projectiles that are tens to hundreds of kilometers in diameter. They occurred early in our solar system history, more than 3.9 billion years ago, when collisions with large objects were more frequent. The largest impact

simple crater

A large, bowl-shaped cavity in the ground or on a planetary object, typically caused by an explosion or the impact of a meteorite. Simple craters are generally up to 2 km (1.2 mi) in diameter.

complex crater

A type of large impact crater that forms at diameters greater than 2 km on Earth, where the collapse and modification of the transient crater is extensive and may result in a central uplift and ring structures.

central peak structure

Part of a complex crater that has an uplifted center area formed by rebound of the crater floor during a meteorite impact event.

peak-ring structure

A complex impact crater with a roughly circular ring or plateau, possibly discontinuous, that surrounds the center of the crater and is inside the outer rim.

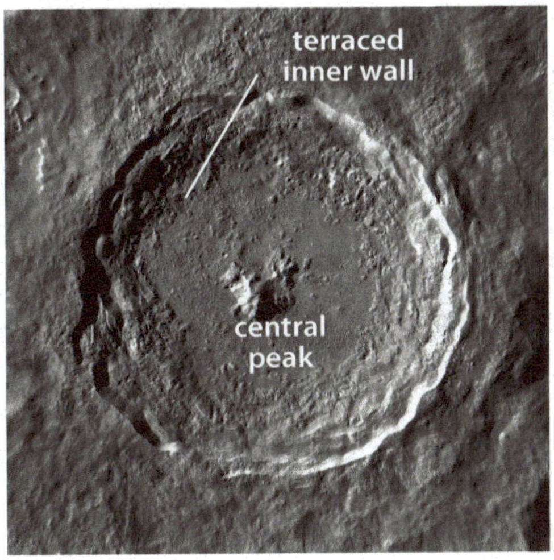

Source: NASA.

Figure 7.25 Tycho crater on the Moon is about 108 million years old, with a central peak that rises 1,600 meters (1 mile) above the crater floor and terraced inner walls.

Figure 7.26 Vredefort crater in South Africa.

Source: NASA.

crater identified on Earth is the Vredefort crater in South Africa, which is 160 km (100 mi) in diameter (**Figure 7.26**). It would have been more than 300 km (186 mi) in diameter when it formed about two billion years ago, from the impact of a 10 to 15 km (6 to 9 mile) diameter asteroid. Much of the crater has eroded away since then, leaving only the remnants of the uplifted central area, called the Vredefort Dome.

Comets

comet

Body of icy and rocky material that moves around the Sun; most comets originate from the Oort Cloud.

A **comet** is the equivalent of a dirty snowball, or icy dirtball, and contains a nucleus, or central core, which averages 5 to 10 km (3 to 6 mi) in diameter (**Figure 7.27**). The core consists mainly of frozen CO_2 and water ice, along with ammonia, methane, and dust. They may also have a small rocky inner core. Comets are thought to be made out of leftover material that initially formed the Solar System. As they near the Sun, ice on the surface of the comet nucleus warms up, sublimates, and turns into a gas, forming a cloud around the nucleus called a coma. The coma may be up to 15 times Earth's diameter. Surrounding the coma is a hydrogen cloud, which is millions of kilometers in diameter but is composed of a very sparse envelope of hydrogen atoms. Charged particles from the Sun's solar winds interact with the gas

Figure 7.27 A comet is made up of an inner core of solid material, called the nucleus. When it reaches about 5 AU from the Sun, the nucleus begins to heat up and sublimate, forming a cloud of gas and ions of the coma and hydrogen cloud. At about 1 AU the tails start to form. The plasma tail is swept back by solar wind, and the dust tail is pushed by sunlight.

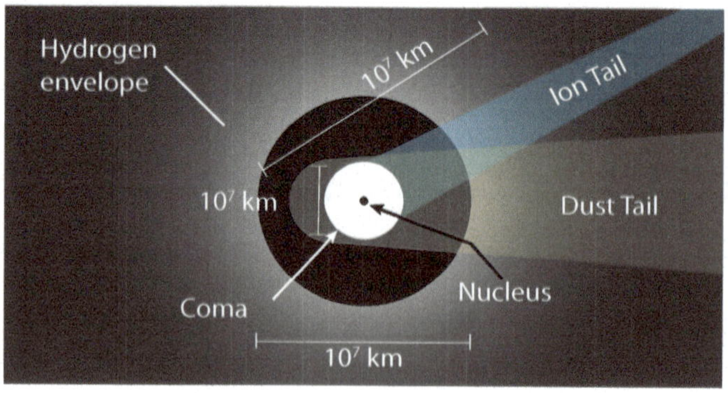

© Kendall Hunt Publishing Company.

© Kendall Hunt Publishing Company.

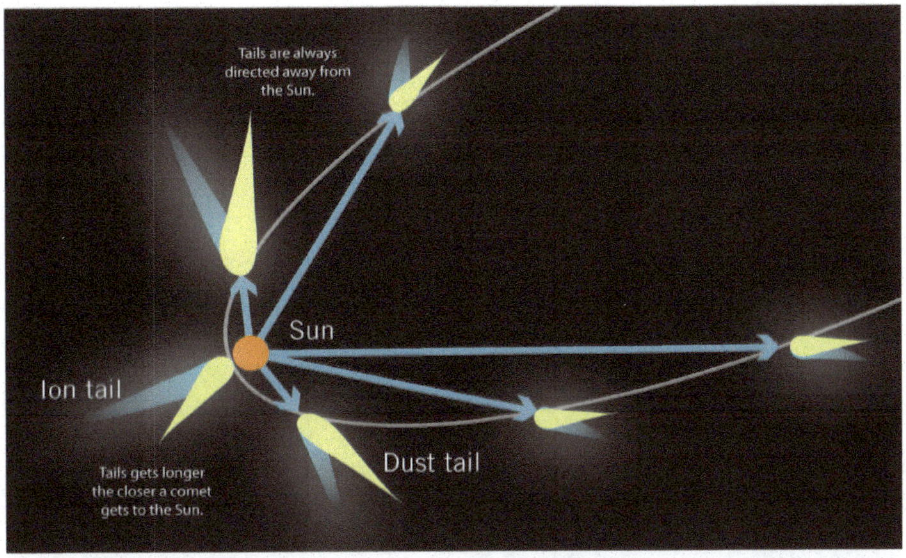

Figure 7.28 Comet tails form as the comet comes into the solar system and begins to warm up from the energy of the Sun, and as the comet gets closer to the Sun the tails get longer. The tails always point away from the Sun—gassy ion tails point straight away from the Sun at every direction whereas the dust tail curves slightly with the orbital path of the comet.

of the coma and ionize some of it, forming a light blue ion tail that always points directly away from the Sun, and a grey dust cloud made up of small particles pushed out of the coma by solar radiation. The dust tail, made of smoke-sized dust particles, can be 10 million km (6 million mi) long, often forms a curved tail, and the particles that are left behind on the comet's track later form meteor showers when Earth passes through this dust trail during orbit (**Figure 7.28**). Some of these occur regularly: the Perseid meteor shower happens every year in mid-August when Earth passes through the dust trail of the Swift-Tuttle Comet, and the Orionid shower in October is from the dust trail of Halley's Comet. The ion tail is several hundred million kilometers long, and can be up to an astronomical unit in length, or equal to the distance between the Earth and the Sun (**Figure 7.27**). The orbits of comets are often very ellipitical and they are invisible except when they are near to the Sun. After about 500 passes most of the ice and gas has been lost from the comet and what is left is a small rocky object that looks very much like an asteroid. Up to half of the near-Earth asteroids might be the dead cores of old comets.

Oort Cloud
A far-reaching, spherically shaped body of material located between 50,000 and 100,000 AU from the Sun; the major source of comets.

The source of 90% of comets is the **Oort Cloud**, named for the Dutch astronomer Jan Oort, who discovered it in 1950. It is a large, spherical region in space that stretches from 20,000 AU to 50,000 AU from the Sun. The Oort Cloud is the outermost part of our Solar System (**Figure 7.29**). Within the Oort Cloud are several trillion icy bodies. If the orbit of a body is disturbed by the passing of another object, the icy body can leave the Oort Cloud and come under the gravitational attraction of the Sun and the larger outer planets. This causes the icy body to begin a journey into our Solar System; the body is now a comet, speeding toward the Sun (**Figure 7.30**).

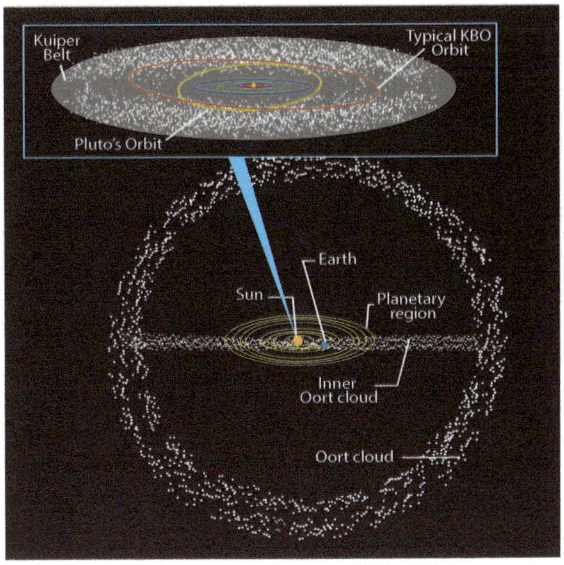

© Kendall Hunt Publishing Company.

Figure 7.29 Kuiper Belt and Oort cloud map.

Long-period comets come from the Oort Cloud and have very elliptical orbits and return periods that can range from 200 years to thousands or millions of years. Short-period comets have a return period of less than 200 years,

© MarcelClemens/Shutterstock.com.

Figure 7.30 Hale-Bopp comet passing Earth in 1997, the blue ion tail and the yellow dust tail are both visible.

and they usually orbit in the same plane and direction that the planets orbit around the Sun. They pass outside the region of the outermost planets into the **Kuiper** (KY-per) **Belt**, a region lying just outside the orbit of Pluto, which tends to be the source of short-term comets. The gravitational effect of Neptune on objects in the Kuiper Belt causes a body to occasionally move out of its orbit and head toward the Sun.

Kuiper Belt

A region beyond the orbits of Neptune and Pluto that is a source of short-period comets.

Halley's Comet

The most well-known of all comets is Halley's Comet. It is the only known short-period comet that is visible to the naked eye from Earth, and the only naked-eye visible comet that might appear twice during a human lifetime. The earliest recorded observations of the comet were made more than 2,000 years ago in China. It may have been recorded as early as 467 B.C. in ancient Greece and China. Subsequent observation and study showed the comet to have a very predictable return rate. The comet was studied by Edmond Halley, an English astronomer who was later named Royal Astronomer. He predicted its return in 1758 based on an observed return cycle of 75 or 76 years. The comet was named in his honor following his death in 1742. It last appeared in the inner Solar System in 1986 and will next appear in 2062 (**Figure 7.31**).

Other calculations revealed that Halley's Comet was the same one that appeared in 1066 before the Battle of Hastings. Its appearance then was interpreted by the Saxons as a bad sign prior to being defeated by the Normans (**Figure 7.32**).

Spacecraft observations of Halley's Comet in 1986 indicated that its core is about 15 km long and 8 km wide (9 by 5 mi). This nucleus contains icy chunks that are thought to be covered with carbon compounds, causing the core to appear dark.

Figure 7.31 **Halley's Comet orbit brings it into the inner solar system once every 75 years or so. Its last close approach was in 1986 and it will return in 2062. As Earth passes through dust from previous paths of Halley's Comet each year, we see two meteor showers—the Orionids in late October, which appear to radiate from the constellation Orion, and the Eta Aquarids in early May, which radiate from the constellation Aquarius.**

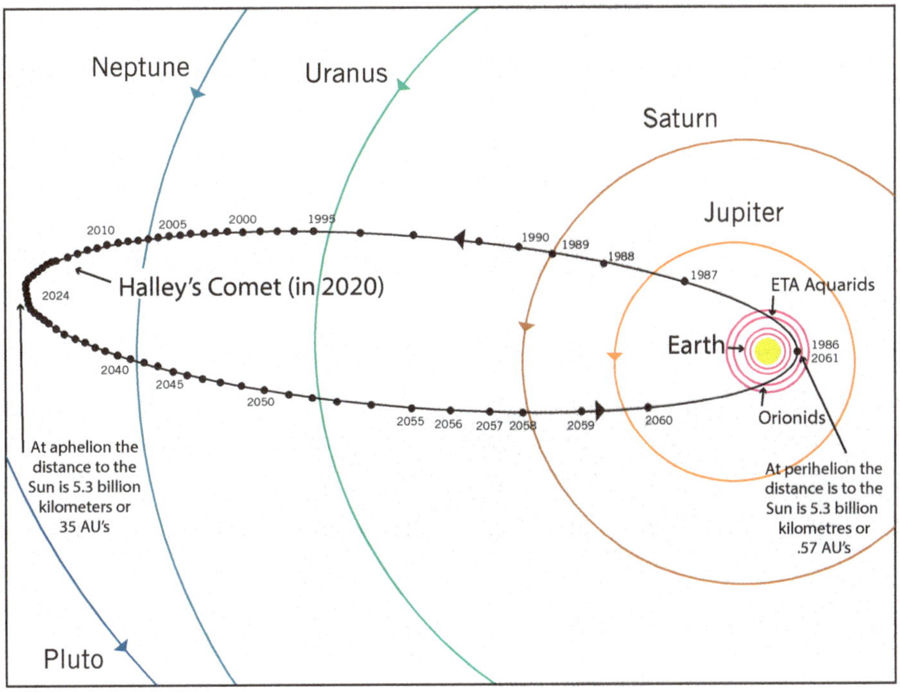

© Kendall Hunt Publishing Company.

Comet Shoemaker-Levy 9

In March 1993 Gene and Carolyn Shoemaker along with David Levy discovered the comet bearing their name (this was the ninth comet the team discovered). It was soon determined that the comet was in orbit around Jupiter and had the potential to strike the planet. Jupiter's huge gravitational forces broke the comet into at least 21 separate pieces that collided with the planet over a 6-day period beginning on July 16, 1994 (Figure 7.33). This event represented the first time we could observe the collision of a planet by a large extraterrestrial object in our solar system (Figure 7.34). The spectacular series of events was described by the Shoemakers and Levy to a worldwide television audience of millions of viewers throughout the world. It should be noted that following the untimely death of Gene Shoemaker in Australia in 1997, Carolyn Shoemaker has continued to discover asteroids, and has found more than 800 asteroids and 32 comets. She is a remarkable scientist who began her career in astronomy in 1980, at the age of 51.

© jorisvo/Shutterstock.com.

Figure 7.32 **This image from the 11th century Bayeux Tapestry shows King Harold during the Norman invasion of England with Halley's Comet in the upper left, portending bad omens.**

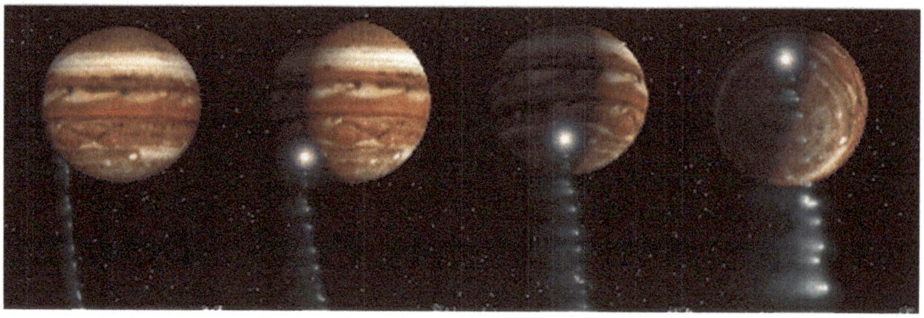

Figure 7.33 From the 16th to 22nd of July, 1994, 21 fragments of Comet Shoemaker-Levy 9, some up to 2 km (1.2 mi) across, collided with the surface of Jupiter. This is the first collision of two solar system bodies ever observed. This artist's rendition shows the collision of fragment Q with Jupiter from different perspectives.

Source: NASA.

Source: NASA.

Figure 7.34 The largest impact was from fragment G. The impact spot from this fragment was over 12,000 km (7,456 mi) across and equivalent to 6 million megatons of TNT, about 600 times the world's nuclear weapons.

Photo by Vokrug Sveta.

Figure 7.35 A photograph of trees blasted down in the Tunguska explosion of 1908, taken by an expedition to the area in 1929. Trees were blown over in a radial pattern away from the site directly below where the airburst was thought to have taken place.

A Recent Collision With Earth

Tunguska, Siberia, experienced an explosion of catastrophic proportions on June 30, 1908. It is the largest impact event on Earth in the recorded history of civilization. Scientists have hypothesized that a fragment of a comet or asteroid from 60 to 190 m (200 to 620 ft) in diameter exploded about 5 to 10 km (3 to 6 mi) above the ground surface, generating a shockwave that flattened more than 2,000 square km (1,250 square mi) of thick forest and knocked down over 80 million trees (Figure 7.35). There are no known human casualties as a result of this event.

Effects of Impacts

Mass Extinction Events

An extinction event is a time in Earth history with widespread and rapid drop in the biodiversity or number of species on Earth. We know of five large extinction events, the worst of which eliminated about 90% of all species living on Earth at the time, and researchers have discovered that there may have been meteorite impact events that contributed to each of these. The boundary between the Cretaceous and Paleogene at about 65.5 million years ago is one of the three largest mass extinction events in the last 500 million years, and coincides with the Chicxulub meteorite impact in Mexico. This extinction event resulted in the disappearance of about 76% of all fossilizable species on Earth, including the non-avian dinosaurs. The asteroid was about 10 kilometers in diameter and impacted the area that is now the Yucatan Peninsula in Mexico at 20 km (13 mi) per second (Figure 7.36). The transient crater was 100 km (62 mi) wide and 20 to 30 km (13 to 19 mi) deep, and the final crater is about 180 km (110 mi) wide and 2 km (1.2 mi) deep. Ejected target rocks were thrown up into the upper atmosphere and space, where many of them fell back to the ground. The re-entry created a brief

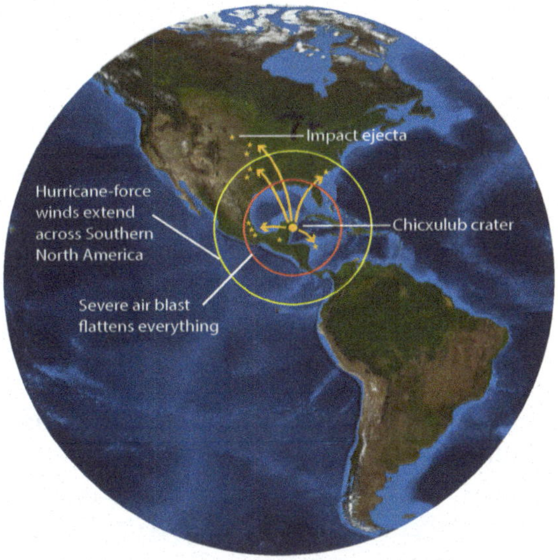

© Kendall Hunt Publishing Company.

Figure 7.36 The Chicxulub Basin stretches for approximately 180 km (112 mi) and is the result of the impact of the Chicxulub asteroid 65 million years ago.

National Science Foundation.

Figure 7.37 This outcrop on Interstate 25 near Raton Pass, Colorado, US, contains the clay layer formed during the Chicxulub impact event. The white layer marks the impact; below that layer, pollen and spores from Cretaceous plants are abundant and above that they are absent.

pulse of infrared radiation, killing any exposed organism. Based on the amount of soot found in the global debris layer from that time, the entire terrestrial biosphere may have burned, blocking out the Sun and creating a nuclear winter effect. The dust cloud would have also blocked the Sun for up to a year. The target rocks the crater formed in had high levels of sulfur, which was injected into the stratosphere, reducing the amount of sunlight reaching Earth's surface by up to 50% and causing acid rain. Sea surface temperatures dropped by up to 7 °C (12.6 °F) for decades after the impact. The impact itself would have caused some of the largest earthquakes ever generated—some greater than magnitude 11.0—as well as tsunamis that radiated across the Gulf of Mexico, Caribbean, and North Atlantic Oceans. Winds stronger than 1,000 km per hour (621 mph) blew across the area surrounding the impact site. The layer of impact material from Chicxulub is found globally and contains high levels of iridium, which is rare on Earth but abundant in asteroids (Figure 7.37).

Key Terms

amino acids *(page 204)*
asteroid *(page 195)*
astronomical unit *(page 196)*
Big Bang Theory *(page 194)*
brecciated *(page 205)*
central peak structure *(page 215)*
comet *(page 216)*
complex crater *(page 215)*
kiloton *(page 209)*
kinetic energy *(page 210)*
Kuiper Belt *(page 218)*
lamellae *(page 206)*
Main Asteroid Belt *(page 195)*
megaton *(page 209)*
meteor (page 198)
meteorite *(page 198)*
meteorite fall *(page 200)*
meteorite find *(page 200)*
meteoroid *(page 198)*
Oort Cloud *(page 217)*
peak-ring structures *(page 215)*
simple crater *(page 215)*
transient crater *(page 214)*

Summary

The universe originated in a Big Bang about 13.8 billion years ago, and our Sun and Solar System formed about 4.6 billion years ago. The oldest inclusions in meteorites are thought to be the first solids formed in our Solar System and represent pre-solar material. The inner planets are the Earth-like terrestrial planets of Mercury, Venus, Earth, and Mars, and the outer, Jovian gas giant planets are Jupiter, Saturn, Uranus, and Neptune. The Main Asteroid Belt is located between Mars and Jupiter and is made up of tens of millions of asteroids. Asteroids that pass near Earth's orbit are called Near Earth Asteroids and are divided into Amor, Aten, and Appollo groups based on their orbital paths. If an asteroid crosses Earth's orbit and is greater than 150 m (490 ft) in diameter it is considered potentially hazardous.

Meteoroids are pieces of debris in space, meteors are objects that disintegrate in Earth's atmosphere, and meteorites are objects that survive the passage through the atmosphere and land on Earth's surface. Meteorites are divided into different groups based on composition. Original unmodified Solar System material that has accreted together forms stony meteorites called chondrites. Solar System bodies that have had melting and differentiation generate iron meteorites from fragments of metal cores, stony-iron meteorites from areas near the core-mantle boundary, and stony achondrite meteorites from differentiated crust. Meteorites represent some of the oldest material in our Solar System as well as fragments of other nearby planets like Mars and the Moon.

Each year an average of one meteorite per square kilometer hits Earth. They range in size from a few millimeters to several meters in diameter (up to about 10 ft) and can be identified by their physical characteristics: irregular shape, black fusion crust, surface textures, metal content, internal texture, and magnetic properties.

Meteorite impact craters form when large meteorites hit Earth's surface. The first stage of crater formation is the initial contact and compression stage which occurs when the meteorite first touches the surface and lasts for a fraction of a second in all but the largest impact events. Once all of the energy of the meteorite has been transferred to the target material, the excavation stage starts. This is when the actual crater forms as shock waves push the target material down and out to generate a bowl-shaped depression. The modification stage starts once the shock waves are no longer able to deform the target material and gravity begins to shape the crater. Simple craters are up to about 4 km (2.5 mi) in diameter and have a bowl shape, larger craters are more complex in shape and can have a central uplifted peak or rings inside the outer rim as the outer edge collapses under the effects of gravity. The largest crater on Earth is the Vredefort in South Africa and would have been more than 300 km (186 mi) in diameter when it formed about 2 billion years ago. Impact events on Earth can have significant global effects including major mass extinction events.

Comets are icy rocky masses with a central rocky core that originate in the outer reaches of the solar system in the Oort Cloud or Kuiper Belt and pass through the inner Solar System. As they travel close to the Sun the icy material sublimates to gas, forming a coma and hydrogen cloud around the nucleus. As the comet interacts with charged particles from the solar wind, it forms a blue ion tail and also leaves a dust trail. As Earth passes through old comet trails, the old dust material left behind enters our atmosphere and generates meteor showers.

References and Suggested Readings

Makishima, A. (2017). *Origins of the Earth, Moon, and life: An interdisciplinary approach.* Elsevier.

Norton, O. R., & Norton, D. S. (1998). *Rocks from space: Meteorites and meteorite hunters.* Mountain Press.

Smith C., Russel, S., & Benedix, G. (2010). *Meteorites.* Firefly Books.

Websites for Further Reference

https://asteroidday.org/

https://b612foundation.org/

https://psyche.asu.edu/

https://www.passc.net/EarthImpactDatabase/

Questions for Thought

1. Explain the difference between a meteor and a comet.
2. How is a meteor different from a meteorite?
3. Distinguish between the asteroid belt and the Oort Cloud.
4. What is the basic concept associated with the Big Bang theory?
5. What is the relationship of the age of the Sun and its Solar System as measured against that of the Earth?
6. Give two major differences between the inner planets and the outer planets.
7. What can cause an asteroid to change its travel path?
8. Why are relatively few impact features on Earth well preserved?
9. Describe two types of tail structures associated with comets.
10. What was significant about Comet Shoemaker-Levy?
11. Explain the significance of the Chicxulub impact crater in Yucatan, Mexico.
12. Describe how meteorite impact craters form and the difference between simple and complex craters.
13. Describe how meteorites are classified.

8 Weather Hazards

© Cammie Czuchnicki/Shutterstock.com.

Supercell thunderstorms are the least common type of storm but are the most likely to produce severe weather including strong winds, large hail, and tornadoes. They are fuelled by a persistent and strong rotating updraft called a mesocyclone.

Lightning Is a Killer

Lightning is a large-scale spark or discharge of electrical energy that occurs within the atmosphere or between the atmosphere and the surface of the Earth. It forms a highly conductive plasma channel within the air where the current flows from an electrically charged area to another, to try to re-equilibrate the electrical charge. The air around the plasma can heat up to 25,000 °C (45,000 °F), which is more than 4 times hotter than the surface of the Sun. There are 50 to 100 lightning strikes worldwide every second.

© Cloudscrest Images/Shutterstock.com.

Figure 8.0.1 **A severe thunderstorm with lightning passes over Perth, Australia.**

Severe weather is any meteorological activity with the potential to cause damage, serious social disruption, or loss of human life. All places on Earth have the potential for severe weather, making this a common global hazard. The types of severe weather that might occur depend on the latitude, altitude, topography, and atmospheric conditions of the place at that time (**Figure 8.1**). There are roughly 16 million thunderstorms worldwide each year, and at any given moment there are 2,000 thunderstorms in progress across the planet. About 10% of these reach severe levels.

Weather is the state of the atmosphere at a particular time and place, and this refers to things like temperature, clouds, sun, wind, rain, snow, and other events (**Figure 8.2**). Weather refers to short-term variations in the atmosphere at a particular place. Thunderstorms are a normal result of the interaction of different air masses, and aren't classified as severe weather unless they reach certain specific conditions. In the United States, the National

© NAntoine/Shutterstock.com.

Figure 8.1 **Severe weather events are a global occurrence and include high winds, hail, intense precipitation, thunderstorms, downbursts, tornadoes, blizzards, ice storms, and dust storms, depending on the location and other conditions.**

© Everett-Art/Shutterstock.com.

Figure 8.2 **Weather is the short-term condition of a particular place at a particular time. Thunderstorms generally need to meet specific criteria before they are classified as severe weather.**

Weather Service classifies intense thunderstorms into one of three different categories. "Approaching severe" thunderstorms have hail between 13 to 25 mm (0.5 to 1 inch) diameter or winds between 80 to 93 km/h (50 to 58 mph). At this stage, these storms are assigned a Significant Weather Alert. To be classified as "Severe," thunderstorms must have hail with a diameter between 25 to 51 mm (1 to 2 inches), winds from 93 to 121 km/h (58 to 75 mph), or generate an F1 tornado. These storms are assigned a Severe Thunderstorm Warning. "Significant Severe" thunderstorms are the most intense, and these have hail larger than 51 mm in diameter (2 inches), winds greater than 120 km/h (75 mph), or generate a tornado stronger than F1. These storms are also assigned a Severe Thunderstorm Warning. Countries such as Canada, Australia, Germany, Mexico, and the United Kingdom have similar criteria for severe thunderstorms but the specifics of of hail size, wind speed, and the presence of a tornado vary from country to country. For example, in New Zealand a severe thunderstorm needs to have hail larger than 2 cm (0.79 in), wind speed higher than 110 km/h (68 mph), and a minimum rainfall rate per hour of 2.5 cm (0.98 in). In the United States, high rainfall rates trigger a separate flash flood warning.

Earth rotates on its axis roughly once every 24 hours, or once a day. The Earth spins on its axis from west to east, and this rotation produces an apparent deflection of any objects moving in a straight path relative to the Earth's surface, such as airplanes, wind, and ocean currents. The effect is termed the **Coriolis Effect**, after an early 19th-century French mathematician who proposed its existence. The effect is that any object in the Northern Hemisphere appears to move to the right (clockwise) when viewed from the ground below (on the spinning Earth) and an object in the Southern Hemisphere appears to move to the left (counterclockwise) when viewed from below (**Figure 8.3**). There is actually no physical force involved, because

Coriolis Effect

An imaginary force that appears to be exerted on an object moving within a rotating system. The apparent force is simply the acceleration of the object caused by the rotation. Along the equator, there will be no such rotation.

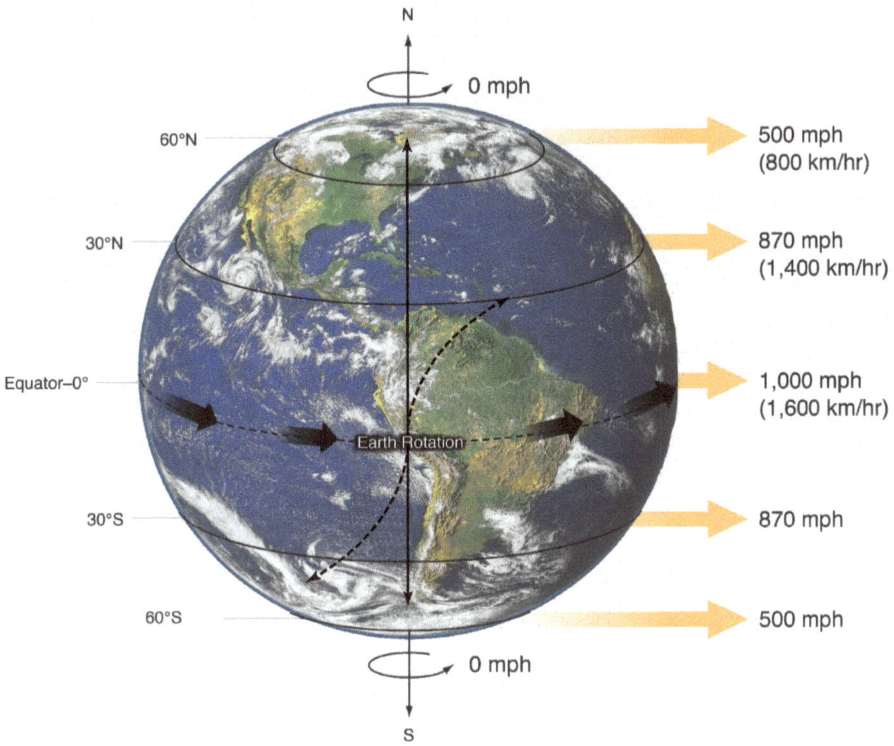

© Sandy Stifler/Shutterstock.com. Adapted by © Kendall Hunt Publishing Company.

Figure 8.3 The Coriolis Effect is the apparent change in an object's path as it moves above the rotating Earth. This causes an apparent deflection of an object moving in a straight line, if an object is moving on a north-south path, it will appear to be deflected to the right in the Northern Hemisphere and to the left in the Southern Hemisphere. The Coriolis Effect is zero at the equator and gets stronger as you approach the poles.

it is just the ground moving a different speed than an object moving above the Earth's surface, and the effect is proportional to the speed of Earth's rotation at different latitudes, becoming more extreme as you move toward the poles and decreasing to zero at the equator. A plane flying in a straight line from the equator toward the North will appear to take a curved path when viewed from the ground above, because while the plane is flying in a straight line, the Earth is rotating below it. For the planet's atmosphere, it means that the winds appear to be deflected counterclockwise in the Northern Hemisphere and clockwise in the Southern Hemisphere.

Atmospheric Circulation

air mass

A large body of air of considerable depth which are approximately homogeneous horizontally. At the same level, the air has nearly uniform physical properties, especially temperature and moisture.

The atmosphere acts like fluid and moves readily. Several factors produce movement in the troposphere. Differential heating of the surface of the Earth and the overlying air generate warm and cool **air masses** that move across the surface. The upward motion of warm air into the upper troposphere is caused by convection. The equatorial regions receive the greatest amount of solar radiation (**Figure 8.4**). Rising warm air moves toward the poles in the upper portion of the troposphere. This air contains large amounts of water vapor derived from the evaporation of ocean waters near the equator. The area near the equator, where warm air is rising, is one of low pressure because of the buoyant lifting effect of the rising air.

Colder temperatures in the upper troposphere cause condensation of this water vapor, which returns to Earth as precipitation. The air in the upper troposphere is now cold and depleted of moisture, so it begins to sink in a region near 30° to 35° North and South of the equator. This region is called the subtropical high-pressure zone. High pressure exists in those

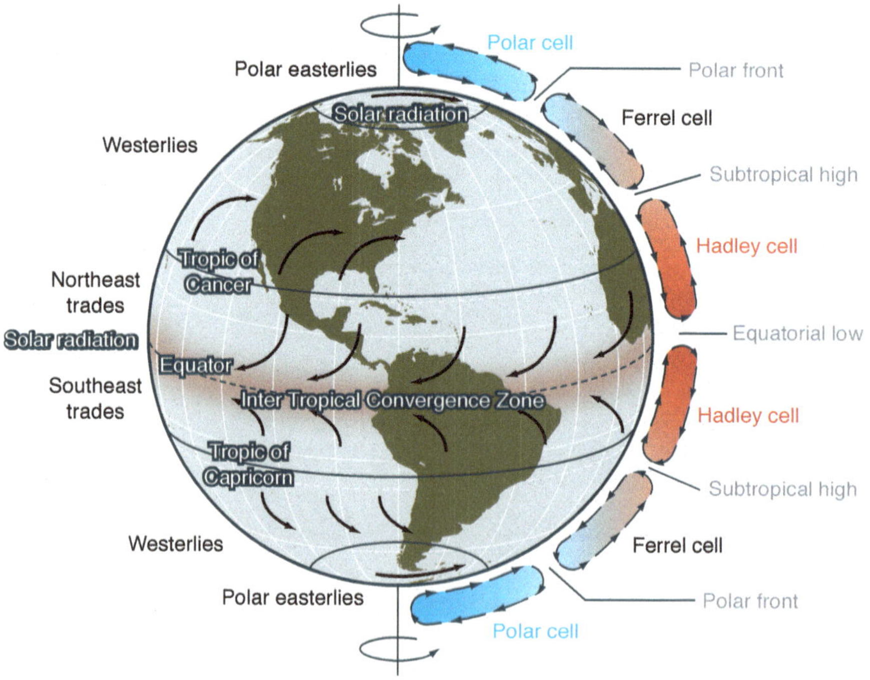

© Kendall Hunt Publishing Company.

Figure 8.4 Large amounts of solar radiation near the equator cause heating of the surface and atmosphere. Rising air then moves poleward where it descends in the subtropical regions near 30 degrees north or south of the equator.

areas where the cooler, dry air is descending back to the surface. As the air descends, it is compressed and begins to heat up, so the surface air is warm and dry. These conditions form many of the mid-latitude deserts on Earth, such as the deserts of North America, the Middle East, and Saudi Arabia in the Northern Hemisphere, and in Australia in the Southern Hemisphere. These descending winds are also deflected by the Coriolis Effect and produce the prevailing westerlies.

Air Masses

Large-scale movement of the atmosphere often involves expansive air masses. North America is affected by these masses, as they often move down from regions around the North Pole or off the Pacific Ocean. Polar air masses are typically cold and dry, while those coming off the Pacific and Gulf of Mexico are warm and moist. General movement is from west and northwest to east and southeast. These directions are driven by Coriolis forces and the prevailing west-to-east **jet streams** that traverse the country (Figure 8.5). Occasionally these two different air masses will collide; the result is often very unstable weather conditions.

jet stream
A high-speed, meandering wind current, generally moving from a westerly direction at speeds often exceeding 400 kilometers (250 mi) per hour at altitudes of 15 to 25 kilometers (10 to 15 mi).

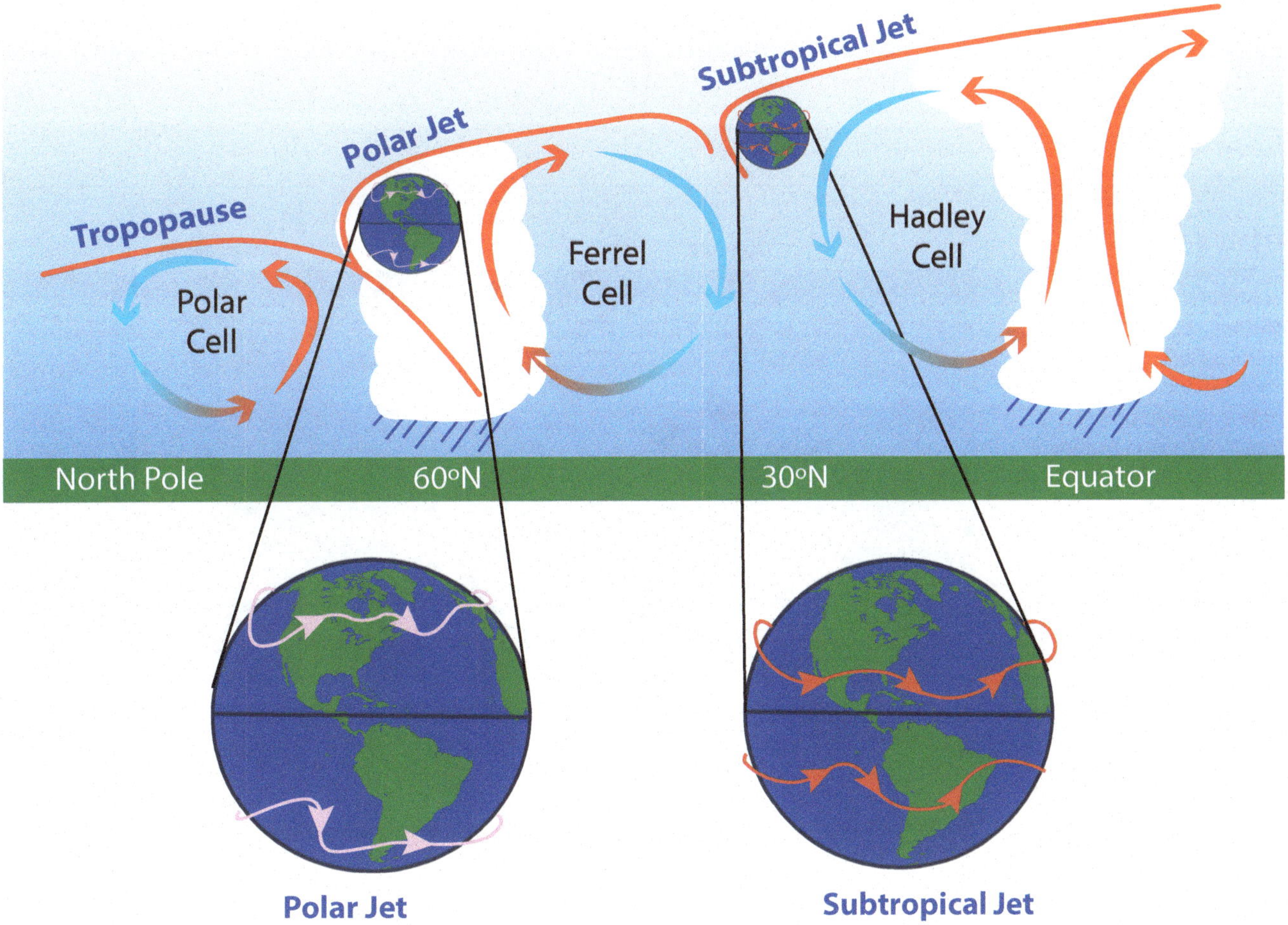

© Kendall Hunt Publishing Company.

Figure 8.5 Jet stream diagram.

Low-Pressure Conditions

In an area where less dense air rises due to heating, the upward force generates an area of low pressure. This lower pressure zone draws air in from surrounding areas of higher pressure (Figure 8.6). It typically indicates bad weather. Air near the ground is drawn into the rising low-pressure center, where it moves upward, expanding and cooling as it rises. As it cools, it condenses water which leads to higher humidity, rain, and clouds. In the Northern Hemisphere this rising air rotates in a counterclockwise manner and in the Southern Hemisphere the zone of low pressure rotates in a clockwise manner. These low pressure systems are termed **cyclones**.

cyclone

An area of low atmospheric pressure having is counterclockwise circulation in the Northern Hemisphere and clockwise motion in the Southern Hemisphere.

High-Pressure Conditions

After air has risen and cooled off, it becomes more dense and then begins to sink. This is an area of high pressure where the air moves down and then diverges and moves out at the surface. This air circulation pattern typically indicates fair weather. Near the surface, air at the center of an anticyclone is forced away from the zone of high pressure by a downward current of air from higher altitudes. The air compresses and heats up as it moves down, dropping in humidity and resulting in few clouds and sunny weather. In the Northern Hemisphere this sinking air rotates in a clockwise motion and in the Southern Hemisphere the zone of high pressure rotates in a counterclockwise manner. These high pressure systems are termed **anticyclones** (Figure 8.6).

anticyclone

An area of high atmospheric pressure having clockwise circulation in the Northern Hemisphere and counterclockwise motion in the Southern Hemisphere.

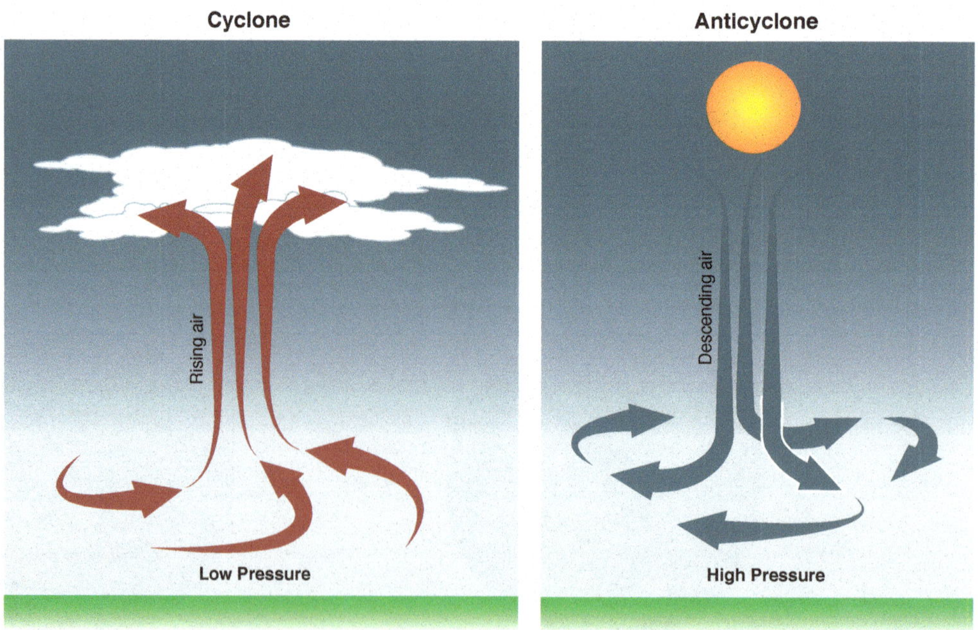

© Kendall Hunt Publishing Company.

Figure 8.6 Rising air produces a low pressure condition on the surface in the Northern Hemisphere, producing a counterclockwise rotation; high pressure produced by colder, descending air generates clockwise winds at the surface.

Role of Land Masses on Global Circulation

The mean height of land above sea level is 840 meters (2,756 ft), and ranges from −418 meters at the Dead Sea to 8,848 meters at the top of Mount Everest (−1,371 ft to 29,029 ft). An idealized circulation pattern for Earth does not take into account land masses (**Figure 8.4**). On several continents there are many mountain ranges that disrupt the flow of air in the atmosphere. The long, relatively linear stretches of mountains such as the Rocky Mountains of North America and the Andes Mountains of South American stretch for thousands of kilometers in a north-south direction. These impedances cause the flow of air to be altered, and thus change the weather and climate associated with the theoretical flow patterns. The Himalaya Mountains of Asia are the highest on Earth, reaching 8,850 m (29,000 ft) above sea level. Air that strikes these peaks is driven higher into the upper troposphere where the moisture is concentrated and returned to Earth as snow. These features also disrupt the normal flow of air around the globe. **Figure 8.7** shows global atmospheric patterns as seen in the clouds. The line of clouds along the equator in the tropical eastern Pacific and Atlantic Oceans is typical of the area where the air is heated, rises, and condenses to form clouds.

Role of Water in Global Climate

Water can exist in one of three states or phases: as a liquid, as a solid (ice), or as a gas (water vapor). When water changes from one state to another, heat is either absorbed or released, as seen in **Figures 8.8a** and **8.8b**. The amount of heat, measured in calories, needed to change 1 gram of water from one phase to another can be measured. For water to be transformed

Source: NASA.

Figure 8.7 This MODIS image, or moderate resolution imaging spectroradiometer, is an instrument on NASA's Terra and Aqua satellites that acquire new images of the Earth's entire surface every 1 to 2 days.

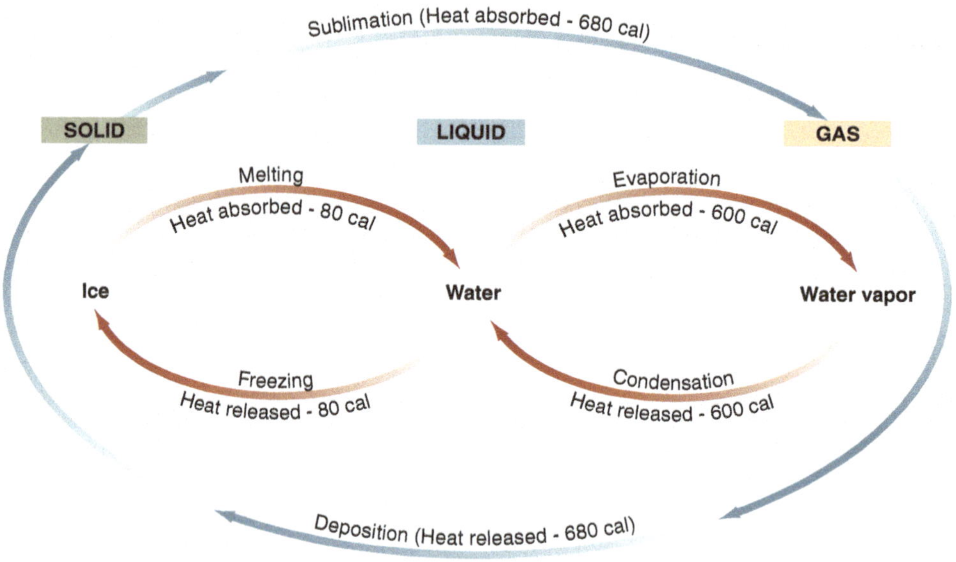

© Kendall Hunt Publishing Company.

Figure 8.8a Heat is absorbed as water changes from a solid to a liquid and then to a gas. Heat is given off when it changes from a gas to a liquid and then a solid. The number of calories shown are those needed to change the state for 1 gram of water.

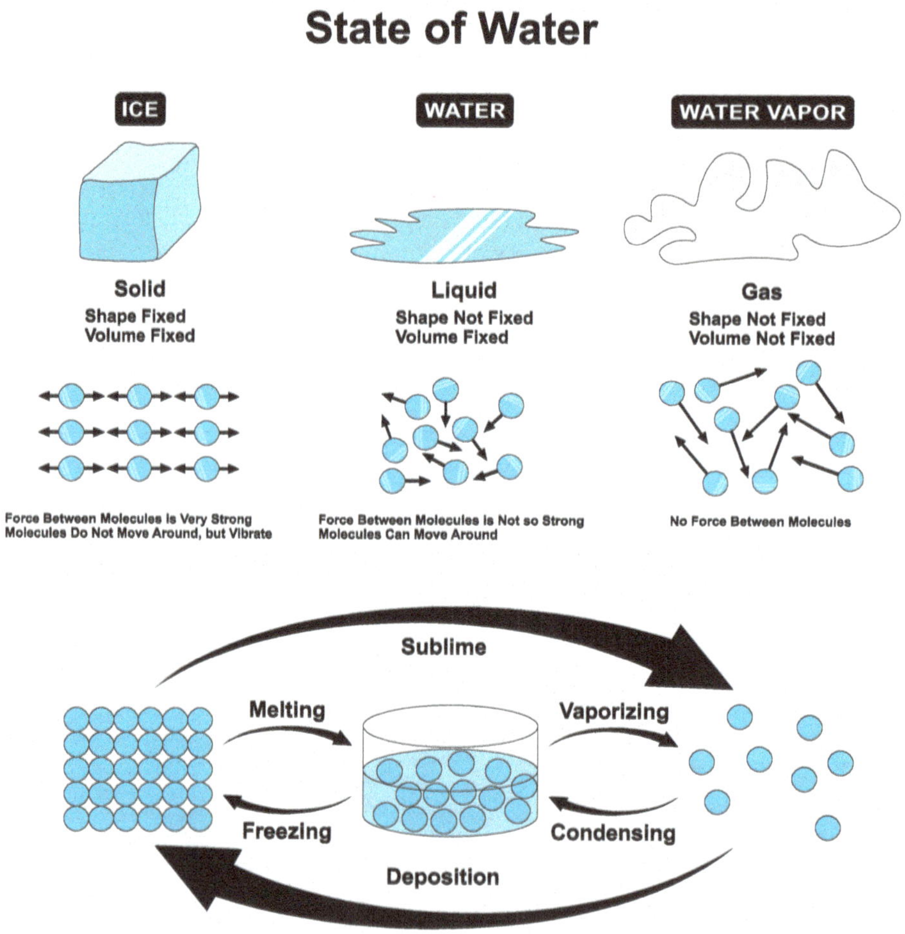

© udaix/Shutterstock.com.

Figure 8.8b Phase transformations of water from solid (ice) to liquid (water) to gas (steam) reflect increasing energy and changes in the arrangement of water molecules in the material at each step.

latent heat of vaporization
Heat stored in water vapor as it changes states from a liquid to a vapor.

latent heat of fusion
Heat released when water freezes to form ice.

sublimation
The process that changes a solid into a gas, bypassing the liquid phase.

latent heat of condensation
Heat released when water vapor absorbs heat to be transformed to water.

from a liquid to a gas, the liquid must absorb 600 calories to evaporate. This is termed the **latent heat of vaporization**. When liquid water freezes, heat is released (80 cal per gram). This is the **latent heat of fusion**. In a case where the liquid phase is bypassed (the transformation of ice directly into water vapor), 680 calories are absorbed. In this instance, the process is termed **sublimation**. The **latent heat of condensation** (600 cal per gram) is associated with the change from water vapor to a liquid. This is a primary source of energy in cyclonic storms, as countless billions of grams of water vapor at high altitudes condense to form liquid water which then falls as rain. This heat adds to unstable atmospheric conditions and also helps fuel storms.

Ocean water, lake water, and other surface water together cover about 75% of the Earth's surface. Of all commonly occurring substances, water has one of the highest heat capacity. This characteristic means that water requires a large amount of heat in order to increase its temperature. The equatorial regions of Earth receive the greatest amount of solar radiation, so this heat can be stored by the oceans. Evaporation is most effective at latitudes at or near the equator. These processes that absorb and release heat are key to driving the convection process in the atmosphere and the oceans.

Fronts and Mid-Latitude Cyclones

Whenever two different air masses collide with one another, they generally do not mix. A cold air mass with little moisture in it is dense and will not intermix with a warm, less dense, moisture-laden air mass. In North America cold air masses descend from the North Pole regions toward the equator. As these cold air masses pass through the mid-latitudes (30° to 45° North), they often collide with warm air masses that have moved northward from the Gulf of Mexico or off the western edge of the Atlantic Ocean. When a cold air mass collides with warm, moist air, the cold air pushes the warm air to higher altitudes. Cold air is more dense and so it is able to force warm air out of the way quickly. Condensation takes place as the moist air encounters colder temperature (Figure 8.9a). A line of high rising, vertical clouds results.

© stihii/Shutterstock.com.

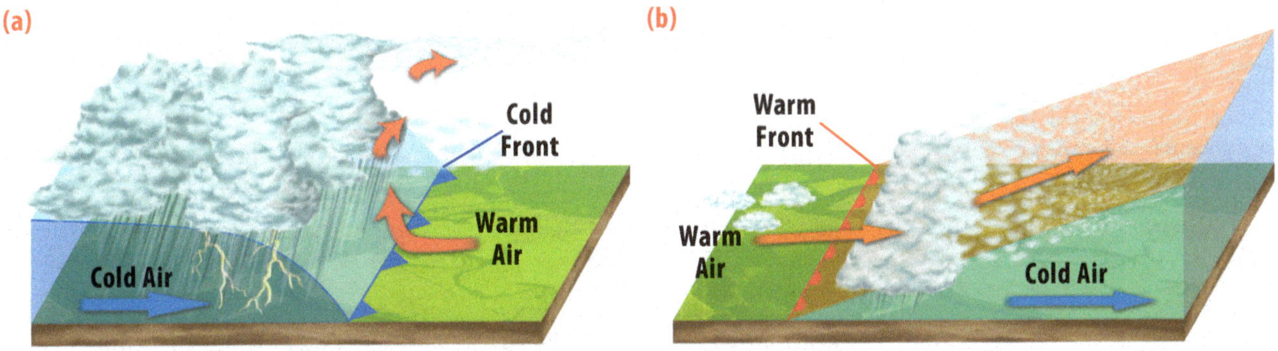

Figure 8.9 When a warm front overrides a cold air mass, the warmer air is spread out along a long distance, producing clouds. When a cold air mass encounters warm air, the warmer air is forced upward, causing condensation to produce thick clouds and the possibility of thunderstorms and lightning.

When warm air collides with cold air, the warm air rises above the denser cold air and smears up and out along a long surface. The warmer air is pushed up to a higher altitude so condensation occurs. This elongated string of condensation produces cloudy conditions that extend over hundreds of kilometers (Figure 8.9b).

© Svetlana Lukienko/Shutterstock.com.

Figure 8.10 Cumulus clouds are the early stage of thunderstorm development. As more updraft moves moisture to high altitudes, the storm enters the mature stage.

Elements of Hazardous Weather

The collision of air masses of different temperatures can create hazardous or severe weather. These weather events are often relatively short-lived in duration, but can be very destructive to property and can cause a significant loss of life if populated areas are affected. Most violent weather begins as picturesque cumulus clouds that eventually are pushed higher in the atmosphere, where their dynamics change (Figure 8.10).

Thunderstorms

Thunderstorms typically form in warmer months of spring, summer, and fall when there is more heat added to the atmosphere, but in some warmer places they can occur at any time of year. There are three basic things that are necessary to develop a thunderstorm: moisture, an unstable atmosphere, and some way to trigger the movement of the atmosphere. Moisture produces the thunderstorm clouds and precipitation. In warm, humid air, there's generally enough moisture present to start a thunderstorm if other ingredients are also present. Atmospheric instability allows the warm, moist air to rise up from the surface. Rising air produces clouds, and rapidly rising air produces thunderstorms. When the atmosphere is unstable, the air near the ground can become buoyant and rise rapidly through the atmosphere. In general, the warmer the air near the Earth's surface and the colder the air above it, the more unstable the atmosphere will be. Moisture near the ground surface can also contribute to atmospheric instability. There also needs to be some kind of a trigger to start air motion. This can be an air mass front, changes in elevation forcing air up, heating from the Sun, or cooling of upper-level air. Once a thunderstorm starts, it can continue to generate triggers for additional thunderstorms. Thunderstorms have three stages of formation: the developing or towering cumulus stage, the mature stage, and the dissipating stage.

thunderstorm
A local storm produced by a cumulonimbus cloud and accompanied by lightning and thunder.

The developing stage involves the formation of cumulus clouds which are puffy clouds that resemble large cotton balls or heads of cauliflower that are pushed upward by a rapid **updraft** of rising warm air (Figure 8.11). Little or no rain forms as the clouds are coalescing at elevations of 5 to 7 kilometers (3 to 4.3 mi).

updraft
A small-scale current of rising air. If the air is sufficiently moist, then the moisture condenses to become a cumulus cloud or an individual tower of a towering cumulus.

A continuation of the updraft pushes moisture to higher altitudes where condensation takes place and precipitation begins to fall to the surface. Once the warm air cannot rise any further it begins to spread out and forms an anvil shape (Figure 8.11). The falling rain drags surrounding air with it and generates downdrafts. This combination of updraft and downdraft forms shear within the thunderstorm. The storm now has reached the mature stage, when it becomes its most violent. Strong downdrafts can generate high winds; hail and heavy rain can fall, and intense **lightning** can develop. If sufficient rotation exists within the storm, tornadoes can be spawned.

lightning
A visible electrical discharge produced by a thunderstorm. The discharge may occur within or between clouds, between the cloud and air, between a cloud and the ground or between the ground and a cloud.

When the amount of downdraft exceeds the rising updraft, the storm reaches the dissipation stage. Descending cold air intercepts warm air near

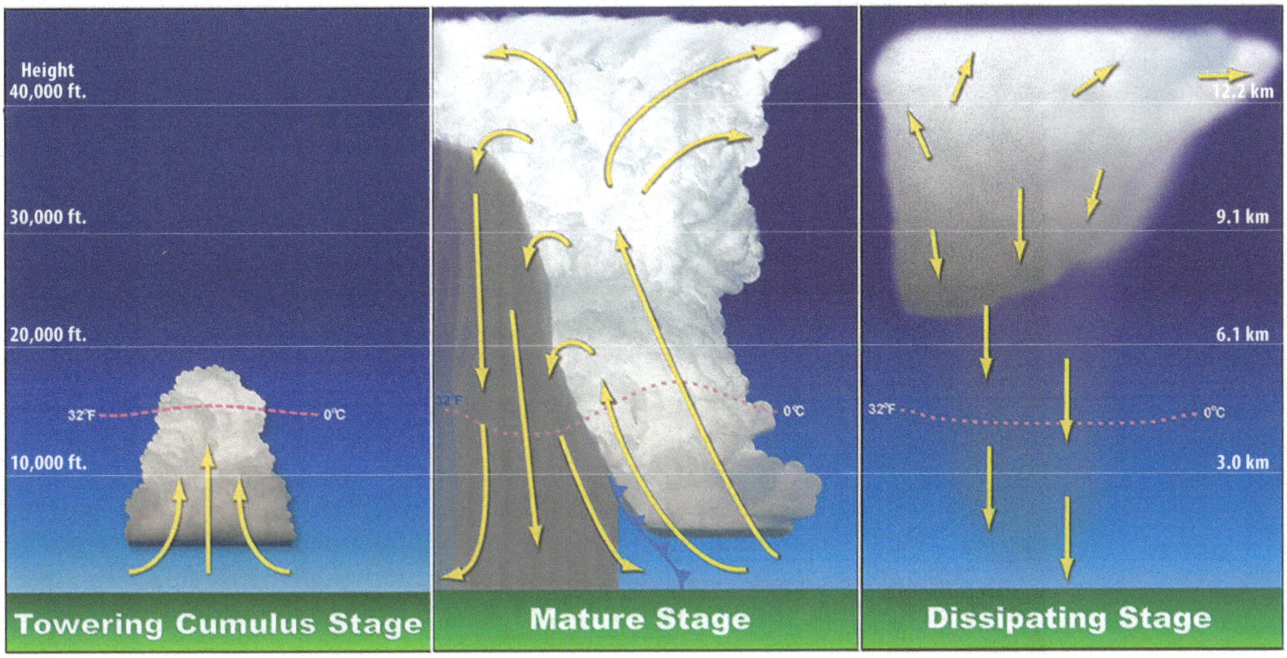

Figure 8.11 Stages of development of a thunderstorm. The developmental or towering cumulus stage begins as warm, moist air rises. The mature stage is characterized by updrafts and downdrafts. The dissipating stage occurs when the upward movement of air has ceased.

the base and prevents the warm air from rising to fuel the storm. Precipitation is in its final stages, although lightning can still be a threat.

Thunderstorms are of four main types. An isolated or single-cell storm is often a short-lived event, lasting maybe 20 or 30 minutes. They are also called air-mass thunderstorms and form from a single main updraft. Their isolated nature generally prevents them from becoming very severe as they do not have sufficient energy to become very large. Some storms can be classified as a severe thunderstorm if they have winds that are at least 93 km per hour (58 mph), hail of 25 mm (0.98 inch) in diameter or greater, or if they have a funnel cloud or tornado. Severe thunderstorms can occur from any kind of storm but are most often produced from multicell clusters, multicell lines, and supercell thunderstorms.

The most common occurrence of severe thunderstorms is as a multicell cluster. Several storm cells move along as a unit, with each cell representing a different phase in the life cycle of the multi-storm system. This configuration typically has a mature cell near the center of the cluster with dissipating cells on the downwind edge of the cluster. Each cell in this group may only last 20 or 30 minutes but the entire group of cells could be active for hours.

A squall line or multicell line is an elongated line of thunderstorms that can be hundreds of kilometers in length and generally form along or ahead of a cold front (Figure 8.12). These lines are known for having strong downdrafts ahead of the line, with large hail, frequent

Figure 8.12 A squall line in Texas, US.

© Minerva Studio/Shutterstock.com.

Figure 8.13 **A supercell thunderstorm over the Great Plains, US.**

lightning, and heavy rainfall accompanying the strong winds, along with possible tornadoes.

Supercell thunderstorms are large, last for 2 to 4 hours, and are characterized by the presence of a rotating updraft or mesocyclone (**Figure 8.13**). These kinds of storms are the least common but have the potential to be the most severe type of thunderstorm. Supercells are characterized by having extremely strong updrafts of about 112 km/hour and up to 280 km/hour in some cases and a significant amount of rotation (69.5 to 174 mph). Hail can exceed 5 cm (2 inches) in diameter due to the extreme updrafts pushing the moisture repeatedly upward. Violent downdrafts are common, and strong to violent tornadoes are most commonly associated with supercell thunderstorms. The extreme updrafts prevent precipitation from falling through the center of the storm. The structure of a supercell is characterized by an overshooting top formed by the powerful updraft in the mesocyclone (**Figure 8.14**). As the rising warm air reaches the troposphere, it spreads out into an anvil-like shape once it reach about 15,000 to 21,000 meters, and juts out in front of the storm. Wall clouds form when humid air is pulled into the updraft and condenses, they appear to descend from the base of the supercell. The forward flank downdraft is the area of most intense precipitation.

The National Oceanic and Atmospheric Administration reports that the typical thunderstorm is 30 kilometers (18.6 mi) in diameter, and lasts about 30 minutes. Thunderstorms can occur during any month, but the largest number occur during the summer months, due to the increased heating of the ocean and atmosphere, particularly in the Gulf of Mexico. All states in the United States experience these storms but the largest number occurs in Florida, where thunderstorms take place an average of 80 to 100 days

Figure 8.14 **The structure of a supercell.**

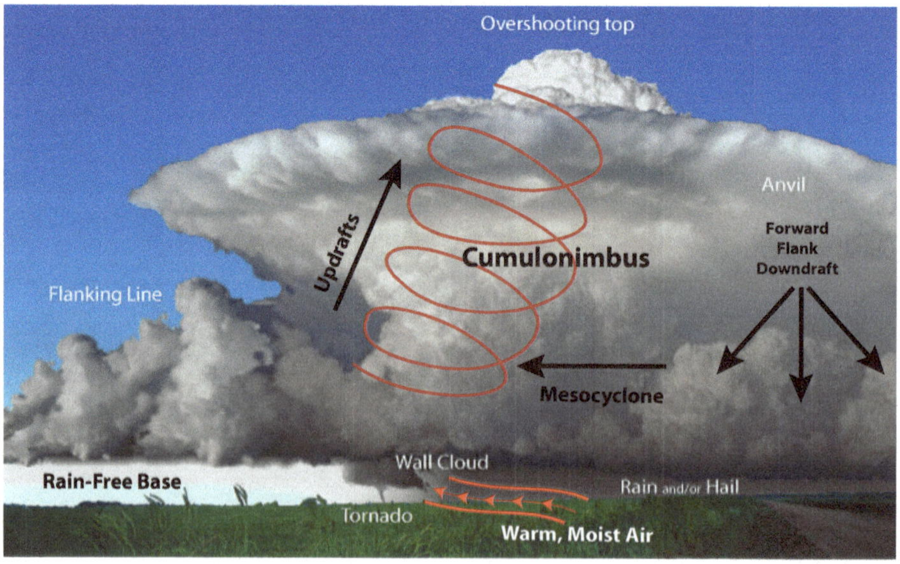

Source: NOAA and National Weather Service

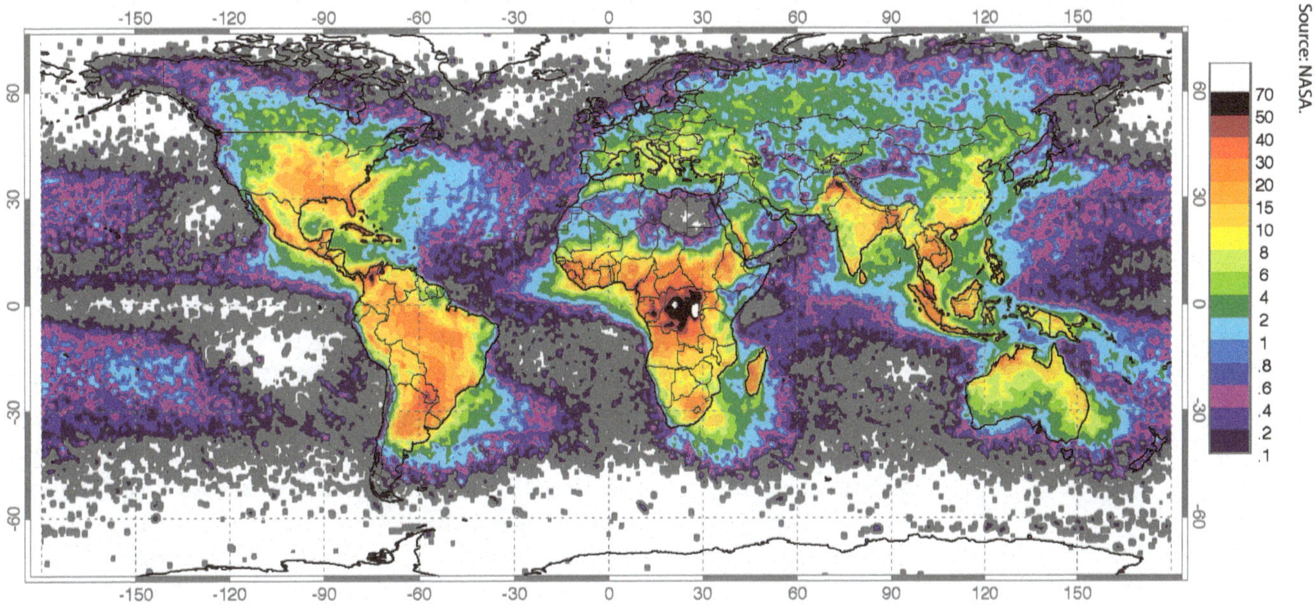

Source: NASA.

Figure 8.15 **The world's most intense thunderstorms were identified by NASA using data from the Tropical Rainfall Measuring Mission (TRMM) satellite, and were located in the Andes Mountains of Argentina. Other areas with powerful storms were southern parts of the Sahara, northern Australia, and parts of the Indian subcontinent. Amazonia, Southeast Asia, and Indonesia experienced frequent storms, but they were rarely severe. Thunderstorms that produce the most lightning are shown here, with the color scale representing average annual number of lightning flashes per square kilometer (0.6 mi).**

per year (Figure 8.15). Venezuela's Lake Maracaibo, located along the Andes Mountain, is the largest lake in South America and the site of, on average, 297 thunderstorms a year. The storms there are known as the Never-Ending Storm, and it is the stormiest place on Earth. The Andes Mountains force air up where it cools and condenses, and that cold air then interacts with the warm, moisture-laden lake air to generate near-constant thunderstorms. Africa is the stormiest continent, with 283 of the 500 most stormy locations on Earth.

hail
Solid, spherical ice precipitation that has resulted from repeated cycling through the freezing level within a cumulonimbus cloud.

Hail

As thunderstorms move air vertically in updrafts, moisture is pushed to higher altitudes, where it condenses out and freezes. Most hail takes on a spherical layered shape and becomes progressively larger as water continues to freeze to the outer surface. At some point, the weight of the frozen sphere becomes too much for the updraft to hold and the particle falls to Earth as **hail** (Figure 8.16). Grapefruit-size hail has been reported in storms in the Midwestern United States, but modern weather warning systems usually allow people to find shelter. The deadliest hailstorm on record was in Moradabad, India, in 1888, where hailstones the size of oranges killed 246 people. While human deaths are rare, large hail can kill livestock unable to find shelter, as well as destroy agricultural crops and can damage roofs and automobiles.

© Rosliak Nataliia/Shutterstock.com.

Figure 8.16 **Hailstones can fall at speeds up to 170 km/hour (105 mph). Bigger hailstones will typically fall faster.**

What Causes Lightning?

Lightning results from a sudden electrostatic discharge, and commonly occurs during thunderstorms. The charged areas temporarily re-equilibrate through the discharge of charged particles in a flash of bright light. This discharge can also be a lightning strike if it interacts with an object on the ground. The discharge is extremely hot and a lightning bolt can heat the surrounding air to temperatures more than four times hotter than the surface of the Sun. As the lightning bolt passes through the air and heats it, the air expands and vibrates rapidly, which is what generates the shock wave that we hear as thunder.

Benjamin Franklin was fascinated by storms and lightning. In June 1752 he conducted experiments with his kite and discovered that lightning was a form of electricity. Soon after that he developed the lightning rod that was placed on top of many buildings of the period. Lightning is a natural phenomenon that is present in large hurricanes, volcanic eruptions, extremely intense wildfires, heavy snowstorms, and most commonly thunderstorms.

The electrical imbalance that causes lightning is generated as a cloud grows, when water droplets in the bottom of the cloud are carried upwards in updrafts and interact with ice crystals from the top of the cloud that are being pushed down in downdrafts. As these particles bump into each other electrons are stripped off of them. The electrons move toward the bottom of the cloud and the positively charged particles move toward the top of the cloud, with the atmosphere between them acting as an insulator within the cloud itself (**Figure 8.17**).

On Earth's surface underneath a thunderstorm, positive charges collect and move along the surface with the storm. Eventually the electrical force between the negatively-charged cloud base and the positively-charged ground surface creates lightning (**Figure 8.18**). The charge differences within a storm tend to be much stronger, explaining why about 75 to 80% of lightning occurs within storm clouds.

Types of Lightning

Ground flashes involve the ground or something attached to the ground being hit. Lightning can originate from the cloud, and strike the positively-charged surface of the Earth. Lightning can also start on the ground and move from the ground upward to the cloud. Lightning bolts travel about 100,000 kilometers per second (62,137 miles per second) and have an estimated width of 1 to 2 cm (0.4 to 0.8 inches).

Following a bold of lightning, thunder travels out at the speed of sound, roughly 1,236 km per hour (770 mph). The sound is the result of rapidly expanding gases that were superheated by the lightning. Air adjacent to a bolt is heated to 10,000 °C. This superheated air expands and creates the rumbling sound that can be used to estimate the distance to the lightning strike. Thunder is generally heard within 25 km (15.5 mi) of a storm. Sound travels roughly 343 m per second (1,125 feet/sec). Begin counting when you see a lightning flash and count until you hear the thunder, then divide your number of seconds by 5 to get the approximate distance to the lightning

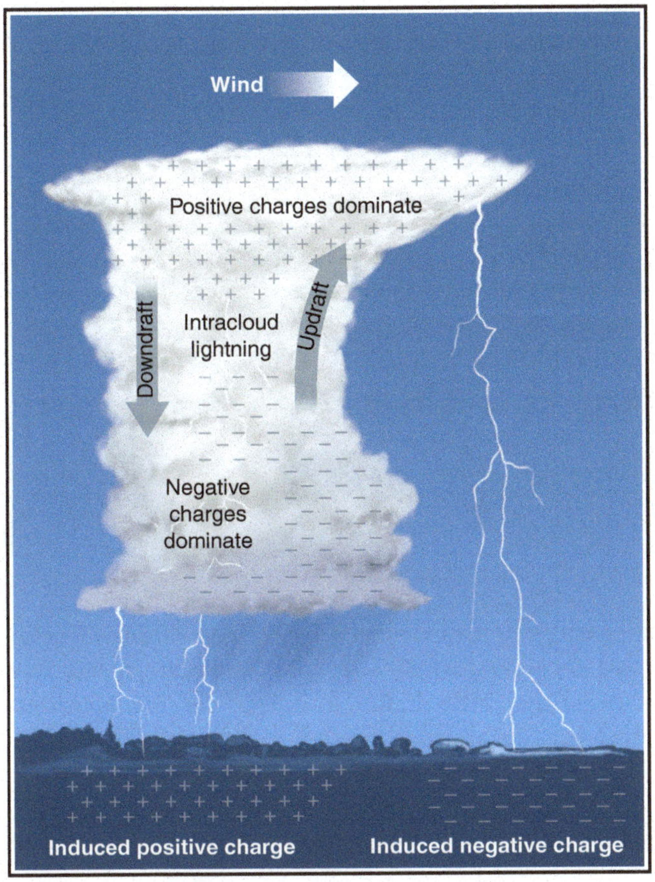

© Kendall Hunt Publishing Company.

Figure 8.17 Electrical charges in a thundercloud are separated with positive charges in the top and negative charges toward the bottom. Once the energy is strong enough, the opposite charges attract and connect to create lightning.

Source: NOAA, Shane Lear.

Figure 8.18 Lightning is a common feature of thunderstorms.

in miles. To calculate in kilometers, divide by 3 because sound travels a kilometer every 2.9 seconds.

Effects of Lightning on Humans

Lightning strikes have a mortality rate of from 10 to 30% and up to 80% of survivors have long-term injuries from high voltage induced nerve and muscle damage. Over the last 20 years an average of about 50 people were killed in the United States and about 300 more injured by lightning strikes. A study completed by the National Weather Service and the National Severe Storms Laboratory in the mid-1990s found that men accounted for 84% of the 3,239 deaths and 9,818 injuries caused by lightning between 1959 and 1994. Only flash floods and river floods caused more weather-related deaths during that period.

Globally, there are about 24,000 deaths and 240,000 injuries a year from lightning. While deaths caused by lightning are rare, there are areas that have higher rates than others. Central Africa west of Kilimanjaro has a greater incidence of lightning strikes than any other large region on Earth.

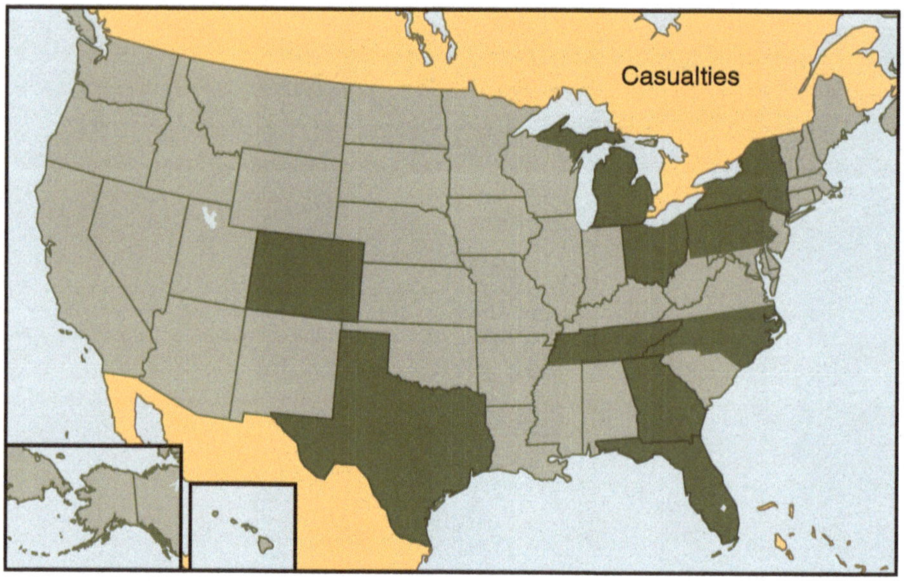

Source: NWS.

Figure 8.19 Of the ten states with the highest number of lightning casualties (deaths and injuries combined), Florida has the highest number.

In this region, mountains cause uplift in moist air from the Atlantic Ocean and generate year-round thunderstorms. The rural population has few warning systems, poor buildings, little shelter, and poor medical facilities. In 2011, 18 children and a teacher were killed when lightning hit their primary school in Uganda. The estimate death rate per million people per year in Malawi is 84, the highest rate in the world. In developing nations it is 6, and in developed nations it is 0.3.

Documented cases exist of lightning striking up to 15 km from a thunderstorm. If you can hear thunder, you are potentially vulnerable to being struck. There is no safe option to avoid lightning outside, so the best response is to go inside immediately. If you can take shelter in a car or other vehicle this will also help protect you, just make sure not to touch any metal surfaces. If you're caught outside in the open, you might be able to recognize signs of an electrical charge building up, signifying a possible imminent lightning strike. If you feel a tingling sensation on your skin, the hairs on your neck or arms stand up, you see a dull, blue-white glow from metal objects nearby, or you hear an unexplained buzzing, clicking, or crackling sound, you may be in the proximity of a strike that is about to occur. Don't lie flat on the ground. Most people are hit by lightning that has struck the ground nearby and passed up through them, so you want to limit the amount of your body surface area that is touching the ground. Instead, you must get as close to the ground as you can without actually touching the ground with more than your feet. Crouch down to make a direct strike of your body less likely. Put your head between your knees and cover your ears. This does two things, it keeps your head and brain from being your highest point and most likely strike point, and limits potential hearing damage. If lightning strikes you or close to you, the sound is loud enough to cause permanent hearing damage. Balance on the balls of your feet and touch your heels together. This reduces your surface contact with the ground and if lightning strikes the ground near you, it may only come up one foot and cross over into the other, limiting its path through your body.

Tornadoes

The intense atmospheric dynamics of a thunderstorm can produce a **tornado**, a rapidly rotating column of air that ranks as the most violent type of naturally occurring weather condition. These funnel-shaped clouds, which extend down from the base of severe thunderstorms, may or may not make contact with the ground. The appearance of a tornado ranges from light gray in color that contains mainly moisture droplets to pitch black with a high amount of dust, dirt, and debris. The siphoning effect of the updrafts carries material high into the atmosphere and spreads it across the landscape as the tornado moves along.

tornado

A rotating column of air usually accompanied by a funnel-shaped downward extension of a cumulonimbus cloud and having a vortex several hundred yards in diameter whirling destructively at speeds of up to 600 kilometers per hour (350 miles per hour).

How Do Tornadoes Form?

The most common location for the formation of tornadoes is a supercell thunderstorm. These are highly organized, extremely intense storms that derive their energy from the strong mesocyclone updrafts that rotate and become tilted (Figure 8.20). These supercells contain vertical air currents that stretch 15 km (9 mi) in diameter and reach altitudes of 15 km (9 mi). Rotation within the cloud system produces a mesocyclone which can be seen on Doppler radar (Figure 8.21). A tornado represents a very small extension of the larger roational cell.

Researchers do not fully understand how tornadoes form, in spite of years of lab research and observations in the field. A supercell must exist and have a well-defined rotating updraft. Winds traveling in two opposite directions or at different speeds at different altitudes can produce a shear force similar to placing your flat palms together and sliding them in opposite directions. This shearing force produces a horizontal rotation that gets transformed into a vertical rotation by updrafts in the storm. Warm, moist air near the surface is lifted and cooled. The temperature differences in the air at altitude adds to the energy of the storm.

Fewer than 20% of supercells spawn tornadoes. Scientists are undertaking large-scale studies to learn more about these phenomena. Vortex2, the Verification of the Origins of Rotation in Tornadoes Experiment 2 (a collaborative endeavor involving governmental agencies and universities) is examining the conditions that form tornadoes in the Midwest. This is the

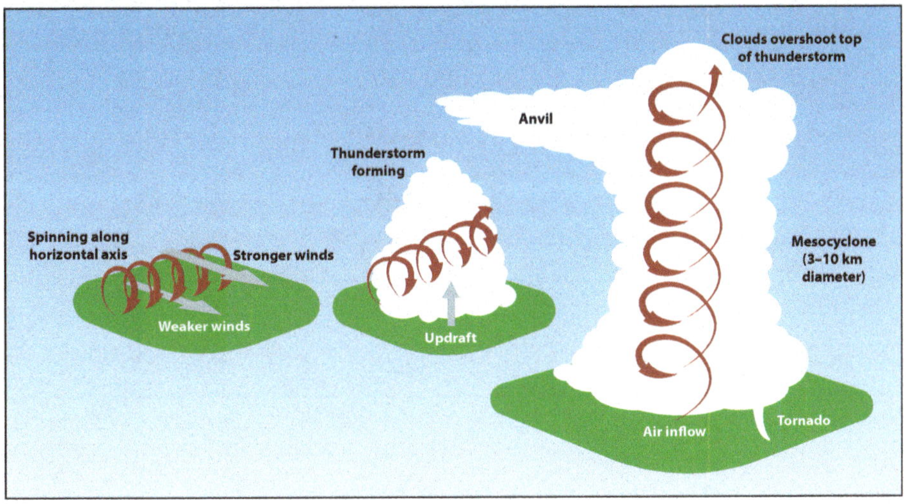

© Kendall Hunt Publishing Company.

Figure 8.20 The rotational vortex of a tornado lifts dust and debris from the surface high into the funnel cloud.

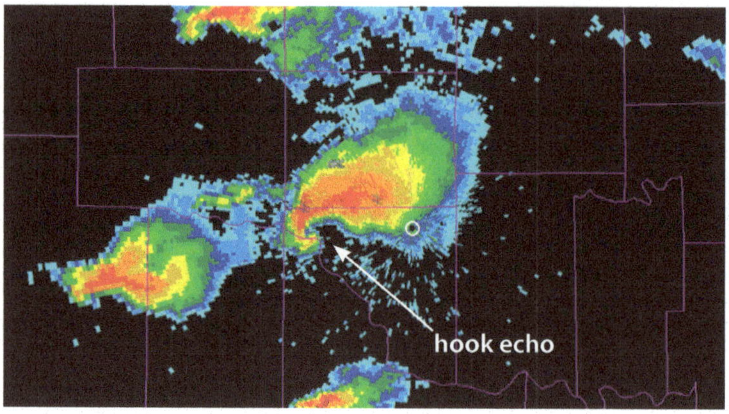

Source: NOAA.

Figure 8.21 Hook echo on Doppler radar image of supercell thunderstorm.

largest tornado research project in history to explore how, where, and why tornadoes form. A fleet of 10 mobile radars and groups of scientists drove over 24,000 km (15,000 mi) across the Midwestern United States chasing thunderstorms that could spawn tornadoes for scientific research.

Once a tornado forms, the rotational vortex spins in a counterclockwise direction (less than 1% rotate clockwise). Most tornadoes form at the trailing end of a thunderstorm, stretching down from the cloud base. Tornadoes move along with the thunderstorm at velocities ranging from close to stationary to 100 km per hour (62 mph). The rotational speed of winds can reach more than 500 km per hour (310 mph). As tornadoes stretch down the base in contact with the ground they can jump up so that they appear to skip along the surface.

Tornado Intensity

Wind velocity values are used to classify the intensity of tornadoes. These velocities are either measured directly or extrapolated from the damage the storms produce. Twenty-eight different measures go into the determination and assignment of the value for a given tornado. The Enhanced Fujita Scale is used to assign a value to a given tornado (Table 8.1). The scale was originally set up in 1971 by Professor Ted Fujita, a world-renowned researcher who worked at the University of Chicago. His research into severe storms

Table 8.1 The Enhanced Fujita Tornado Scale

Fujita Scale			Operational EF-Scale		
F Number	3 Second Gusts (km/h)	3 Second Gust (mph)	EF Number	3 Second Gusts (km/h)	3 Second Gust (mph)
0	60–110	45–78	0	90–130	65–85
1	120–170	79–117	1	135–175	86–110
2	180–240	118–161	2	180–220	111–135
3	250–320	162–209	5	225–265	136–165
4	330–410	210–261	4	270–320	166–200
5	420–510	262–317	5	> 320	Over 200

Source: http://www.ncdc.noaa.gov/oa/satellite/satelliteseye/educational/fujita.html

and weather led to the discovery of phenomena such as microbursts and downbursts, sudden violent blast of air that produces damaging results.

Frequency of Tornado Occurrences

Tornadoes occur everywhere across the world. They have been documented on every continent except Antarctica. However, the highest number of tornadoes occur in the United States, where all 50 states have experienced a tornado at some point in time. Because tornadoes are directly related to thunderstorms, they are more common in those regions struck by thunderstorms. Because clashes of cold and warm, moist air generate unstable atmospheric conditions, the Midwest has the greatest number of tornadoes (**Figure 8.22a**). Warm, moist air from the Gulf of Mexico drawn up into the Midwest collides with cold air from Canada (**Figure 8.22b**). The result creates severe thunderstorm and tornado conditions.

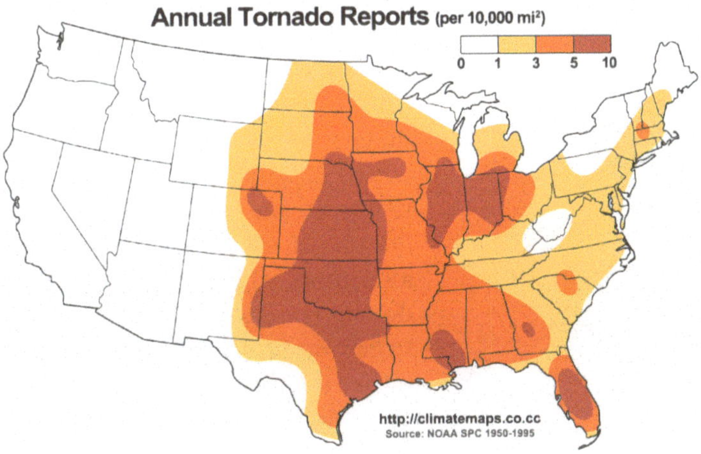

NOAA. Source: Mamatus/Wikimedia Commons.

Figure 8.22a Tornadoes have formed in every state, but are most common in the Midwest.

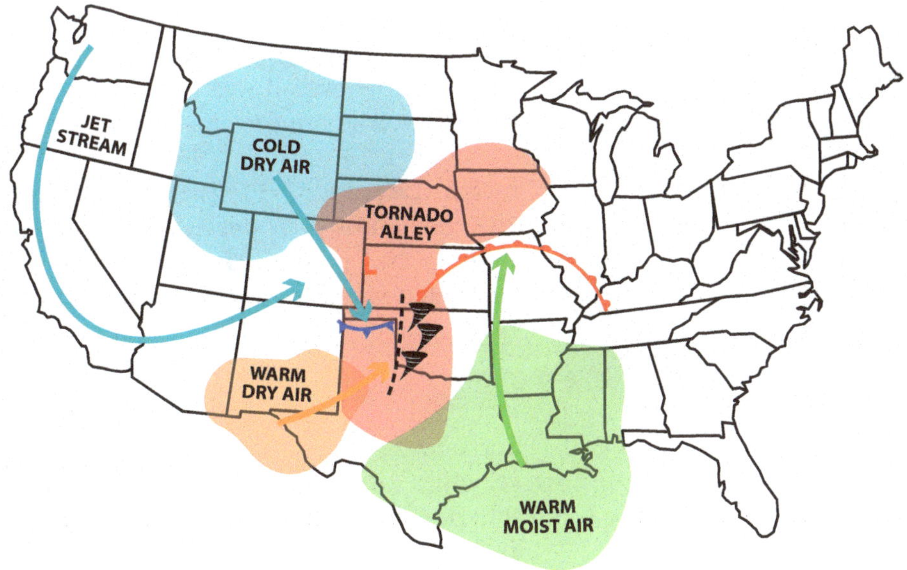

© Kendall Hunt Publishing Company.

Figure 8.22b Tornadoes form in the continental United States because of the confluence of warm, moist air and cold, dry air over the broad, flat central Midwestern plains.

Figure 8.23 Tornadoes are most likely to occur in the late afternoon, when land temperatures are at their highest and thunderstorms are most likely to develop.

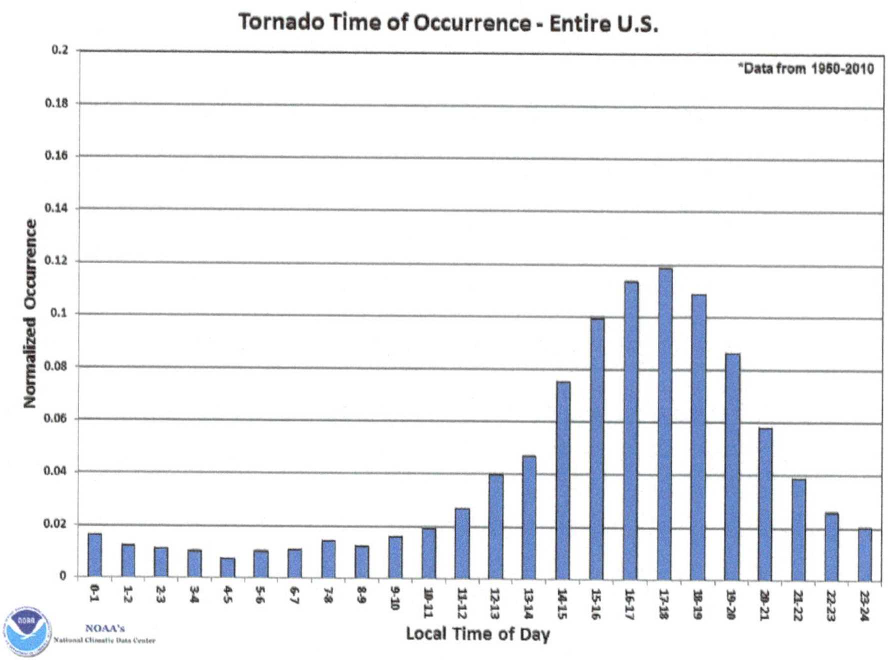

Source: NOAA.

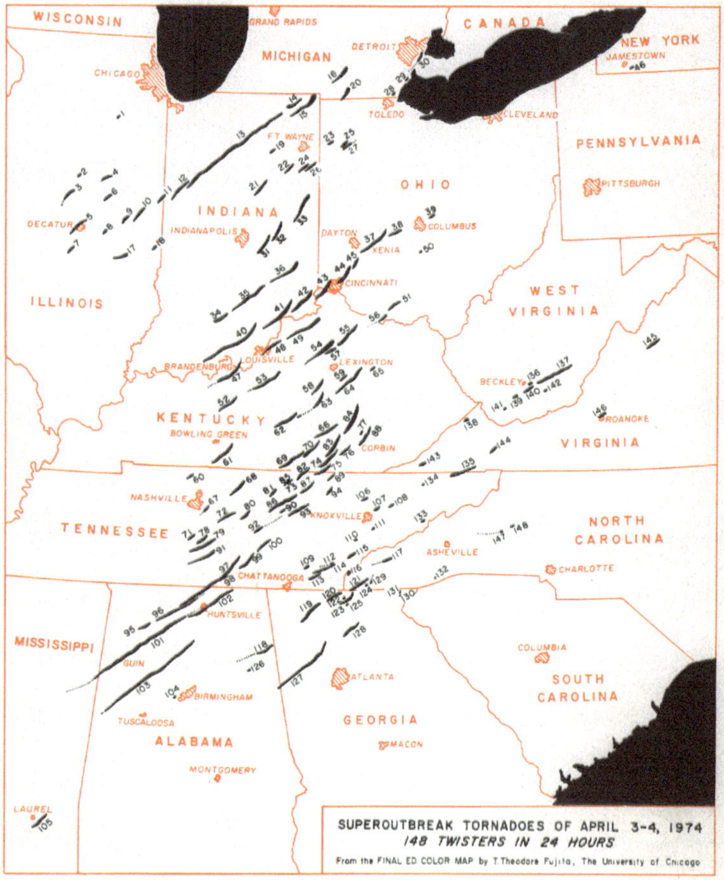

Source: NOAA.

Figure 8.24 The super outbreak of tornadoes across the Mississippi Valley region on April 3 and 4, 1974, generated 148 tornadoes.

The highest number of tornadoes occur during the spring and summer months, when clashing air masses of different temperatures and moisture content collide. Most tornadoes tend to occur in the afternoon, after temperatures have risen during the day (**Figure 8.23**). Long weather fronts often spawn swarms of tornadoes rather than isolated events. A megaswarm of tornadoes occurred on April 3 and 4, 1974 (**Figure 8.24**). Plots of the paths of the tornadoes showed the general northeast movement of tornadoes. This is normal for tornadoes in the United States, as they are driven by the overall weather pattern and westerlies. One-hundred forty-eight tornadoes struck 13 states in a 16-hour period. Included were six F-5 twisters. At the end 307 people lost their lives and more than 6,000 were injured. Property damage amounted to more than $600 million (in 1974 dollars).

Tornado Damage

Tornado damage is caused by high wind velocities and a large difference in atmospheric pressure between the tornado and its surroundings. The rotating winds can destroy weak structures, and the extremely low pressure inside the tornado generates strong pressure differences between the inside and outside of buildings. This pressure difference causes roofs to be lifted and removed. The high winds pick up smaller objects including small structures, animals, people, cars, and especially mobile homes, and can carry these objects up to several kilometers. The debris picked up by the winds becomes rapidly moving projectiles that could be lethal when hurled against a human body.

Homes in the Midwest have storm cellars that allow families to go underground to wait out the high winds. As seen in the opening image of this chapter, an EF5 tornado can turn a community into a flattened mass of rubble. Winds from a less violent EF3 storm can still have enough energy to strip trees of their leaves and throw building roofs into the cleared area (Figure 8.25).

Figure 8.25 An EF3 tornado hit Mena, Arkansas, on April 11, 2009. The tree was stripped of all its leaves and caught an aluminum roof from a nearby building.

Tornado Prediction and Warning

Tornadoes cannot be predicted with precision. However, when strong thunderstorm activity is detected, a **tornado watch** is generally issued for all areas that may fall in the path of the thunderstorm. **Doppler radar** can detect rotating motion within a thunderstorm and when this is detected, or a tornado is actually observed, a **tornado warning** is issued for all areas that may fall in the path of the thunderstorm. Tornado safety (from the Federal Emergency Management Agency, FEMA) includes the following.

If at home:

- Go at once to the basement, storm cellar, or the lowest level of the building.
- If there is no basement, go to an inner hallway or a smaller inner room without windows, such as a bathroom or closet.
- Get away from the windows.
- Go to the center of the room. Stay away from corners, because they tend to attract debris.
- Get under a piece of sturdy furniture such as a workbench or heavy table or desk and hold onto it. Use arms to protect head and neck.
- If in a mobile home, get out and find shelter elsewhere.

If at work or school:

- Go to the basement or to an inside hallway at the lowest level.
- Avoid places with wide-span roofs such as auditoriums, cafeterias, large hallways, or shopping malls.
- Get under a piece of sturdy furniture such as a workbench or heavy table or desk and hold onto it. Use arms to protect head and neck.

tornado watch
This is issued by the National Weather Service when conditions are favorable for the development of tornadoes in and close to the watch area. Their size can vary depending on the weather situation.

Doppler radar
Radar that can measure radial velocity, the instantaneous component of motion parallel to the radar beam (i.e., toward or away from the radar antenna).

tornado warning
This is issued when a tornado is indicated by the WSR-88D radar or sighted by spotters; therefore, people in the affected area should seek safe shelter immediately.

If outdoors:

- If possible, get inside a building.
- If shelter is not available or there is no time to get indoors, lie in a ditch or low-lying area or crouch near a strong building. Be aware of the potential for flooding. Use arms to protect head and neck.

If in a car:

- Never try to out-drive a tornado in a car or truck. Tornadoes can change direction quickly and can lift up a car or truck and toss it through the air.
- Get out of the car immediately and take shelter in a nearby building. If there is no time to get indoors, get out of the car and lie in a ditch or low-lying area, away from the vehicle. Be aware of the potential for flooding.

After the tornado:

- Help injured or trapped persons. Give first aid when appropriate. Don't try to move the seriously injured unless they are in immediate danger of further injury. Call for help.
- Turn on radio or television to get the latest emergency information.
- Stay out of damaged buildings. Return home only when authorities say it is safe.
- Use the telephone only for emergency calls.
- Clean up spilled medicines, chemicals, gasoline, or other flammable liquids immediately.
- Leave the building if you smell gas or chemical fumes.
- Take pictures of the damage—both to the house and its contents—for insurance purposes.
- Remember to help your neighbors who may require special assistance—infants, the elderly, and people with disabilities.

Inspecting utilities in a damaged home:

- Check for gas leaks—If you smell gas or hear a blowing or hissing noise, open a window and quickly leave the building. Turn off the gas at the outside main valve if you can and call the gas company from a neighbor's home.
- If you turn off the gas for any reason, it must be turned back on by a professional.

Mitigation of Potential Tornado Damage

Tornadoes can occur wherever thunderstorms occur—basically anywhere. Therefore it is financially impossible to prepare every possible locality for a tornado. The best preparation is to ensure that people are aware of warning systems and have taken some precaution, such as building a storm cellar if they live in an area that is likely to experience tornadoes. Building codes can also require more sturdy construction. However, if a structure is hit directly by an EF 5 tornado it will certainly be severely damaged, if not totally destroyed. Warnings broadcast by governmental agencies must be timely, accurate, and heeded by those for whom they are sent.

Other Severe Weather Phenomena

Derechoes

Derechoes are named after the Spanish word for straight, or direct, reflecting the strong straight winds generated in these storms which is in contrast to tornadoes. They were first named after a significant derecho which swept across Iowa, United States, on July 31, 1877. They can generate hurricane-force winds, tornadoes, heavy rain, and flash floods. They move rapidly in the direction of storm movement, and the extremely strong winds remain sustained for hours or even days and may increase in strength as they progress. These are a warm-weather phenomenon and occur in areas of strong atmospheric instability and strong vertical wind shear. Most derechoes are associated with self-perpetuating thunderstorm systems. They occur with squall lines and wind speeds can reach 225 km/h (140 mph). Derechoes can strike with little warning, unlike thunderstorms which can typically be heard approaching. Derechoes can severely damage above-ground utilities, especially electrical systems. The 2011 Derecho that struck Chicago, Illinois left more than 860,000 people without electricity, and the August 10–11, 2020, derecho that hit Iowa, United States, affected 1.9 million customers, some of whom were without power for weeks. Almost 66% of the state's crops were destroyed, residential and commercial property was damaged or destroyed, resulting in an estimated $11 billion US dollars in damage. The chief executive officer of an affected electrical infrastructure company described it as the "equivalent of a 40-mile wide tornado that rolled over 100 miles of the state" [of Iowa], which is a path of destruction 64 kilometers wide and more than 160 km long.

derecho

A derecho is a widespread, long-lived, straight-line wind storm that is associated with a fast-moving group of severe thunderstorms with wind speeds equivalent to tornadoes and hurricanes.

Nor'easters

A **nor'easter** is an extratropical cyclonic storm that forms in a region away from the tropics. It has some of the characteristics of a tropical storm, winds that rotate in a counterclockwise direction, and a low pressure center. These storms that affect the east coast of the United States form off the south Atlantic coast. The counterclockwise winds come out of the northeast (hence the name). Because these usually occur in the fall and winter months, they are less intense than hurricanes because the colder ocean waters do not provide a large amount of thermal energy. However, damage can be extensive (Figure 8.26).

nor'easter

A strong low-pressure system with winds from the northeast that affects the mid-Atlantic and New England states between September and April. These weather events are notorious for producing heavy snow, copious rainfall, and tremendous waves that crash onto Atlantic beaches, often causing beach erosion and structural damage.

Winter Blizzards and Severe Weather

Severe winter weather is often accompanied by high winds, significant snowfall, and cold temperatures. These conditions can produce a **blizzard**, which the National Weather Service defines as having sustained 35 mph (56 kph) winds that lead to blowing snow, and causes visibilities of ¼ mile or less, lasting for at least 3 hours (Figure 8.27). Although no specific temperatures are associated with a blizzard, the high winds often produce sub-zero wind chill conditions. These conditions form when a ridge of high-pressure interacts with a low-pressure system; this results in the horizontal movement of air from the high-pressure zone into the low-pressure area.

blizzard

A severe winter storm that has the following conditions that are expected to prevail for a period of 3 hours or longer: sustained wind or frequent gusts to 35 miles an hour or greater and significant falling and/or blowing snow (i.e., reducing visibility frequently to less than one fourth mile).

FEMA.

Figure 8.26 A nor'easter hit New England in April 2009. Damage in Saco, Maine, was extensive along the shoreline.

FEMA, photo by Michael Rieger.

Figure 8.27 A blizzard hits Denver, Colorado, in December 2006 with more than 2 feet of snow.

Other winter weather condition watches and warnings that are issued by the National Weather Service include:

- **Winter Storm Watch.** Conditions are favorable for hazardous winter weather conditions including heavy snow, blizzard conditions, or significant accumulations of freezing rain or sleet. These watches are issued by the Weather Service Forecast Office in Chicago and are usually issued 12 to 36 hours in advance of the event.
- **Winter Storm Warning.** Hazardous winter weather conditions that pose a threat to life and/or property are occurring, imminent, or likely. The generic term, winter storm warning, is used for a combination of two or more of the following winter weather events: heavy snow, freezing rain, sleet, and strong winds.
- **Heavy Snow Warning.** Snowfall of 6 inches or more in 12 hours or less, or 8 inches or more in 24 hours or less.
- **Lake Effect Snow Warning.** Lake effect snowfall of 6 inches or more in 12 hours or less, or 8 inches or more in 24 hours or less. The source of moisture for these storms comes from large bodies of water, such as the Great Lakes. Michigan, northern Ohio and Pennsylvania, and western New York state are often hit with heavy snow that developed from the Great Lakes.

Heat Waves

heat wave

A prolonged period of excessively hot weather and high humidity.

Several significant heat waves have occurred throughout the world over the past 30 years. A **heat wave** is a prolonged period of excessively hot weather, which may be accompanied by high humidity. There is no universal definition of a heat wave, as the term is relative to the usual weather in the area. Temperatures that people from a hotter climate consider normal can be termed a heat wave in a cooler area if they are outside the normal climate pattern for that area. Increased humidity adds to the effect of the elevated temperatures to create deadly conditions.

The term *heat wave* is applied both to routine weather variations and to extraordinary spells of heat which may occur only once a century. Severe, prolonged heat waves have caused catastrophic crop failures, thousands of deaths from hyperthermia, and widespread power outages due to increased use of air conditioning (**Figure 8.28**). In the 40-year period from 1936 through 1975, nearly 20,000 people were killed in the United States by the effects of heat and solar radiation. In the disastrous heat wave of 1980, more than 1,250 people died, mainly in the central and southern Plains. Extreme heat now causes more deaths in United States cities than all other weather events combined, and heat waves such as the one that hit the Northern Hemisphere in 2018 resulted in dozens of deaths across the US and Canada as well as Japan. In the US, a high-pressure system locked in a dome of heat that resulted in temperatures over 90 °F in 44 of 50 states. Heat waves are especially deadly when nighttime temperatures remain high—the human body can't recover from the effects of extreme heat if the air temperature doesn't drop below 80 °F at night. The top four hottest years on record are 2016, 2017, 2015, and 2018. The continuing rise in global temperatures will result in more extreme heat in the future, and an estimated 150 Americans will die every summer day due to this extreme heat by 2040, totaling about 30,000 heat-related deaths each year.

© Jasper Sujiten/Shutterstock.com.

Figure 8.28 Heat waves can result in extreme drought conditions and crop failure.

In the summer of 2003 Europe was hit by a heat wave that killed more than 37,000. Much of the heat was concentrated in France, where nearly 15,000 people died. Elderly people were the most affected. In July 2006, the United States experienced a massive heat wave, and almost all parts of the country recorded temperatures above the average temperature for that time of year. Temperatures in some parts of South Dakota exceeded 115 °F (46 °C), causing many problems for the residents. Also, California experienced temperatures that were extraordinarily high, with records ranging from 100 to 130 °F (38 to 54 °C). On July 22, the County of Los Angeles recorded its highest temperature ever at 119 °F (48.33 °C) (**Figure 8.29**).

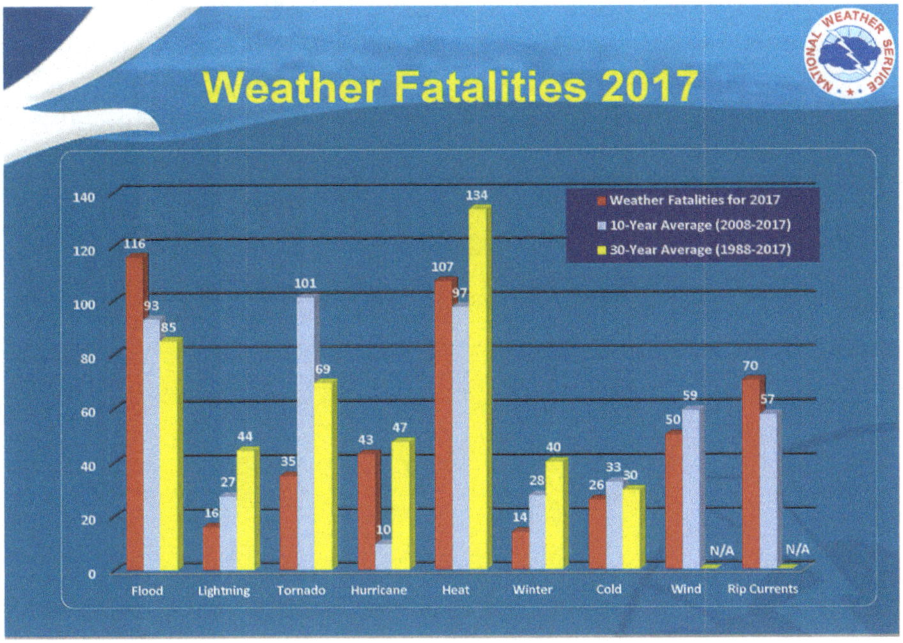

Source: NOAA.

Figure 8.29 Deaths attributable to weather-related events show a wide range of values.

Drought and Famine

Drought is an extended period lasting months or years when a region receives much less than normal precipitation, resulting in a significant shortage of water. Both surface water and groundwater supplies are drastically reduced. This shortage rapidly reduces the ability of people to grow crops and to sustain life, and the ecosystem of the area is impacted to the point that it begins to deteriorate.

© Everett Historical/Shutterstock.com.

Figure 8.30 A dust storm in Elkhart, Kansas, in May 1937.

The onset of a drought is caused by changes in the upper-level air flow. In the United States it is associated with a prolonged ridge of high pressure that pushes dry air down to the surface. This sinking air is warmed by compression and begins to reduce the humidity even more. The dry air reduces moisture in the soil, reducing the ability to grow crops. In the heartland of the United States these conditions began to develop in the early 1930s and lasted for several years. The result was severe dust storms throughout the central part of the United States, giving the name "Dust Bowl" to the region (**Figure 8.30**). Widespread crop failures resulted in the malnutrition of hundreds of thousands of people. Numerous farms were abandoned and there was a mass migration to the West, especially California, where former farmers hoped to reestablish the way of life they had known previously. These events came on the heels of the Great Depression, making living conditions and survival very problematic for thousands of people.

Key Terms

air mass *(page 228)*
anticyclone *(page 230)*
blizzard *(page 247)*
Coriolis Effect *(page 227)*
cyclone *(page 230)*
derecho *(page 247)*
Doppler radar *(page 245)*
hail *(page 237)*
heat wave *(page 248)*
jet stream *(page 229)*
latent heat of condensation *(page 232)*
latent heat of fusion *(page 232)*
latent heat of vaporization *(page 232)*
lightning *(page 234)*
nor'easter *(page 247)*
sublimation *(page 232)*
thunderstorm *(page 234)*
tornado *(page 241)*
tornado warning *(page 245)*
tornado watch *(page 245)*
updraft *(page 234)*

Summary

The Coriolis effect results from the spin of the planet and appears to cause objects to be deflected as they move through the atmosphere—this is because as they are moving in a straight line, the planet's surface is spinning underneath them. In the Northern Hemisphere, the Coriolis effect results in an apparent clockwise movement to the right.

The equator receives large amounts of solar radiation, warming air masses and causing them to rise. As they encounter colder regions in the upper atmosphere, the now cool air condenses water vapor and begins to sink near 30 to 35 degrees north and south of the equator. This sinking cool air warms again, absorbing surrounding moisture and forming mid-latitude deserts. Large-scale movement of air masses forms weather patterns. Where less dense air is rising due to heating, the upward force generates an area of low pressure. When cool air sinks to the surface as it is cooled, it generates an area of high pressure. Winds generated by the air masses move outward from the center in a spiral fashion. Topographic features such as mountain ranges interact with air masses and disrupt the global atmospheric circulation pattern. Phase changes in water, from solid to liquid and gas, either release or absorb heat, depending on the phase state shift. When air masses collide they interact to generate fronts and thunderstorms.

Hazardous weather and thunderstorms require an updraft to maintain the energy feeding the storm. There are four main types of thunderstorms: single-cell storms, multicell storms, squall lines, and supercell thunderstorms. They are capable of generating hazards such as hail, lightning, and tornadoes. Hail forms in updrafts that support frozen particles until their weight becomes greater than the updraft to keep aloft. Lightning results from a sudden electrical discharge and is generated when regions of positively charged ions and negatively charged ions, often within clouds but also between clouds and the ground, re-equilibrate their charge differences. Tornadoes form from a rotating vortex called a mesocyclone that forms the core of supercell thunderstorms. Only about 20% of supercell storms spawn tornadoes, which are classified by wind intensity in the Enhanced Fujita Scale which ranges from 0 to 5.

Other additional weather hazards are nor'easters (extratropical cyclonic storms), blizzards (severe winter storms), heat waves (prolonged periods of hot weather), and drought (lower than normal precipitation lasting for months or years).

References and Suggested Readings

Burt, C. C. (2004). *Extreme weather: A guide and record book.* W. W. Norton & Company.

Collins, A. (2006). *Violent weather—Thunderstorms, tornadoes, and hurricanes.* National Geographic Society.

Sandlin, L. (2013). *Storm kings—The untold history of America's first tornado chasers.* Pantheon Books.

Schneider, B. (2012). *Extreme weather: A guide to surviving flash floods, tornadoes, hurricanes, heat waves, snowstorms, tsunamis and other natural disasters.* St. Martin's Press.

Uman, M. (1986). *All about lightning.* Dover Publications.

Williams, J. (1997). *USA Today the weather book.* Vintage Books.

Websites for Further Reference

https://droughtmonitor.unl.edu/

https://www.cdc.gov/disasters/tornadoes/index.html

https://www.noaa.gov/weather

https://www.ready.gov/thunderstorms-lightning

Questions for Thought

1. Distinguish between low pressure and high pressure conditions. What types of weather are associated with each?
2. Explain how changes in Earth's eccentricity affect global climate.
3. How does hail form?
4. Why do most thunderstorms form in the central and southeast parts of the United States?
5. What are the general weather conditions associated with a drought?

9 Flooding

Source: David M. Best.

Intense rainfall can produce significant runoff. This park in Scottsdale, Arizona, was flooded by an intense thunderstorm more than 20 km (12 mi) away.

The Role of Water

Water, considered by many to be the "staff of life," is an essential part of processes that are found on Earth. More than 70% of Earth's surface is covered by water, whether in liquid or solid form, as well as being in the atmosphere. The continental regions on which people live have lakes, rivers, and streams, and a small amount of surface covered by ice. When these areas are affected by water-laden storms, the additional water can create flooding that can be very disruptive, dangerous, and costly. Oftentimes this occurs without warning. People who live in some regions, such as low-lying areas along coastlines, are aware of the potential for problems, while those who live in arid climates often do not think about the potential for disastrous events. Everyone should recognize that water has the ability to generate a natural disaster with potentially catastrophic consequences.

All climates, especially arid ones, are prone to flooding. Every year, numerous floods affect millions of people throughout the world. Weather systems can stall over an area and generate copious amounts of rainfall, causing streams and rivers to overflow their banks when too much water enters a drainage basin. Rapid melting of snow and ice also create abnormal water discharge. Areas that are most affected tend to lie toward the lower portions of a river system, as the topography there is less steep and the increased downstream flow cannot readily move water away from the affected region.

Floods can also be the result of poor urban planning or dam construction that causes water to flow unexpectedly in places where it was not meant to go. Coastal regions become flooded when maritime storms land ashore. The strong winds of hurricanes, cyclones, and typhoons generate storm surges that bring massive amounts of water inland, flooding low-lying areas. Tsunami, although relatively rare, rapidly push seawater past beach zones, inflicting severe damage to communities in the path of the waves. Floods can also result from volcanic eruptions that melt snow and ice atop a volcano, the condition that contributed to the lahars of Mount St. Helens (see Chapter 3).

The National Weather Service reports that flooding in the United States annually causes an average of 82 deaths and almost $8 billion in damages. In many low-lying areas, home and landowners cannot purchase flood insurance and the losses they might sustain are not covered. Often the lower socioeconomic classes reside in these topographically lower locations and suffer immense losses in terms of property and human life. Although floods can be caused by storms, tsunamis, and volcanoes, this chapter will address flooding associated with streams.

Stream Processes

hydrologic cycle
The cycle that moves water and water vapor among the oceans, land, and atmosphere through evaporation, condensation, precipitation, transpiration, and respiration.

Earth's hydrosphere includes all water at or near the surface in addition to what is contained in the atmosphere. Water, which can be a liquid, a gas (water vapor), or a solid (ice), moves around in the **hydrologic cycle**—a continuous circulation of water around the globe. Solar energy drives this cycle; it stretches from the equator to the poles, where water movement is obviously slower but nevertheless part of the cycle.

© stihii/Shutterstock.com.

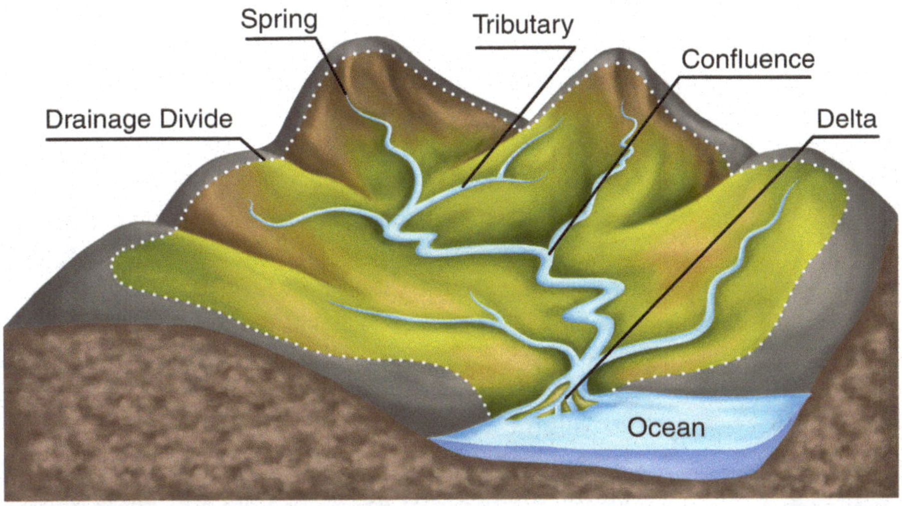

Figure 9.1 Streams flow from high to low elevations. In the upper portions small streams form the tributary system. Water flows downslope to feed other larger streams or it flows into a lake, sea, or ocean.

Geologists define a **stream** as a body of water that flows within a confined channel on the surface. When several streams or **tributaries** join to produce a larger flowing body, the term **river** is then applied (Figure 9.1). Water that falls onto Earth's surface can produce **runoff**, which travels along the surface and ends up in a channel, or it can infiltrate into the ground, where it temporarily resides in underground collection areas before eventually finding its way into a stream system. In arid climates, infiltrated water often evaporates and reenters the hydrologic cycle as water vapor.

stream
A body of water that flows downhill under the influence of gravity and lies within a defined channel.

tributary
A stream that flows into another larger stream.

river
A large stream.

runoff
Water that flows across a surface and into a stream or other body of water.

Drainage Basins

A **drainage basin** or **watershed** is an area on the surface that collects water that flows into a stream and forms a drainage pattern. Drainage divides, which are often ridges or a series of high points that divide downhill slopes, separate one watershed from its adjacent neighbors (Figure 9.2). The amount of water collected by a drainage system is dependent on the amount of precipitation, the size of the area, and the subsurface characteristics that control infiltration and runoff. As long as the amount of water flowing into streams and falling on the watershed is carried away by the existing system, water remains in the channels and does not present a problem.

drainage basin
An area that drains water to a given point or feature, such as a lake.

watershed
See drainage basin.

Source: USGS.

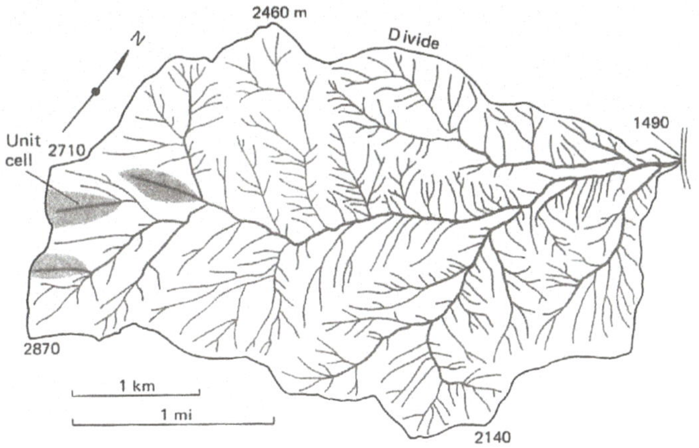

Figure 9.2 A watershed is outlined by its divide, which separates it from adjacent watersheds. Elevations are in meters. Notice how the stream patterns collect water and channel it into a larger stream downhill.

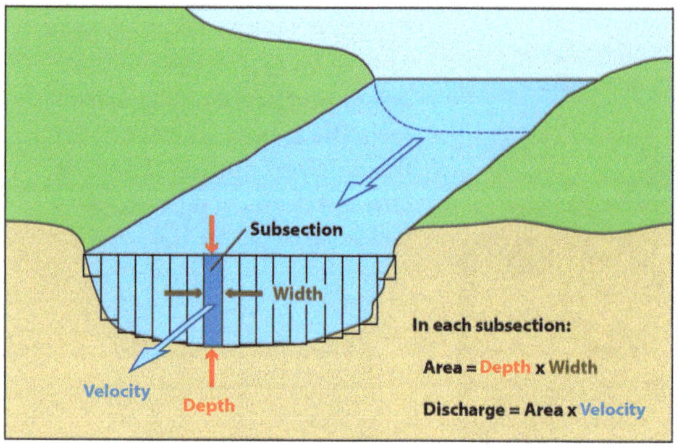

Source: USGS.

Figure 9.3 Discharge measurements are made by determining the discharge in each subsection of the channel cross section and then by summing these discharge amounts, the total discharge is calculated.

Think of a stream configuration as the transportation system that is moving water to some end point, either a temporary one, such as a lake, or the ultimate end point—a sea or an ocean. A stream begins to cut a channel into the landscape, thereby creating the "highway" that is allowing water to move through an area. Each channel has a cross-sectional view, in which we see the width and depth of the channel. This area, coupled with the length of a particular stream segment and its drop in elevation, defines the volume or how much water is contained (and moved) by the stream and at what velocity (Figure 9.3). **Discharge** is the volume of water that flows downstream past a given point in a given period of time. We measure this volume in cubic feet per second (cfs) or cubic meters per second (cms), and the amount can vary widely depending on weather, surface conditions, and stream characteristics. The world's largest rivers move enormous volumes of water each second (Table 9.1). All of these rivers have very large watersheds and many of the rivers flow through regions that have wet, temperate climates.

discharge

The volume of water flowing through a stream channel in a given period of time, usually measured as cubic feet per second or cubic meters per second.

floodplain

The flat area alongside a stream that becomes flooded when water exceeds the banks of the stream.

Floodplain

A stream channel is bordered by its banks and the area to either side, which is termed the **floodplain**. When water spills over the banks, it is no longer moving in its channel. As it spreads out, the velocity of the water decreases rapidly, causing any sediment carried by the stream to be deposited. This process is repeated every time the stream floods. The continual buildup of sediment along the banks creates natural levees that increase in height, thereby deepening the stream channel. Repetition of this process permits the stream to carry more water than before, because its cross-sectional area

Table 9.1 World's 10 Largest Rivers by Discharge

TERRESTRIAL HAZARDS

River	Country	Average Discharge at Mouth (Cubic Feet per Second)
Amazon	Brazil	7,500,000
Congo	Congo	1,400,000
Yangtze	China	770,000
Brahmaputra	Bangladesh	700,000
Ganges	India	660,000
Yenisey	Russia	614,000
Mississippi	US	611,000
Orinoco	Venezuela	600,000
Lena	Russia	547,000
Parana	Argentina	526,000

Source: http://www.waterencyclopedia.com/Re-St/Rivers-Major-World.html.

has increased. When water spills over onto the floodplain, it does not tend to drain back into the main river channel because the natural levees now act as a dam. This water slowly moves downhill on the floodplain, creating a yazoo tributary. Eventually this water finds a site where it can flow back into the main stream or off the floodplain (Figure 9.4).

Figure 9.4 This image of the Mississippi River at Burlington, Iowa, shows the floodplain on either side of the river. Periodic flooding of the rivers covers the floodplain and impacts the cities of Burlington and West Burlington.

Streams provide water, transportation, food, irrigation, and soil to people living nearby. The finer silt carried by streams is deposited to create arable farmland, such as in the delta region of the Nile River in Egypt. Before the High Aswan Dam was completed in 1970, these rich farmlands were frequently flooded, destroying crops, homes, and killing people, but also depositing valuable silt that enriched farming areas near the mouth of the river. Construction of dams lessens the occurrence of major floods, but the dams hold back valuable silt, rendering many downstream regions unsuitable for growing crops.

FACT BOX **Equivalent Measures and Weights of Water at 4 °C.**

1 gallon = 0.134 cubic foot = 8.35 lb (3.79 kg)
1 cubic foot per second (cfs) = 7.48 gal/sec = 62.4 lbs/sec (28.3 kg/sec)
= 449 gals/min

Types of Floods

Regional River Floods

Flooding that is related to seasonal rains or snow melts, or a combination of rain falling on snow, produces a large volume of water that cannot be handled by existing stream systems. Such flooding is common in wetter climates that have generally larger, more established rivers. Floods can occur anywhere along the length of a stream. In the upstream regions, such floods are caused by intense rainfall or snow melt over a watershed that flows into smaller streams and tributaries. When several watersheds feed into a larger stream, the volume of water can be immense and produce widespread flooding.

These vast amounts of water can have an effect on the landscape. One foot of water covering 1 acre is termed an **acre foot**. This amounts to 325,851 gallons (1.233 million liters) of water, so for every inch of rain falling on an acre, there are 27,154 gallons (102,778 liters) that can flow across the surface. If 1 acre foot of water moves into a river, it will contribute 43,560 cubic feet (1,233 cu meters) of flow to the stream's volume. Generally, these intense

acre foot
The amount of water that covers one acre to a depth of one foot; equivalent to 325,851 gallons of water.

(a)

(b)

Figure 9.5 **The Reedy River in Greenville, South Carolina. (a) normal flow conditions in March; (b) following a June thunderstorm.**

rainfalls are so rapid that very little water percolates into the subsurface, producing stream flows that become rushing torrents (**Figures 9.5a–b**). As the gradient of the stream flattens out and the downstream river channels become wider, the velocity drops and the water becomes calmer. However, the volume of water continues to increase, especially if more streams are present in the system.

Many watersheds consist of hundreds or thousands of acres, resulting in massive amounts of water moving downslope and downstream. The lateral and down cutting erosive power of the water causes dimensions of the stream channel to increase. Upstream regions generally have fairly steep slopes and the longitudinal profile of the stream shows a rapid drop in the elevation of the stream (**Figure 9.1**). Therefore water will tend to move downslope quickly and generate large-scale flooding.

As higher elevation streams move water downstream, they join with other streams and increase the size of the trunk stream (in a fashion similar to the trunk of a deciduous tree having many branches that feed into the main trunk). If a widespread rainstorm covers several different watersheds and feeds water into the trunk stream, the downstream region can experience flooding. Widespread saturation of the ground prevents water from percolating into soils, so the water must flow under the force of gravity to lower elevations. This downslope flow is not instantaneous, so there is often a lag that gives some warning to communities at the lower end of a drainage area.

Downstream communities generally have some preventive measures in place in preparation for recurring floods. Increasing the height of levees and deepening the river channel help the river to handle a greater flow, thus preventing flooding. However, these measures are not always successful, particularly if there is a flow that greatly exceeds the capabilities of the infrastructure to handle the runoff. The result can be flooding so severe and widespread that the governor of the state requests presidential declaration of the region as a disaster area so that it can receive federal help (**Figure 9.6**). As we see from the map in **Figure 9.6**, very few regions in the United States are exempt from flooding over a period of several decades.

Other natural hazards play a role in changing the drainage regime of an area. The 1991 eruption of Mount Pinatubo, a volcano in the Philippine Islands, deposited massive amounts of ash across the countryside. Within a few months, copious rainfalls struck the region and moved the ash downslope as sediment. When flooding occurred in the streams, deposits were created along the banks and formed natural levees that increased the depth of the

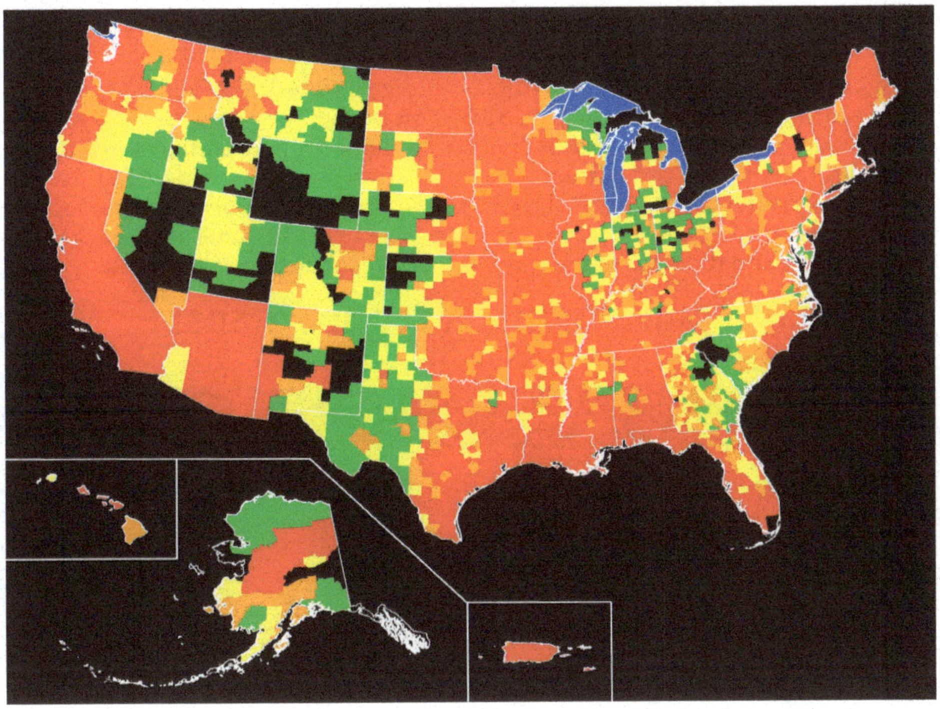

Federal Emergency Management Agency. Photo by Michael Baker, Jr.; National Atlas, and the USGS.

Figure 9.6 **Presidential disaster declarations related to flooding in the United States, shown by county: Green areas represent one declaration; yellow areas represent two declarations; orange areas represent three declarations; red areas represent four or more declarations between June 1, 1965, and June 1, 2003. Map not to scale.**

channels. After several episodes of this redistribution of the ash, streambeds were flowing at a level higher than the original surface. When the streams experienced later flooding, the water easily flowed into the lower lying areas adjoining the raised stream channels. Removal of ground cover increases surface runoff that causes sediment to affect stream flow.

Flash Floods

Flash floods occur with little or no warning. They are usually caused by torrential rainfall that takes place over a very short time, often associated with severe, localized thunderstorms or from a series of storms continually soaking an area. Although flash floods can occur anywhere, they tend to be more devastating in two areas: (1) in mountainous areas, where steep slopes funnel water into narrow streams, and (2) in desert regions, where normally dry or low flow streambeds are quickly transformed into raging torrents. Rainfall does not have any chance of infiltrating into the subsurface and is often flowing in a rapid, sheet-wash manner across impervious surfaces to collect in dry stream beds. The intense nature of flash floods makes them capable of causing extensive damage and loss of life. When people drive their vehicles through normally dry washes and streambeds that are filled with fast-moving water, their vehicles begin to float, are pushed along by the flow, or are overturned, trapping the victims inside. The force with which fast-moving, sediment-laden water hits a surface is great (Box 9.1)

Winter Climate-Driven Floods

Ice-jam floods are a problem in regions where rivers freeze and then begin to thaw or receive surface water from rainfall or nearby melting. In the winter, some regions receive excessive snowfall, which rests on frozen ground. Rivers that normally drain the snowmelt freeze and cannot move any water

ice-jam flood
A flood, usually in the spring, that results from broken pieces of river ice blocking the flow of a river, thereby flooding areas adjacent to the river.

BOX 9.1 **How One State Deals With Not-So-Smart Drivers**

The State of Arizona has enacted the Stupid Motorist Law (Arizona Revised Statutes 28-910), which imposes a fine of up to $2,000 on drivers who have to be rescued from a flooded area. In spite of this law, people continue to drive around barricaded crossings and attempt to get through flooded roadways. A lack of adequate storm culverts causes water to flow across low points on streets and highways. The depth of water, even when it is flowing across a roadway that might be familiar to the driver, is very uncertain and the force of the water is much greater than one realizes.

DO NOT DRIVE THROUGH FLOODWATERS!

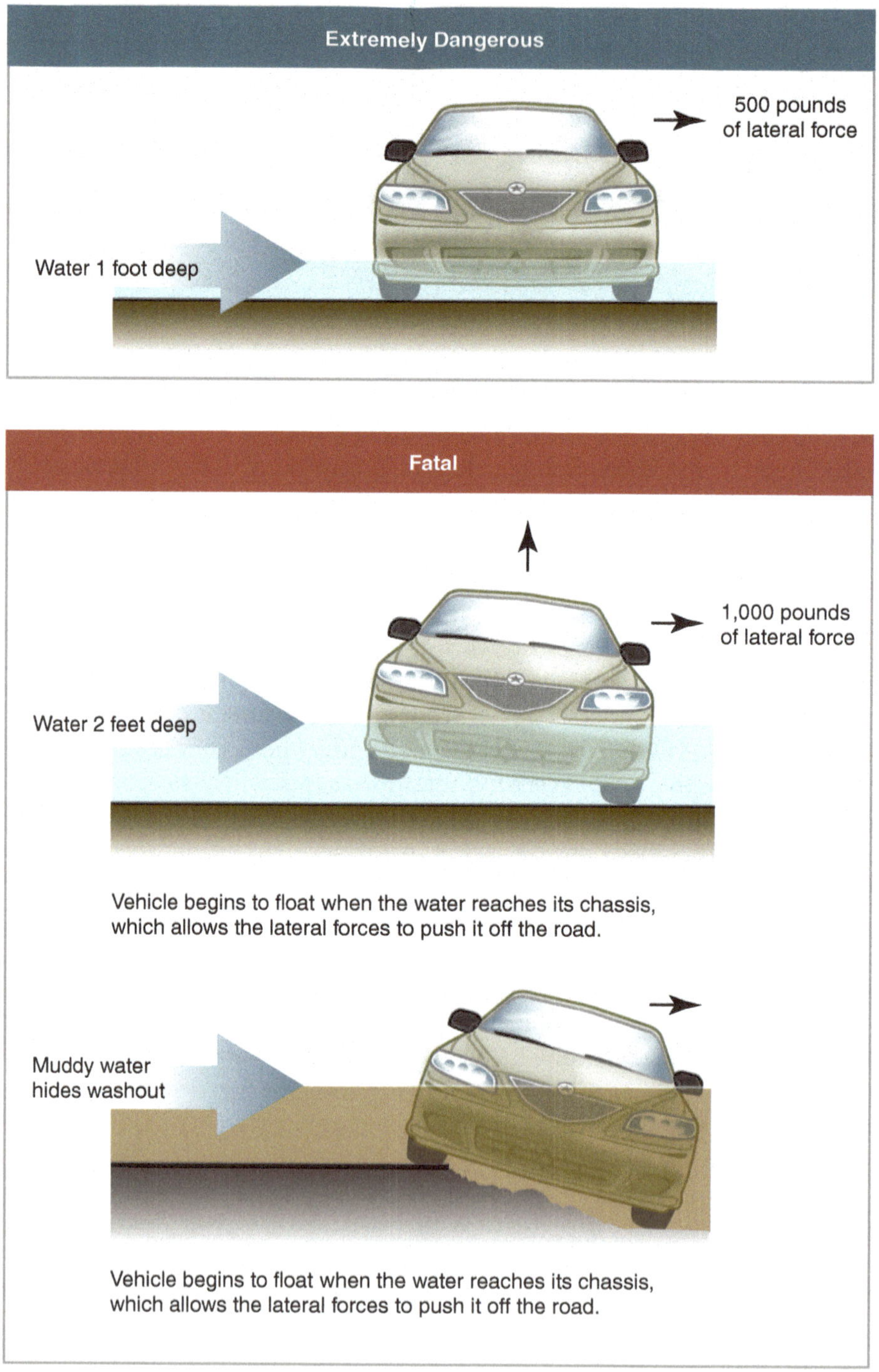

Source: USGS.

downstream. Ice builds up whenever a small period of melting occurs and then refreezing takes place, thus making the drainage situation worse.

During the winter of 1996 and 1997, the watershed of the Red River of the North, which forms the state boundary of North Dakota and Minnesota, was besieged by excessive rainfall and a series of blizzards. Precipitation totals were more than three times normal and cold weather early in the winter froze the soil, preventing any percolation of water. In early April the region experienced record low temperatures along with another 25 to 30 cm (10 to 12 in) of snow. Within a 10-day period, daytime temperatures swung from single digits to highs in the upper 50s. Rapid melting occurred that produced extensive flooding.

Because the Red River of the North flows north—one of few rivers in the United States to do so—the water was draining into an area where the river was still frozen. The surrounding farmland had a very low gradient (less than a few inches per mile), so the water had nowhere to go. The result was that almost 4.5 million acres (18,210 sq km) were covered in water. Grand Forks, a city of about 48,000 on the banks of the river, was flooded (Figure 9.7) and more than 24,000 homes were destroyed. The downtown area was under several meters of water. News photos recorded fires burning that destroyed several key buildings in the downtown area. Farms were under water, major crops were lost, and more than 120,000 cattle died. Losses amounted to several billion dollars. The damage continued into Canada, where more than $800 million worth of property was destroyed.

Source: USGS.

Figure 9.7 Sorlie Bridge, which connects Grand Forks, North Dakota, with East Grand Forks, Minnesota, lies under water in the April 1997 floods. Losses exceeded $3.5 billion in these cities.

In early February 2021, a glacier in the Himalaya Mountains of northern India broke apart and destroyed a hydroelectric dam. More than 150 people died or were missing, including many who worked at the dam. A combination of ice and flood water moved through the region, producing major damage to buildings and the landscape. Global warming has been attributed to the cause of this event.

Dam-Related Floods

Vaiont Dam, Italy, 1963

Completed in 1961, the Vaiont Dam, the world's sixth highest dam, was built to generate hydroelectric power for northern Italy (Figure 9.8). The region behind the 262 m (860 ft) high dam included steep hillsides that were underlain by sedimentary rocks, including shales, in the Dolomite region, about 100 km north of Venice. In October 1963, after the reservoir filled naturally, abnormal rainfall caused a block of approximately 270 million cubic meters (9.4 billion cu ft) to detach itself from one wall and slide into the lake at velocities of up to 30 m/sec or 65 mph. This generated a wave 100 m high that went over the top of the dam and blasted down the valley below. More than 50 million cubic meters (13.2 billion gallons) of water shot downstream. More than 2,500 people died in the disaster, which was very preventable. During planning

© R. Carnovalini/Getty Images.

Figure 9.8 Vaiont Dam in northern Italy was topped by a massive wave generated by a landslide that went into the impounded reservoir. The resulting flood killed more than 2,500 people in villages along the downstream reach of the Vaiont River, in northern Italy.

and construction of the dam, geologists and engineers failed to note the potential of rock slippage. Fortunately the dam itself did not fail.

Buffalo Creek, 1972

During a 5-year period (1972 to 1977), several major dams failed in the United States. In 1972, a privately controlled, slag-heap dam on Buffalo Creek in West Virginia gave way following excessive rainfall, resulting in 125 deaths and more than $50 million in damage. After the main dam broke, water rushed downstream and destroyed two more dams. In a matter of minutes more than 1,000 people were injured and 4,000 were left homeless. Interestingly, the U.S. Department of the Interior had warned officials in the State in 1967 that the potential for a disaster existed, but nothing was done to address the issues.

Teton Dam, 1976

On June 5, 1976, the Teton Dam on the Teton River in eastern Idaho, near the city of Rexburg, collapsed. The earthen dam was originally designed to serve as a multipurpose structure, providing irrigation water to agricultural land in the area, hydroelectric capabilities, recreational opportunities, and flood control. The dam was the subject of lawsuits in federal court during its planning and early construction, but it was completed and put into service in January 1976. It failed on the first filling of the impounded reservoir. In the end, the federal government paid out almost $400 million in damage claims, and the dam was not rebuilt. The final toll was 11 lives lost and 13,000 head of livestock drowned. Flaws in the design of the dam were considered the source of the failure.

Flood Severity

Some floods are accompanied by a slow rise in water level; others produce sudden raging torrents of water, silt, and other debris. Slow floods are generally forecast to happen whereas rapid ones provide little or no warning for people downstream. The degree of flooding is a function of the amount of water involved, the level it reaches, the expanse the water covers, and the slope of the land, which controls how rapidly an area might drain following an event. In addition to the damage caused by water, many floods deposit thick layers of silt and other debris. One of the major problems for homeowners is the pervasive nature of the mud that is left behind after a house is flooded.

Although flash floods are rapid and impart a great deal of damage, they are usually confined to relatively small areas. Large-scale flooding associated with major rivers or coastal areas affects huge areas and often is in localities that do not drain well, causing the water to linger for weeks. For anyone who has experienced a flood, the end results are personally devastating. Recovery time can extend into months or even years after a major event.

When flooding occurs in wet, humid regions, water takes a longer time to evaporate. Molds and mildew begin to grow and usually become a health hazard. Extensive flooding can also cause cholera and other water-related diseases to spread rapidly among the survivors of these disasters. More detail about these diseases is provided in Chapter 13, which discusses biological hazards as they relate to some types of natural disasters.

Evidence From the Geologic Record

Flooding has occurred on Earth since water began falling from the sky. As the continents began to form several billion years ago, water was an important part of the erosion that began to wear these landmasses down. Eventually, stream patterns formed that allowed water to move more efficiently downhill. Water can flow in a channelized system or across the surface as unconfined surface flow or sheet wash.

Dubiel and others (1991) reported that exposed continental land lay astride the equator during the Triassic Period (about 225 million years ago). Paleoclimate models showed a maximum effect of monsoonal circulation in the atmosphere. Their study pointed to climatic conditions similar to present-day moisture and circulation patterns of Asia and the Indian Ocean. We could employ the principle of uniformitarianism ("the present is the key to the past") and conclude that the possibility existed for cyclonic storms to develop as they do now, thus leaving their mark on coastal and adjacent low-lying regions. Many of those sedimentary environments are now situated well above sea level as a result of later activity that uplifted much of the continents through plate tectonics.

Examples From the Historic Record

Floods have affected Earth's surface for millions of years, just like the other catastrophes we cover in this book. Historians have documented the occurrence of floods worldwide during recorded history. From their reports we learn that no part of the globe is exempt from flooding.

Yanosky and Jarrett (2002) have been able to analyze tree rings for past records of floods to determine the frequency and magnitudes with which they occurred. In some instances, they could determine the date of a flood within a few weeks. One technique they used was to notice where trees were injured by stream debris. By counting the rings, they were able to determine the year, and sometimes the season, when the major flooding took place. Such studies are helpful in determining the recurrence interval for various levels of floods in given areas.

Thousands of major floods have occurred throughout history. Some have killed many people, some have damaged or destroyed cities and surrounding areas. Sometimes the floods could have been prevented or at least the damage minimized. Several noteworthy floods have occurred in the United States in the past 125 years.

Johnstown Flood

On May 31, 1889, an earthen dam failed, flooding the town of Johnstown, Pennsylvania. The dam, which was built in 1853 to hold water for the Pennsylvania Canal, had not received maintenance for several decades. After the canal changed owners, the spillway was altered and a grating was installed to prevent fish from escaping from the lake. This led to a buildup of debris that eventually clogged the outlet.

© Everett Historical/Shutterstock.com.

Figure 9.9 **Ruins of the Cambria Iron Works following the failure of the Johnstown dam in May 1889.**

In the early spring of 1889 the region received significant snowfall that was later followed by heavy rains that melted the snowpack. With a watershed measuring 657 sq mi (1,725 sq km), several billion gallons of water flowed into the lake. By May 30 water was passing over the dam, and the following day the dam gave way with a flood crest of almost 12 m (40 ft). The estimated flow was approximately that of the water passing over Niagara Falls. Numerous debris flows resulted.

The town of Johnstown, located 25 km (15 mi) down the valley, was obliterated (Figure 9.9), and 2,209 people perished. Debates still continue about whether the townspeople were warned of the coming water or if they simply ignored the warning, having heard similar ones in the past when nothing happened. The event is memorialized at the Johnstown Flood National Memorial, part of the U.S. National Park system.

Source: USGS.

Figure 9.10a **Destruction was severe in portions of Big Thompson Canyon due to the July 31, 1976, flooding.**

Big Thompson River Floods, 1976 and 2013

The Big Thompson River has its headwaters in Rocky Mountain National Park, located in the central Rocky Mountains of Colorado. The river, with a length of 125 km (75 mi), flows through several cities and towns and eventually passes through Big Thompson Canyon. This steep-walled canyon, a scenic area measuring 40 kilometers, is traversed by US Highway 34.

On July 31, 1976, a severe thunderstorm developed over a portion of the river's watershed. In less than 5 hours as much as 30 centimeters of rain fell, rapidly moving downstream. The resulting flash flood lasted several hours and produced a wall of water more than 6 meters high (Figure 9.10a). The water moved at 6 meters per second (14 miles per hour), destroying more than 450 homes and businesses, as well as 400 vehicles. Campers along the river were warned of the impending conditions, but many failed to act. One hundred forty-three people died, making this event the worst loss of life in Colorado history.

Raul Valenzuela, United States Geological Survey.

Figure 9.10b **Extreme water velocities caused extensive flooding and damage in Big Thompson Canyon in September 2013.**

In September 2013 a similarly prolonged rainstorm struck the area. Fortunately, the loss of life was much less. However, flooding from this event destroyed a high-water marker set up by the USGS to signify the 1976 flood (Figure 9.10b).

Great Flood of 1993 in the US Midwest

An unusual series of events produced the Great Flood of 1993, which resulted in the most catastrophic flooding the United States has ever experienced. At its peak, more than 15% of the contiguous United States was affected, more than 50,000 homes were damaged or destroyed, and water covered more than 400,000 square miles (1.04 million sq km) in Illinois, Iowa, Kansas, Minnesota, Nebraska, North Dakota, South Dakota, and Wisconsin. The infrastructure of America's heartland was disrupted and its economy was upset, affecting millions of people. For more than 2 months there was no barge traffic on the Mississippi and Missouri Rivers and railroad traffic stopped in the Midwest, one of the hubs of rail transportation in the United States. Because many highways and bridges were flooded, truck traffic came to a stop or had to be rerouted, delaying the delivery of many of the nation's goods.

Larson (1995) provides a good background for examining the conditions that led to the Great Flood of 1993. Flood forecasts are made by 12 National Weather Service River Forecast Centers located throughout the country. Components of the models used to forecast flood potential include temperatures, precipitation amounts, and the soil moisture content of various watersheds. Stream runoff is used to produce a flow model, which uses data from a unit hydrograph. The unit hydrograph represents 1 inch of runoff from a rain storm that evenly covers a defined headwater drainage basin over a given amount of time, usually taken as 6 hours. Data from different upstream drainage basins are combined to determine the amount of flow for a selected downstream location, allowing hydrologists to estimate the potential flooding severity.

The flooding that occurred in the Midwest in 1993 has been attributed by some researchers to El Niño, which is an oscillation of the ocean-atmosphere system located in the tropical Pacific Ocean. From January to June 1993, the upper reaches of the Mississippi River drainage basin received more than 1.5 times the normal rainfall (Figure 9.11). The continuing rains in July caused portions of North Dakota, Iowa, and Kansas to record more than four times their normal rainfall amounts. Rains fell often in very intense storms that passed over the region. Normally, floods only last a few days or perhaps a week or two at most. However, these prolonged rains produced supersaturated soils, causing subsequent rain to run off into the drainage basins and produce unprecedented flooding.

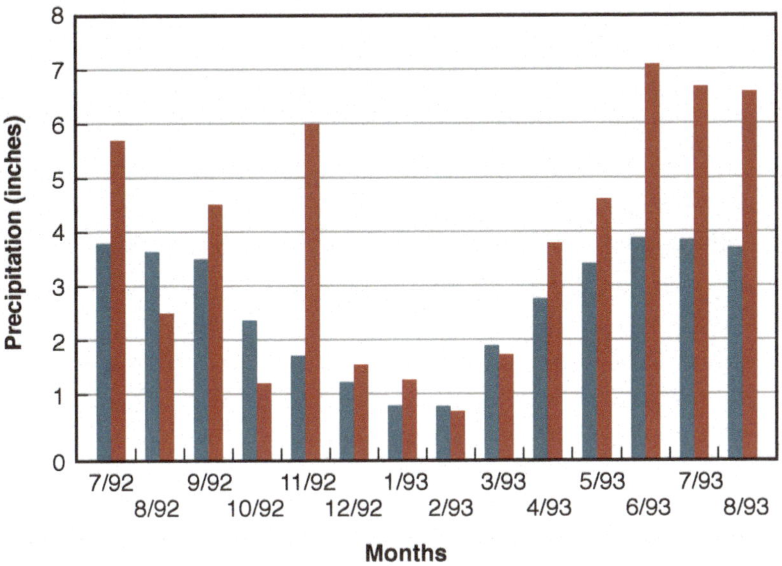

© Kendall Hunt Publishing Company.

Figure 9.11 **Comparison of observed and average monthly precipitation totals for the Upper Mississippi River Basin.**

Figure 9.12 A comparison of the river flows of the Illinois, Mississippi, and Missouri Rivers, August 1991 (normal conditions) and August 19, 1993 (full flood conditions).

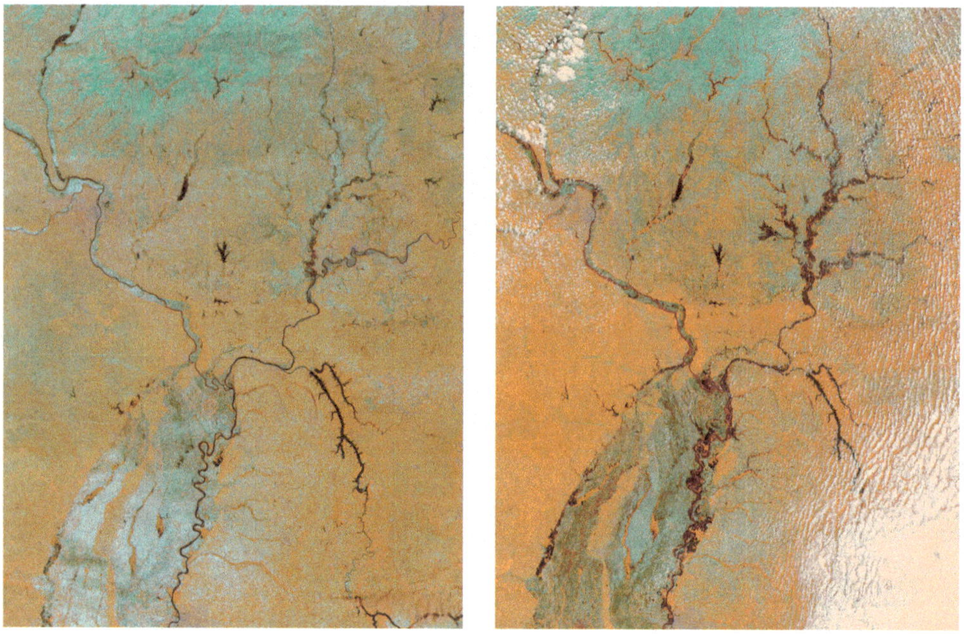

(a) (b)

Source: Earth Observatory, NASA.

FEMA, Andrea Booher.

Figure 9.13 Major flooding occurred in Jefferson City, Missouri, during the 1993 floods.

The Great Flood of 1993 lasted for months, with many locations experiencing flood stage conditions for 5 or 6 months. The duration of the regional flooding was caused by an extremely unusual set of weather patterns that kept the region very wet (Figure 9.12). Saturated soils could not absorb any rain, thereby causing it to run off into streams incapable of carrying the volume of water. Rainfall amounts set new all-time records for many recording stations, as the amounts recorded approximate those expected in a 75- to 300-year event (Larson, 1995).

With the continual rains and high humidity, very little evaporation occurred. The Mississippi River remained above flood stage in St. Louis, Missouri, for 146 days between April 1 and September 30, 1993 (Figure 9.13). In Cape Girardeau, a city located along the Mississippi River in southeastern Missouri, floodwaters crested 5 m (16 ft) above **flood stage**. The river remained at flood stage in Cape Girardeau for 126 days. The floods of 1993 resulted in 50 deaths and damage estimates in the Midwest ranged between $15 and $20 billion.

flood stage

The point in time when a body of water, such as a river, rises to a level that causes damage to the adjacent areas.

Each year in the United States major rainfall events occur. In recent years many parts of the country have received record amounts of rainfall, often as the result of tropical storms and hurricanes. These episodes had numerous devastating results.

Since 2000 more than 50 major episodes of flooding have occurred. Among the most recent events are those that took place in Hawaii in 2018. A series of thunderstorms in April produced record rainfall on the islands of Kauai and Oahu when more than 5.5 inches (140 millimeters) of rain fell. Four months later, Hurricane Lane produced 50 inches (1,475 millimeters) of rain over Mauna Loa. Flooding throughout the state created more than

$250 million of damage. In 2021 severe flooding returned to Maui, Kauai, and Oahu as the result of winter storms.

In 2019 the conterminous United States had the wettest winter on record. Some areas had flooding that exceeded records set by the Great Mississippi River flood of 1927. Flooding, the most-costly natural hazard, affected the Mississippi and Missouri River basins, including Nebraska and western Iowa which were heavily damaged in the early stages. Predicted to last through the spring of 2019, the flooding eventually cost several billion dollars in terms of property, farm, and cattle losses.

Other countries have also had major flooding in recent years. One region that is often adversely hit with flooding is the area including Indonesia and surrounding islands. In March 2019, Indonesia's easternmost province of Papua had nearly a month's worth of rain in 7 hours, triggering flash flooding and landslides. The heavy rainfall killed 89 people, with dozens of others missing causing many remote villages to be out of contact. Landslides destroyed roads and bridges, severely impacting rescue efforts. More than 7,000 people were displaced from their homes. In 2020 and 2021 Indonesia and Timor to the east experienced massive flooding, resulting in major damage and loss of life (Figure 9.14).

© Maharani affah/Shutterstock.com.

Figure 9.14 Extensive flooding is often an annual occurrence in Indonesia.

Many countries worldwide have experienced unusual rainfall in the recent past. Europe has had major flooding with portions of Austria, France, Germany, Great Britain, Italy, the Netherlands, and Spain being inundated in 2018, 2019, and 2021 (Figure 9.15). India, China, Vietnam, and Australia have also experienced floods of epic proportions since 2018. Hundreds of lives have been lost in the areas.

© Lisa S./Shutterstock.com.

Figure 9.15 Flooding in Steyr, Austria, in July 2019.

Global climate change has affected numerous countries with disastrous and often deadly results. In July 2021 eastern Tennessee and western North Carolina had massive flooding, resulting in the loss of several dozen lives.

Human Interactions and Flooding

Streams will naturally flood at some point during their existence. Whether flooding occurs regularly or on a very periodic time scale, water will flow over the banks and onto a floodplain. However, mankind has introduced new parameters into the flooding process that create flooding conditions more frequently than the natural processes.

Construction of Levees and Channels

Construction of natural levees, although meant to allow a stream to flood in a natural way less often, upsets the normal balance of a stream system. Increased height of levees causes more water to flow in the deeper channel, so when it does occasionally break through, major problems develop. Water cannot easily flow back into the main channel.

Construction of concrete channels to contain flow through urban areas only speeds up any excess water from heavy rainfall or snowmelt that enters the channels. The Los Angeles River has been channelized as the river passes through the city of Los Angeles. More than 60% of its watershed is covered by impervious material (asphalt and concrete), so infiltration is greatly reduced. Water is sent to lower elevations in the channel system. There are times when too much water accumulates at the lower reaches of these channels, and flooding results in extensive, unexpected damage.

When flooding occurs, sandbags are often used to contain the flow of water. Bags are placed along the banks of a river as temporary, "quick-fix" levees that hold most of the water in a channel. Bags are also used in low-lying areas to slow the flow of water into buildings and their doorways or low windows. Sandbags work well in the short term for localized flooding but do not help with major overbank conditions. Sometimes large sheets of plastic are placed over the bags but these sheets are not continuous and will allow leaks. All of these techniques are insufficient in solving the larger problem of streams overflowing their banks.

Flood Control

When water is out of control, it creates major problems. Several techniques have been devised to help control and channel potential floodwaters. Channelization can involve the clearing or dredging of an existing channel to speed up the flow. Another method is to construct artificial cutoffs that shorten the length of a stream, increasing its velocity and gradient. The result is that more water flows through an area with higher velocities. The goal is to reduce the chances of flooding, but channelization has produced mixed results.

Recall that natural levees form along rivers and deepen the channel. Piling extra earth atop natural levees or putting earth alongside a river creates artificial levees that deepen the channel and increase the volume of water in the stream channel. However, these are only temporary solutions that eventually prove unsuccessful in controlling floods.

Dams help to reduce flooding by storing water and releasing it in a controlled manner. In addition to the impounded water, however, the dams collect sediment that normally would have been transported downstream to enrich floodplain farmland or to create riparian habitats along the stream. Dams create an imbalance in the ecology of the river environment, a fact that was not recognized until recently. On the other hand, the thousands of dams in the United States also generate hydroelectric power, provide water for agricultural irrigation, and serve as recreation areas.

Predicting Floods

The occurrence of floods depends on many variables, most of which we have little control over. Weather-related floods can be forecast with some detail in the short term, but over longer periods of time we cannot foresee their arrival. However, researchers can analyze flood frequency over longer periods of time, provided adequate data exist.

The USGS has a system of gauging stations situated all over the country that provide real-time data. Stream flow is constantly monitored and the data are available for researchers and scientists to analyze. For the data to be useful, researchers attempt to determine some type of recurrence pattern or interval for flooding events. The term *100-year flood* is a standard time period used to describe the recurrence interval of floods. As Table 9.2 shows, the "100-year recurrence interval" means that a flood of that magnitude has a 1% chance of occurring in any given year. In other words, the chance that a river will flow as high as the 100-year flood stage in any given year is 1 in 100. Statistically, each year begins with the same % chance that a 100-year event will occur.

Even though a 100-year flood happens in one given year, its occurrence does not imply that another 100 years will pass before the water level would be that high again. Many variables are at work controlling the amount of moisture in an area so the event can indeed occur at any time, even twice or more often within a given year. Should the 100-year flood begin to become more common, it will be necessary to reexamine all the data and redefine the 100-year flood, taking into account changes in the occurrence of such a flood level.

Table 9.2 Recurrence Intervals for Various Levels of Floods

Recurrence Interval, in Years	Probability of Occurrence in Any Given Year	Percent Chance of Occurrence in Any Given Year
100	1 in 100	1
50	1 in 50	2
25	1 in 25	4
10	1 in 10	10
5	1 in 5	20
2	1 in 2	50

Key Terms

acre foot *(page 257)*
discharge *(page 256)*
drainage basin *(page 255)*
floodplain *(page 256)*
flood stage *(page 266)*
hydrologic cycle *(page 254)*
Ice-jam flood *(page 259)*
river *(page 255)*
runoff *(page 255)*
stream *(page 255)*
tributary *(page 255)*
watershed *(page 255)*

Summary

Rivers and streams play a key role in the movement of surface water. Depending on the terrane across which they travel, and weather conditions, they can produce floods. Floods occur every day somewhere on Earth. When a drainage system receives more water than it can handle, the excess water flows over riverbanks and across the landscape, flooding farmland and communities. Drainage systems that develop in watersheds serve to remove water that falls in the area. As streams flow downhill, they gather more water and have a higher flow in the lower elevations. If the amount of water exceeds the capacity of the stream channel, flooding occurs.

The past several decades have seen significant flooding worldwide. Excessive rainfall throughout the globe has produced billions of dollars of property damage and caused many people to die.

Regional floods result from prolonged rainfall or excessive snowmelt. Such floods usually last several days or weeks and create widespread damage to a region. Usually the loss of human life is small in these events. Flash floods form rapidly. They provide very little warning and are the primary cause of loss of human life. Floods caused by ice dams or the breaking of artificial dams have widespread devastating effects. Failure of artificial dams is often traced to engineering problems. The effects of historic floods throughout the world indicate that many regions are affected by the influx of water. Low-lying areas suffer the most because floodwaters are slow to recede.

Humans have attempted to intervene in preventing floods but generally the results are mixed. Channelization of streams and overland flow move water through an area more rapidly than normal but can produce unexpected effects if too much water ends up in the drainage system. The construction of dams for flood control is often a temporary solution as the dams tend to fill with stream sediment that eventually reduces their capacity to impound the amount of water for which they were originally designed. Impounded sediment can upset the downstream ecosystems both in the river and along the banks.

The ability of researchers to predict floods is helpful in forecasting potential problems in a given area. However, the fickle nature of weather systems and climate change makes the prediction of major events very difficult. Due to an increase in flooding, it is necessary to use recent information when attempting to forecast future patterns of floods and related events.

References and Suggested Readings

Benito, G., Baker, V. R., & Gregory, K. J. (Eds.). (1998). *Palaeohydrology and environmental change.* John Wiley.

Collier, M., & Webb, R. H. (2002). *Floods, droughts, and climate change.* University of Arizona Press.

Dubiel, R. F., Totman Parrish, J., Parrish, J. M., & Good, S. C. (1991). The Pangaean megamonsoon—Evidence from the Upper Triassic chinle formation, Colorado plateau. *Palaios, 6,* 347–370.

Goodell, J. (2017). *The water will come—Rising seas, sinking cities, and the remaking of the civilized world.* Little, Brown and Company.

House, P. K., Webb, R, H., Baker, V. R., & Levish, D. R. (Eds.). (2002). *Ancient floods, modern hazards—Principles and applications of paleo-flood hydrology.* American Geophysical Union.

Kandel, R. (2003). *Water from Heaven.* Columbia University Press.

Larson, L. W. (1995). The great USA flood of 1993. http://www.nwrfc.noaa.gov/floods/papers/oh_2/great.htm

Mayer, L, & Nash, D. (Eds.). (1987). *Catastrophic flooding.* Allen and Unwin.

McCullough, D. (1987). *The Johnstown flood.* Simon and Schuster.

Meyer, S. (2013). *Adapting to flooding and rising sea level.* Rosen Central.

Renton, J. J. (2011). *Physical geology across the American landscape* (3rd ed.). Kendall Hunt Publishing Company.

Sene, K. (2013). *Flash floods—Forecasting and warning.* Springer.

Welky, D. (2011). *The thousand-year flood—The Ohio-Mississippi disaster of 1937.* University of Chicago Press.

Williams, G. (2013). *Washed away—How the great flood of 1913, America's most widespread natural disaster terrorized a nation and changed it forever.* Pegasus Books.

Yanosky, T. M., & Jarrett, R. D. (2002). Dendrochronologic evidence for the frequency and magnitude of paleofloods. In P. K. House, R. H. Webb, V. R. Baker, & D. R. Levish (Eds.)., *Ancient floods, modern hazards—Principles and applications of paleoflood hydrology.* American Geophysical Union.

Websites for Further Reference

https://photolib.noaa.gov

https://pubs.usgs.gov/circ/2003/circ1245

https://water.usgs.gov

https://waterdata.usgs.gov/nwis

https://www.noaa.gov

https://www.weather.gov

Questions for Thought

1. How does the size of a drainage basin affect the amount of discharge in a stream?
2. How does the cross-sectional area of a stream affect its velocity?
3. What factors lead to flash floods? What would you do if you encountered a flash flood?
4. What are ice-jam floods and during what part of the year are they most likely to occur?
5. Give two examples of how poor planning resulted in disastrous failures of dams.
6. Explain the role that levees play along major rivers.
7. How predictable are floods?
8. What would be the effects of a 2-foot flood in your hometown? Where would such a flood come from and what would be the long-term effects? Look at your hometown's website to see if any plans are in place for such a disaster.
9. What effects do changes in climatic conditions have on the occurrence of floods?
10. How does human activity and development affect localized flooding?
11. Explain why coastal flooding takes a long time to recede.

10 Coastal Regions and Land Loss

© Bruce Ellis/Shutterstock.com.

Morris Island Lighthouse, near Charleston, South Carolina, was built about 400 meters (1,300 ft) inland in 1876. By 1938 changes in nearshore currents had eroded the land so that the lighthouse was at water's edge. Today it is about 450 meters (1,500 ft) from the shoreline.

The Role of Oceans on Earth

Oceans cover 70% of Earth's surface. The three largest, the Pacific, Atlantic, and Indian, cover more than 100 million square miles (259 million sq km). All oceans are key to our weather and they, along with the atmosphere, serve as the main mechanism for heat transfer around the globe. Solar radiation provides the necessary energy to heat the water.

Ocean currents redistribute heat from the equatorial regions toward the poles. Thousands of kilometers of shoreline are the interface between the continents and the oceans. This contact zone is under constant stress, as it is being eroded by the energy of waves and storms. As sea levels continue to rise, we must be aware of the potential for increased flooding and erosion of coastal regions, areas that are home to several billion people on Earth.

The relentless attack by these waters will have devastating effects on the world's population in coming decades, as coastal areas demonstrate that the oceans are a dynamic part of our existence. Within our lifetimes there will undoubtedly be significant changes to coastal areas that were once thought to be stable.

Coastal regions, where the land meets the ocean or a large lake, are attractive places to live or vacation. Because large bodies of water moderate temperatures, coastal areas are cooler in summer and warmer in winter than land that is located farther inland. Humans have always been attracted to the coastal regions, to take advantage of the milder climate, abundant seafood, easy transportation, recreational opportunities, and commercial benefits. It is no wonder that the coastlines have become heavily urbanized and industrialized. Approximately 40% of the world's population lives within 100 kilometers of ocean coasts. Counties in the United States that lie directly on the ocean shoreline are about 10% of the total land area (excluding Alaska) but have 39% of the total population. The population density of these areas is more than six times greater than that of inland counties.

Increasing coastal development is of major concern to geologists because they know that coastlines are among Earth's most geologically active and fragile environments. Water in oceans and lakes is constantly in motion due to winds, tides, currents, and, occasionally, tsunami. Therefore, coasts are dynamic and constantly changing from interactions between the energy in the water and the land. These coastal changes occur on two very different time scales. Short-term change (years to decades) is largely due to coastal erosion from waves, storms, and coastal flooding, while long-term changes (hundreds to thousands of years) are due to slower sea-level variation that causes a landward shift in the coastline. These natural processes posed no problem until people began to live along coasts.

Hurricanes and coastal storms are major hazards affecting most coasts in the United States, due to the high energy waves they bring to these areas. As these coasts become increasingly developed, they are highly vulnerable to these natural hazards, and storm damage continues to rise dramatically. Most of the coastline along the Western United States is affected by multiple hazards including landslides caused by cliffs continually being undermined by large waves. Today, coastal erosion affects businesses, homes, public facilities, beaches, cliffs, and bluffs (cliffs along lakes) built close to the water's edge. It is estimated that within the next 60 years, coastal erosion may claim one out of four structures within 150 meters of the coastline of the United States.

Even though the coasts are dynamic environments, predicting future coastal change is often difficult, because of the many variables inherent in world climate, weather, nature of the coastline, and human activity. However, scientific studies show evidence that sea level will continue rising, and that storms will become more common and powerful in the coming years. If our present patterns and rates of development along coasts continue, then we are on a collision course with more disasters and catastrophes. It is therefore important to understand the nature of coastal processes and their inherent natural hazards as future development is planned along the coastal zone.

© LedyX/Shutterstock.com.

Figure 10.1 The shoreline serves as a boundary between the hydrosphere, geosphere, and atmosphere, where all three meet and interact.

Coastal Processes

Coastal Basics

The **coastline** is a unique boundary where the geosphere, atmosphere, and hydrosphere meet and the systems interact (**Figure 10.1**). At this boundary, dynamic processes of erosion and deposition are constantly at work shaping and reshaping the landscape. Coastal processes active along the coasts are the result of interactions within the climate system and the solar system. Coastal surf and storms result from interactions between the atmosphere and the hydrosphere, with the Sun as the ultimate source of energy driving them. Wave activity that derives from blowing winds is the most important process acting along lake and marine coastlines. Gravity is also an important source of energy in producing rising and falling tides and currents that mostly affect oceanic coastlines. Tides are produced by gravitational interactions between Earth, the Sun, and Moon. Thus waves and tides are important processes in bringing energy to the coasts for erosion of the land and the transport and deposition of sediment along the shorelines.

coastline
Unique boundary where the geosphere, atmosphere, and hydrosphere meet and the systems interact.

Waves

Wave Generation

A **wave** is energy in motion that is the result of some disturbance. The energy that causes waves to form in water is called a disturbing force. For example, a rock thrown into a still lake will create waves that radiate in all directions from the disturbance. Mass movement into the ocean, such as coastal landslides and calving glaciers (which creates icebergs) produce waves commonly known as splash waves. Sea floor movements change the shape of the ocean floor and release tremendous amounts of energy to the entire water column and create very large waves. Examples

wave
Energy in motion that is the result of some disturbance that moves over or through a medium with speeds determined by the properties of the medium. Ocean waves are usually generated by wind blowing across the water surface.

Figure 10.2 **Wave form with characteristics. Crests and troughs alternate across the wave form. Note that water motion dies off at a depth of about one half the wavelength.**

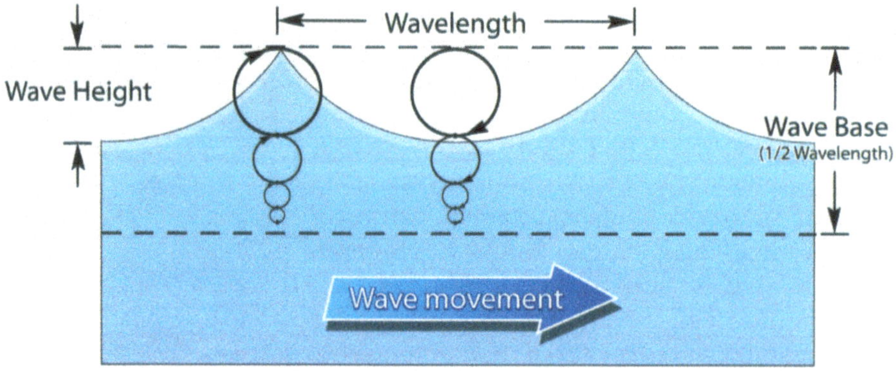

© Kendall Hunt Publishing Company.

crest

The highest point of a wave.

trough

The low spot between two successive waves.

wave height

The vertical distance between the crest and adjacent trough of a wave.

wave period

The time it takes for one full wavelength to pass a given point.

wave speed

The velocity of propagation of a wave through a liquid, relative to the rate of movement of the liquid through which the disturbance is propagated.

wave steepness

The measured ratio of wave height to wavelength.

break

When a wave steepness exceeds 1/7, the wave becomes too steep to support itself and it breaks, or spills forward, releasing energy and forming whitecaps often observed in choppy waters or the surf area along a beach.

include underwater avalanches, volcanic eruptions, and fault movement, all of which can generate a tsunami (Chapter 5). Human activities can also generate waves, such as when ships travel across a body of water and leave behind a wake, which is a wave. Wind blowing across a body of water disturbs surface waters and generates most waves that we commonly see on the surface of the oceans or large lakes. Wind-generated waves represent a direct transfer of kinetic energy from the atmosphere to the water surface. In all these cases, some type of energy release creates waves; however, wind-generated waves provide most of the energy that reaches land and shapes and modifies the coastlines.

Wave Characteristics

When wind blows unobstructed across the water, it deforms the surface into a series of wave oscillations. The highest part of the wave is the **crest**, and the lowest part between crests is the **trough** (Figure 10.2). The vertical distance between the crest and the adjacent trough is the **wave height**. The horizontal distance between any two similar points of the wave, such as two crests or two troughs, is the wavelength. The **wave period** is the time it takes for one full wavelength to pass a given point. The **wave speed** is the rate at which the wave travels and is equal to the wavelength divided by the period. **Wave steepness** is the ratio of wave height to wavelength. If the wave steepness exceeds 1/7, the wave becomes too steep to support itself and it **breaks**, or spills forward, releasing energy and forming whitecaps, often observed in choppy waters or in the surf zone along a beach. A wave can break anytime the 1:7 ratio is exceeded, either in the open ocean or along the shoreline.

Wave Motion

Waves are a mechanism by which energy is transferred along the surface of the water. Waves can travel many kilometers from their place of origin. Waves generated in Antarctica have been tracked as they traveled over 10,000 kilometers (6,250 mi) through the Pacific Ocean before finally expending their energy a week later on the shores of Alaska. But it is important to note that the water itself does not travel this great distance. The water is merely the medium for the waveform (the energy) to travel through, similar to earthquake seismic waves as they travel through solid rock.

In open water, water particles pass the energy along by moving in a circle (Figure 10.3). This is known as **circular orbital motion**. This motion can be observed easily by observing a floating object as it bobs up and down and sways back and forth as the wave passes. The object itself does not travel along with the wave. From the side, the object can be viewed moving in a circular orbit with a diameter at the surface equal to the wave's height. Beneath the surface, the orbital motion of the water particles diminishes downward with depth. At a depth equal to approximately one half the wavelength, there is no movement associated with surface waves. This bottom depth of orbital motion is known as the **wave base**. Submarines can avoid large ocean waves by submerging below the wave base. Even seasick scuba divers can find relief by submerging into the calm water below the wave base. If the water depth is greater than the wave base, the waves are called deep-water waves and do not contact the ocean floor. The motion of water in waves is therefore distinctly different from the motion of water in currents, in which water travels in a given direction and does not return to its original position.

© Kendall Hunt Publishing Company.

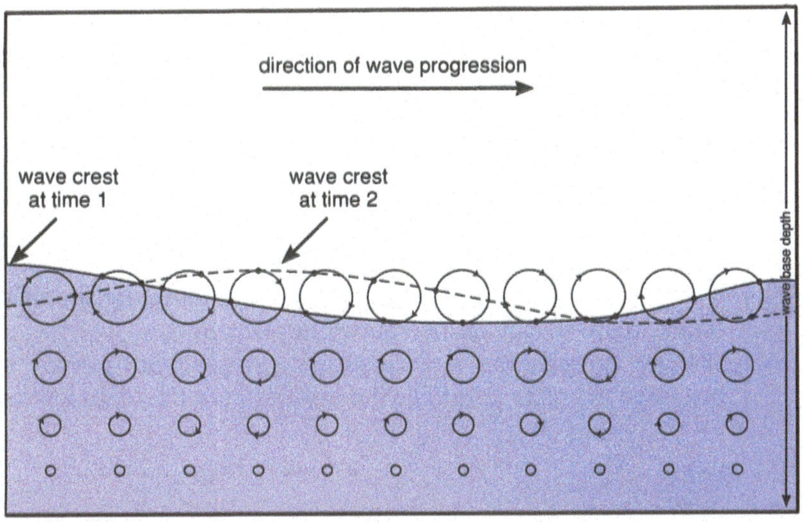

Figure 10.3 Circular motion of a wave form in open water. Individual particles of water rotate in a circle, producing an overall lateral movement of a wave.

circular orbital motion
The movement of a particle by moving in a circle.

wave base
Depth equal to one half the wavelength where there is no movement associated with surface waves.

Wave Energy

As wind blows over the surface of a body of water, some of its energy is transferred to the water. The mechanism of energy transfer is related to frictional drag resulting from one fluid (the air) moving over another fluid (the water). Waves that crash along the coast in the absence of local winds are generated by offshore winds and storm events, sometimes thousands of kilometers away. The energy of a wave depends on its height and length. The higher the wave's height, the greater the size of the orbit in which the water moves. The height, length, and period of a wave depend on the combination of three factors: (1) wind speed (the stronger the wind speed, the larger the waves), (2) wind duration (the longer time the wind blows, the more time the wind can transfer energy to the water, and the larger the waves), and (3) **fetch**, the distance over which the wind blows (a longer fetch allows more energy to be transferred and form larger waves). If we compare waves on a lake to those on the ocean, when wind speed and duration are the same, the waves will be higher on the ocean because the fetch is far greater than on a lake. When the maximum fetch and duration have been reached for a given wind speed, waves will be fully developed and will grow no further. This is

fetch
The distance over which the wind blows across open water.

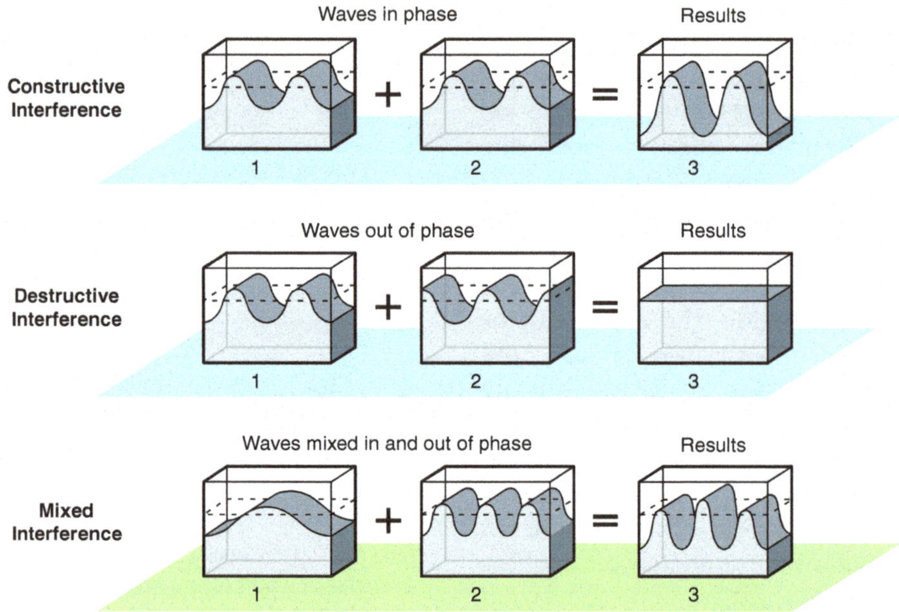

© Kendall Hunt Publishing Company.

Figure 10.4 **Interference patterns in water. Constructive interference creates larger waves, destructive interference reduces the waves to a flat surface, and mixed interference generates a mixture of small and large waves.**

because they are losing as much energy breaking as whitecaps as they are receiving from the wind.

In areas where storm waves are generated, waves will have different lengths, heights, and periods. When the wind stops, or changes direction, waves will separate into waves of uniform length called **swells**. Swells moving away from the storm center can travel great distances before the energy of the wave is released by breaking and crashing onto the coast.

swell

Wave of uniform wavelength moving away from a storm center. They can travel great distances before the energy of the wave is released by breaking and crashing onto the coast.

Interference Patterns and Hazardous Rogue Waves

When swells of deep water waves from different storms run together, the waves interfere with one another to produce different interference patterns (**Figure 10.4**). The interference pattern produced when two wave systems collide is the sum of the disturbance that each would have created individually. Constructive interference occurs when two waves having the same wavelength come together in phase (meaning crest to crest and trough to trough). The wave height will be the sum of the two, and if it becomes too steep, the wave may break forming whitecaps. Destructive interference occurs when waves having the same wavelength come together out of phase (meaning the crest of one will coincide with the trough of the other). If the wave heights are equal, the energies will cancel each other and the water surface will become flat. If the waves are traveling in opposite directions, the waves will return to their normal heights once they travel through the interference area.

It is common for mid-ocean storm waves to reach 7 meters (22 ft) in height, and in extreme conditions such waves can reach heights of 15 meters (50 ft) or more. However, solitary waves called **rogue waves** can reach enormous heights and can occur when normal ocean waves are not unusually high. The word rogue means unusual, and in this case the waves are unusually large—monsters up to 30 meters in height (approximately the height of a 10-story building)—that can appear without warning. Rogue waves appear to be caused by an extraordinary case of constructive

rogue wave

Large solitary wave caused by constructive wave interference that usually occurs unexpectidly amid waves of smaller size.

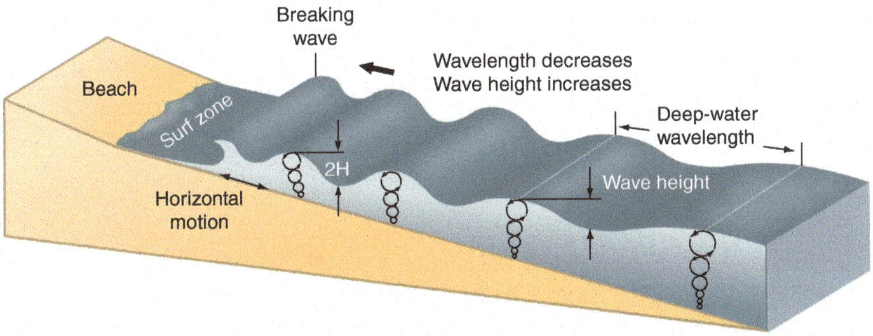

©Kendall Hunt Publishing Company.

Figure 10.5 **Wave hitting the shore with the bottom of the wave intersecting the sea floor. Waves form breakers and move toward the beach as the surface of the water moves faster than the water underneath.**

interference that can be very destructive and have been popularized in movies such as *The Poseiden Adventure* and *The Perfect Storm.*

In 1942 during World War II, the RMS *Queen Mary* was carrying 15,000 American troops near Scotland during a gale and was broadsided by a 28-meter high rogue wave and nearly capsized. The ship listed briefly about 52 degrees before the ship slowly righted herself.

Waves Reach the Shore: Shallow-Water Waves and Breakers

When deep-water waves approach shore, the water depth decreases and the wave base starts to intersect the seafloor. At this point the wave comes in contact with the bottom and the character of the wave starts to change (**Figure 10.5**). In this zone of shoaling (shallowing) waves, they grow taller and less symmetrical. Because of friction at the bottom, the wave speed decreases, but its period remains the same, and thus, the wavelength will decrease. The circular loops of water motion also change to elliptical shapes, as loops are deformed by the bottom. As the wave moves farther shoreward, the wavelength shortens considerably and the wave height increases. The increase in wave height, combined with the decrease in wavelength, causes an increase in wave steepness. With continued forward motion at the top of the wave and friction at the bottom, the front portion of the wave cannot support the water as the rear part moves over, and the wave breaks as surf. Here in the **surf zone**, actual forward movement of water itself occurs within the wave as all the water releases its energy as a wall of moving, turbulent surf known as a **breaker**.

There are three main types of breakers (**Figure 10.6**). **Spilling breakers** form on shorelines with gentle offshore slopes and are characterized by turbulent crests spilling down the front slope of the wave. **Plunging breakers** form on shorelines with steeper offshore slopes and have a curling crest that moves over an air pocket. Plunging breakers are prized waves for surfing. **Surging breakers** form when the offshore slopes abruptly and the wave energy is compressed into a shorter distance and the wave surges forward right at the shoreline.

After the breaker collapses, a turbulent sheet of water, the **swash**, rushes up the slope of the beach (called the swash zone). The swash is a powerful surge that causes landward movement of sediment (sand and gravel) on the beach. When the energy of the swash is dissipated, the water flows by gravity back down the beach toward the surf zone as **backwash**. Therefore, as a wave approaches the shore, it breaks, and the stored energy in the wave

surf zone
The nearshore zone of breaking waves.

breaker
A wave in which the water at the top and leading edge falls forward producing foam.

spilling breaker
Forms on shorelines with gentle offshore slopes and are characterized by turbulent crests spilling down the front slope of the wave.

plunging breaker
Forms on shorelines with more steep offshore slopes and have a curling crest that moves over an air pocket.

surging breaker
Forms when the offshore slopes abruptly and the wave energy is compressed into shorter distance and the wave surges forward right at the shoreline.

swash
A turbulent sheet of water that rushes up the slope of the beach following the breaking of a wave at shore.

backwash
The flow of water down the beach face toward the ocean from a previously broken wave.

Figure 10.6 Spilling, plunging, and surging breakers. Sea floor topography plays a key role in the type of breaker that forms.

Fundamentals: Breakers

There are three types of breakers:

- **Spilling breakers** break gradually over considerable distance.
- **Plunging breakers** tend to curl over and break with a single crash. The front face is concave, the rear face is convex.
- **Surging breakers** peak up, but surge onto the back without spilling or plunging. Even though they don't "break," surging waves are still classified as breakers.

© Kendall Hunt Publishing Company.

is expended in the surf and swash zones, causing erosion, transport, and deposition of sediment along the coast (Figure 10.7).

Wave Refraction

When waves approach an irregular shoreline, or at an angle to the shore, the wave base will initially encounter shallower water areas first and begin to slow down before the rest of the wave does resulting in **wave refraction**, or bending of the wave. Refraction of waves approaching coastlines concentrates wave energy on protruding **headland** areas and dissipates energy in the bays (Figure 10.8). The concentrated energy on headlands erodes them into cliffs and causes deposition of sediment in the bays; thus, headlands erode faster than bays due to stronger wave energy. The result of wave refraction is to erode headlands and smooth out the coastline. The eroded sediments are deposited offshore, on beaches, or in bays.

wave refraction

The process by which the part of a wave in shallow water is slowed down, causing it to bend and approach nearly parallel to shore.

headland

A steep-faced irregularity of the coast that extends out into the ocean.

Tides

Tides produce short-term fluctuations in sea level on a daily basis. **Tides** are the rising and falling of Earth's ocean surface caused by the gravitational attraction of the Moon and the Sun on the Earth. Because the Moon is closer to the Earth than the Sun, it has a greater effect and causes the Earth's

tide

The periodic rising and falling of the water that results from the gravitational attraction of the Moon and Sun acting on the rotating Earth.

water to bulge toward it, while at the same time a bulge occurs on the opposite side of the Earth due to inertial forces. These different bulges remain stationary while Earth rotates and the tidal bulges result in a rhythmic rise and fall of the ocean surface, which is not noticeable in the open ocean but is magnified along the coasts. The changing tide produced at a given location is the result of the changing positions of the Moon and Sun relative to the Earth, coupled with the effects of Earth rotation and the local shape of the sea floor.

© Dmitry Naumov/Shutterstock.com.

Figure 10.7 Swash forms as a breaking wave moves up the flat surface along the shore. Receding water produces the backwash.

The regular fluctuations in the ocean surface result in most coastline areas having two daily high tides and two low tides as sea level rises and falls along the shore. A complete tidal cycle includes a **flood tide** that progresses upward on the shore until high tide is reached, followed by an **ebb tide** falling off the shore until low tide is reached and exposing the land once again. Tidal ranges between high and low tides along most coasts range about 2 meters (6 ft). However, in narrow inlets tidal currents can be strong and cause variations in sea level up to 16 meters (53 ft). High and low tides do not occur at the same time each day but instead are delayed about 53 minutes every 24 hours. This is because the Earth makes a complete axis rotation in 24 hours, but at the same time, the Moon is orbiting the Earth in the same direction, which means the Earth must spin an additional 53 minutes for the same point on Earth to be directly beneath the Moon again (and thus in the bulge at its highest). This explains why the Moon rises in the sky about 53 minutes later each day, and in the same manner, why the tides are also about 53 minutes later each day.

flood tide

The incoming or rising tide; the period between low water and the succeeding high water.

ebb tide

That period of tide between a high water and the succeeding low water; falling tide.

Because the Sun also exerts a gravitational attraction on the Earth, there are also monthly tidal cycles that are controlled by the relative position of the Sun and Moon to the Earth. Although the Sun's gravitational pull on

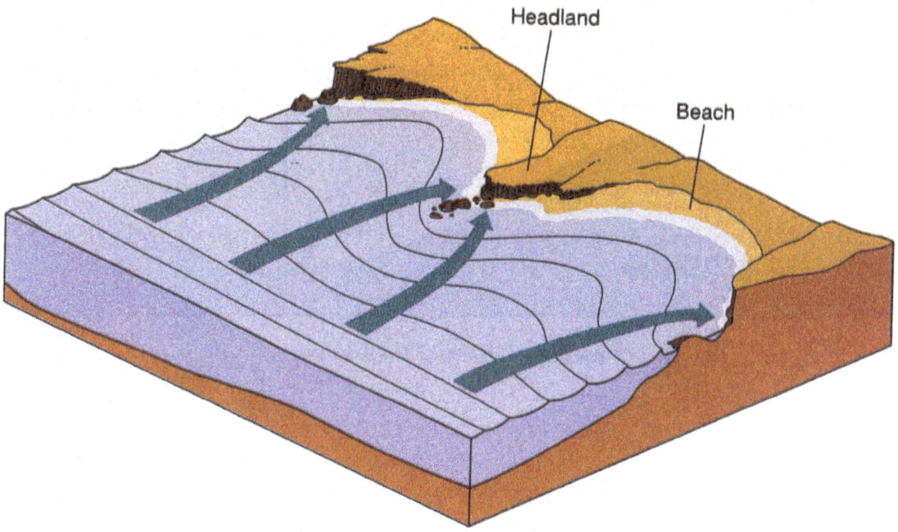

© Kendall Hunt Publishing Company.

Figure 10.8 Refraction of wave energy around a headland. Incoming wave energy is bent or refracted toward the headland due to changes in the sea floor topography that cause the wave to change speed. This produced a zone of high energy that erode the protruding headland.

Figure 10.9 The tidal ranges are affected by the position of the Moon with respect to Earth. New and full moons produce spring tides, while first and third quarter moons produce neap tides.

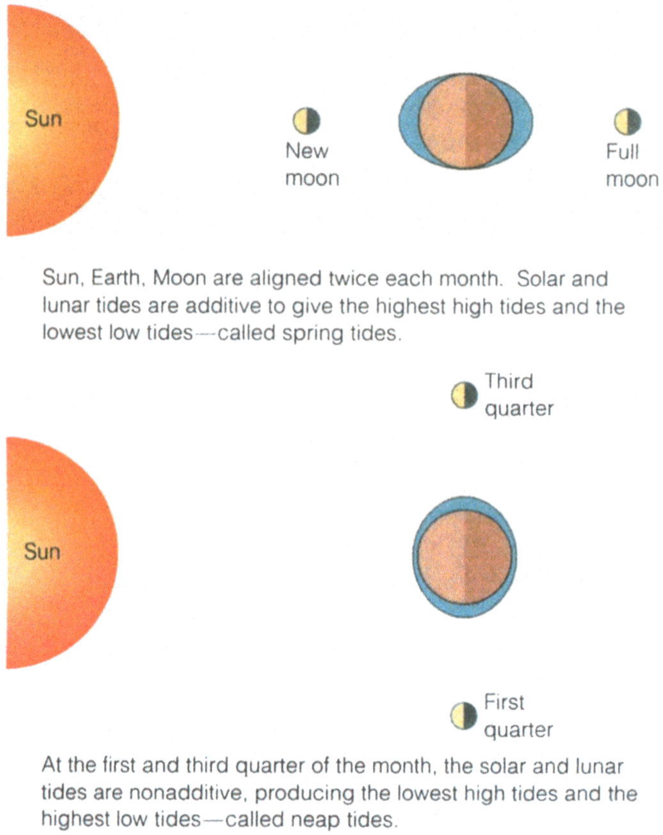

© Kendall Hunt Publishing Company.

the oceans is smaller than the Moon's, it does have an effect on tidal ranges (the difference in elevation between high and low tides). The largest variation between high and low tides, called **spring tides**, occurs when the Sun and the Moon are aligned on the same side of the Earth (new moon) or on opposite sides of the Earth (full moon) (**Figure 10.9**). Here the gravitational attractions of the Moon and Sun amplify each other and produce higher and lower tides. The lowest variation between high and low tides, called **neap tides**, occur when the Moon is at right angles relative to the Earth and Sun (quarter moons).

spring tide

The highest high and the lowest low tide during the lunar month. The exceptionally high and low tides that occur at the time of the new moon or the full moon when the Sun, Moon, and Earth are approximately aligned. Contrast with neap tide.

neap tide

A tide that occurs when the difference between high and low tide is least; the lowest level of high tide. Neap tide comes twice a month, in the first and third quarters of the moon. Contrast with spring tide.

The timing when hazards, such as storms or tsunami events, strike the shoreline, especially at spring tides phases, is very important. The intensity of a disaster is magnified when rising water from a storm surge or tsunami arrives at the same time as the highest high tides. The combination of these events often sends the destructive power of the water farther inland, producing much more catastrophic results.

Coastal Erosion

The erosion of the coastlines is due to the constant battering of waves which causes the land to retreat. Most coastal erosion occurs during intense storms when waves and storm surges are more energetic. The rate of wave erosion varies greatly along coasts of different compositions but is typically rapid along sandy coasts and slower along rocky coasts. Water weathers and erodes coastlines by processes of hydraulic action, abrasion, and corrosion.

© Kendall Hunt Publishing Company.

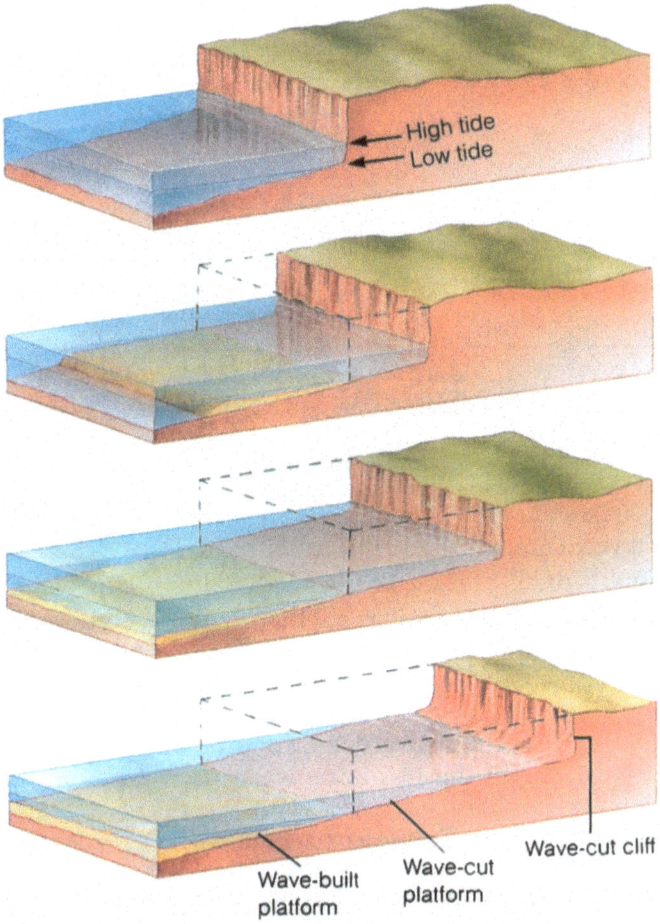

Figure 10.10 Wave-cut platforms and wave-cut cliffs are formed by active erosion produced by incoming tides.

The force of water alone, called hydraulic action, is an effective erosional process. Breaking waves exert a tremendous force on the shores by direct impact of the water and are very effective on cliffs composed of sediment or fractured rocks. A large wave 10 meters (33 ft) high striking a 10-meter high cliff produces four times the thrust energy of a space shuttle's three main orbiter engines. A wave striking a cliff drives water into cracks or other openings in the rock and compresses air inside. As this happens, the water and air combine to create hydraulic forces on the surrounding rock that is large enough to dislodge rock fragments or large boulders. Repeated countless times, hydraulic action wedges out rock fragments from cliff faces which fall to the bottom of the cliff or sea bed. The debris can be picked up and used for another erosive wave action–abrasion. Loose sand is easily moved by wave action and by currents that run parallel to the shoreline.

Landforms of Erosional Coasts

Along coasts where erosion dominates, rocky coastlines are common. Since erosion occurs at sea level, abrasion and hydraulic action undercut exposed bedrock forming a wave-cut cliff. As the cliff continues to erode, it leaves behind a flat or gently sloping **wave-cut platform** (Figure 10.10). Farther offshore a wave-built platform can be formed by transported sediments that are deposited as the water moves seaward. Locally, refraction of waves onto a narrow headland can cut a cave into the rock which may eventually

wave-cut platform

A gently sloping surface produced by wave erosion, extending far into the sea or lake from the base of the wave-cut cliff.

Figure 10.11 Arches represent the remains of a once-protruding headland that reached out into the ocean. The sea stack in the distance seen through the arch is the remnants of a collapsed arch.

Figure 10.12 Waves strike a rugged shoreline in the Channel Islands National Park, California.

sea arch
An opening through a headland caused by erosion.

sea stack
An isolated rock island that is detached from a headland by wave action.

longshore drift
The net movement of sediment parallel to the shore.

beach drift
The movement of sand along a zigzag path along the beach parallel to shore due to successive waves on the beach.

longshore current
A current that flows parallel to the shore just inside the surf zone. It is also called the littoral current.

erode all the way through the headland forming a **sea arch**. When the sea arch collapses, a small portion of the headland may be isolated from the retreating sea cliff and remain as a **sea stack** (Figure 10.11). As waves continue to batter the rocks, eventually the sea stacks crumble. The overall effect is to produce a rugged shoreline that is constantly being hit by incoming waves (Figure 10.12).

Coastal Sediment Transport

Sediment that is created by the abrasive and hydraulic action of waves, or sediment brought to the coast by streams, is picked up by the waves and transported. One of the most important processes of sediment transport within the shoreline area is longshore drift. Sediment is also transported by rip currents and tidal currents.

Longshore Drift and Currents

Longshore drift is the net movement of sediment parallel to the shore. The process starts when waves approach the shore at an oblique angle. Waves striking the shore at an angle, as opposed to straight on, will cause the wave swash to move sediment along the beach at an angle. The backwash brings the sediment directly down the beach slope, under the influence of gravity. This has the net effect of gradual movement of the sediment along the shore by the swash and backwash processes. The swash of the incoming wave moves sand up the beach in a direction perpendicular to the incoming wave crests and the backwash moves the sand down the beach perpendicular to the shoreline. Thus, with successive waves, the sand will move along a zigzag path along the beach parallel to shore. This process is known as **beach drift** (Figure 10.13).

A related process, known as a **longshore current**, develops in the surf zone and a little farther out to sea (Figure 10.14). The movement of swash and backwash in and out from the shore at an angle creates turbulent water in the surf that transports sediments along the shallow bottom in the same

© Kendall Hunt Publishing Company.

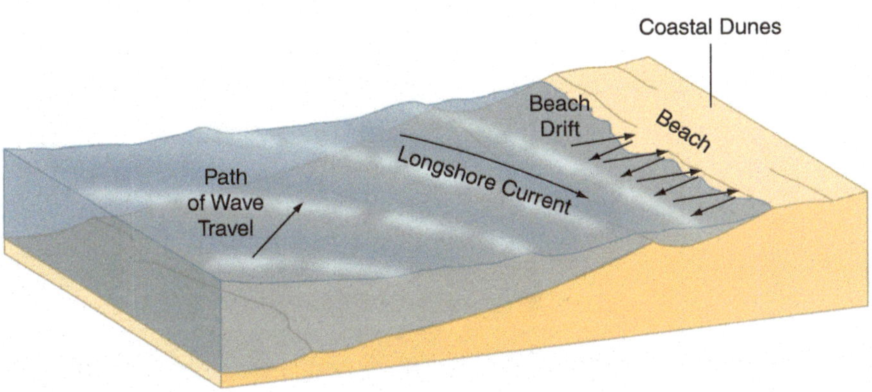

Figure 10.13 Beach drift is created by incoming waves hitting the beach at an angle moving sand that returns to sea in a direction perpendicular to the beach. Longshore current is the overall movement of water parallel to the beach but in the same general direction as the incoming wave direction.

© Kendall Hunt Publishing Company.

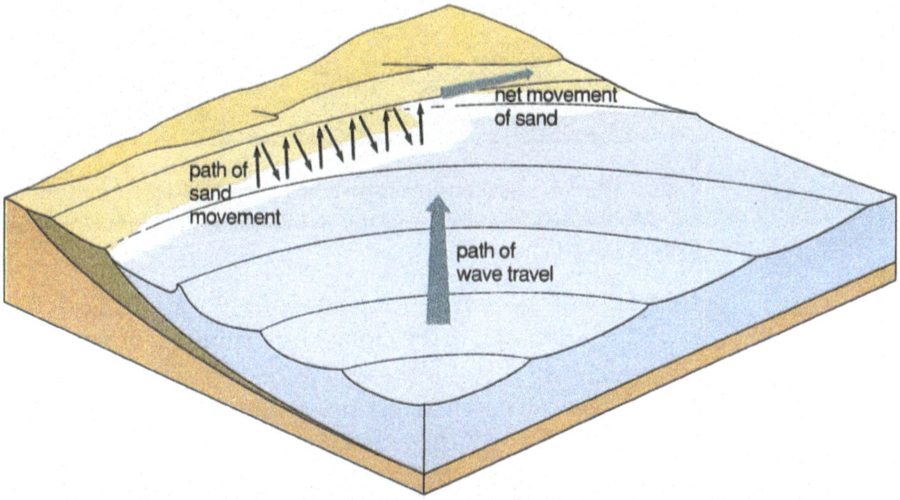

Figure 10.14 Longshore drift creates a net movement of sand along the beach and shallow water zone.

direction as the longshore drift. Substantially more sediment is transported along many beaches as a result of longshore currents than beach drift. Thus longshore currents and beach drift work together as longshore drift to transport huge amounts of sediment along a coast (**Figure 10.14**). At Sandy Hook, New Jersey, approximately 2,000 tons of sediment per day move past any given point on the beach. No wonder beaches are often referred to by coastal geologists as rivers of sand.

Hazardous Rip Currents

There are times when waves can pile large volumes of water on the beach that are much greater than normal. The only way this water can return to the ocean is to ebb back in a channel through the surf zone. This creates a **rip current** that typically flows perpendicular to the shore and the strong surface flow can have sufficient force to be a hazard to swimmers (**Figure 10.15a**). It is often incorrectly called a "rip tide" or "riptide." However, the occurrence is not related to tides. Rates of return flow can range from 0.5 meter (1.6 ft) per second to as much as 25 meters (8.25 ft) per second. The position of rip currents can shift along the beach during the day as differing amounts of water are pushed up onto the shore.

rip current
Movement of water back into the ocean in narrow zones through the surf zone.

Often two characteristics are present that allow us to identify a rip current. As the water is receding toward the ocean, its force counteracts the incoming force of the waves, thereby canceling out the waves. Thus a relatively smooth

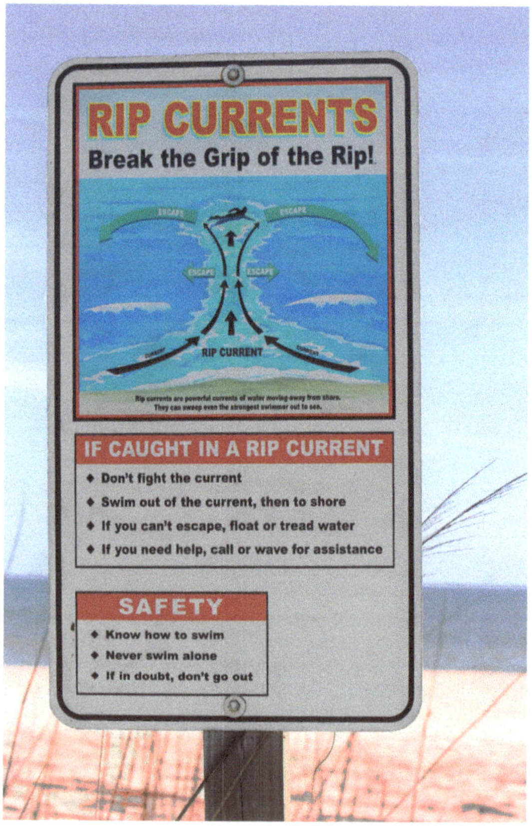

© Chet Mitchell/Shutterstock.com.

Figure 10.15a Warning signs are posted in areas where rip currents are likely to form.

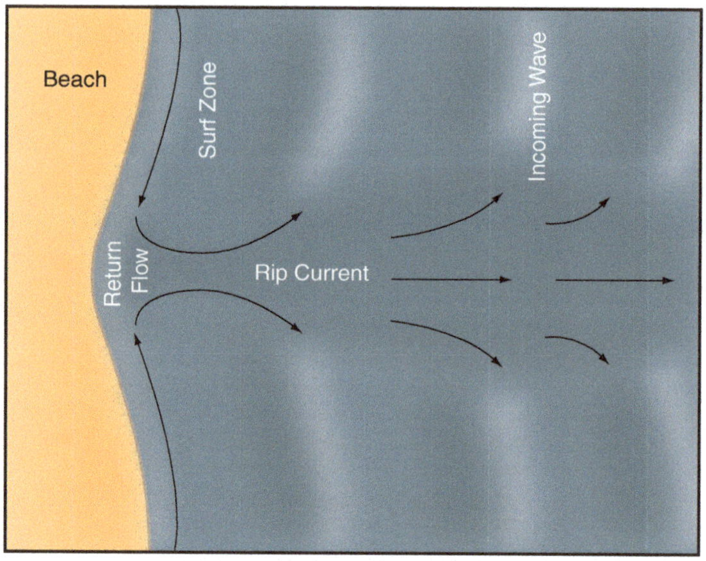

© Kendall Hunt Publishing Company.

Figure 10.15b Rip currents form when too much water is pushed up onto the beach. Its return to the sea causes a rapid movement of water seaward.

surface will be flanked by incoming waves. Also the receding water can carry along large amounts of sand and silt, which will discolor the water. It is advisable to look for the existence of a rip current before heading into the water. Such currents can be extremely dangerous, dragging swimmers away from the beach and leading to death by drowning when they attempt to fight the current and become exhausted. The United States Lifesaving Association reports that rip currents cause approximately 30 deaths annually in the United States, mainly on unguarded beaches. Over 80% of all rescues by beach lifeguards are due to rip currents, totaling almost 24,000 in 2017, a decrease of almost 40% in the past decade, as swimmers are paying more attention to posted warnings (Figure 10.15b).

If a swimmer is caught in a rip current, one should not try to swim directly back to shore but rather swim parallel to the shoreline in order to get out of the current. Rip currents can be between 15 and 45 meters (50 and 150 ft) wide. If you see a person caught in a rip current, yell at them to swim parallel to the shore and you should move along the shoreline in a direction that leads them out of the current.

Coastal Sediment Deposition and Landforms

Sediment transported along the shore is deposited in areas of lower wave energy and produces a variety of landforms. Common landforms include beaches and barrier islands. Erosion of headlands and sea cliffs is the source of some sediment, but probably no more than 5 to 10% of the total. The primary source of sediment is that transported to the coast by rivers that drain the continents. This sediment is then redistributed along the shore by longshore drift.

© Kendall Hunt Publishing Company.

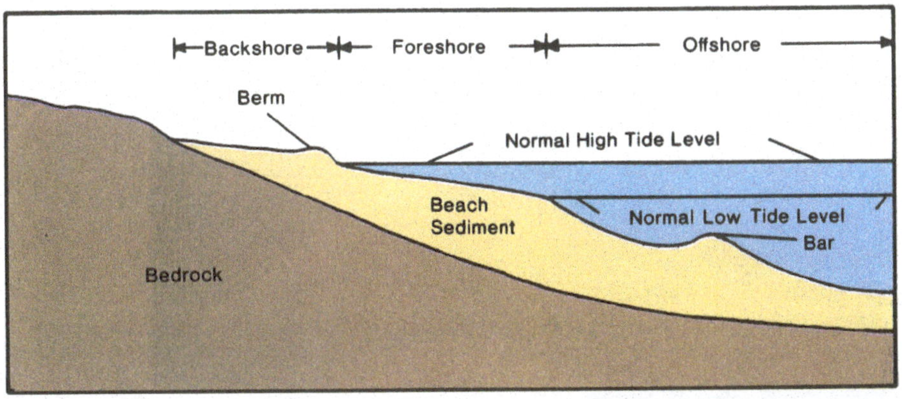

Figure 10.16 The topographic features of a beach area show the different "shore" zones that extend from the land out to sea.

Beaches

A **beach** is an accumulation of unconsolidated sediment along part of the coastline that is exposed to wave action. Beaches are formed from the wave-washed sediment along a coast, and represent interconnected zones of onshore and offshore sediment accumulation. Most beaches can be described in terms of three geomorphic zones (Figure 10.16). The **offshore zone** is the portion of beach that extends seaward from the normal low tide level. Strong backwash currents usually transport some sediment off exposed portions of the beach and deposit it offshore as submerged offshore bars. The **foreshore zone** represents the area between normal low and high tide levels. The **backshore zone**, which is commonly separated from the foreshore by a distinct ridge, called a **berm**, is the part of the beach extending landward from the high tide level to the area reached only during storms. Sediments in this zone are frequently redistributed by wind to form sand dunes (Figure 10.17). Dunes serve as a line of defense against stormwater that can overrun the beach zone. Salt-resistant grasses are often planted to stabilize the dunes. Behind the backshore may be a zone of cliffs, marshes, or additional sand dunes.

Source: David M. Best.

Figure 10.17 Grasses stabilize sand dunes that serve as a break for storm waters.

Even though beaches are areas of deposition, they are in a constant state of change, and dynamically responding to variations in the energy of waves and currents. The effect that waves have depends on their strength and duration. Destructive waves, occur on high energy beaches and are typical of winter storms (Figure 10.18). They reduce the quantity of sediment present on the beach by carrying it out to offshore bars under the sea. Constructive, weak waves are typical of low energy beaches, and occur usually during summer months. These are the opposite of destructive waves because they increase the size of the beach by removing sand from the offshore bars and piling it up onto the berm. This strong and weak wave activity alternates seasonally at most beaches. The weak wave activity produces a high and

beach
An aggregation of unconsolidated sediment, usually sand, that covers the shore.

offshore zone
The portion of beach that extends seaward from the low tide level.

foreshore zone
The area located between the normal low and high tide levels.

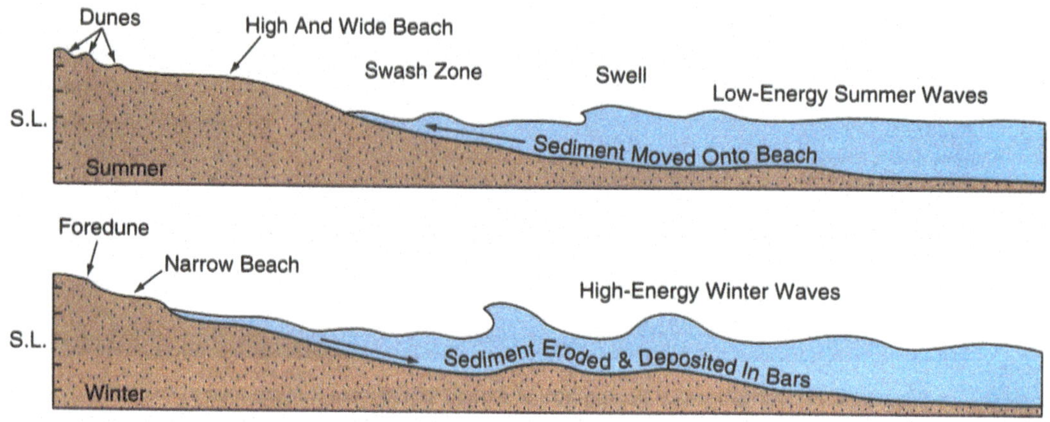

© Kendall Hunt Publishing Company.

Figure 10.18 **Beach profiles during the summer and winter differ in that the lower energy of the summer allows the beach to become enriched with sand. Winter storms erode most of the summer buildup and reduce the size of the beach.**

backshore zone
The part of the beach extending landward from the high tide level to the area reached only during storms.

berm
A low, incipient, nearly horizontal or landward-sloping area, or the landward side of a beach, usually composed of sand deposited by wave action.

barrier island
A long, usually narrow accumulation of sand, that is separated from the mainland by open water (lagoons, bays, and estuaries) or by salt marshes.

emergent coastline
A coastline which has experienced a fall in sea level, because of global sea level change, local land uplift, or isostatic rebound.

submergent coastline
A coastline which has experienced a rise in sea level, due to a global sea level change or local land subsidence.

wide sandy beach at the expense of the offshore bars. The strong wave activity produces a narrow beach during the winter months and builds prominent offshore bars.

Barrier Islands

A **barrier island** is a long, narrow offshore island of sediment running parallel to the coast and separated from it by a lagoon (Figure 10.19). These islands range between 15 and 30 kilometers (9 and 18 mi) long and from 1 to 5 kilometers (0.6 to 3 mi) wide. The tallest features are wind-blown sand dunes that reach heights up to 10 meters (33 ft). Barrier islands are common features along the Atlantic and Gulf coasts of the United States which form the longest chain of barrier islands in the world. However, barrier islands are dynamic coastal features as they grow parallel to the coast by longshore drift and are often eroded by storm surges that often cut them into smaller islands. Despite their transient nature, many barrier islands are heavily populated with homes and resorts. Even several major cities occupy barrier islands, including Miami Beach, Atlantic City, and Galveston.

Emergent and Submergent Coastlines

While sea level fluctuates daily because of tides, long-term changes in sea level have also occurred. Such changes in sea level result from uplift or subsidence along a coastline. Many coastal geologists classify coasts based on changes that have occurred in the past with respect to sea level. This commonly used classification divides coasts into two categories: emergent and submergent. An **emergent coastline** is a coastline that has experienced a fall in sea level, because of global sea level change, local land uplift, or isostatic rebound. Emergent coastlines are identifiable by the coastal landforms which are now above the high tide mark, such as raised beaches or raised wave-cut benches (marine terraces) (Figures 10.20a–b). Alternatively, a **submergent coastline** is one that has experienced a rise in sea level, due to a global sea level change or local land subsidence. Submergent coastlines are identifiable by their submerged, or "drowned" landforms, such as drowned valleys and fjords.

United States Geological Survey.

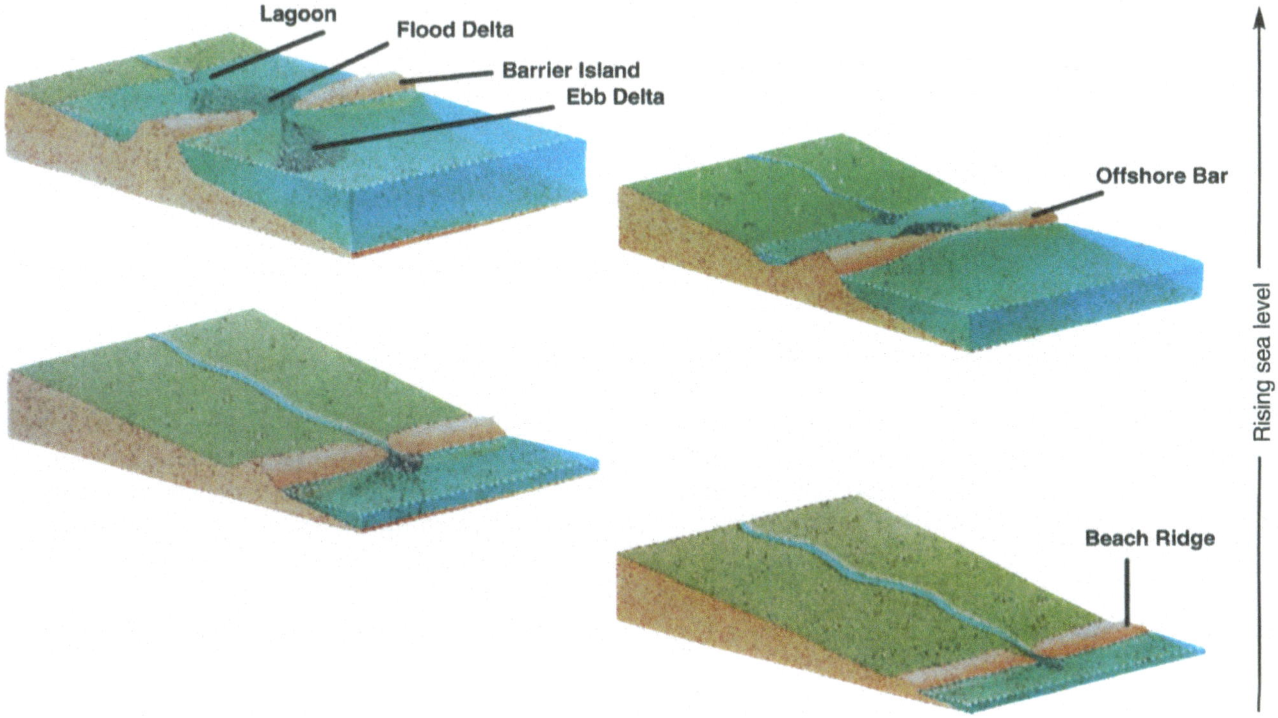

Figure 10.19 **Barrier islands formed from sand deposited along low energy beachs. These islands provide protection of the shoreline during storms and periods of high energy. During periods of very high water these islands are overrun with water.**

The type of coast produced is controlled mainly by tectonic forces and meteorological conditions (climate and weather). Tectonic processes can cause a coastline to rise or sink while lithospheric isostatic adjustment can depress or elevate sections of a continent. The tectonics at active plate margins can produce uplift or subsidence of a coast. In the Northwestern United States, the coastline is slowly rising due to subduction of the Juan de Fuca plate beneath the North American plate. During the 1964 Alaska earthquake, large sections of the coast rose and other parts subsided beneath sea level from a single event.

During glacial periods, large continental ice sheets can displace the lithosphere into the plastic asthenosphere. About 18,000 years ago, a huge continental glacier covered most of Scandinavia, causing the land to sink

© Kendall Hunt Publishing Company.

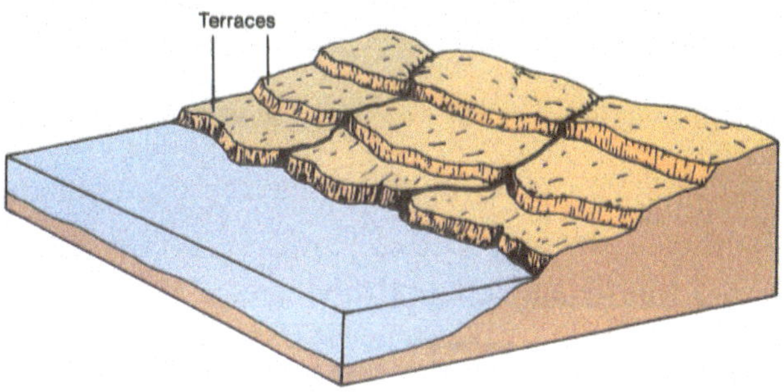

Figure 10.20a **An emergent coastline often has wave-cut terraces that represent older period of higher sea level.**

From Physical Geology by Dallmeyer. Copyright © 2000. Reprinted by permission of Kendall Hunt Publishing Company.

Figure 10.20b **This wave-cut terrace is part of an emergent shoreline at Baker's Point, California.**

isostatically. As the lithosphere sank, the displaced asthenosphere flowed southward, causing the region around modern-day Netherlands to rise. After the ice melted, the process was reversed and the asthenosphere flowed back north from below the Netherlands to Scandinavia. Today, Scandinavia is rebounding and the Netherlands is sinking. During the same glacial episode in North America, Canada was depressed by ice, and asthenosphere rock flowed southward. Today the asthenosphere is flowing back north and much of Canada is rebounding while much of the United States is now sinking.

eustatic
Global changes in sea level.

Global changes in sea level can also occur. Such global sea level changes are called **eustatic** changes and can occur by three mechanisms: the growth or melting of glaciers, changes in water temperature, and changes in the volume of the mid-ocean ridges. During glacial periods large amounts of water that evaporated from the oceans became stored on the continents as glacial ice. This caused sea level to become lower, resulting in global land emergence. Similarly, when glaciers melt, water flows back into the oceans and sea level rises globally, causing land submergence.

Changes in the volume of mid-ocean ridges can also affect sea level. Growth of a mid-ocean ridge displaces seawater upward. If lithospheric plates spread slowly from the ridge they create a narrow mountain ridge system that displaces relatively small amounts of seawater, resulting in lower sea level. In contrast, rapidly spreading plates produce a high-volume ridge system that displaces more water upward, resulting in higher sea level. At times during Earth's history, sea floor spreading has been relatively rapid, and as a result, global sea level has been higher.

Coastal Hazards: Living With Coastal Change

From the discussion of coastal areas it is apparent that diverse and complex processes are at work continually changing the coastal landscapes. Vast areas of coastal land have been lost since the mid-1800s as a result of natural processes and human activities (Table 10.1). The natural causes that have the greatest influence on coastal land loss are relative sea level rise, erosion from frequent storms, and reductions in sediment supply; whereas the most important human activities are sediment excavation, river modification, and coastal construction. Any one of these causes may be responsible for most of the land loss at a coast, or the land loss may be the result of several of these factors acting at the same time.

From a hazard point of view, coastal erosion is the most widespread and continuous process affecting the world's coastlines and contributing to land loss destruction. Global warming and sea level rise are slow-onset hazards that greatly contribute to the erosional process. However, catastrophic, rapid-onset events play a very significant role both for coastal erosion and human suffering. These include erosion and destruction from storms, landslides, and tsunami.

Table 10.1 Common Physical and Anthropogenic Causes of Coastal Land Loss

NATURAL PROCESSES	
Agent	**Examples**
Erosion	• waves and currents • storms • landslides
Sediment Reduction	• climate change • stream avulsion • source depletion
Submergence	• land subsidence • sea-level rise
Wetland Deterioration	• herbivory • freezes • fires • saltwater intrusion
HUMAN ACTIVITIES	
Agent	**Examples**
Transportation	• boat wakes, altered water circulation
Coastal Construction	• sediment deprivation (bluff retention) • coastal structures (jetties, groins, seawalls)
River Modification	• control and diversion (dams, levees)
Fluid Extraction	• water, oil, gas, sulfur
Climate Alteration	• global warming and ocean expansion • increased frequency and intensity of storms
Excavation	• dredging (canal, pipelines, drainage) • mineral extraction (sand, shell, heavy mins.)
Wetland Destruction	• pollutant discharge • traffic • failed reclamation • burning

Source: United States Geological Survey.

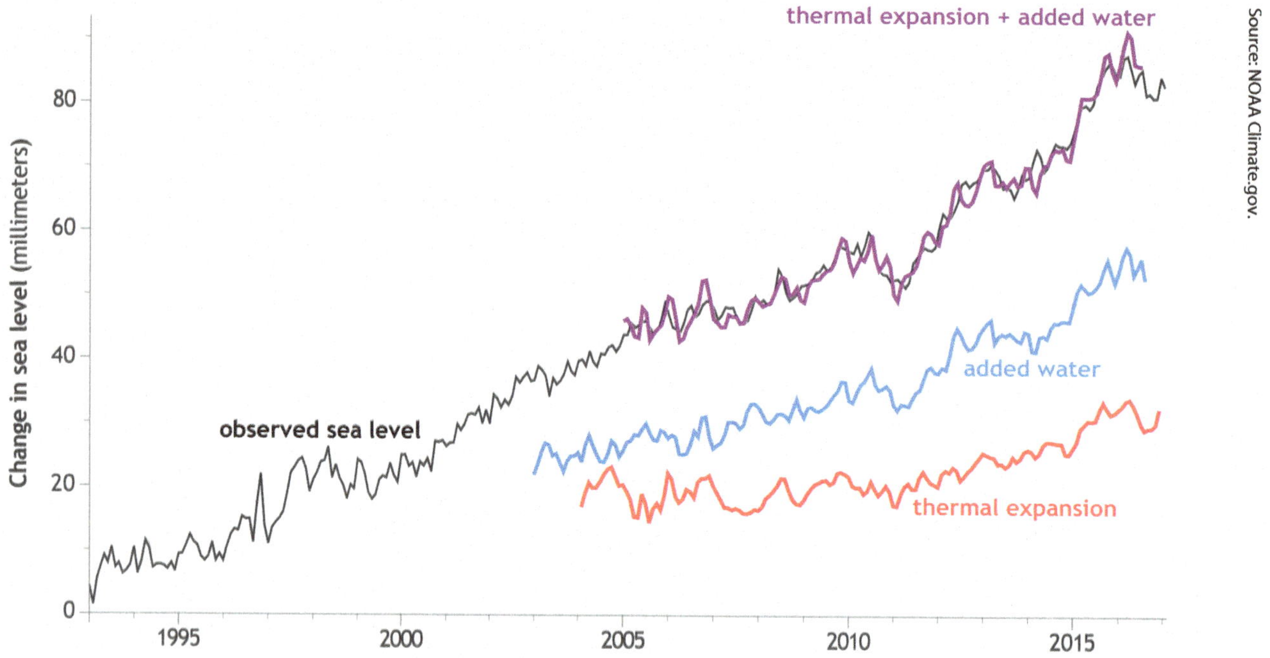

Figure 10.21 Observed sea level as measured by satellite altimeter recordings since 1993 (black line). Added water (glacial melting; blue line) and thermal expansion (red line) match observed sea level very well.

Coastal Land Loss by Global Sea-Level Rise and Subsidence

A significant amount of coastal erosion presently plaguing today's coastlines is the result of gradual but sustained global sea level rise (Figure 10.21). Most of this rise is from the melting of polar continental ice sheets, coupled with expansion of the water itself as global temperatures increase. The sea level rise is currently estimated at about 0.3 meter (1 foot) per century. Although this amount does not sound very threatening, additional factors increase the risk. First, the slope of many coastal areas is very gentle so that a small rise in sea level results in a far larger inland advance of the coast than steeper sloping areas. The vulnerability of coastal regions along the Eastern Seaboard of the United States shows a wide range of potential risks (Figure 10.22). Estimates of the amount of coastline retreat in the United States due to sea level rise of 0.3 meter (1 foot) would be 15 to 30 meters (50 to 100 ft) in the northeast, 65 to 130 meters (215 to 430 ft) in California, and up to 300 meters (1,000 ft) in Florida. Second, the documented rise of atmospheric carbon-dioxide levels (discussed in Chapter 8) suggests that global warming from the increased greenhouse-effect will melt the glaciers more rapidly, as well as warming the oceans more, thereby accelerating the rise of sea level. Some estimates put the anticipated rise in sea level at 1 meter by the year 2100, which could affect as much as 50% of the population of the country. In addition to increased beach erosion, it would also flood 30 to 70% of coastal wetlands in the United States that protect shores from storm flooding events.

The most widely assessed effects of future sea level rise are coastal inundation (submergence), erosion, and barrier island migration. The USGS estimates the primary impacts of a sea level rise on the United States to be: (1) the cost of protecting ocean communities by pumping sand onto beaches and gradually raising barrier islands in place; (2) the cost of protecting developed areas along sheltered waters through the use of levees (dikes) and bulkheads; and (3) the loss of coastal wetlands and undeveloped lowlands. The total cost for a 1 meter (3 ft) rise is estimated to be \$270–475 billion, ignoring future development.

coastal submergence
The permanent flooding of coastal areas by global sea level rise or land subsidence.

Coastal submergence refers to permanent flooding of the coast caused by either a rise in global sea level or subsi-dence of the land, or both. At many coastal sites, submergence is the most important factor responsible for land loss and as sea level rises, or the land subsides, it will inundate present unprotected low-lying coastal areas and cities such as Boston, New York, Miami, New Orleans, and Los Angeles. Submergence also accelerates coastal beach erosion and landslides because it facilitates greater inland penetration of storm waves. In addition to accelerated land loss, coastal submergence causes intrusion of salt-water into coastal freshwater aquifers.

Coastal Land Loss by Erosion: Impacts From Storms and Landslides

Superimposed on the slower sea level rise are shorter duration water fluctuations caused by storm events. The most damaging coastal storms for the United States are tropical cyclones (hurricanes) and extratropical cyclones (winter storms) that form around low-pressure cells (see Chapter 11 for discussions of these storms).

Hurricanes form in the tropics during summer to early fall months and migrate northward and westward into temperate regions of the Atlantic and Gulf coasts. The extratropical storms occur mostly in winter (like nor'easters) and can cause erosion of the coastline at much higher rates than normal. Although each type of storm is unique, there are several factors common to all storm types which include strong winds, generation of large waves, and elevated water levels known as a storm surge.

Storm Impacts

Strong storms bring more energy to the coastline causing higher rates of erosion (Figure 10.23). Higher erosion rates are due to several factors:

- Wave velocities are higher during storms and thus larger particles can be carried in suspension causing sand on beaches to be picked up and moved offshore. This leaves behind coarser grained particles such as pebbles and cobbles, thus reducing the width of sandy beaches;
- Storm waves reach higher levels onto the coast and destroy and remove structures and sediment from areas not normally reached by normal waves;

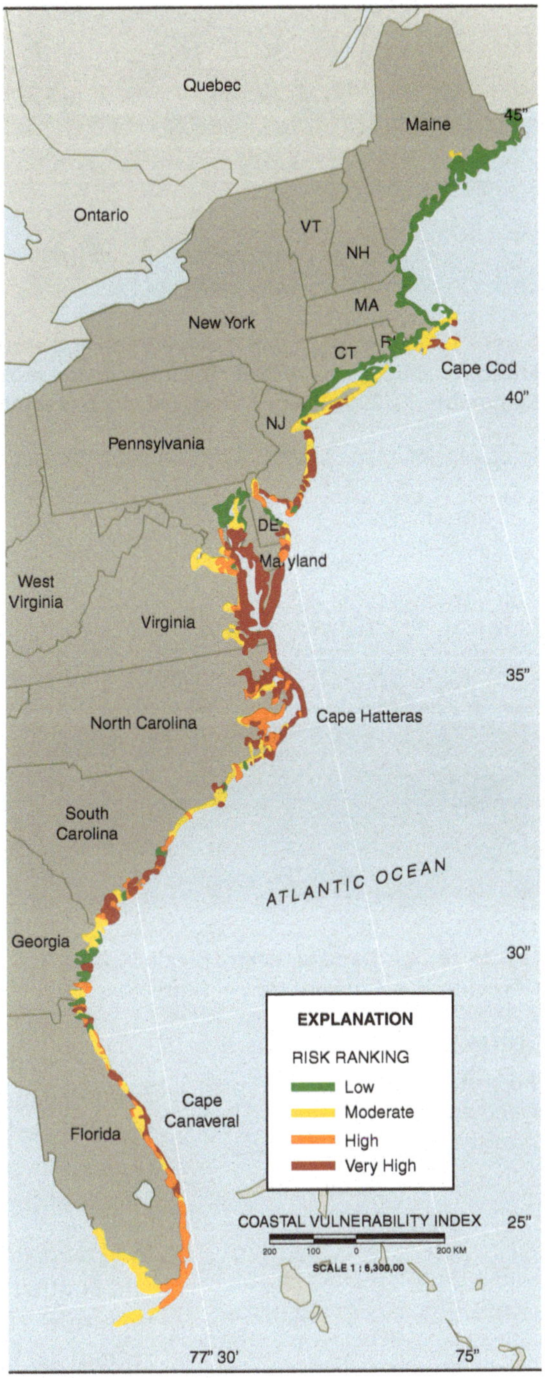

From Thieler and Hammar-Klose (2000). United States Geological Survey.

Figure 10.22 Map of the Coastal Vulnerability Index (CVI) for the US East Coast showing the relative vulnerability of the coast to changes due to future rises in sea level. Areas along the coast are assigned a ranking from low to very high risk, based on the analysis of physical variables that contribute to coastal change.

Figure 10.23 Upper figure shows erosion and property damage near Floridana Beach, Florida, caused by Hurricane Frances on September 4, 2004. The lower image shows the same area following the arrival of Hurricane Jeanne on September 25, 2004. Jeanne produced much greater beach erosion.

Figure 10.24 Damage along a pier in St. Mary's, GA, caused by Hurricane Irma in September 2017.

- Wave heights increase during a storm and crash higher onto cliff faces and rocky coasts. Larger rock debris or debris from destroyed structures is flung against the rock causing rapid rates of erosion by abrasion;
- Hydraulic action increases as larger waves crash into rocks. Air and water occupying fractures in the rock become compressed and thus the pressure in the fractures is increased which causes further fracturing of the rock.

storm surge
An onshore flood of water created by a low pressure storm system.

Storm surge is responsible for about 90% of all human fatalities and damage during storms (Figure 10.24). A **storm surge** is an onshore flood of water created by a low pressure storm system. The surge is caused primarily by the strong winds of the storm blowing over the sea surface and causing the surface water to pile up above sea level. Low pressure at the center of the storm also elevates the surface water upward and enhances the height of the mound of water. As the storm nears land, the shallower sea floor prevents the piled-up water from collapsing and it floods inland as a deadly storm surge. Storm surges are at their highest and most damaging when they coincide with high tide (especially at the high tides during the spring tide cycles), combining the effects of the surge and the tide.

When Superstorm Sandy struck the coastal areas of the Mid-Atlantic states and southern New England in October 2012, its landfall coincided with a high tide. This caused extreme flooding and beach erosion throughout the region. Sections of the Atlantic City Boardwalk were destroyed in addition to hundreds of homes being damaged or destroyed. Billions of tons of beach sand were eroded.

Dune and Beach Recession

High storm-generated waves erode large quantities of sediment from dune and beach areas. From March 5–8, 1962 ,a major coastal storm, known as the "Ash Wednesday" storm, moved northward and became stalled against the middle Atlantic coast through five high tides. The documented erosion that occurred at Virginia Beach, Virginia, showed that 30% of the beach and dune sand was removed. The crest of the dunes at Virginia Beach was reduced from an elevation of 4.9 to 3.4 meters thus enabling future storm surges to rise over the dunes and flood inland areas.

Dune and Beach Breaching and Overwash

Large amounts of sediment can be eroded from a beach and dunes during a major storm with some migrating along the shore by accelerated longshore currents and some moved offshore. In addition, storm waves may wash sediment through low areas between the dunes of islands and onto the back side of it. This **overwash** is important because it maintains the barrier island's width as its front is eroded but can be devastating to homes and other structures. Overwash fans are lobe-shaped deposits eroded from the ocean side of a shore and deposited in the bays and lagoons behind barrier islands.

overwash
A deposit of marine-derived sediments landward of a barrier system, often formed during large storms; transport of sediment landward of the active beach by coastal flooding during a tsunami, hurricane, or other event with extreme wave action.

El Niño Effects on Coastlines

Along the Pacific coast, winter storms and unusual oceanographic conditions such as El Niño cause the most erosion and land loss. Approximately every 4–5 years, El Niño conditions cause warm surface water of the Pacific Ocean to flow eastward piling up water along the west coast of North and South America. The elevated water levels and the unusually strong storms

BOX 10.1 Barrier Island Breaching

In the United States, barrier islands are found along the Gulf Coast and Atlantic Ocean coastline where they serve as a break between the sea and land. Examples of these islands include Padre Island, Texas, islands off the coast of Georgia, and a major set off the coast of North Carolina. Several barrier islands off the coast of Louisiana have been eroded as much as 30 meters (100 ft) due to a rise in sea level in the area.

These features are formed by the deposition of sand transported to the sea by rivers (**Figure 10.1.1**). Because the material is unconsolidated, the islands are dynamic and change their size and shape over time. The major factor that alters these features is an increase in the force of sea water driven ashore by storms. The storms cut through the unconsolidated sand of a barrier island to produce a tidal inlet or tidal delta. These features are often short-lived as they close naturally in a few weeks as the result of longshore drift deposits moving sediment along the shoreline.

The Outer Banks of North Carolina are a major series of barrier islands which are frequently affected by hurricanes and tropical storms (**Figure 10.1.2**). In September 2003, the islands were subjected to the forces of Hurricane Isabel which hit the coast with winds of 105 mph (170 kph). Waves as high as 8 meters (25 ft), coupled with a storm surge of 1.8 to 2.4 meters (6 to 8 ft), breached the island, forming two inlets that were each more than 600 meters wide and 5 meters deep (**Figure 10.1.3**). One inlet was named Isabel Inlet to recognize the effect of the storm.

Several sections of State Highway 12, which runs the length of Hatteras Island, were destroyed. In addition, more than 30 beach houses and several motels were knocked off their pilings and foundations. One hundred thirty kilometers north in Dare County, North Carolina, surge flooding and strong winds damaged several thousand houses. In the end the storm caused more than $450 million in damage in North Carolina, with a total damage assessment in the United States of $5.5 billion. Fortunately, no deaths or injuries were reported in the Outer Banks.

© makasana photo/Shutterstock.com.

Figure 10.1.1 Driving north on North Carolina Highway 12 large sand dunes and the Atlantic Ocean (not seen) are on the right.

Source: NASA.

Figure 10.1.2 The Outer Banks of North Carolina are a major system of barrier islands between the land and Atlantic Ocean.

Source: USGS

Figure 10.1.3 Hatteras Island was breached by Hurricane Isabel, producing two new inlets along the barrier beach.

during El Niño events cause extensive flooding and erosion of beaches and cliffs. In 1983, an unusually strong El Niño caused torrential rainfall, rapid beach erosion, and massive landslides along the Pacific coast of the United States. Land loss was concentrated along the southern California coast where numerous expensive homes built on cliffs were damaged or destroyed.

United States Geological Survey, Center for Coastal and Watershed Studies.

United States Geological Survey, Center for Coastal and Watershed Studies.

Figure 10.25 **Coastal region near Ventura, California. Left image shows the coast following the El Niño storm of October 1997; right image shows a definite change in the coastal morphology following the April 1998 El Niño storm.**

Since 1900 at least 30 events have occurred. Two major El Niño storms hit California in October 1997 and six months later in April 1998. This series of storms was the largest of the 30 events and had significant effects on the weather in the central and eastern Pacific Ocean regions. Coastal areas were heavily eroded by these two events (Figure 10.25). Another set of storms struck the west coast of the United States in 2014 to 2016.

Source: David M. Best.

Figure 10.26 **The small community of La Conchita, California, lies along the coast of the Pacific Ocean. Bluffs that overlie the town have collapsed twice since 1995 due to heavy rainfall that saturated the ground.**

Landslides and Cliff Retreat

Coastal landslides occur where unstable slopes fail and land is displaced downslope (Figure 10.26). Some of the fundamental causes of slope failures that lead to land loss are: (1) slope oversteepening, (2) slope overloading, (3) shocks and vibrations, (4) water saturation, and (5) removal of natural vegetation. Sea level rise can elevate waves so they can erode and undercut cliffs at higher elevations, initiating mass movements. Cliffs may stay relatively stable and then retreat several meters in a single storm event which makes building structures near the edge of cliffs especially risky during times of sea level rise. More discussion of landslides and their mechanisms is found in Chapter 6.

Coastal Land Loss by Human Activities

There is increasing evidence that recent land losses in many coastal regions are largely anthropogenic, as the changes are attributable to human alteration of the coastal environment. Land losses indirectly related to human activities are difficult to quantify because they promote alterations and imbalances in the primary factors causing land loss such as sediment budget, coastal processes, and relative sea level changes. Human activities causing land loss are transportation networks that tend to increase erosion, coastal construction projects that typically increase deficits in the sediment budget, subsurface fluid extraction and climate alterations that accelerate submergence, and excavation projects that cause direct losses of land.

There are countless examples of human interference with coastal processes. The beach at Miami Beach must be restored periodically by sand pumped from offshore. In southern Louisiana the land is subsiding as sea level rises, thus causing loss of natural coastal wetlands. Many California beaches are eroding due to the damming of rivers for irrigation and flood control. The river-supplied sediment that normally replenishes the beaches is being trapped in reservoirs behind dams. Since this sediment is not being supplied to the ocean, longshore currents cannot resupply the beaches with sediment. Instead, longshore currents carry the existing sediment in the downdrift direction, resulting in significant erosion of the beaches.

Eliminating wetlands for development and agriculture removes the natural flood protection and storm-swollen estuaries now flood barrier islands from the bay side as storms move inland. Whenever beach dunes are removed, the most effective barrier to storm waters has been lost and over washing can occur. Over the last 10,000 years, most of the state of Louisiana has formed from the deposition of sediments by the flooding of the Mississippi River. Humans, however, have prevented the river from flooding by building levee systems that extend to the mouth of the river. As previously deposited sediments become compacted they tend to subside. Since no new sediment is being supplied by Mississippi River flooding, the subsidence results in a relative rise in sea level. This, coupled with a current rise in eustatic sea level, is causing coastal Louisiana to erode at an incredible rate and experience more flooding.

Tsunami

As discussed in Chapter 5, a tsunami is a shallow water sea wave generated by earthquakes, volcanic eruptions, meteorite impacts in the ocean, or landslides. Tsunami can cause coastal flooding and catastrophic destruction thousands of kilometers from where they were generated. Such waves can have wave heights up to 30 meters or more, and have great potential to wipe out large coastal cities.

Evidence From the Geologic Record

Since the oceans formed over four billion years ago, they have influenced the coastlines of the newly forming continents. Sea level has risen and fallen repeatedly in the geologic past, and its coastlines have subsequently submerged and emerged throughout Earth's history. The rock record shows countless examples of seas transgressing and regressing over the continents. Marine fossils found thousands of kilometers inland from the present coast attest to these fluctuations in past sea levels.

At times during Earth's long history, tectonic movements arranged the continents into very different configurations from those of today. When there were large amounts of land near the poles, the rock record shows unusually low sea levels during past ice ages due to large ice sheets forming on the continents. During times when the land masses clustered around the equator, ice ages had much less effect on sea level. However, over most of geologic time, long-term sea level has been higher than today. Only at the Permian-Triassic boundary about 250 million years ago was long-term sea level lower than today.

The world's present coastlines are not the result of present-day processes but were affected by the rise of sea level caused by the melting of the Pleistocene glaciers beginning between 15,000 and 20,000 years ago (Figure 10.27). The rising sea flooded large parts of the low coastal areas, which are now part of the continental shelf, and moved the coastlines inland. Sea-level has risen about 130 meters since the peak of the last ice age about 18,000 years ago. It was during this time of very low sea level that there was a dry land connection between Asia and Alaska over which humans migrated to North America over the Bering Land Bridge. However, for the past 6,000 years (a few centuries before the first known written records), the world's sea level has been gradually approaching the level we see today.

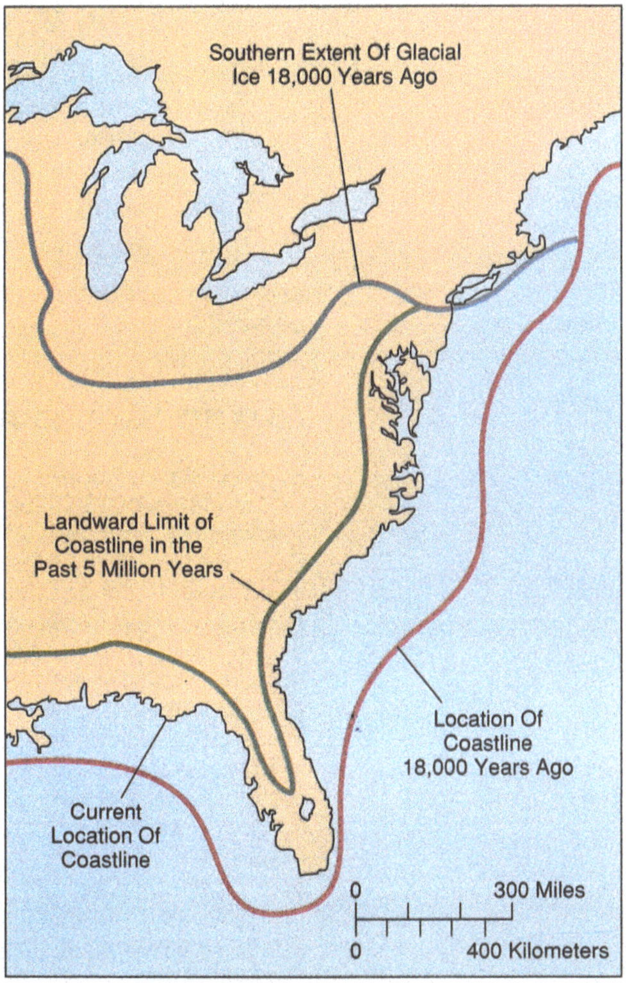

© Kendall Hunt Publishing Company.

Figure 10.27 **Coastlines have fluctuated about 400 km over the past five million years. Notice that sea level was much lower in the past 18,000 years.**

Lessons From the Historic Record and the Human Toll

The increase in coastal populations together with rising sea level and intense storms combine to make coastal erosion and flooding very costly and life-threatening. Since the early 1900s, property damage in the United States has been on the rise. However, the death tolls have generally decreased (at least until Hurricane Katrina) because of advanced warnings and evacuations to the weather-related storm events. Unfortunately, people continue to build more structures along migrating shorelines and are unaware of the dynamic balance between erosion and deposition (refer to Box 10.2). Nearly all human intervention with coastal processes interrupts natural processes and thus can have an adverse effect on coastlines.

Coastal environments are a delicate setting in which conflicts arise when humans and natural processes attempt to coexist. With about 60% of people in the United States living within 150 km (100 mi) of a shoreline, it is inevitable that problems will develop. These problems are all too common when cyclonic storms hit a coastline. The rising waters and wind cause large amounts of damage to property, and unfortunately take people's lives. Examples are well known to those who live in these areas, and others who live elsewhere become aware of the issues when large storms hit. The case of Hurricane Katrina striking the Gulf Coast region of the United States in late August 2005 brought these problems to everyone's attention.

Direction of Longshore Current

Barrier islands and beaches, since they consist of unconsolidated sediment, and sea cliffs, since they are susceptible to landslides due to undercutting, are difficult to protect from the erosive action of the waves. Human construction methods can attempt to prevent erosion, but cannot always protect against abnormal conditions. In addition, other problems are sometimes caused by these engineering structures.

BOX 10.2 One Case Study Made By the United States Geological Survey

OCEAN CITY, MARYLAND: AN URBANIZED BARRIER ISLAND

Relative recent changes in the shoreline have occurred in populated regions of the United States. USGS Circular 1075 provides a good summary of the events that have involved the shoreline near Ocean City, Maryland. For more than a century Ocean City, Maryland, has been a popular beach resort for vacationers from the Northeast and Mid-Atlantic States (Figure 10.2.1). During the Roaring 1920s, several large hotels and a boardwalk were built to accommodate visitors and development continued slowly until the early 1950s. Then a period of rapid construction began that lasted almost 30 years. Concerns about the coastal environment were raised in the late 1970s and led to Federal and State laws to limit dredging and filling of wetlands. The resort is built on the southern end of Fenwick Island, one of the chain of barrier islands stretching along the East Coast (Figure 10.2.2). The Great Hurricane of 1933 (before names were assigned to hurricanes) opened the Ocean City Inlet by storm-surge overwash from the bay side. To maintain the inlet as a navigation channel, two large stone jetties were constructed by the U.S. Army Corps of Engineers. These jetties helped stabilized the inlet, but they have drastically altered the sand-transport processes near the inlet. The net longshore drift at Ocean City is southerly; it has produced a wide beach north of the jetty, but Assateague Island, south of the inlet, has been starved of sediment. The result is a westerly offset of more than 500 meters in the once-straight barrier island.

The most damaging storm to hit Ocean City within historic times was the Five-High or Ash Wednesday northeaster of early March 1962 which caused severe erosion and flooding along much of the middle Atlantic Coast. For 2 days, over five high tide cycles, all of Fenwick Island, except the highest dune areas was repeatedly washed over by storm waves superimposed on a storm surge measuring 2 meters (6.6 ft) high. Property damage in Ocean City was estimated at $7.5 million. Given the dense development of the island over the last 30 years, damage from a similar storm would today be hundreds of millions of dollars.

© Racheal Grazia/Shutterstock.com.

Figure 10.2.1 The beaches at Ocean City, Maryland, are a popular vacation spot for many people who live in the heavily populated Atlantic Seaboard of the Eastern United States. Large amounts of construction are at risk along the beaches of the United States from cyclonic storms.

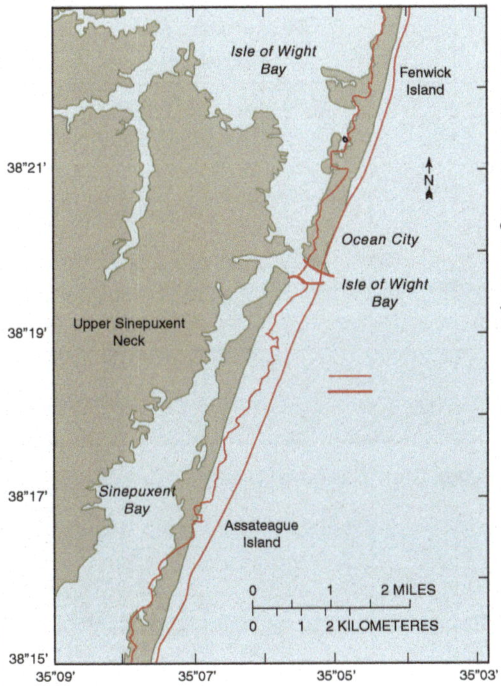

Source: United States Geological Survey.

Figure 10.2.2 The natural sediment transport along the Fenwick Island-Assateague Island region has been altered by the construction of two large jetties at Ocean City Inlet. The landward shift of the southern barrier island was caused by a change in the longshore currents in the area.

Protection of the Shoreline

Shoreline protection can be divided into two categories: hard stabilization in which solid structures are built to reduce wave action, and soft stabilization which mainly refers to adding sediment back to a beach as it erodes.

Hard Stabilization

Two types of hard stabilization are often used. The first type interrupts the flow of sediment along the beach. These structures include **groins** (Figures 10.28 and 10.29) and **jetties** (Figure 10.30), built at right angles to the beach to trap sand and widen the

Source: David M. Best.

Figure 10.28 This groin along the English Channel is constructed along a beach that consists of pebbles and cobbles rather than sand.

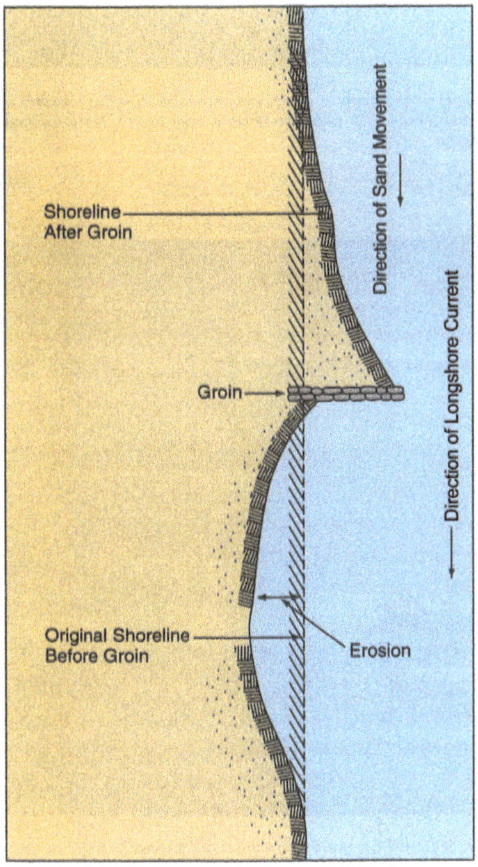

© Kendall Hunt Publishing Company.

Figure 10.29 The effects of constructing a groin on a beach are that portions of the beach undergo erosion while others down drift experience erosion.

Figure 10.30 Jetties are extensions of pre-existing channels or river channels. Notice how depositional patterns change and sand begins to migrate around the jetty, eventually becoming a hazard to the entrance to the river.

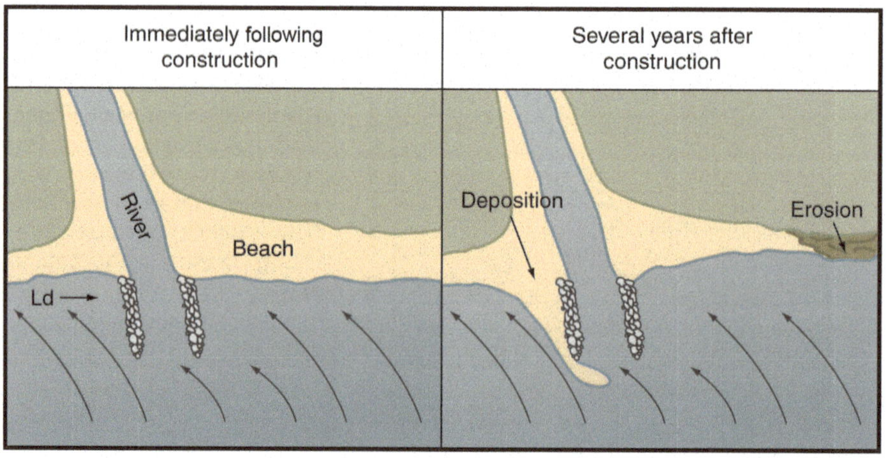

© Kendall Hunt Publishing Company.

Figure 10.31 This rock sea wall is built to absorb the energy of storm waves coming off the ocean. Unfortunately this feature removes the possibility of having a recreational beach.

From Physical Geology by Dallmeyer. Copyright © 2000. Reprinted by permission of Kendall Hunt Publishing Company.

groin

Solid structure built at an angle from a shore to reduce erosion from long shore currents, and tides.

jetty

A structure extending into the ocean to influence the current or tide in order to protect harbors, shores, and banks.

seawall

Massive structure built along the shore to prevent erosion and damage by wave action.

breakwater

Structure built offshore and parallel to shore that protects a harbor or shore from the full impact of waves.

beach. The second type interrupts the force of the waves. **Seawalls** are built parallel to the coastline to protect structures on the beach (Figures 10.31) by allowing waves to crash against them and preventing water from running up the beach. **Breakwaters** serve a similar purpose, but are built offshore parallel to the beach (Figure 10.32), again preventing the force of the waves from reaching the beach and any structures.

While hard stabilization usually works for its intended purpose, it does cause sediment to be redistributed along the coast. A breakwater, for example, causes wave refraction, and alters the flow of the longshore current. Sediment is trapped behind the breakwater, and the waves become focused on another part of the beach where they can cause significant erosion (Figure 10.32). Similarly, because groins and jetties trap sediment, areas in the downdrift direction are not resupplied with sediment by the longshore current, and beaches are eroded and become narrower in the downdrift direction.

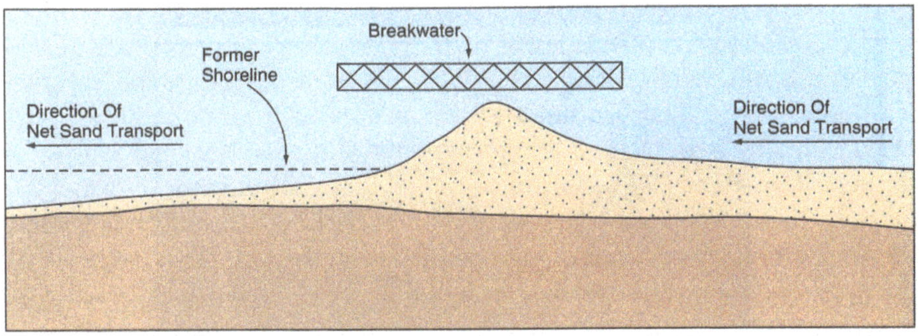

© Kendall Hunt Publishing Company.

Figure 10.32 The construction of a breakwater offshore and parallel to the shoreline creates a buildup of sand that extends from the shoreline out to the breakwater.

Soft Stabilization

Soft stabilization is primarily accomplished by adding sediment to the shoreline, which is called **beach nourishment**. This is usually done by dredging sediment from offshore and pumping it onto the shoreline. Adding sediment is necessary when erosion removes too much material. However, because the erosive forces are still operating, additional sediment will need to be periodically replenished at continuing expense (Figure 10.33). Less invasive approaches include construction of access walkways and steps to reduce erosion of fragile dunes, as well as planting and protection of well-rooted vegetation. One approach to recovering sand dunes is to use discarded Christmas trees to form wind breaks (Figure 10.34).

beach nourishment
A soft stabilization technique primarily accomplished by adding sediment to the shoreline.

Abandonment

Protecting ourselves from coastal hazards, such as beach erosion and coastal flooding, requires long-term strategic planning. Should we continue to attempt to defend the coasts from rising seas and ever larger storms? Or do we recognize the awesome power of natural processes and strategically abandon the coast? Both methods are very costly and decisions will most likely require passage of governmental regulations. Presently, public policies are encouraging the development of hazardous areas by providing federal flood-insurance and disaster relief programs which encourage homeowners and businesses to rebuild after a disaster.

© Kochneva Tetyana/Shutterstock.com.

Figure 10.33 The replenishment of beach sands is key to maintaining the health of the shoreline. Each year millions of cubic yards of sand are added to coastal regions to maintain the balance between the ocean and the land.

Source: David M. Best.

Figure 10.34 Discarded Christmas trees are helping reform sand dunes along beach areas.

Many coastal geologists believe that instead of rebuilding our eroding beaches, we should return more of them to public use after they are devastated by a storm. Examples of such public coastal resources are the National Seashores, such as Cape Cod in Massachusetts, Padre Island in Texas, Hatteras in North Carolina, and Point Reyes in California. National Seashores are kept in as natural a condition as possible for all to enjoy. Storms will continue to damage their roads, parking lots, bathhouses, and concessions, but these are inexpensive to rebuild compared to the cost of restoring high-rise hotels, condominiums, and beach homes.

Key Terms

backshore zone *(page 288)*
backwash *(page 279)*
barrier island *(page 288)*
beach *(page 287)*
beach drift *(page 284)*
beach nourishment *(page 303)*
berm *(page 288)*
break *(page 276)*
breaker *(page 279)*
breakwater *(page 302)*
circular orbital motion *(page 277)*
coastal submergence *(page 293)*
coastline *(page 275)*
crest *(page 276)*
ebb tide *(page 281)*
emergent coastline *(page 288)*
eustatic *(page 290)*
fetch *(page 277)*
flood tide *(page 281)*
foreshore zone *(page 287)*
groin *(page 302)*
headland *(page 280)*
jetty *(page 302)*
longshore current *(page 284)*
longshore drift *(page 284)*
neap tide *(page 282)*
offshore zone *(page 287)*
overwash *(page 295)*
plunging breaker *(page 279)*
rip current *(page 285)*
rogue wave *(page 278)*
sea arch *(page 284)*
sea stack *(page 284)*
seawall *(page 302)*
spilling breaker *(page 279)*
spring tides *(page 282)*
storm surge *(page 294)*
submergent coastline *(page 288)*
surf zone *(page 279)*
surging breaker *(page 279)*
swash *(page 279)*
swell *(page 278)*
tides *(page 280)*
trough *(page 276)*
wave *(page 275)*
wave-cut platform *(page 283)*
wave base *(page 277)*
wave height *(page 276)*
wavelength *(page 276)*
wave period *(page 276)*
wave refraction *(page 280)*
wave speed *(page 276)*
wave steepness *(page 276)*

Summary

Coastlines are among Earth's most geologically active environments and are continually changing because of the dynamic interaction between the water and the land. Coastal changes occur on two very different time scales. Short-term change is largely due to seasonal severe storm events, while long-term changes are due to slow sea-level rise. Waves, longshore currents, and tidal currents interact with the geologic structures and plate tectonic processes of the coast to shape coastlines into a multitude of landforms.

Waves are formed by wind forces acting on the water surface and transferring energy in a circular motion through the water. Motion in a wave ceases at a depth equal to one half its wavelength. As a wave enters shallow water, its wave base intersects the bottom and the wave begins to break. Most of the energy carried within a wave is then released within the surf zone by breakers. Sediment is moved along the shoreline by longshore drift and into and out of tidal inlets by tidal currents. Beaches are accumulations of sediment along the shoreline that is carried to the oceans by rivers and streams.

Coastal storms and hurricanes are inherent natural hazards along most shorelines in the United States. As population and development continues to increase in those areas, storm damage has risen dramatically. Many coasts are affected by multiple hazards such as storm erosion, landslides, tsunami, and coastal flooding. Scientists see evidence that sea level is rising, that storms and hurricanes will continue to occur as frequently as they do today, and that their magnitude will increase as global temperatures rise.

Human interference with natural coastal processes and landform development is causing serious problems. Seawall construction protects structures against wave attack, but causes beach erosion. Groins cause the updrift beach to grow, but the downdrift beaches erode. Breakwaters and jetties prevent erosion of beaches in some areas, but their effect is to cause sand to move to other positions along the beach and waterfront. Hard and soft stabilization of beaches is only a short-term solution in the overall process.

References and Suggested Readings

Aldersey-Williams, H. (2016). *The tide—The science and stories behind the greatest force on Earth.* W. W. Norton & Company.

American Museum of Natural History. (2014). *Ocean—The definitive visual guide.* Dorling Kindersley, Ltd.

Bush, D. M., Pilkey, Jr., O. H., & Neal, W. J. (1996). *Living by the rules of the sea.* Duke University Press.

Coch, N. K. (1995). *Geohazards, natural and human.* Prentice Hall.

Douglas, B. C., & Peltier, W. R. (2002). The puzzle of global sea-level rise. *Physics Today, 55*(3): 35–41.

Earle, S. A. (2009). *The world is blue—How our fate and the oceans are one.* National Geographic Society.

Hobbs, C. H. (2012). *The beach book—Science of the shore.* Columbia University Press.

Hutchinson, S., & Hawkins, L. E. (2005). *Oceans—A visual guide.* Firefly Books, Inc.

Komar, P. D. (1997). *The Pacific Northwest Coast.* Duke University Press.

Morton, R. A. (2003). *An overview of coastal land loss: With emphasis on the Southeastern United States.* U.S. Geological Survey, Open File Report 03-337.

Neal, W. J., Pilkey, O. H., & Kelley, J. T. (2007). *Atlantic coast beaches—A guide to ripples, dunes, and other natural features of the seashore.* Mountain Press.

Pilkey, O. H., Neal, W. J., Graham Cooper, J. A., & Kelley, J. T. (2011). *The world's beaches—A global guide to the science of the shoreline.* University of California Press.

Pilkey, O. H., Pilkey-Jarvis, L., & Pilkey, K. C. (2016). *Retreat from a rising sea—Hard decisions in an age of climate change.* Columbia University Press.

Trujillo, A. P., & Thurman, H. V. (2017). *Essentials of oceanography* (12th ed.). Prentice Hall.

Williams, S. J., Dodd, K., & Cohn, K. K. (1990). *Coasts in crisis.* U.S. Geological Survey Circular 1075.

Websites for Further Reference

https://coast.noaa.gov/
https://marine.usgs.gov/
https://pubs.usgs.gov/circ/c1075/change.html
https://pubs.usgs.gov/of/2003/of03-337/
https://solidearth.jpl.nasa.gov/PAGES/sea01.html
https://www.fema.gov/protecting-homes/atlantic-ocean-coastal-information
https://www.noaa.gov/oceans-coasts

Questions for Thought

1. How are waves that affect coastlines generated?
2. Explain longshore currents and how they cause sediment transport on beaches.
3. Describe what happens to waves as they move into shallow water.
4. What are the main causes of coastal erosion?
5. What are the greatest coastal hazards?
6. What are barrier islands, and how do they form?
7. Describe ways in which the relative elevation of land and sea may change.
8. What is the present trend in global sea level and what effect does it have on coastal regions?
9. Describe the motion of a water particle as a wave passes.
10. Why are beaches often referred to as rivers of sand?
11. What is wave refraction, and how does it affect coast erosion and deposition?
12. What influence do tides have on coastal processes and hazards?
13. Why is a combination of a high tide and an incoming storm surge so destructive?

11

Hurricanes, Cyclones, and Typhoons

Source: NASA. Photo by Joshua Stevens.

Hurricane Jose is poised to strike New England while Hurricane Maria approaches Puerto Rico, September 19, 2017.

Why Study Cyclonic Storms?

These events occur throughout the globe at all times of the year in both the Northern and Southern Hemispheres. They generate destructive conditions that affect our communities, landscape, and daily lives. In the United States these storms, whether lower grade tropical storms or more intense hurricanes, affect more than half the population of the country. Of the 10 most populated states, six of them (Texas, Florida, New York, Pennsylvania, Georgia, and North Carolina) lie in the paths of these events. Although not all of these states are impacted every year, when hurricanes do come ashore, their fury is extremely damaging and disruptive. In addition to property being destroyed, commerce and travel can come to a halt, producing economic upheaval. Since 2005, evidence has made us acutely aware of the forces of nature that alter our well-being. If you have not been affected by these events, perhaps you know people who have. As more people move to warmer regions of the country, including those mentioned above, they will come to understand the role that storms play in their lives.

cyclonic storm
A generic term that covers many types of weather disturbances that are typified by low atmospheric pressure and rotating, inwardly directed winds.

Large-scale **cyclonic storms** occur in many areas of the globe, generated by warm, oceanic waters that create rising air circulations that lift moisture and heat from the water surface into rotating storms. Few regions on Earth are spared from these weather phenomena. Both the geologic and human impacts can be long-lasting when these storms encounter continental regions.

Hundreds of cyclonic storms have hit coastal and inland regions along the Pacific, Indian, and North Atlantic oceans, as well as the Caribbean Sea and Gulf of Mexico. Only a few specific storms are discussed in this chapter.

Areas of Cyclonic Storm Generation

Wind is the movement of air. It can be affected by changes in air pressure and the Earth's rotation. On a global scale the differences in heating near the equator and the cooling at the poles produce large circulation cells that

Figure 11.1 A sequence of satellite images shows the position of Hurricane Andrew as it approaches and moves across Florida and the Gulf of Mexico. The storm image to the right was taken August 23, 1992; the center image is its position on August 24, and the image on the left was taken August 25, 1992. Notice the changes in the size of the eye of the storm.

© Kendall Hunt Publishing Company.

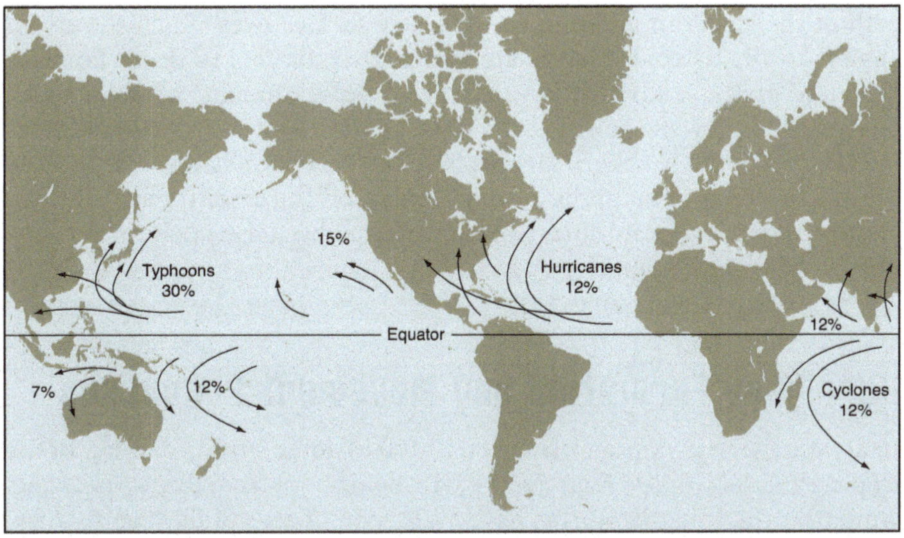

Figure 11.2 Global map showing where storms first form. The majority form in the Northern Hemisphere. Strong wind shear and lack of weather disturbances do not allow cyclonic storm formation in the areas east and west of South America.

drive large weather systems across Earth's surface. In some cases these are gentle breezes, but at other times winds can produce storms that become major natural disasters. In areas of low atmospheric pressure, which are associated with stormy conditions, rotational forces cause winds to create a swirling vortex. As the rotation intensifies, a cyclonic disturbance forms and moves through the Earth's atmosphere, with the potential of becoming a major cyclonic storm (**Figure 11.1**).

Three different terms are used to categorize these cyclonic storms, based on where they first develop. **Cyclones** are associated with the Indian Ocean and South Pacific Ocean and affect South Asia, Australia, and the east coast of Africa. **Typhoons** are generated in the western and southern Pacific, affecting islands in the Pacific Ocean as well as southeast Asia, China, and Japan. Cyclonic storms in the western portions of the North Atlantic Ocean, the Caribbean Sea, and the eastern Pacific off Mexico and Central America are termed **hurricanes** (**Figure 11.2**). Unless we refer to a specific storm in a portion of the globe other than the North Atlantic and adjoining regions, we will refer to cyclonic storms as hurricanes.

Many variables are involved in the formation of these storms. They have a range of sizes and intensities as well as different forward velocities, so each storm is unique in terms of its characteristics and behavior. However, some traits are common to all cyclonic storms.

- They initially form in only a few areas over the eastern and southwest Pacific Ocean, eastern North Atlantic Ocean, and north Indian Ocean.
- Storms form between the latitudes of 5° and 20° N or S.
- They develop more frequently during the summer months of their respective hemispheres, but they can occur in any month.

The movement of cyclonic storms across Earth's surface is influenced by Earth's rotation and the jet stream. This rotation produces the Coriolis effect, which generates a force that causes storms in the Northern Hemisphere to curve to the right. The reverse movement is seen in the Southern Hemisphere. The physics of fluid dynamics also controls movement within the low-pressure zones. Areas at risk include the shorelines lying in the

cyclone

A rotating mass of low pressure in the atmosphere that covers a large area; the warm air mass rotates counterclockwise in the Northern Hemisphere and clockwise in the Southern Hemisphere. The term is strictly applied to large low pressure storms in the Indian Ocean and South Asia.

typhoon

A cyclonic storm that forms in the central and western Pacific Ocean. Refer to cyclone and hurricane.

hurricane

Term applied to cyclonic storms that occur in the North Atlantic Ocean, Caribbean Sea, Gulf of Mexico, or eastern Pacific Ocean. Minimum wind velocity is 74 miles per hour. Refer to cyclone and typhoon.

path of the storm in addition to the water surface over which the storm passes. Low-lying coastal areas are frequently subjected to major flooding produced by the landward movement of massive amounts of water being pushed ashore by the storm's winds. In general, east-facing coastal areas are affected in the Northern Hemisphere. An exception is the west coast of Mexico, which is often hit by storms coming off the eastern Pacific Ocean and the western side of India, which is affected by storms that form in the northern Indian Ocean.

Hurricane Formation and Movement Processes

In an average year, approximately 80 to 100 cyclonic storms develop in the tropical and subtropical regions on Earth, being born from the warm waters lying near the equator. Storms have a life span of several days up to a few weeks, during which time they can intensify from a small tropical depression into a major atmospheric event that can affect millions of people.

tropical disturbance
The beginning stage of a cyclonic storm lasting at least 24 hours, originating in the tropics or subtropics; clouds and moisture become organized and a vertically rotating wind mass creates atmospheric instability.

tropical depression
A slow-forming cyclonic storm with sustained surface winds of 38 miles per hour (62 kph) or less.

Hurricane Conditions

For a **tropical disturbance** to evolve into a **tropical depression** and eventually a hurricane, several conditions must be met:

- Warm, ocean surface water (exceeding 28 °C, 82 °F), along with high humidity, and unidirectional, constant winds
- Rotating winds produced by thermal heating and from the Coriolis effect in the lower latitudes
- Low atmospheric pressure near the equator that creates an atmospheric, tropical depression (**Figure 11.3**) that can develop into more intense storms.

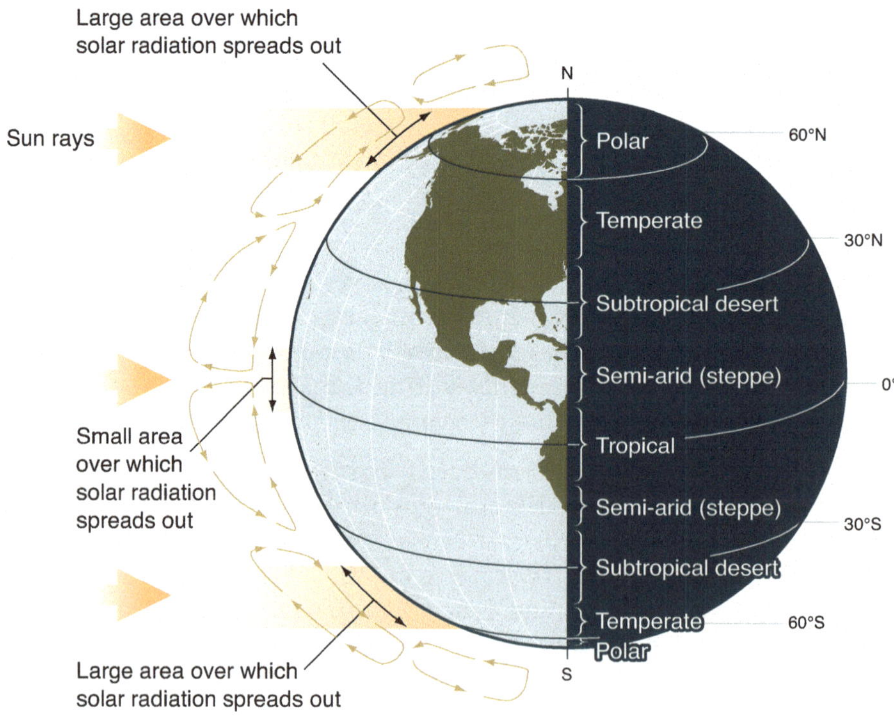

© Kendall Hunt Publishing Company.

Figure 11.3 Global atmospheric circulation patterns. Note the formation of low pressure centered on the equator as heated air rises into the atmosphere and descends in the subtropical regions.

Cyclones form in tropical and subtropical latitudes, generally between 5° and 20° north or south of the equator. This region on Earth receives the highest amount of solar radiation, which warms ocean waters that evaporate into the atmosphere (Figure 11.3). Seawater must be a minimum temperature of 28 °C to be warm enough for heat and moisture to contribute to the formation of a storm. The resulting low-pressure conditions begin to rotate as the heat and moisture rise. Circulation is counterclockwise in the Northern Hemisphere and clockwise in the Southern Hemisphere.

The National Hurricane Center defines the progression of storm development as shown in Table 11.1. Over a period of days before the system reaches the threshold of a **tropical storm**, wind velocities increase and the storm becomes better defined; at this time it is assigned a name by the National Hurricane Center. Once hurricane conditions are reached, the high winds and torrential rainfall produce a significant threat to the environment and the communities the storm might impact. In the Northern Hemisphere, ocean conditions are at an optimum for storm development from August through October, when the largest number of tropical storms occur in the North Atlantic Ocean (Table 11.2). This period is near the end of summer as ocean waters need the summer months to collect enough solar radiation and heat to generate storms. Historically the greatest number of hurricanes that make landfall in the United States occurs in September, also the stormiest month.

Thermal energy—energy that comes from heat drawn from surface waters—is what drives hurricanes. As this energy rises into the atmosphere, it rotates and produces unstable conditions aloft. Further rotation concentrates energy toward a central column and the velocity increases (the spinning ice skater effect). The ocean surface becomes more agitated, which increases the heat and moisture in the atmosphere.

tropical storm

A cyclonic storm with sustained surface winds between 39 miles per hour to 73 miles per hour (62 kph to 117 kph). At this level of activity the system is assigned a name to identify and track it.

thermal energy

The amount of energy in a system that is related to the temperature of its constituents; for hurricanes this is heat originally taken from the oceans.

Table 11.1 Definitions of Tropical, Low-Pressure Weather Systems (Storms begin as disturbances and can become hurricanes.)

System Designation	Conditions
Tropical disturbance	Organized mass of convectional air and thunderstorms with partial rotation present; generally 100 to 300 nautical miles in diameter; forms in the subtropics or tropics.
Tropical depression	Closed circulation with sustained winds of 38 mph (33 kt or 62 kph) or less.
Tropical storm	Sustained winds (1-minute measurement, 10 m above water) of 39 mph (63 kph or 34 kts) up to 64 kts. A name is now assigned to the storm.
Hurricane	Sustained winds of 74 mph (119 kph or 64 kts) or more (see Box 11.1).

Source: National Hurricane Center.

Table 11.2 **Monthly Occurrence of Tropical Storms and Hurricanes in the North Atlantic Ocean, for the Period 1851 to 2020**

TOTAL AND AVERAGE NUMBER OF TROPICAL STORMS BY MONTH						
	Tropical Storms		Hurricanes		US Landfall Hurricanes	
Month	**Total**	**Average**	**Total**	**Average**	**Total**	**Average**
January-April	6	*	2	*	0	*
May	27	0.1	4	*	0	*
June	89	0.5	30	0.2	19	0.11
July	116	0.7	54	0.3	30	0.16
August	365	2.2	242	1.4	81	0.47
September	491	2.9	346	2.0	115	0.66
October	295	1.7	177	1.0	56	0.31
November	64	0.4	39	0.2	5	0.03
December	9	*	4	*	0	*
Year	1,456	8.6	869	5.2	306	1.75

* Less than 0.05.
Source: National Oceanic and Atmospheric Administration.

BOX 11.1 **Why Do Hurricane-Force Winds Start at 64 Knots?**

Beaufort Wind Scale	
Force 0	Calm
Force 1	Light Air
Force 2	Light Breeze
Force 3	Gentle Breeze
Force 4	Moderate Breeze
Force 5	Fresh Breeze
Force 6	Strong Breeze
Force 7	Near Gale
Force 8	Gale
Force 9	Strong Gale
Force 10	Storm
Force 11	Violent Storm
Force 12	Hurricane

Contributed by Neal Dorst

In 1805–1806 Commander Francis Beaufort RN (later Admiral Sir Francis Beaufort) devised a descriptive wind scale in an effort to standardize wind reports in ships' logs. His scale divided wind speeds into 14 Forces (soon after pared down to 13) with each Force assigned a number, a common name, and a description of the effects such a wind would have on a sailing ship. And since the worst storm an Atlantic sailor was likely to run into was a hurricane, that name was applied to the top Force on the scale.

During the 19th century, with the manufacture of accurate anemometers, actual numerical values were assigned to each Force level, but it wasn't until 1926 (with revisions in 1939 and 1946) that the International Meteorological Committee (predecessor of the World Meteorological Organization, an agency of the United Nations) adopted a universal scale of windspeed values. It was a progressive scale with the range of speed for Forces increasing as you go higher. Thus Force 1 is only 3 knots in range, while Force 11 is 8 knots in range. So Force 12 starts out at 64 knots (74 mph, 33 m/s).

There is nothing magical in this number, and since hurricane force winds are a rare experience, chances are the committee that decided on this number didn't do so because of any real observations during a hurricane. Indeed, the Smeaton-Rouse wind scale in 1759 pegged hurricane force at 70 knots (80 mph, 36 m/s). Just the same, when a **tropical cyclone** has maximum winds of approximately these speeds, we see the mature structure (eye, eyewall, spiral rainbands) begin to form, so there is some utility with setting hurricane force in this neighborhood.

Source: http://www.aoml.noaa.gov/hrd/tcfaq/.

Hurricane Eye

Eventually an **eye** forms at the center of the rotating vortex. The eye is a region of calm air and relatively clear conditions characterized by a column of descending cool, dry air (Figure 11.4). The column is bordered by the **eye wall**, a thick mass of moisture-laden clouds spinning at a high velocity, and represents the most violent portion of a hurricane. Upward rotational movement in the eye wall lifts warm, moist water to higher altitudes, where condensation occurs and heat is given off.

Rotational forces throw most of the air and moisture outward, but some air flow is directed toward the center of the eye, where it is heated by compression. This warmer air can now absorb additional moisture, so most water vapor in the eye will be absorbed into the air. Any clouds in the eye dissipate, producing a zone devoid of clouds. The rising air creates lower pressure along the storm-ocean water boundary, which causes more moisture and heat to be drawn upward and the process continues. In the early stages of storm development, the eye is not well-defined but has better definition as overall rotation of the entire storm system increases (Figure 11.5). Once formed, the eye ranges in diameter from 5 to 40 miles (8 to 65 km), while the entire storm can stretch across 250 to 400 or more miles (400 to > 650 km).

tropical cyclone

A low-pressure system having a warm center that developed over tropical (sometimes subtropical) water and has an organized circulation pattern. The magnitude of its winds defines it as a disturbance, depression, storm, or hurricane/typhoon.

eye

The central core of a cyclonic storm, normally relatively small and lacking clouds, moisture, and wind.

eye wall

The boundary between the eye of a cyclonic storm and the innermost band of clouds.

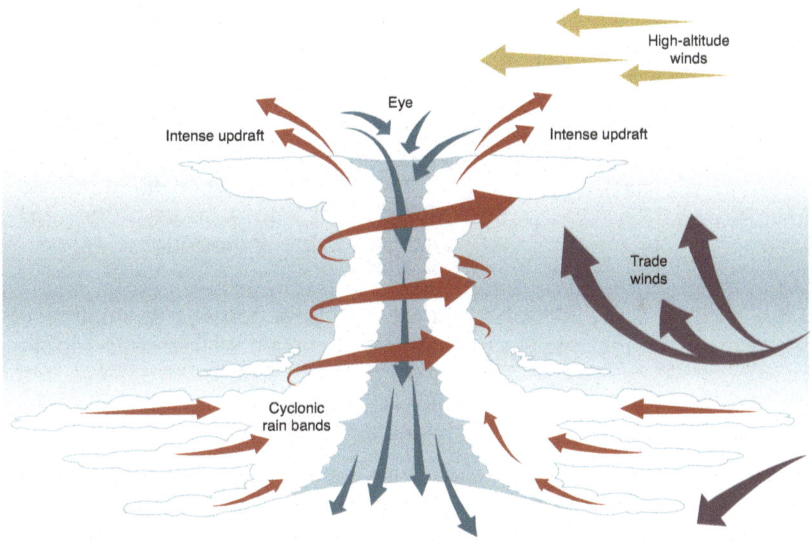

Source: National Geophysical Data Center, NOAA.

Figure 11.4 Generation of the eye wall and circulation pattern of a hurricane.

Measuring the Size and Intensity of a Hurricane

Hurricanes have a huge amount of thermal energy inside them. The National Oceanic and Atmospheric Administration (NOAA) reports that an "average" hurricane has at least 1.3×10^{17} Joules per day or 1.5×10^{12} watts of energy, which is equivalent to about half the world's entire electrical generating capacity. Because of the extensive size of a hurricane and its slow forward speed, winds blow for many hours in a given location. At first winds are blowing from one direction and then as the storm passes directly over an area, the winds diminish to almost none at all as the eye passes overhead. For the uninformed, this might signal the end of the storm. After the eye passes, winds increase again—but from the opposite direction—and will continue to blow until the storm moves out of the area.

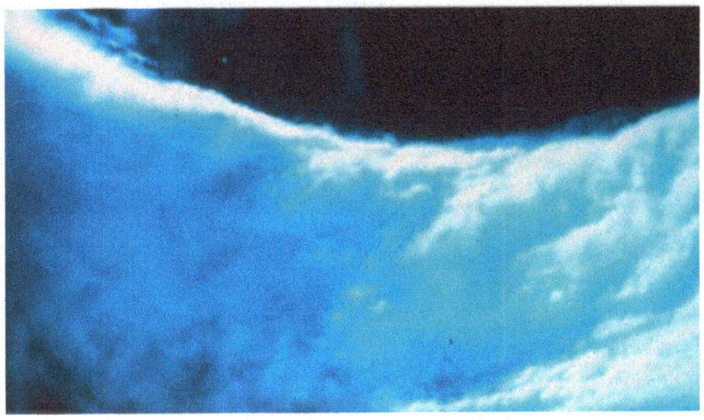

Source: The National Oceanic and Atmospheric Administration.

Figure 11.5 Photo taken in the eye of Hurricane Katrina, showing the eye wall. Note the clear conditions aloft and the rotating cloud pattern.

Tropical storms, and those that grow into hurricanes, extend over very large areas. Gale-force winds (those exceeding 39 mph or 73 kph) can reach out from the center of the storm 300 to 400 miles. Hurricane-force winds, which exceed 74 mph, can extend out up to 100 miles or more from the center. Hurricane Sandy in 2012 is the hurricane that covered the largest area. Its gale-force winds were felt more than 1,000 miles (1,600 km) from the storm center. Fortunately most of those winds were situated over open ocean to the east of the coastline of the mid-Atlantic states and New England. However, those winds created a massive storm surge that affected the entire region.

Hurricane forecasters assign storms to one of five categories, based on their wind speeds (Table 11.3). This designation provides an estimate of the potential property damage and flooding expected from the storm.

Table 11.3 The Saffir-Simpson Scale Used to Categorize Intensities of Hurricanes

Category	Winds	Effects
1	74–95 mph (119–153 kph)	No real damage to building structures. Damage primarily to unanchored mobile homes, shrubbery, and trees. Also, some coastal road flooding and minor pier damage.
2	96–110 mph (115–177 kph)	Some roofing material, door, and window damage to buildings. Considerable damage to vegetation, mobile homes, and piers. Coastal and low-lying escape routes flood 2 to 4 hours before arrival of center. Small craft in unprotected anchorages break moorings.
3	111–130 mph (179–209 kph)	Some structural damage to small residences and utility buildings, with a minor amount of curtain wall failures. Mobile homes are destroyed. Flooding near the coast destroys smaller structures with larger structures damaged by floating debris. Terrain continuously lower than 5 feet above sea level may be flooded inland 8 miles or more.
4	131–155 mph (211–249 kph)	More extensive curtain wall failures, with some complete roof structure failure on small residences. Major erosion of beach. Major damage to lower floors of structures near the shore. Terrain continuously lower than 10 feet above sea level may be flooded requiring massive evacuation of residential areas inland as far as 6 miles.
5	Greater than 155 mph (249 kph)	Complete roof failure on many residences and industrial buildings. Some complete building failures with small utility buildings blown over or away. Major damage to lower floors of all structures located less than 15 feet above sea level and within 500 yards of the shoreline. Massive evacuation of residential areas on low ground within 5 to 10 miles of the shoreline may be required.

Table 11.3 The Saffir-Simpson Scale Used to Categorize Intensities of Hurricanes (*continued*)

RECENT EXAMPLES			
Category	**Sustained Winds (MPH/KHP)**	**Description**	**Examples**
1	74–95 (118–153)	Minimal	Barry 2019 LA
2	96–110 (154–177)	Moderate	Gert 2017 East Coast of United States
3	111–130 (179–209)	Extensive	Humberto 2019 Caribbean Region
4	131–155 (211–249)	Extreme	Katrina 2005 LA | Laura 2020 LA
5	>155 (>249)	Catastrophic	Camille 1969 MS | Andrew 1992 FL

Source: National Hurricane Center, formulated in 1969 by Herbert Saffir, a consulting engineer, and Dr. Bob Simpson, director of the National Hurricane Center.

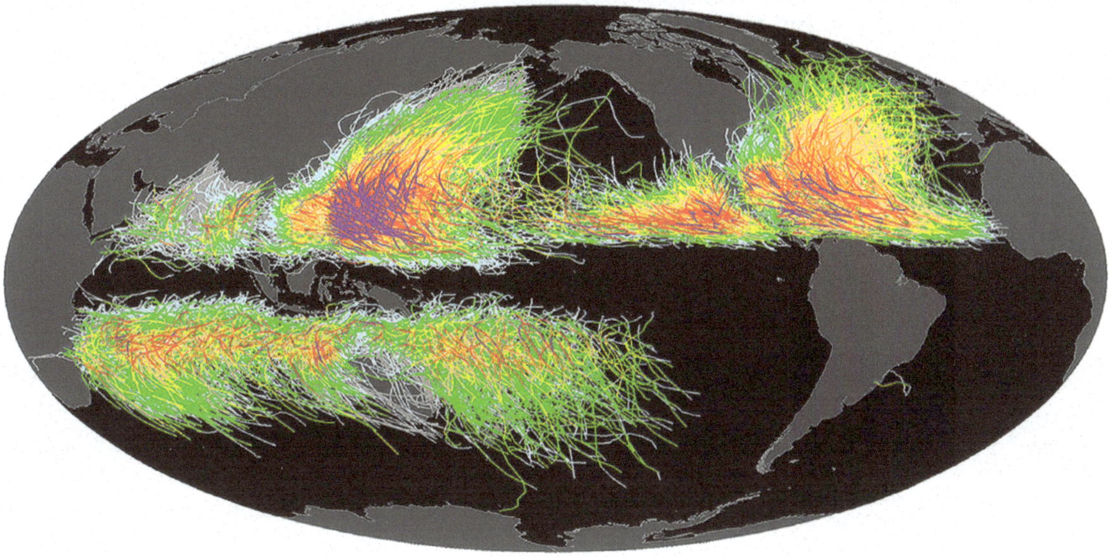

Source: NOAA.

Figure 11.6a Global tropical storm activity shows much stronger activity in the Northern Hemisphere.

Causes of Damage

Wind Action

Forward motion of the entire mass of a tropical storm is created by global circulation patterns prevalent in the area of the storm. In **Figure 11.6a** we see several interesting features. As mentioned earlier, there are no tropical storms that occur near the equator and no storms exist on the oceanic sides of South America. The strongest storms are in the western Pacific Ocean. Very few of the storms in the Southern Hemisphere become major storms. Australia is the primary land mass that is hit by these storms. The two sections of the continent that are affected lie along the western and northeastern coasts. Intensity levels are generally much less than those experienced in the northern Pacific and Atlantic Oceans.

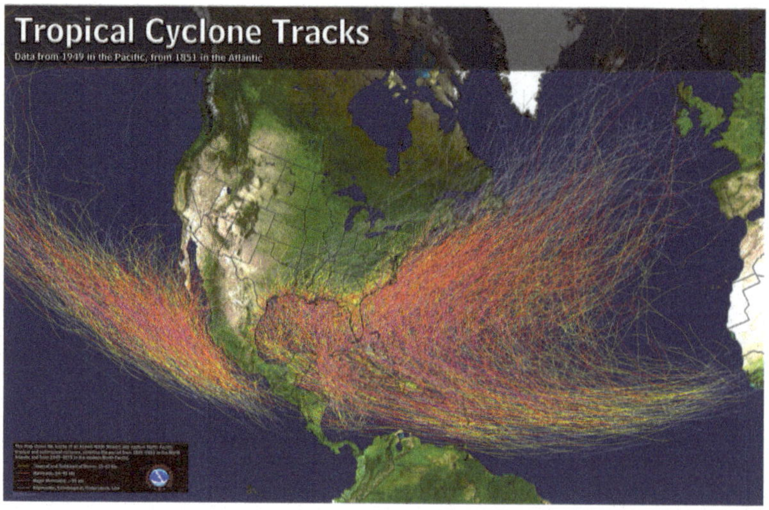

National Weather Service.

Figure 11.6b **Activity in the North Atlantic over the past 150 years shows the prevailing movement of storms as they travel westward and then to the northeast.**

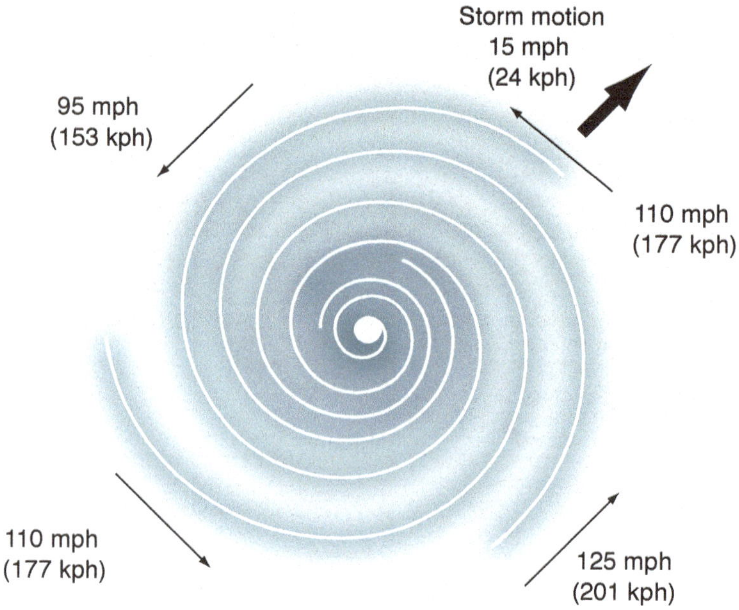

© Kendall Hunt Publishing Company.

Figure 11.7 **The strongest winds are on the right side of the storm. This value considers both the forward motion of the storm as well as the wind speed around the eye.**

In the North Atlantic Ocean, the prevailing westerlies push storms from their regions of origin in the east toward the west. The science of fluid mechanics also helps explain much of the movement of these storms. As a mass of fluid (in this case, water-laden, rotating air) moves from the equator toward the poles, it experiences increased rotation from a state of no rotation at the equator to a maximum value at the poles. In the Northern Hemisphere, there is a maximum rotation on the west side of a storm and a minimum value on the east side. This difference in the rotational strength, along with the Coriolis force, generates a deflection of the storm toward the north. Thus, hurricanes in the North Atlantic Ocean curve northward as they progress to the west (**Figure 11.6b**). In its earliest stages, the forward motion of a hurricane ranges from almost 0 mph to more than 10 or 12 mph (0–20 kph). When storms move out of the latitudes of the trade winds (about 30° north and south of the equator), they increase their velocity to about 15 to 20 mph (25 to 35 kph). This forward motion, when coupled with the counterclockwise circulation of winds makes the northeast side of a storm the most devastating, as the two wind velocities are added together (**Figure 11.7**).

As storms get closer to the coast of the United States, they often encounter continental weather systems that have moved across the Midwest. These large weather fronts push the dissipating cyclonic low to the northeast and sometimes entrain the low pressure into its primary wind pattern, dragging it along and breaking it up. Tropical storms (which were previously hurricanes) have been recorded to move as fast as 50 mph (80 kph) across the mid-Atlantic and New England states northeastward into the North Atlantic, where colder waters fail to provide the energy needed to sustain the system.

Wind damage is perhaps the most obvious product of a hurricane. Photos and video images show buildings being torn apart, roofs flying through the air, and trees and limbs being bent. In many coastal areas that lie in the

FEMA, photo by David Gatley.

Figure 11.8 **Hurricane-force winds can easily destroy weak, freestanding structures.**

FEMA, photo by David Gatley.

Figure 11.9 **Tall pine trees, which have shallow roots systems, are among the first trees to be blown over due to saturated soils.**

paths of hurricanes, building standards have improved to lessen the effect of wind damage, but major storms are capable of producing severe wind damage. Roofs are very susceptible as winds catch under shingles or corrugated metal and rip these off. Roofs on older buildings are frequently blown off as a single unit as they are often not attached securely to the sides and internal walls of buildings. Isolated or freestanding structures are especially affected if they lack walls (Figure 11.8). Roofs on new construction in coastal communities are now required to be tie-bracketed to interior and exterior walls to help maintain some structural integrity for the building when strong winds strike. Tall trees are often toppled because the ground holding their root systems becomes saturated and can't hold the trees in place (Figure 11.9). Occasionally trees are snapped off, but that is mainly the result of sudden gusts of higher velocity winds. Although the effects of wind are readily seen in an area hit by a hurricane, there is a more costly and devastating agent of destruction at work.

Storm Surge and Flood Hazards

Once a hurricane strikes land, it begins to lose its strength as it is no longer over the warm waters that feed it. Within a short period (usually less than 1 day), the hurricane is downgraded to a tropical storm or depression. However, these reduced winds, along with intense rainfall, will continue to affect areas in the storm's path. Extensive flooding often results from this stage of the storm. In June 1972, Hurricane Agnes developed off the Yucatan Peninsula of Mexico, gained strength as it passed over the warm waters of the Gulf of Mexico, and hit land in the panhandle of Florida. After it made landfall, it was downgraded to a tropical storm or depression and moved across Georgia and the Carolinas. It then joined another cyclonic low in the westerlies and regained strength as it went over the ocean to the east of Virginia. As it swung back inland, it was picking up moisture from the Atlantic Ocean and proceeded to drench the Mid-Atlantic states with intense rains (Figure 11.10). Portions of Maryland, northeastern Pennsylvania, upstate New York, and northern Virginia received more than 15 inches of rain. There were 122

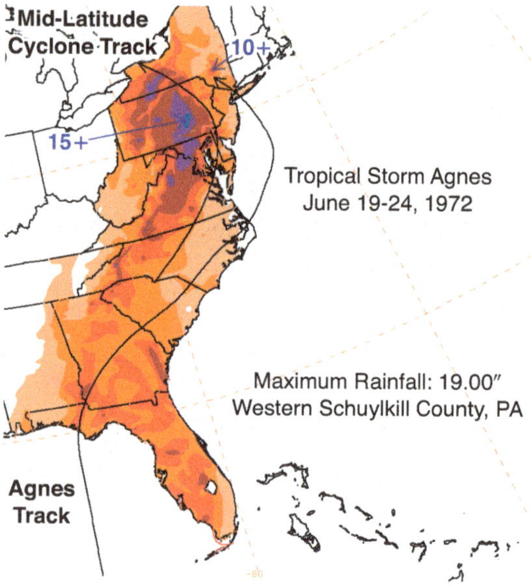

Source: NOAA.

Figure 11.10 Rainfall map for tropical storm Agnes. The thin, solid line shows the location of the center of the storm as it moved northward.

recorded deaths in the United States, the majority caused by drowning and mudslides.

Seawater from storm surges produces the most extensive damage of a hurricane, especially when the surge strikes at high tide. The momentum of this water is difficult to recognize because the wind and driving rain catch everyone's attention. Within hours, rising waters can appear. The storm surge, which can be 3 to 5 m (10 to 15 ft) or higher, inundates the shore and adjacent inland areas. The surge rises as the hurricane makes landfall. The rise in water level is usually gradual but relentless (Figures 11.11 and 11.12). The greatest recorded storm surge was associated with Tropical Cyclone Mahina that struck Australia in 1899. Reports stated a surge in excess of 13 m (42 ft). Dolphins were reported to be lying on the ground 14 m (45 ft) above sea level!

Because coastal areas are relatively flat and often have low spots, floodwaters recede very slowly and the subsequent flooding can continue for several days or weeks. Flooding kills many more people than the wind. Hurricane Camille in 1969 was one of only three category 5 hurricanes to hit the continental United States mainland (the other two were the unnamed Florida hurricane of 1935 and Hurricane Andrew in 1992, since upgraded by a review of its activity). Most of the 256 deaths associated with Camille were a result of flooding along inland portions of the Gulf Coast states. One hundred thirteen people died in Virginia due to flooding caused by extreme rainfall. From 1970 to 1994, 59% of the 589 deaths in the United States were by drowning in rainwater that falls from an average hurricane at the rate of 2,300 cu meters per second. When measured against the average flow of the St. Lawrence River in New England and eastern Canada (a flow of 6,900 cu meters per second), an average hurricane produces that much water (in just 3 seconds). If rain falls in areas where streams funnel the runoff into developed areas, devastating conditions develop rapidly.

© Danny E. Hooks/Shutterstock.com.

Figure 11.11 Onshore winds push water over low-lying areas and produce beach erosion.

© John Orsbun/Shutterstock.com.

Figure 11.12 Large wind-driven waves produce significant erosion.

NOAA. LCDR Mike Moran.

Figure 11.13 **New Orleans experienced severe flooding due to a breeched levee system.**

FEMA. Jocelyn Augustino.

Figure 11.14 **Flooded New Orleans neighborhood near inundated roadway. Numerous rooftop rescues were carried out.**

Hurricane Katrina, one of the strongest storms to hit the United States in the last 100 years, killed more than 1,800 people, most of whom drowned in floodwaters that covered Alabama, Louisiana, and Mississippi. New Orleans, a city of almost 500,000 people before the onslaught of Katrina, is situated below sea level, surrounded by a system of levees that were constructed to hold back the Mississippi River and Lake Ponchartrain. Several major breaches in the levees resulted in rapid flooding that covered the city with more than 20 ft (7 m) of water (Figure 11.13).

Analysis of the region after Katrina passed led investigators to realize that the levees had not been sunk deep enough into the ground to withstand the enormous forces inflicted by the wind and water. Also, floodwaters undercut the bottom edges of the levees, thereby removing the material they rested on and they collapsed. Hurricane Katrina struck New Orleans as a category 3 storm, having lost some of its earlier punch when it was a category 5 storm over the central Gulf of Mexico. The relentless winds and storm surge battered the levees that were in place to protect the city. Once the levees broke, the entire city was flooded and remained in those conditions for several weeks until the water could be pumped out (Figure 11.14). Low-lying areas, such as the Ninth Ward, were inundated to depths exceeding 20 ft (6 m).

© Steven J. Taylor/Shutterstock.com.

Figure 11.15 **Beach erosion is rapid in areas without any grasses or other vegetation to hold sand in place.**

United States Geological Survey.

Figure 11.16 This house once sat on sand dunes along the beach in the Outer Banks of North Carolina. Severe beach erosion associated with Hurricane Dennis in August 1999 removed large amounts of sand and isolated many structures along the shoreline.

Any time there is a storm surge, there can be removal of sand from the beaches (Figures 11.15 and 11.16). This material is carried inland by the water and deposited once the surge abates. Thick sand deposits cover streets and yards and become a hazard (Figure 11.17). Beaches must be rebuilt to maintain the natural balance between the ocean and streams that normally supply the sand. One novel way to rebuild beaches is to use old Christmas trees to serve as wind breaks, allowing sand to reform coastal dunes (Figure 11.18).

The Historic and Recent Records and the Human Toll

Hurricanes have been reported by sailors since the first ships sailed the Pacific and Atlantic Oceans. Prior to that, the indigenous peoples of the Pacific Ocean and other regions all had stories of storms that were passed along through their oral histories. As countries took to the seas, sailors experienced the forces of nature when ships sailed through and around storms. Records of storms that have affected the United States have been kept for more than a century and detail some of the most destructive natural disasters to hit the US mainland.

Source: Mark Wolfe, Federal Emergency Management Agency.

Figure 11.17 Storm surge from Hurricane Frances in September 2004 pushed sand several hundred meters inland at Fort Pierce Beach, Florida.

Source: David M. Best

Figure 11.18 Discarded Christmas trees are used to reestablish eroded beach dunes on the North Carolina coast.

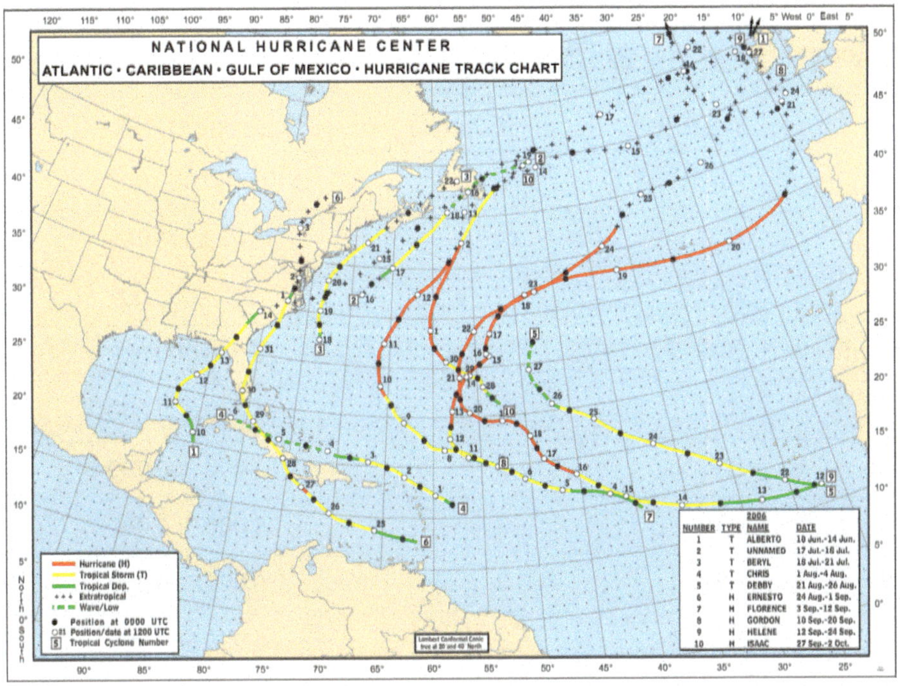

Source: NOAA.

Figure 11.19 Tropical storm tracks in the North Atlantic Ocean for the 2006 season. No hurricanes hit the United States and the season had only 10 named storms, a marked contrast to 2005, when 27 named storms formed.

Often conflicting reports are given in terms of the number of deaths and the amount of property damaged by tropical storms and hurricanes. These data vary because of different reporting techniques that include a wide range of interpretations. We have attempted to provide reliable data that represent reports provided by US government offices or by reputable research agencies, but differences will be encountered if you use a variety of sources.

The United States, certainly a country with a well-developed economic base and a means to deal with the ravages of hurricanes, is not exempt from the destruction that results when major storms hit the coastline. The National Oceanic and Atmospheric Administration (NOAA) reports that the average annual damage from hurricanes in the mainland United States is $4.9 billion. Damage increases exponentially with rising winds, so a category 4 hurricane can produce up to 250 times the damage of a category 1 storm. NOAA reported that in 2005 the country experienced 27 named storms, of which 15 reached hurricane status. Four hurricanes reached category 5, and five named storms developed in July—a record on both counts. This significant activity in terms of number and intensity of storms fortunately did not carry over into 2006, as some forecasters had feared. Figure 11.19 shows that no hurricanes made landfall in 2006.

Table 11.4 shows the ten most costly hurricanes in the United States. Texas and Florida have been very vulnerable to these storms. Note that in 2005 and 2017 these two states were hit by major events. Several factors contribute to the high loss of property. Resurgent construction and population growth frequently follows a devastating storm as people think another storm will not come along to inflict damage again, but obviously it happens. Data available from NOAA shows that in the period 1985 to 2015, almost 25% of catastrophic losses related to hurricanes occurred in Texas and Florida. These figures were calculated before the 2017 hurricane season when Hurricanes Harvey and Irma struck those regions with devastating results. As mentioned earlier, numerous hurricanes have struck the continental United States. Two major events that have done so recently are discussed in Boxes 11.2 and 11.3.

Table 11.4 Ten most costly hurricanes in the United States to 2020

Rank	Hurricane	Year	Category	Damage (USD)
1	Katrina (SE FL, LA, MS)	2005	3	$179,000,000,000
2	Harvey (TX, LA)	2017	4	139,000,000,000
3	Maria (PR, USVI)	2017	4	100,000,000,000
4	Sandy (Mid-Atlantic, NE US)	2012	1	78,000,000,000
5	Irma (FL)	2017	4	54,000,000,000
6	Andrew (SE FL, LA)	1992	5	51,000,000,000
7	Ike (TX, LA)	2008	2	38,000,000,000
8	Ivan (AL, NW FL)	2004	3	29,000,000,000
9	Wilma (S FL)	2005	3	26,000,000,000
10	Michael (W FL, GA, SC, NC, VA)	2018	5	26,000,000,000

Source: NOAA; note figures are rounded

BOX 11.2 Superstorm Sandy 2012

Most major hurricanes that affect the Caribbean and continental United States occur between mid-August and mid-October each year. Late October 2012 saw the formation of a hurricane that was unusual due its late formation and its severity. Hurricane Sandy was the 18th named storm of 2012.

Within a 6-hour period on October 19, a small tropical disturbance in the Caribbean developed into a tropical storm. Five days later it became a hurricane, the 10th of that year. Between October 24 and 27, Sandy's strength oscillated between being a Category 2 storm, a tropical storm, and back to a Category 1 hurricane as it moved northward. Its first encounter with the East Coast of the United States was when it came ashore near Atlantic City, New Jersey, on the evening of October 29. Unfortunately, this landfall coincided with high tides associated with a full moon, driving the storm surge approximately 20% higher than normal. Extreme damage resulted from flooding and widespread power outages.

New York City was subjected to extensive flooding that filled tunnels and the subway system. Battery Park at the southern end of Manhattan experienced a surge of 4.2 meters. A major loss of electricity darkened Lower Manhattan and surrounding areas. Within a few days more than eight million people in the area were without electric power. The effects were far-reaching, as gas stations and other businesses closed, airlines canceled almost 15,000 flights, and normal operations throughout the immediate region ceased. In the end, the storm cost almost $78 billion in damaged property and an estimated $10 to $30 billion in lost business revenue and wages.

FEMA, Jocelyn Augustino.

Figure 11.2.1 Damaged boardwalk at Rockaway, New York following Hurricane Sandy.

BOX 11.3 Late Season Superstorms

As of mid-October 2018, the Atlantic hurricane season had experienced 15tropical depressions, including two major hurricanes (Category 3 or higher): Florence and Michael. As reported by the National Hurricane Center, Hurricane Michael was the third most intense hurricane to strike the mainland United States, behind the Labor Day hurricane of 1935 and Hurricane Camille in 1969.

As Hurricane Michael traveled across the warm waters off the west coast of Florida, it rapidly intensified into a Category 4 hurricane. Its extreme winds and storm surge devastated several cities in the Florida panhandle and then weakened as it moved northeastward toward the mid-Atlantic states. More than 50 people were killed. Estimated insurance losses totaled more than $25 billion.

Source: US Army Corps of Engineers.

Figure 11.3.1 Mexico City, Florida, experienced extreme damage due to the winds and storm surge of Hurricane Michael in October 2018.

In 2020 seven named storms formed in October and November. Four of these became major hurricanes which was twice the previous record for that late in the season. The 2020 season marked the fifth Atlantic hurricane season since 1966 to have 10 hurricanes by October 20 (previous years were 1969, 1995, 2005, and 2017). Note the general increase in activity and intensity of these storms in the past 50 years.

The naming of North Atlantic and Caribbean storms began in 1953, and each year the World Meteorological Organization prepares lists of names to be used for potential storms each year (see Box 11.4). Through 2020, 92 names have been retired, as explained in Box 11.5, as these names still evoke unfortunate memories for people affected by those storms' destructive power. Note that more than 60% of the retired names are for hurricanes that have occurred since 2000. Undoubtedly, climate change has contributed to the increase in more intense storms.

In the past century, hundreds of thousands of people have died worldwide in cyclones, typhoons, and hurricanes (Table 11.5). The densely populated regions of South Asia and the Far East have been subjected to catastrophic storms that hit coastal areas where many people live because of their need to be near the sea for their livelihoods and the availability of easy, cheap transportation. Unfortunately, it is the very sea these people rely upon that takes its toll and often wipes out entire communities and villages in the course of a few hours.

BOX 11.4 Hurricane Names for North Atlantic Ocean from National Hurricane Center

Experience shows that the use of short, distinctive given names in written as well as spoken communications is quicker and less subject to error than the older more cumbersome latitude-longitude identification methods. These advantages are especially important in exchanging detailed storm information between hundreds of widely scattered stations, coastal bases, and ships at sea.

Since 1953, Atlantic tropical storms have been named from lists used by the National Hurricane Center. They are now maintained and updated by an international committee of the World Meteorological Organization. The original name lists featured only women's names. In 1979, men's names were introduced and they alternate during a given year with the women's names. Six lists are used in rotation. Thus, the 2020 list will be used again in 2026. Note that no names begin with the letters Q, U, X, Y, or Z.

In the event that more than 21 named tropical cyclones occur in the North Atlantic basin in a season, additional storms will take names from the Greek alphabet: Alpha, Beta, Gamma, Delta, and so on. This occurred in 2005 and again in 2020, which produced the most tropical storms in the North Atlantic and Caribbean regions in recorded history. After 2020 Greek letters will no longer be used to label storms, but an alternate list will be provided, if needed. If a storm forms in the off-season, it will take the next name in the list based on the current calendar date. For example, if a tropical cyclone formed on December 28, it would take the name from the previous season's list of names. If a storm formed in February, it would be named from the next season's list of names.

2020	2021	2022	2023	2024	2025
Arthur	Ana	Alex	Arlene	Alberto	Andrea
Bertha	Bill	Bonnie	Bret	Beryl	Barry
Cristobal	Claudette	Colin	Cindy	Chris	Chantal
Dolly	Danny	Danielle	Don	Debby	Dorian
Edouard	Elsa	Earl	Emily	Ernesto	Erin
Fay	Fred	Fiona	Franklin	Francine	Fernand
Gonzalo	Grace	Gaston	Gert	Gordon	Gabrielle
Hanna	Henri	Hermine	Harold	Helene	Humberto
Isaias	Ida	Ian	Idalia	Isaac	Imelda
Josephine	Julian	Julia	Jose	Joyce	Jerry
Kyle	Kate	Karl	Katia	Kirk	Karen
Laura	Larry	Lisa	Lee	Leslie	Lorenzo
Marco	Mindy	Martin	Margot	Milton	Melissa
Nana	Nicholas	Nicole	Nigel	Nadine	Nestor
Omar	Odette	Owen	Ophelia	Oscar	Olga
Paulette	Peter	Paula	Philippe	Patty	Pablo
Rene	Rose	Richard	Rina	Rafael	Rebekah
Sally	Sam	Shary	Sean	Sara	Sebastien
Teddy	Teresa	Tobias	Tammy	Tony	Tanya
Vicky	Victor	Virginie	Vince	Valerie	Van
Wilfred	Wanda	Walter	Whitney	William	Wendy

Source: NOAA.

BOX 11.5 Retired Hurricane Names 1954-2017 From the National Hurricane Center

The NHC does not control the naming of tropical storms. Instead, a list of names has been established by an international committee of the World Meteorological Organization. For Atlantic hurricanes, there is actually one list for each of 6 years. In other words, one list is repeated every 7th year. The only time that there is a change is if a storm is so deadly or costly that the future use of its name on a different storm would be inappropriate for obvious reasons of sensitivity. If that occurs, then at an annual meeting by the committee (called primarily to discuss many other issues) the offending name is stricken from the list and another name is selected to replace it.

Below is a list of retired names for the Atlantic Ocean, Caribbean Sea, and the Gulf of Mexico. There are, however, a great number of destructive storms not included on this list because they occurred before the hurricane naming convention was established in 1953.

In April 2021 the World Meteorological Organization decided to no longer use Greek letters to name storms when more than 21 occurred in a given year. A supplemental list of names will be provided each year.

List of Retired Names by Year

				1954 Carol Hazel	**1955** Connie Diane Ione Janet	**1956**	**1957** Audrey	**1958**	**1959**
1960 Donna	**1961** Carla Hattie	**1962**	**1963** Flora	**1964** Cleo Dora Hilda	**1965** Betsy	**1966** Inez	**1967** Beulah	**1968** Edna	**1969** Camille
1970 Celia	**1971**	**1972** Agnes	**1973**	**1974** Carmen Fifi	**1975** Eloise	**1976**	**1977** Anita	**1978**	**1979** David Frederic
1980 Allen	**1981**	**1982**	**1983** Alicia	**1984**	**1985** Elena Gloria	**1986**	**1987**	**1988** Gilbert Joan	**1989** Hugo
1990 Diana Klaus	**1991** Bob	**1992** Andrew	**1993**	**1994**	**1995** Luis Marilyn Opal Roxanne	**1996** Cesar Fran Hortense	**1997**	**1998** Georges Mitch	**1999** Floyd Lenny
2000 Keith	**2001** Allison Iris Michelle	**2002** Isidore Lili	**2003** Fabian Isabel Juan	**2004** Charley Frances Ivan Jeanne	**2005** Dennis Katrina Rita Stan Wilma	**2006**	**2007** Dean Felix Noel	**2008** Gustav Ike Paloma	**2009**
2010 Igor Tomas	**2011** Irene	**2012** Sandy	**2013** Ingrid	**2014**	**2015** Erika Joaquin	**2016** Matthew Otto	**2017** Harvey Irma Maria Nate	**2018** Florence Michael	**2019**
2020 Dorian Laura Eta Iota									

Source: National Hurricane Center.

Table 11.5 Major Loss of Life Caused by Cyclones in the Western Pacific and South Asia Since 1940

Date	Location	Number Killed	Notes
October 1942	Bengal, India	40,000	
October 1960	East Pakistan	6,000	
May 1963	East Pakistan	22,000	
May/June 1965	East Pakistan	47,000	
December 1965	Karachi, Pakistan	10,000	
November 1970	East Pakistan (Bangladesh)	300,000	
November 1977	Andhra Pradesh, India	20,000	
May 1985	Bangladesh	15,000	
April 1991	SE Bangladesh	139,000	
October 1999	Orissa state, India	9,600	10 million homeless
October 2007	SE Bangladesh	3,000	1.4 million homes destroyed or damaged
May 2008	Myanmar (formerly Burma)	138,000	2 million homeless
November 2013	Philippines	8,000	

Source: World Almanac.

Deaths in the United States attributed to hurricanes have not been on the scale of those in other parts of the world, where storms strike areas that have high population densities, poor communication, and few means to evacuate people out of harm's way. Since 1900 more than 18,000 people have died in the United States because of hurricanes and more than 65% of those deaths occurred in the most deadly storms to hit the mainland (Table 11.6). The vast majority of deaths are caused by water driven onshore by storm surge; winds cause very few fatalities.

As we see in **Box 11.6**, the United States continues to experience major catastrophic storms. In the past several decades, improvements in storm forecasting and communications have reduced the loss of life, as more people heed the warnings and leave areas that are likely to be affected. Although fatalities are not on the order they have been in the past, property damage increases due to coastal development and population migration to those areas. Oftentimes, after a storm destroys coastal homes, people rebuild knowing that their property can be affected again.

Table 11.6 **The Deadliest Hurricanes in the United States since 1900.**

Ranking	Hurricane	Year	Category	Deaths
1	TX (Galveston)	1900	4	8,000*
2	FL (Lake Okeechobee)	1928	4	2,500–3,000
3	KATRINA (LA/MS/AL)	2005	4	1,836
4	FL (Keys)/S TX	1919	4	600†
5	NEW ENGLAND	1938	3	600
6	AUDREY (SW LA/N TX)	1957	4	416
7	FL (Keys)	1935	5	408
8	NE US	1944	3	390††
9	LA (Grand Isle)	1909	4	350
10	LA (New Orleans)	1915	4	275
11	TX (Galveston)	1915	4	275
12	CAMILLE (MS/LA)	1969	5	256
13	FL (Miami)/MS/AL/Pensacola	1926	4	243
14	DIANE (East Coast US)	1955	1	184
15	SE FL	1906	2	164

* May have been as high as 10,000 to 12,000.
† Over 500 of these lost on ships at sea; 600–900 estimated deaths.
†† Includes 344 of these lost on ships at sea.

Source: National Hurricane Center.

As shown in **Figures 11.2** and **Figure 11.6a**, cyclonic activity is active in the southern hemisphere over the Indian Ocean which lies east of Africa and south of India. In mid-March 2019, Cyclone Idai struck Mozambique, Malawi and Zimbabwe, creating one of the worst natural disasters to occur in the southern hemisphere (**Figure 11.20**). Winds up to 100 miles per hour (160 kph) created a major disaster for the region. Extensive flooding, reaching depths of 6 meters, inundated more than 3,000 square kilometers (1,200 sq mi). Almost 3 million people were affected. Waterborne diseases,

Source: NOAA.

Figure 11.20 Add Title?????

including cholera, dysentery, and malaria, caused many to become seriously ill. Another key problem was that no drinkable water was available. Although a major relief effort was mounted, it took months for life to show significant improvement. Some areas experienced hardships for more than a year, given the poor conditions that existed before the storm.

Storm Season 2021

The first storm to form in the North Atlantic Ocean in 2021 was Tropical Storm Ana, which formed on May 22, 10 days before the official start of the hurricane season. As reported by the National Hurricane Center, six named storms formed, which was twice the 30-year average. Two major hurricanes, Grace and Ida, developed. On average a major hurricane forms in August every 1 or 2 years. Thus, activity in August 2021 was well above the norm.

On August 29, the 16th anniversary of the 2005 arrival of Hurricane Katrina, Hurricane Ida struck the Gulf Coast as a Category 4 event. This marked the fourth hurricane to hit Louisiana since August 2020, and was the third major hurricane to do so as its winds were measured at 150 miles per hour (one recorded gust was measured at 172 mph). Although Ida had produced stronger winds than Katrina, the storm surge generated by Ida was not as devastating. The death toll of Ida was minimal, as there were fewer levee failures, and better preparations than that which followed Katrina.

Ida rapidly intensified just before hitting the coast of Louisiana. After it came ashore it rapidly was downgraded to a tropical depression. However, it continued to produce massive amounts of rain as it coursed its way through the Mississippi and Ohio Valleys. When it moved over New England it generated record flooding in many areas, particularly in New York and New Jersey. The storm continued to linger into early September.

Early September saw the formation of Hurricane Larry. This storm reached hurricane status in the far eastern portion of the North Atlantic Basin, a rare event. Usually, storms do not become hurricanes until they are over the middle of the western edge of the basin where ocean temperatures are much warmer.

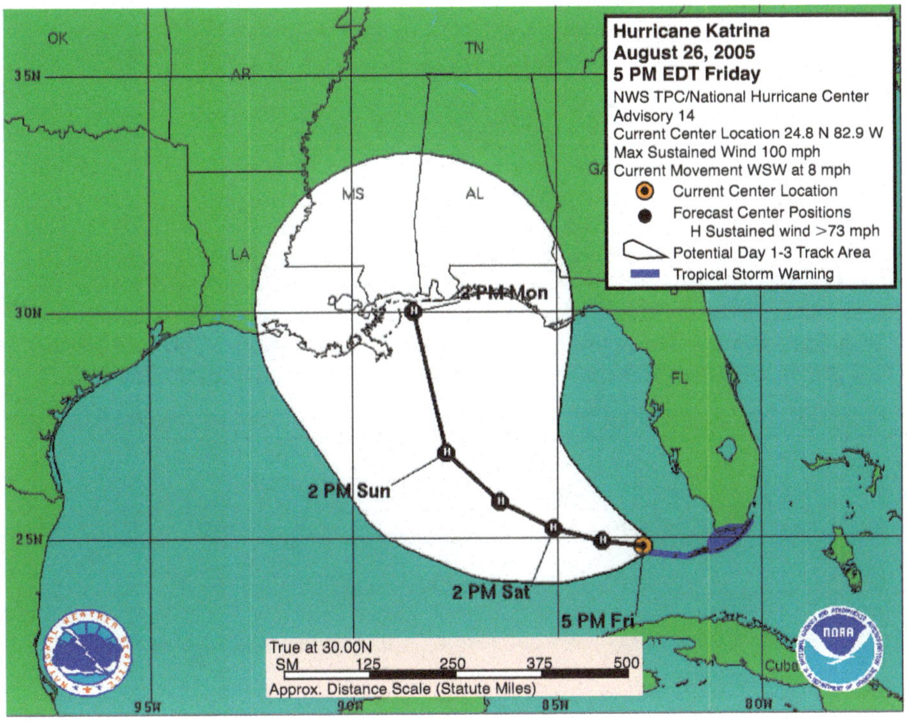

Source: NOAA.

Figure 11.21 The projected storm track for Hurricane Katrina, published 3 days before landfall in the early hours of August 29, 2005. The forecast proved to be very precise in terms of where the storm came ashore.

Hurricane Prediction and Mitigation

Numerous variables control the path of a hurricane, which makes the task of predicting its behavior and intensity a challenge. In the past 20 years, researchers have developed sophisticated computer models that provide a relatively reasonable estimate of the short-term path of a storm. Predictions for Hurricane Katrina were extremely precise (Figure 11.21).

Models depend on adequate data to provide accurate forecasts, but data collection is sparse on the ocean surface. Most information is gained by upper-atmospheric reconnaissance flights by observational aircraft and satellite imagery. The National Hurricane Center in Miami, Florida, is the central clearinghouse for all observations in the North Atlantic, Gulf of Mexico, and the Caribbean regions. Reconnaissance flights by hurricane tracking aircraft ("hurricane hunters") provide real-time information about the wind speed, forward motion, location, and other key hurricane measurements. In addition, recording instruments are dropped from planes that penetrate the center of the storm. The instruments send back such information as wind velocity and moisture content readings that are linked to precise locations given by GPS coordinates within the storm.

BOX 11.6 The Record-Setting Hurricane Season of 2020

As reported by the National Hurricane Center, in 2020 the North Atlantic Ocean, Gulf of Mexico, and Caribbean Sea regions experienced the highest number of tropical storms on record. Thirty named storms formed, two of which occurred before the official start of the season on June 1. This was the second year on record that Greek letters had to be used to designate storms (2005 was the first time). Thirteen storms reached hurricane status; six of those developed into major hurricanes (winds exceeding 115 mph or 185 kph). In addition to the record number of storms, 26 of the 30 named storms formed earlier than their past counterparts having the same alphabetic listing. For example, Hurricane Hanna formed 10 days before Hurricane Harvey which had previously been the earliest named storm beginning with the letter H.

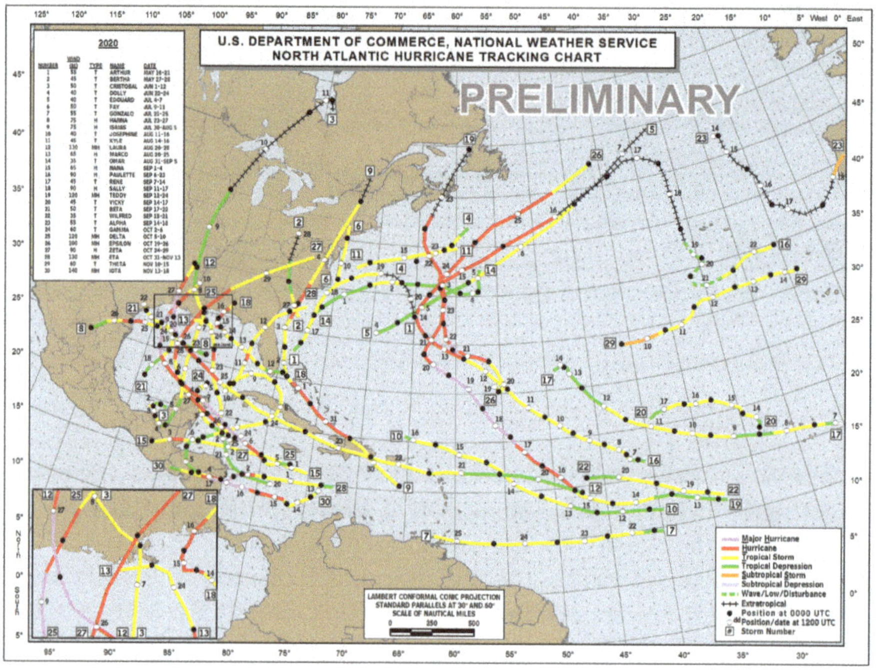

Source: NOAA.

Figure 11.6.1 **Storm tracks for the 2020 season.**

As seen in **Figure 11.6.1**, the Gulf of Mexico was especially hard hit, as the warmer waters in the Gulf provided the heat source necessary for those storms to intensify. Interestingly, an above average level of activity was centered in the north Atlantic Ocean.

In late August, Hurricane Laura expanded in size and intensity as it traversed the Gulf of Mexico. It struck the mainland along the Louisiana coast as a Category 4 hurricane, devastating the city of Cameron, which has been hit by four major hurricanes since 1957. These events resulted in the reduction of the population of Cameron, as the community dropped from approximately 1,900 people to fewer than 400 in 2018. Lying at 1.3 meters (4 ft) above sea level, the area experienced major flooding due to a storm surge of 4.5 meters (15 ft). Severe damage from winds were measured at 150 miles per hour (240 kph), making Laura the strongest storm to ever strike Louisiana. Other areas affected included Texas, Arkansas, and Mississippi. Six weeks later, Hurricane Delta struck the same region as a Category 2 event.

There were several interesting occurrences during the 2020 tropical storm season. For the sixth consecutive year, named storms formed before the usual seasonal start date of June 1. Hurricane Paulette, which lasted for 16 days, had a strange path that took it through the middle of the ocean before it moved due south and then to the east. Tropical Storm Alpha formed in the northeastern portion of the North Atlantic Ocean and struck Portugal in mid-September. Hurricane Iota, the last storm of the season, formed on November 13 and had a straight-line path toward the east coast of Central America. It rapidly became a Category 5 event, the latest such storm in history.

One measure of cyclonic storm activity is the Accumulated Cyclone Energy (ACE), which measures the combined strength and duration of tropical storms and hurricanes. This value has increased over the past 20 years. The ACE values for the very active storm seasons of 2005, 2017, and 2020 ranked second, seventh, and tenth, respectively. Tropical storm activity in 2020 was 75% above the long-term average.

Of particular note is the fact that four of the last six hurricanes in 2020 became major hurricanes, marking the first time in history that such strong storms developed that late in the season. Two of them formed in November, which marks the official end of the hurricane season in the North Atlantic Ocean. This marked the 10th consecutive year that the United States experienced eight or more billion-dollar disasters due to hurricanes.

Because researchers now understand hurricanes better than they did 20 years ago, they can provide relatively reliable computer models of a storm's behavior, fewer lives have been lost in recent years. The unfortunate exception is what occurred with Hurricane Katrina. Many people did not heed the warnings to evacuate and far too many people in New Orleans did not have the necessary resources to leave the city and its low-lying areas. As a result, more than 1,800 died, mostly from drowning related to the storm surge and broken levees that were supposed to protect the city.

The Federal Emergency Management Agency (FEMA) has published a series of guidelines to assist in planning for natural disasters. One document, "Against the Wind—Protecting Your Home from Hurricane Wind Damage," outlines ways to strengthen a home by using reinforced construction techniques. Suggestions are given to build stronger roofs and to ensure that walls and gables are braced and tied down to lessen their chances of being blown down. However, many homes in regions affected by hurricanes have older homes and buildings that do not employ these newer concepts, so often these less-sturdy buildings are destroyed. Mobile homes, a very common housing option in storm-prone areas, offer no protection whatsoever against sustained hurricane-force winds and rising floodwaters.

Attempts can be made, such as the ongoing effort by the Army Corps of Engineers in New Orleans, to improve our artificial defenses against Mother Nature. However, in spite of all our attempts to save our homes, communities, and other possessions, in the end these efforts will often prove costly and rather futile.

The ever-changing dynamics of hurricanes and tropical storms have been evident recently. Box 11.7 tells the tale of two significant events that have affected the same region in a major way over a period of 19 years.

BOX 11.7 Flooding Caused in North Carolina by Two September Hurricanes

The east coast of the United States has been hit with numerous hurricanes and tropical storms for centuries. The areas within 20 to 30 kilometers of the shoreline are generally typified by low elevations above sea level and the presence of rivers and streams that flow into the Atlantic Ocean and coastal sounds. When these storms strike the coast, one major result can be severe flooding of these low-lying areas.

In September 1999, Hurricane Floyd struck the North Carolina coast with winds of 110 mph, along with a 10-to-15-foot storm surge. This occurred at high tide. Rainfall amounts ranged between 15 and 20 inches, which caused most of the flooding. Only 1 week prior to this, Hurricane Dennis had dropped as much as 10 inches of rain on eastern North Carolina. The entire region experienced severe flooding, causing several billion dollars damage.

FEMA Dave Saville.

Figure 11.7.1 An area in Edgecombe County, North Carolina, that was flooded by Hurricane Floyd in September 1999.

In early September 2018, Hurricane Florence was headed to the coast of North and South Carolina, where it made landfall on September 11 near Wilmington, North Carolina. Freshwater flooding was severe due to as much as 30 inches of rainfall in some locations. The city of Wilmington, population of 119,000, was totally cut off and became an island for several days. The overall damage estimate resulting from Florence is $38 billion, making it the sixth most costly storm to hit the United States.

Major environmental concerns included more than 3 million chickens and turkeys killed, along with approximately 5,500 hogs. Floodwater carried countless tons of animal manure into the surface water, creating serious health hazards. In addition, coal ash from several power plants was carried into the floodwaters.

USCG PO3 Zachary Hupp.

Figure 11.7.2 U.S. Coast Guard personnel search flooded neighborhoods in eastern North Carolina for stranded residents.

Key Terms

cyclone (page 309)
cyclonic storm (page 308)
eye (page 313)
eye wall (page 313)
hurricane (page 309)
thermal energy (page 311)
tropical cyclone (page 312)
tropical depression (page 310)
tropical disturbance (page 310)
tropical storm (page 311)
typhoon (page 309)

Summary

Cyclonic storms are generated in tropical and subtropical latitudes where warm ocean waters provide the thermal energy and moisture necessary for atmospheric circulation to expand into storms. Rotational forces in the atmosphere cause these disturbances to move across Earth's surface, gaining momentum. Once a well-defined circulation system has developed, the storm rotates around the eye, where vertically directed winds put more moisture and heat into the center of the storm. Once winds reach 74 miles per hour (118 kilometers per hour), the storm is classified as a hurricane. Hurricanes are classified in terms of wind strength—from Category 1 to Category 5.

In the United States the months of August, September, and October have the largest number of hurricanes. The areas most affected lie along the coastal regions of the Gulf of Mexico and East Coast of the United States. Although we generally think of the wind damage associated with these storms, flooding caused by storm surges and torrential rainfall is their most damaging feature. In the past 30 years, coastal regions in the South and along the Gulf of Mexico have experienced devastating floods. The 10 most costly storms have occurred during that period, including Hurricane Katrina in 2005, which ranks as the worst natural disaster to occur in the United States in terms of property damage. Loss of life is still a major concern but has decreased somewhat by employing better communication techniques to warn people of an impending disaster.

Cyclonic storms play a key role in generating destructive effects on the shoreline. Storms move massive amounts of sand and other material out to sea; beaches must be rebuilt to maintain the equilibrium between the shoreline and the streams that bring sediment to the oceans.

Improvements in monitoring and modeling hurricanes allow forecasters and researchers to mitigate the impact of these storms by giving communities sufficient warning to prepare for the storms' arrival and by providing better information to engineers and builders so they can strengthen building codes and construct buildings that can withstand the hurricanes' destructive force. However, it is necessary that citizens heed these warnings and exercise good judgment when they are in harm's way.

References and Suggested Readings

Barnes, J. (2019). *North Carolina's hurricane history* (4th ed.). University of North Carolina Press.

Brennan, V. M. (Ed.). (2009). *Natural disasters and public health—Hurricanes Katrina, Rita, and Wilma.* Johns Hopkins University Press.

Burt, C. C. (2004). *Extreme weather—A guide and record book.* W. W. Norton.

Elsner, J. B., & Jagger, T. H. (Eds.). (2009). *Hurricanes and climate change.* Springer.

Fitzpatrick, P. J. (1999). *Natural disasters: Hurricanes.* ABC-CLIO.

Gray, W. M. (1968). Global view of the origin of tropical disturbances and storms. *Monthly Weather Review, 96*(October), 669–700.

Horne, J. (2008). *Breach of faith—Hurricane Katrina and the near death of a great American city.* Random House.

Longshore, D. (2008). *Encyclopedia of hurricanes, typhoons and cyclones* (3rd ed.). Facts on File.

Miles, K. (2014). *Superstorm—Nine days inside Hurricane Sandy.* Dutton Publishers.

Pielke, R. A., Jr., & Pielke, R. A., Sr. (1997). *Hurricanes—Their nature and impacts on society.* John Wiley.

Rosenfeld, J. (1999). *Eye of the storm—Inside the world's deadliest hurricanes, tornadoes, and blizzards.* Plenum Trade.

Websites for Further Reference

http://www.aoml.noaa.gov
https://oceanservice.noaa.gov
https://visibleearth.nasa.gov
https://www.fema.gov
https://www.nhc.noaa.gov
https://www.ncei.noaa.gov

Questions for Thought

1. What are the three types of cyclonic storms and where does each most commonly occur?
2. What factors contribute to the formation of most cyclonic storms near the equator?
3. Describe the development of a cyclonic storm.
4. Describe the conditions that exist in and near the eye of a hurricane.
5. Why does water rather than wind cause the most damage in cyclonic storms?
6. Examine Table 11.4 that shows the amount of damage various hurricanes have generated in the United States. Why do we see that the majority of these entries are from the last 30 years?
7. What factors contribute to the much larger loss of life we see in Table 11.5 versus that in Table 11.6?
8. Examine Figure 11.6a. Explain why there are no cyclonic storms forming in the areas around the equator and on either side of South America.
9. Why has the intensity of cyclonic storms increased over the past 40 years?
10. Explain why hurricanes quickly dissipate after making landfall.

12 Wildfires

Source: David M. Best

Wildfire outside Flagstaff, Arizona, generated a pyrocumulus cloud that severely affected air quality.

> **Wildfires Are Worldwide Events**
>
> Wildfires have existed for millions of years. In today's world they have become a major concern as the global population has spread into regions that provide the fuels for these events. For example, in Europe, Australia, and the United States, wildfires over the past 50 years have destroyed millions of acres of forests and grasslands and have consumed thousands of homes while upsetting numerous lives. Prolonged drought in many areas will continue to cause wildfires to be a natural hazard to be aware of. Due to a long period of suppressing wildfires, fuels have increased, thereby causing fires to be much larger than in the past.

Fire Processes

The discovery of fire as a useful tool by the earliest humans allowed them to have heat and light, and they eventually learned that fire could be used to cook their food and improve their diets. Early man used fire to herd animals and to clear land. It was even used in warfare to drive combatants to their eventual death.

Today, we do similar things with fire but we also have the awareness that when fire gets out of control, it can become a destructive force. A **wildfire**, sometimes referred to as a **forest fire** or **brush fire**, is an unplanned and uncontrolled event involving the rapid oxidation of organic matter. These fires occur in either wildland or agricultural areas that are covered by grasses, brush, or trees. Fires are most common in those parts of the world where moist, warm climates permit vegetation to grow. When these areas experience periods of hot weather, coupled with low humidity, vegetation dries out and becomes fuel for wildfires.

The chemical decomposition of a substance by heat is defined as **pyrolysis**. The chemical products of this process are gases and solid matter that become flammable in the presence of sufficient heat. For a fire to occur, three components—fuel, oxygen, and heat—must be present (Figure 12.1). In this **fire triangle**, **fuel** must be available to burn, **oxygen** has to be present to allow the chemical reaction of oxidation to take place, and sufficient **heat** must exist to allow combustion to begin. Usually the heat source generating the fire comes from an open flame produced by a match, a spark, or a concentrated heat source, such as lightning.

Fires result from the accelerated combination of the elements in the fire triangle. Every flammable fuel has a **temperature of combustion**, which is a measure of the amount of heat necessary to cause the substance to ignite and burn in the presence of sufficient oxygen. The chemical process involves the formation of new bonds between oxygen and the carbon/hydrogen bonds present in the fuel.

Fuel

Wood, a common fuel source in nature, is composed mainly of cellulose, a complex carbohydrate containing hundreds of connected glucose molecules. **Glucose** is an organic substance with the chemical formula $C_6H_{12}O_6$.

wildfire
An uncontrolled fire in a natural setting.

forest fire
A fire that occurs in a forest or stand of trees.

brush fire
A fire that burns primarily brush and material low to the ground.

pyrolysis
The process in which a series of reactions break down complex chemical compounds into simpler ones.

fire triangle
Consists of three parts that cause fire to occur: heat, fuel, and oxygen.

fuel
The part of the fire triangle that actually burns.

oxygen
A basic element that allows oxidation to take place; a necessary part of the fire triangle.

heat
The part of the fire triangle that provides the energy to create fire.

temperature of combustion
The temperature at which a specific type of material will ignite.

glucose
Is an organic substance with the chemical formula $C_6H_{12}O_6$.

This organic molecule, when consumed in the presence of heat and oxygen, is broken down into carbon dioxide and water along with the release of heat:

$$C_6H_{12}O_6 + 6O_2 + \text{heat} \rightarrow 6CO_2 + 6H_2O + \text{heat}$$

Because other chemical compounds are found in complex living material, the combustion process forms many more chemical products besides carbon dioxide and water. Many different gases and solids, along with some liquids, are created as the fuel source is thermally decomposed.

Cellulose is also a major constituent of grasses, representing about 35% of their weight when they are dried out. Because grasses are not as compact as wood and can lose moisture much faster than wood in arid conditions, they are a prime candidate for wildfires. Grasses have a lower combustion temperature and can ignite much more rapidly than wood, producing fast-moving fires of higher intensity and shorter duration. Grasses also have more surface area per unit mass so they tend to ignite rapidly. Grass fires are beneficial, as they provide the occasional clearing-out process that removes the fuel source.

© Kendall Hunt Publishing Company.

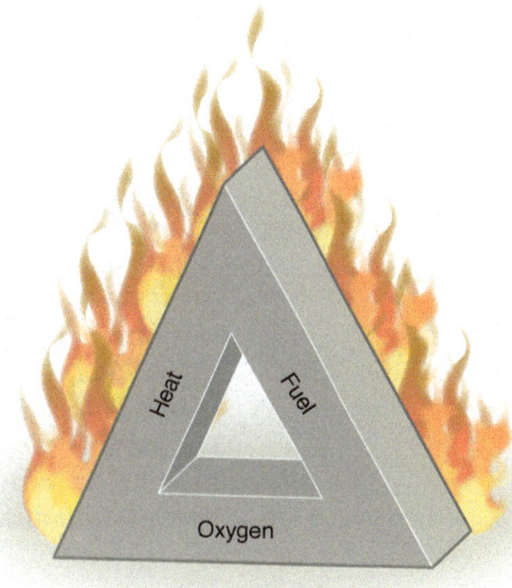

Figure 12.1 **The fire triangle. All three components (fuel, oxygen, and heat) must be present for fire to occur; remove any one and the fire goes out.**

Oxidation

The atmosphere consists of about 21% oxygen. However, only 16% oxygen is required for sustained combustion to take place. This combination of oxygen and fuel is the process of **oxidation**. In the preceding equation it is evident that removal of any component on the left-hand side (cellulose, oxygen, or sufficient heat) will break the fire triangle and put out the fire. The products of the combustion process, carbon dioxide and water, along with heat, reenter the atmosphere and can later be used to form more cellulose.

Water cools the temperature below the ignition point and it temporarily removes oxygen from the process. Retardants and foam remove oxygen along with lowering the temperature below the ignition point.

Heat Transfer

Heat can be transferred by means of **convection**, **radiation**, or **conduction**. Convection involves the rising of heat as the surrounding air becomes less dense. The updraft then draws in surface air located at the edges of the rising column, which then fans the flames. The convective movement is a primary cause of fires in canyons that expand rapidly. Rising, warmer air is funneled up the canyon sides toward the top. This draws in cooler air near the canyon base, thus fanning flames that then produce additional heated air, and the process continues in a continuous cycle.

cellulose

The primary substance that makes cell walls in plant tissue.

oxidation

A process in which oxygen combines with a substance.

convection

Transfer of heat by movement in the atmosphere caused by density differences produced by heating below.

radiation

The process in which heat energy is sent through space in the form of rays.

conduction

Transmission of heat through a material.

Radiant heat is similar to what we encounter when we place our hand over a heating element on a stove that has been turned off. This process is always present in fires and serves to dry out fuel which then can become ignited more easily.

The process of conduction is the direct transfer of heat through a conducting medium, such as the handle on a cast iron skillet. Conduction is not effective in wildfires as wood is a poor conductor of heat.

Wildfire Behavior

Approximately 85% of all wildfires are started by humans either through carelessness, such as leaving campfires unattended or unextinguished, discarding lit cigarettes, or premeditated arson. The majority of these occurs in the Southern United States. The remaining 15% are the result of the more than 4 million lightning strikes that occur daily on Earth. The mountainous parts of the Western United States have the most lightning-generated fires due to warmer climates that dry out the forests, not that the regions have more lightning (Box 12.1)

BOX 12.1 Causes of Wildfires–Humans Versus Mother Nature

Data provided by the National Interagency Fire Center (NIFC) in Boise, Idaho, for the period 2010 to 2020 show that humans caused an annual average of 55,790 wildfires. The greatest number of these (69%) occurred in the Southern and Eastern regions of the United States.

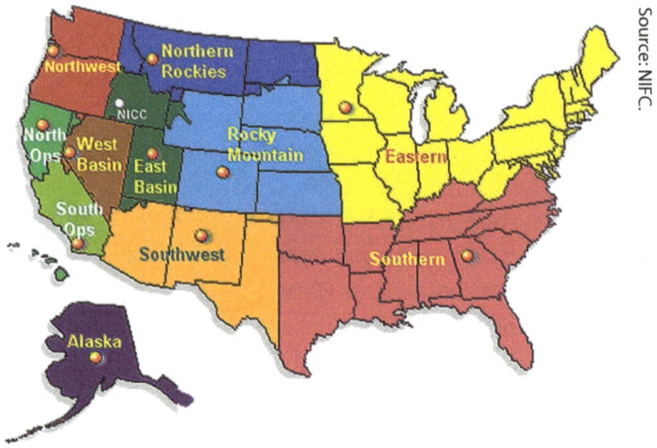

Figure 12.1.1 The United States is divided into 11 fire management regions under the U.S. Forest Service.

The five regions with the most acreage burned due to humans from 2010 to 2020 are:

Southern	9,623,935 acres
Northwest	7,512,565 acres
Northern California	4,444,769 acres
Southwest	3,791,330 acres
Great Basin	3,650,218 acres

During the period 2010 to 2020 lightning caused an annual average of 7,841 wildfires. The five leading regions that were affected were as follows:

Alaska	13,094,362 acres
Great Basin	7,512,265 acres
Northwest	6,910,315 acres
Northern Rockies	4,351,785 acres
Southwest	3,838,400 acres

An interesting observation is that Alaska only experienced an annual average of 214 wildfires as the result of lightning strikes per year, but the wildfires they created destroyed more than three times the acreage of the Great Basin region.

With changes in global climate conditions and weather patterns, wildfires have become larger and more destructive. When they occur near urban areas or in developed sections in the forests, homes can be adversely impacted.

Types of Wildfires

There are three main types of wildfires, depending on the location of the burning material.

1. Ground fires. **Ground fires** consist of slow-burning material on or just below the surface and include plant roots and buried vegetative matter such as leaves or needles (Figure 12.2a). These fires can smolder below the surface so the damage they do is not readily seen. Ash from these and other fires can fill spaces in the soil and create a **hydrophobic** layer that impedes the downward movement of moisture. This condition contributes to erosion and other problems discussed later in the chapter.

2. Surface fires. **Surface fires** consume fuel lying on the surface, such as ground litter, limbs, fallen trees, and grasses (Figure 12.2b). The intensity of surface fires is quite variable depending on how dry the material is and how

ground fire

Fire that moves along the ground and rises several feet above the surface.

hydrophobic

Incapable of absorbing water.

surface fire

Fire that moves along the ground and consumes ground litter.

Source: NASA.

Figure 12.2a Ground fire.

Source: USFS.

Figure 12.2b Surface fire.

© Lumppini/Shutterstock.com.

Figure 12.2c Crown fire.

much fuel is present. Some of these surface fires can become very intense. Many of the wildfires in California are located in areas where fuels have dried out and fires are fanned by **Santa Ana winds**, which result from high pressure over the Great Basin of Nevada. Common from October of each year to the following March, they are produced by cool air in the eastern deserts of California being pushed westward over mountains. At the top, the cold, dry air descends, heating up and generating strong winds that are often associated with wildfires in southern California.

3. Crown fires. In forested areas, fires that begin on the surface often leap into the upper portions of trees to produce **crown fires** (Figure 12.2c). The ladder concept causes low-burning fires to climb up into the crowns of trees, where winds easily fan the flames from tree to tree (Figure 12.3). Unlike surface fires, which can be fought with fire personnel and equipment, crown fires are nearly impossible to control and the results are disastrous (Figure 12.4).

Santa Ana winds

Seasonal winds that originate from high pressure over the Great Basin of Nevada, pushing dry air from east to west across southern California.

crown fire

A fire that moves through the upper portions of trees; these are very difficult to extinguish.

topography

The general configuration of the land surface.

Fire Behavior Variables

Fire behavior is controlled by several variables. Topography, fuel types, and weather each play a major role in how a fire will spread.

Topography

Topography, or the lay of the land, can influence how rapidly a fire moves along the surface. The direction that a slope faces in terms of its exposure to sunlight is critical as the amount of vegetation can be much greater on those slopes receiving a high amount of sunlight. In the northern hemisphere, south-facing slopes produce thicker vegetation, which can become dried out in periods of low humidity by constant exposure to the sun. This is especially true from May to September in the Western United States.

© Kendall Hunt Publishing Company.

Figure 12.3 A progression of fuel heights produces a ladder effect to get the tops of trees burning.

© BGSmith/Shutterstock.com.

Figure 12.4 A wildfire destroyed a once healthy forest in Glacier National Park, Montana.

Healthy forests can be quickly destroyed by wildfires (Figure 12.4). As heated air from a fire rises, it will dry out vegetation at the higher elevations, thereby producing preheated, dry material that can subsequently be burned. A general rule is that the steeper the slope, the faster the fire advances and it spreads upslope. Convection of hot gases causes updrafts that fan the flames from below, causing the fire to spread rapidly.

Fuel Types

Fire behavior is dependent on the type of fuel that is burning. Dried grasses and low shrubs can produce hot fires but these are often short-lived events due to the lack of material. These fires often creep along the ground as low-intensity fires.

Taller trees generate more intense fire activity due to their larger size. Branches long with leaves and needles will ignite and spread the fire upward. When the tops of trees ignite, a crown fire can result. One reason we have seen more large fires in recent years is that for many years the prevailing philosophy was that all fires should be put out immediately. This forest management style resulted in an overall buildup of forest floor fuels, which served to generate large fires once the material was ignited. In recent years, the US Forest Service, National Park Service, Bureau of Indian Affairs, and Bureau of Land Management, stewards of most of the forests and grasslands in the United States, have allowed large fires that were ignited naturally to burn while being monitored. These fires are considered to be in the wildland fire use (WFU) category. Such a designation was used in the summer of 2006, when a large fire on the north rim of Grand Canyon National Park in Arizona was started by a lightning strike. Of the more than 58,000 acres consumed by the fire, almost 19,000 acres were considered as the wildland fire use phase of the incident, while the remaining 39,000 acres burned as a wildland fire.

Governmental fire agencies consider wildland fire use as the management of naturally ignited wildland fires in order to accomplish resource management goals for a given area. These goals include the reduction of likelihood of unwanted fires, the maintenance of natural ecosystems in a given area, thus enhancing the health and safety of firefighters and the public. The purpose of wildland use fires is the same as that of prescribed burns, which are discussed later in the chapter.

Weather

Weather plays an obvious role in the spread of fires. Dry, windy conditions will greatly enhance the spread of wildfires, while moisture-laden storms will help quench a fire. Firefighters are aware of the amount of relative humidity in the air. A higher relative humidity means more moisture is present that can help impede the advance of a fire. As a general rule, relative humidity increases at night, which helps slow the growth of fires. Air temperature is also important as higher daytime temperatures will dry out the fuel.

Rapidly spreading fires are often accompanied by strong winds that tend to flatten out the flames and also transport glowing embers ahead of the fire, making it difficult to fight the fire along a well-defined line. Wind increases the rate of spreading of a fire.

Teams that are sent to fight wildfires are in contact with nearby weather service personnel to keep informed about upcoming conditions. Oftentimes one member of the team assigned to direct firefighting operations serves as the liaison with weather officials.

Besides drying out fuel, drought can also stress trees and reduce their ability to ward off insects because there is less moisture in the trees themselves. This condition makes the trees susceptible to insect damage and diseases. In recent years, the pine bark beetle has killed thousands of acres of trees in North America, particularly conifers such as ponderosa pine and Douglas fir (**Figure 12.5**).

These standing dead trees are then fuel for raging wildfires that can rapidly move through an area, consuming healthy trees as well. Sometimes even without the presence of dead trees, healthy trees will burn, particularly if they are stressed from experiencing dry conditions for several years (**Figure 12.6**).

Many areas in the United States have experienced significant drought conditions over the past 20 or more years. This has greatly reduced the moisture content of trees, grasses, and other vegetation, thus making them primary targets for wildfires. Examination of the data in Table 12.1 for the lower 48 states shows that since 2010, major wildfires have devastated the landscape in the several states in the Central and Western parts of the country. These are areas that do not receive much normal precipitation, so recovery will be very slow.

Firestorms

firestorm

A widespread, intense fire that is sustained by strong winds and updrafts of hot air.

Very large wildfires are capable of producing their own weather by generating massive updrafts that can rise to altitudes of 20 km or more. The upward movement of heated air causes surface winds to be drawn inward, generating a **firestorm**. These self-produced weather systems will have winds in front of the progressing fire and can also generate lightning that can produce more fires. Wind also increases the flow of oxygen to the fire, and the variable nature of wind can make it extremely difficult to forecast where a fire will move.

Fire whirls are a phenomenon sometimes observed at locations of extreme heat. Fire whirls are small tornado-shaped currents that consist of fast, rotating flames. They are capable of carrying embers and other burning debris well above the ground surface and can deposit them ahead of the fire line. Some embers have been carried more than 5 km by wind and convection associated with a fire.

Fires have become larger in the past 20 years although the number of fires has decreased (see **Box 12.2**). Changes in climate have contributed to more dry fuel, which burns faster than wet trees.

Another factor possibly causing larger fires is that the initial response is sometimes slow as the fire's potential growth is assessed. Sudden changes

Source: US Fish and Wildlife.

Figure 12.5 Conifers destroyed by bark beetles. These trees become fuel for fast-moving wildfires.

Source: David M. Best.

Figure 12.6 Results of a wildfire in a young forest that has experienced a decade of drought conditions. The 125-acre fire was started by a thrown radial tire along Interstate 40 outside Flagstaff, Arizona.

in weather conditions, especially in conjunction with thunderstorms, can cause the fires to expand rapidly.

Firefighting Techniques

The response to fighting a wildfire depends on several variables, including its location and severity. Small fires can be handled using nearby resources, including local fire departments. Fires occurring on more open state or federal lands might require larger numbers of specially trained personnel, including support staff to handle a wide range of logistical issues.

Incident Command System

When personnel from a wide range of agencies began to be used in fighting wildfires, problems soon became apparent, especially in the ability of the various agencies to communicate with one another. Problems also developed over territorial control in terms of which group had jurisdiction in a given area.

As a result of fire suppression activity in southern California in the 1970s, the **Incident Command System (ICS)** was initiated. Depending on the magnitude of the fire, one agency assumes overall command of the incident and coordinates fire suppression activity with other responders. The federal government has 17 incident management teams (IMT) located in different regions throughout the United States. These teams consist of experts in all facets of disaster management and are drawn from a wide range of local, state, and federal agencies. They can respond within a day or two to assume oversight of the firefighting when requested.

Incident Command System (ICS)
A system that is used to establish a chain of command for multiple agencies fighting fires.

IMTs are also available to oversee recovery efforts associated with other disasters, such as hurricanes and earthquakes. The experience these teams have allows multiple agencies to interact more efficiently. IMTs are

Table 12.1 Largest Wildfires Greater Than 300,000 Acres in Alaska and the Lower 48 States

Large Fires in Alaska—2010 to 2020

Year	Fire Name	Location	Size in Acres
2015	Tanana Area Fires	Alaska	498,043
2015	Ruby Area Fires	Alaska	421,613
2015	Old Grouch Top	Alaska	307,969

Large Fires in the Lower 48 States—2010 to 2020

Year	Fire Name	Location	Size in Acres
2020	August Complex	California	1,032,649
2017	NW Oklahoma Complex	Oklahoma	779,292
2012	Long Draw	Oregon	557,628
2011	Wallow	Arizona	538,049
2012	Holloway	Nevada	460,850
2018	Mendocino Complex	California	459,123
2018	Martin	Nevada	435,569
2020	SCU Lightning Complex	California	396,624
2014	Buzzard Complex	Oregon	395,747
2020	Creek Fire	California	377,693
2016	Anderson Creek	Oklahoma	367,740
2020	LNU Lightning Complex	California	363,220
2012	Mustang Complex	Idaho	341,448
2020	North Complex	California	318,930
2017	Perryton	Texas	318,156
2012	Rush	California and Nevada	315,577
2011	Rock House	Texas	314,444
2011	Honey Prairie	Georgia	309,200
2010	Long Butte	Idaho	306,113

Source: NIFC.

BOX 12.2 Fire Activity

The occurrence of wildfires has been variable in recent years due to an increase in population and a decrease in precipitation. Prolonged drought conditions in the West and Southwest have allowed fuels to dry out, often leading to major events. There has been a steady increase in the average size of wildfires burned although the number of fires has decreased. Thus the average size of fires has increased from 35 acres per fire in 1985 to 180 acres, more than a 500% increase.

In addition, the number of fires larger than 300,000 acres has increased dramatically in the past 20 years (Table 12.1). Each year the cost of fighting wildfires in the United States has been several billion dollars. In terms of being prepared for fires, we can often see storm systems coming that will alert us to the likelihood of fires starting in forests and grasslands due to lightning. However, fire management officials have no indication where or when human-caused fire will take place, except perhaps during hunting or camping season.

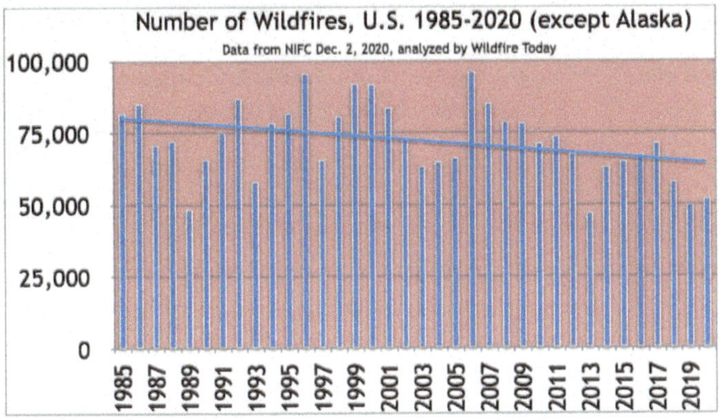

Source: Wildfire today/NIFC.

Figure 12.2.1 Number of wildland fires, 1985 to 2020.

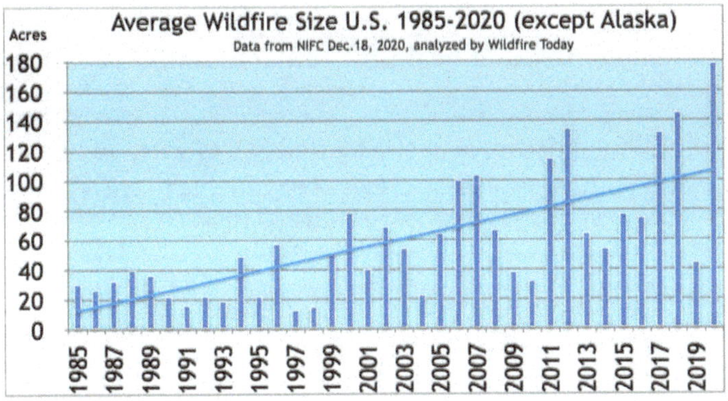

Source: Wildfire today/NIFC.

Figure 12.2.2 Average size of wildfires, 1985 to 2020. Note the steady increase.

networked with numerous federal and state offices to help reduce some of the bureaucratic challenges that often arise during these events.

The job of informing the public about wildfires is often undertaken by a **public information officer (PIO)** for the agency charged with direct oversight of the firefighting operations. The PIO is the contact point through whom other agencies and the general public learn about the progress of the fire and how it is expected to react in the short term. With so many agencies involved in a major fire, it is paramount to have a consistent message being delivered to the general population and other agencies that need information.

public information officer (PIO)
A person who is designated to provide official information regarding disasters.

Source: Bureau of Land Management.

Figure 12.7 Hotshot crews consist of 20 to 22 highly trained and extremely conditioned members.

Source: Bureau of Land Management.

Figure 12.8 Smokejumpers are staging before hiking into a remote area.

Ground Crews

The attack on a fire uses several different techniques. The initial response is to send in firefighters to establish a fire line that controls the movement of the fire on the surface. Often hotshot crews are sent in to make the initial attack. These crews consist of 20 to 22 highly trained and well conditioned firefighters (Figure 12.7). Personnel can come in by trucks or by walking if there are no roads. Helitack firefighting crews can be dropped by helicopter, or if the terrain is extremely rugged and the area is remote, smoke jumpers are called in. These specially trained firefighters parachute into the near vicinity of remote fires and begin the initial assault on a wildfire (Figure 12.8).

There are currently 113 hotshot crews in the United States with 96 of them located in the West and Southwest. These crews are available to travel to any major event anywhere in the country on very short notice.

Aircraft

Occasionally, ground crews are unsuccessful in their attempts to extinguish the fire and it climbs or ladders into the forest canopy. At this point, the firefighters coordinate their efforts with helicopter crews who drop water from the air. Should the fire be moving rapidly in the direction of major stands of fuel, large fixed-wing aircraft are called in to drop retardant just beyond the leading edge of the fire (Figures 12.9 and 12.10).

retardant

A chemical substance used to impede the spread of a fire.

Retardant consists of a dry powder that contains various compounds of ammonium phosphate and water. The phosphate compounds are the retardant; water is simply the carrier for the chemicals. When the retardant coats the fuel, it prevents oxygen from coming in contact with the fuel, thus removing oxygen from the fire triangle. The mixture is also effective as a fire prevention substance even after the water has evaporated. Drops of the slurry coat everything in their path—trees, ground vegetation, and structures. Although it creates a sticky, red-colored surface, it can later be washed off.

Source: USFS.

Figure 12.9 Fixed-wing aircraft dropping fire retardant.

LCPL Daniel Boothe USMC.

Figure 12.10 Heavy-duty helicopter making a water drop ahead of a fire line.

Mapping

The growth of large fires (those exceeding 100 acres; 0.16 sq km) is mapped using global positioning system (GPS) and geographic information system (GIS) techniques. These techniques allow precise location of fire lines and provide the positions of crews working a fire. Very large fires are monitored by flight observers who survey the perimeter of a fire once a day and have continuous recording instrumentation on board that sends the coordinates via satellite to a computer. Large drones are also used to track the position of a fire as well as that of ground crews. A map can then be produced providing an outline of the fire as it changes direction (Figure 12.11).

Evidence From the Geologic Record

Research done by Susan Rimmer of Southern Illinois University has disclosed the preservation of fossil charcoal in near-shore and terrestrial environments dating to the Late Silurian period. An increase in fire activity in the Late Devonian stretching into the Pennsylvanian occurred in Pennsylvania and West Virginia and points to fires that burned mostly surface vegetation. The sources of these fires are unknown. Major fires resulted from the impact of the Chicxulub meteor in the Yucatan Peninsula at the Cretaceous-Tertiary boundary 65 million years ago. The global inferno that resulted produced enough soot to cover the Earth's surface. This soot is preserved in rock units in the Northern and Southern Hemispheres (Wolbach et al., 1990).

Obviously, major volcanic eruptions created extensive fires in the geologic past, but most of the record of these fires has been destroyed or buried beneath the volcanic deposits. During the past 1 million years, there has been a proliferation of plant life that has experienced burning. The ash and carbon that result from those fires is very mobile and not compacted, so it

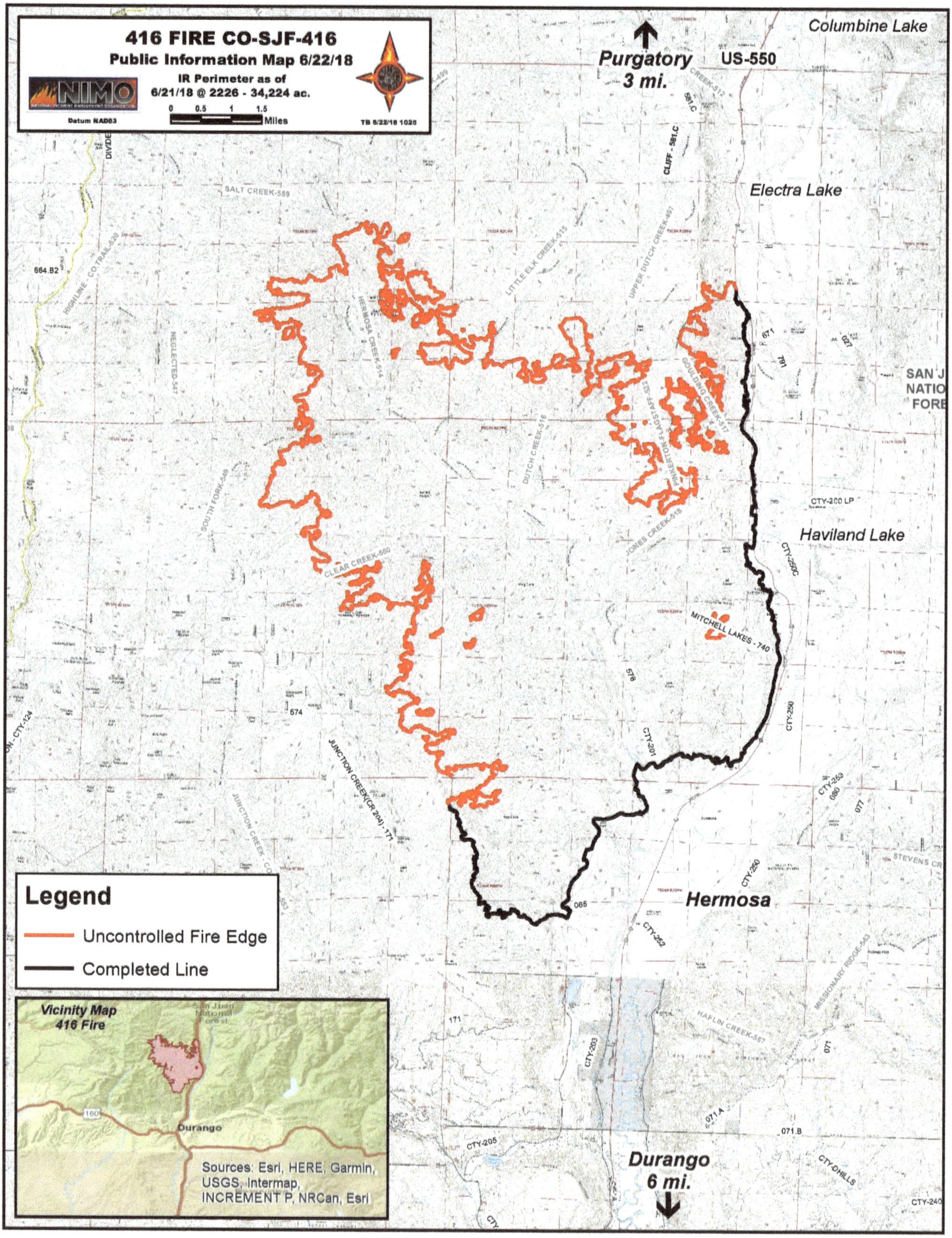

Source: USFS.

Figure 12.11 Map showing the outline of the 416 Fire north of Durango, Colorado, in June 2018.

is seldom preserved. The volcanic eruptions on the island of Hawaii during the spring and summer of 2018 burned numerous acres of the island of Kilauea and destroyed homes as the lava flowed toward the sea.

Lessons From the Historic Record and the Human Toll

Wildfires have been a part of Earth's history, especially since combustible materials began to develop on Earth more than 350 million years ago. Fires that have damaged the largest amount of land are ones that burned in forested regions. However, numerous fires have occurred in populated areas and have destroyed homes, businesses, and the lives of numerous people.

Selected Historical Fires

Peshtigo Fire

The summer of 1871 in the upper Midwest was very dry, with drought conditions spreading across Wisconsin and Michigan. Forested areas had experienced occasional fires, and residents often set fires to clear their land for planting. Conditions were ripe for major conflagrations (fires) to occur.

On October 8, 1871, a massive low-pressure system began moving in from the west (Figure 12.12). Strong winds caused fires in the vicinity of Peshtigo (PESH-tee-goh) to quickly grow together (Figure 12.13), and by late evening they were encroaching on the town of approximately 1,700 people. The fire rapidly overtook the town and surrounding area and within a few days more than 1.2 million acres had burned. Estimates of the loss of life ranged between 1,200 and 2,500 people and the final toll showed 12 towns in the region had been destroyed. This fire caused the greatest loss of human life from a wildfire in the history of the United States. Unfortunately this entire event is seldom referred to as it occurred on the same day as, and within a few hours of, the Great Chicago Fire.

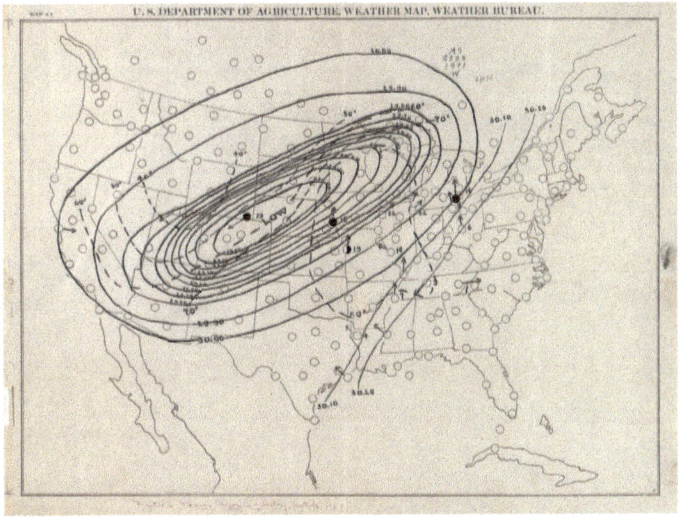

Source: Wisconsin Historical Society, (X3)53832.

Figure 12.12 Weather conditions in the Midwest on October 8, 1871. A strong weather system produced extremely windy conditions throughout Wisconsin, Illinois, and adjoining states.

Source: Wisconsin Historical Society, (X3)53832.

Figure 12.13 Shaded area shows the extent of the Peshtigo Fire. Some estimates placed the loss of life as high as 2,000 people.

Figure 12.14 A Currier and Ives drawing showing the city of Chicago in flames in October 1871.

The Great Chicago Fire

The Great Chicago Fire is well documented to have started in a barn belonging to Patrick O'Leary, although the story of a cow having knocked over a lantern is in dispute. (In 1999 the City Council of Chicago exonerated both the O'Learys and their cow, Naomi.) In 1871, Chicago was a bustling city of more than 330,000 residents and a major center of trade, banking, and farming. The same weather conditions that contributed to the Peshtigo fire caused the fire in Chicago to spread rapidly among the numerous wooden structures that were oftentimes closely spaced. Within a few hours, the fire had raged through the city and jumped the Chicago River to begin burning the downtown area (Figure 12.14). In less than 30 hours, more than 3 square miles (8 sq km) were turned to ash and almost 300 people had lost their lives.

The Great Fire of 1910

The spring and summer of 1910 were hot and dry conditions in the Pacific Northwest, creating conditions that set the stage for major wildfires. Estimates place the number of fires in Washington, Montana, Idaho, and British Columbia between 1,000 and 3,000. Highly resinous conifers served as the source of flammable vapors that spread through the atmosphere and would eventually explode when heated by the fires.

A weather system blew in off the Pacific Ocean on August 20, with winds in excess of 70 miles per hour. These winds fanned several hundred fires that were already burning throughout the region, eventually causing them to form two massive blazes. In 1910 the federal forest service was in its infancy, so its personnel had no experience or preparation to deal with the fires.

In the end, seven towns in Idaho and Montana were totally overwhelmed with fire and burned to the ground. Large sections of other cities near the fires also burned. More than 3 million acres were destroyed and 78 firefighters lost their lives, making it the deadliest wildfire for firefighters in US history.

Tillamook Burns

From 1933 to 1951, a series of large wildfires consumed more than 355,000 acres of the Coast Range in western Oregon. The first fire began in August 1933, when a logger was dragging a downed tree along the ground. The steel cable ignited a log and the resulting fire spread to burn more than 240,000 acres. Numerous burned trees lay on the ground or were still standing. These helped fuel the second fire that took place in 1939 and burned 190,000 acres. Two additional fires, one in 1945 and the second in 1951, consumed more than 210,000 acres, some of which had been previously burned. Much of this area has now recovered because of the high rainfall in the Pacific Northwest that allows trees to grow rapidly.

Yellowstone Fire

Yellowstone National Park, America's oldest national park, was ravaged by major wildfires in the summer and early fall of 1988. The park had a fire policy in place that allowed naturally occurring fires to burn on their own and not be suppressed unless property and human lives were endangered. The rationale was that nature could and would use fires to benefit the landscape. Up until 1988 there had been between 10 and 15 lightning-caused fires each year, which usually burned less than 100 acres each.

The park and surrounding area experienced a dry winter in 1987–1988. The lack of moisture produced large amounts of dry fuel, and stressed trees in the park became susceptible to insect infestation. These conditions, coupled with massive amounts of fuel that had built up from a lack of fires over the previous 80 or so years, allowed lightning to start several dozen fires within the park. In mid-July park officials realized they had to dispatch fire crews to fight these naturally produced fires—a move that was counter to their policy of letting such fires burn unchecked. The fires of Yellowstone National Park became the top priority of national firefighting agencies. More than 9,000 firefighters from across the country battled the fires at a cost of more than $120 million.

In mid-September, rains finally arrived to help quench the fires, but not until significant snowfalls in November were the fires finally put out. The final toll showed more than 1.4 million acres burned in the park and surrounding areas.

The Ravages of Fire on Communities

Fires in Oakland, California in October 1991

It started as a partially extinguished grass fire in the hills about 5 km northeast of downtown Oakland and within a similar distance from Berkeley, home of the University of California at Berkeley. The original fire was 5 acres and was considered under control in the evening of Saturday, October 19.

However, the fire was re-ignited by strong winds late Sunday morning and it quickly jumped two major highways. Frantic efforts by firefighters were for naught as several hundred homes and apartments were rapidly consumed. The fire eventually generated its own unpredictable winds that spread fire in all directions. Later in the evening of October 20, the winds stopped but the firestorm that had resulted proved too much for the army of firefighters battling the widespread blaze (Figure 12.15). The heroic efforts of many had been short-circuited by the lack of a water supply because of downed power lines that prevented pumps from delivering water to the hydrants. The fire ultimately killed 25 people and injured 150 others. Destroyed were 1,520 acres along with 2,449 single-family dwellings and 437 apartment and condominium units. The economic loss was estimated at more than $1.5 billion.

Source: NASA.

Figure 12.15 **Fire in the hills surrounding Oakland, California, October 1991.**

BOX 12.3 Western Fires in 2020

During the late spring, summer and autumn of 2020, portions of the Western United States experienced record wildfire activity. Continually dry conditions throughout the region allowed fires to become rampant, presenting major challenges from Washington to southern California and eastward to Colorado. Peak activity occurred in late July through September, when more than 100 large, uncontained fires were being fought by more than 30,000 firefighters.

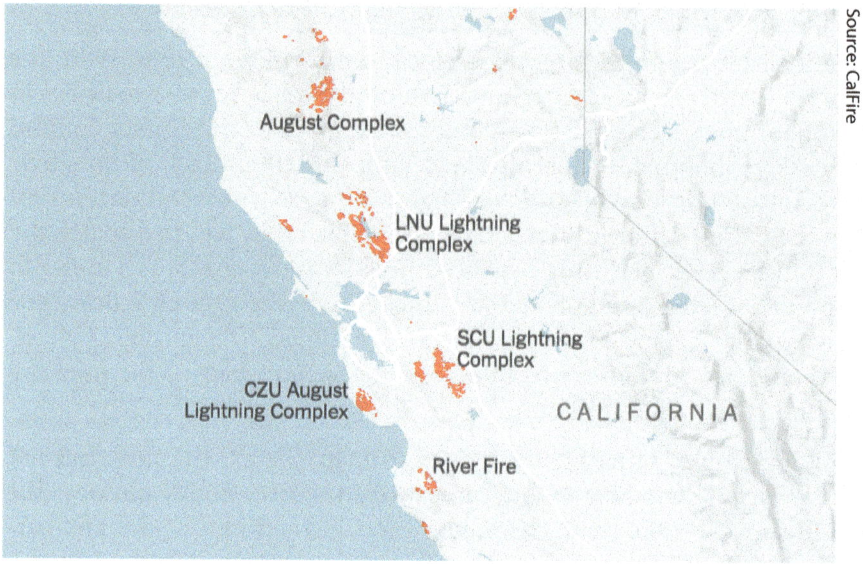

Figure 12.3.1 **Major fires in central California in the summer of 2020 destroyed almost 2 million acres (3,100 sq mi or 8,000 sq km).**

Among the worst were the August Complex, the SCU Lightning Complex, and the Creek Fires in California. The August Complex fire destroyed more than 1.3 million acres (2,030 sq mi or 5,260 sq km). At the same time, Colorado was battling its largest wildfire in history, the Cameron Peak fire. By late October the Cameron Peak fire in northern Colorado, to the northwest and west of Estes Park and Fort Collins, had consumed 208,663 acres (326 sq mi or 844 square kilometers). This represents the largest wildfire in Colorado history, as it was more than 50% larger than the Pine Gulch fire which began 2 months earlier.

Cerro Grande Fire in Los Alamos, New Mexico, in May 2000

As part of a long-range plan to remove fuels from Bandelier National Monument, personnel of the National Park Service started a prescribed burn on the evening of May 4. Changes in local winds spread the fire beyond the containment lines and it was soon declared a wildfire. Within 2 days, strong winds turned it into a major fire. By May 10, winds were carrying embers more than 2 km beyond the defense lines that had been established. Because the towns of Los Alamos and White Rock lay in the immediate path of the fire, more than 18,000 residents of these towns were evacuated. Ground fuels located around homes ignited and caused many of the spot fires to consume these houses. More than 18,000 acres had burned by May 10, destroying 235 homes and threatening the Los Alamos National Laboratory. The laboratory was spared major damage, but by the end, total losses amounted to 48,000 acres of forest and grasslands and more than $1 billion of property. Fortunately, there was no loss of life in this fire.

California Fires of 2017 and 2018

More than 200 wildfires larger than 100,000 acres have occurred in the United States, with 90% of them since 2000, including more than a dozen in California. Both northern and southern California experienced major wildfires in late 2017 into mid-2018. In October the wine country surrounding Calistoga and Santa Rosa had several large fires that devastated the region. More than 8,800 structures were burned and 22 people died. Several wineries experienced major losses of vineyards and stored wines.

In December 2017 the Thomas Fire started near Ojai, a small community to the east of Santa Barbara. Santa Ana winds and extremely dry conditions allowed the fire to become the second largest in California history, burning 281,893 acres and destroying more than 1,000 structures. Within a month of the start of the fire, heavy rains fell on the burn area, creating mudflows and debris flows that destroyed 129 homes and damaged another 300 (Figure 12.16). In the city of Montecito several dozen homes were destroyed by the flows which killed 17 people. The fire was officially declared out in June 2018, having cost more than $205 million to fight.

Source: Mike Eliason, San Bernardino County Fire Department.

Figure 12.16 Homes in Montecito, California, were submerged in debris flows that resulted from the Thomas Fire in December 2017.

The Camp fire began along Camp Creek Road in northern California in early November 2018. The fire spread to the west and destroyed numerous homes in the town of Paradise. The fire began as the resulted of a faulty power line; Pacific Gas and Electric assumed responsibility for the fire and paid out more than $13 billion in claims. A total of 85 people died.

The 2021 Fire Season

The Western United States has experienced severe drought conditions for the past decade. One result is that the forests and grasslands have become much drier than normal, thus making those areas very susceptible to wildfires.

In the late spring fires burned large areas in Arizona. When above average monsoon rains fell on those regions in July, major flooding resulted. The towns of Globe and Miami were inundated with floodwaters that carried debris washed from the forests.

In June and July numerous wildfires burned more than two million acres in Washington, Oregon, and California. On July 13, 2021, the Dixie Fire began burning in northern California, which spread over two national forests, and Lassen Volcanic National Park (a five-county area). Drought, high winds, and hot weather contributed to its very active behavior. At the peak of activity, more than 5,000 personnel were assigned to the fire. Its smoke column reached an altitude of more than 30,000 feet (9.2 km). The Dixie Fire was fully contained in October, after covering more than 963,000 acres (1,505 sq mi). It ranks as the second largest fire in California history.

Another major fire in California, the Caldor Fire, about 100 miles (160 km) southeast of the Dixie fire has threatened the city of South Lake Tahoe. More than 10,000 people were placed on evacuation standby notice. More than 4,000 personnel were needed to combat the fire which was fully contained in mid-October.

Major Fires Caused by Arson

Rodeo-Chediski Fire, East-Central Arizona in 2002

More than 469,000 acres were burned in east-central Arizona when two separate fires converged in late June and early July 2002 (Figure 12.17). One fire, the Chediski, was caused when a lost hiker lighted a signal fire in an area that was extremely dry. The downdraft of the rescue helicopter rapidly expanded the fire. The Rodeo fire was purposefully set by a seasonal firefighter seeking work. When the fires grew together they became the largest fire in the history of Arizona. More than 400 homes were destroyed. At its peak, more than 4,000 personnel, including four Type-1 **Incident Management Teams (IMT)**, which were coordinated by a national area command team, fought the inferno.

Incident Management Team (IMT)

A rapid response team that oversees the operations of large-scale firefighting.

Hayman Fire in 2002

The Hayman fire, the largest in the history of Colorado, was started by a U.S. Forest Service employee who lit a fire in a fire ring in a nonburn zone. The fire eventually destroyed more than 130,000 acres and included a 1-day run that burned more than 60,000 acres. Six hundred structures were destroyed, including 133 homes.

Source: NASA.

Figure 12.17 Rodeo-Chediski fire consumed more than 730 square miles.

Esperanza Fire in California in October 2006

An arsonist set a fire about 25 km west of Palm Springs, California, early on the morning of October 26, 2006. Santa Ana winds quickly spread the fire to more than 24,000 acres in only 18 hours (Figure 12.18). Eventually it covered approximately 40,200 acres before being contained. Five firefighters died when flames were blown over their position along a fire line. More than four dozen homes and outbuildings were destroyed.

Source: NASA.

Figure 12.18 **Notice the direction of the smoke from the Esperanza Fire as Santa Ana winds were blowing to the southwest.**

Southern California Fires of October and November 2007

Tinderbox fire conditions occurred in October 2007 fueling wildfires that swept across more than 500,000 acres of southern California and destroyed more than 2,100 homes. As many as 15 major fires ranged from north of Santa Barbara to the United States–Mexico border, several of which were investigated as possible arson-related. In San Diego County more than 350,000 homes were ordered evacuated, with more than 10,000 people being housed in Qualcomm Stadium, the home of the San Diego Chargers football team.

One fire that covered more than 36,000 acres was started by a 10-year-old boy playing with matches. That particular fire destroyed 21 homes and injured at least three people. In November 2007 fires ravaged homes situated in canyons surrounding Malibu. Santa Ana winds drove the flames that destroyed more than 50 homes and burned almost 5,000 acres.

BOX 12.4 The Deadly Hazards of Fighting Wildfires

SOUTH CANYON FIRE, GLENWOOD SPRINGS, COLORADO, 1994

On July 2, 1994, lightning started a fire several miles west of Glenwood Springs, Colorado, in an area that was extremely dry because of a prolonged drought. High temperatures and low humidity contributed to the tinderbox conditions. Because of other fires in the vicinity, resources were not sent to this fire until it had burned for 2 days. By July 6 several small crews had been assigned to the fire, including two groups of smoke jumpers who had been dropped into the fire zone. During the midafternoon of July 6, winds up to 45 miles per hour resulted in sixty- to 90-meter flames causing the fire to jump a drainage area, where it rapidly overtook and killed 12 members of a hotshot crew along with two members of a helitack crew. Fortunately, 35 firefighters survived the onslaught. The combination of extremely steep terrain and catastrophic weather conditions with an explosive fuel source, along with poorly established escape routes, led to the disaster. The entrapment of these firefighters in this 2,115-acre fire marked the greatest loss of life in a wildfire in more than 40 years. Two primary causes of the personnel disasters in the Mann Gulch fire and the South Canyon tragedies were that the firefighters were working downhill toward the fire and were inattentive to the weather and fire behavior.

Figure 12.4.1 Terrain and chaparral vegetation typical of that found at the South Canyon and Yarnell Hill Fires.

YARNELL HILL FIRE, YARNELL, ARIZONA, JUNE 2013

June 2013 was a typically dry period in central Arizona, but monsoon activity had begun to build toward the latter part of the month. The region around Yarnell, located 100 kilometers northwest of Phoenix, is situated in the Weaver Mountains, which consist of rugged granitic outcrops and interspersed chaparral (Figure 12.4.1). The chaparral, ranging from 1 to 3 meters in height was extremely dry due to drought, which had lasted more than 15 years.

Thunderstorm activity in the late afternoon of June 28 ignited a fire that drew a small crew the next day to contain it. Within 24 hours it had expanded to about 100 acres, and by the early morning of June 30, it covered almost 500 acres. Its rapid growth prompted the assignment of a Type 1 Incident Management Team to oversee operations. Several crews were dispatched, including the Granite Mountain Hotshots, based in nearby Prescott, Arizona. They were on the fire less than 24 hours when they were overrun by extreme fire conditions. The fire was moving at an estimated 10 to 12 miles per hour, being driven by winds of 35 to 40 miles per hour. Their planned escape route to a designated safety zone was cut off. They deployed their fire shields but 19 firefighters died. Only their lookout survived.

The South Canyon and Yarnell Hill fires bring home the fact that more than 950 personnel have died fighting wildfires in the United States. There have been instances when more than 10 firefighters died in single burnovers due to rapidly changing conditions, inadequate communication, and a lack of awareness of the terrain and the fuel conditions in which they were operating. With each unfortunate episode the governmental agencies conduct thorough reviews and strive to improve field operations.

Minimizing Wildfire Hazards and Their Aftermath

Land Clearing

Areas of urban expansion often have an interface with forested land. Many municipalities have put plans in place that require the removal of vegetation around homes and other structures. This causes its own problem, however, as it increases soil erosion, and runoff can be up to several hundred times greater than the normal amount after surface fuels are cleared. Unfortunately, most homeowners living in residential neighborhoods and forested areas like having trees and vegetation near these buildings. The establishment of defense zones surrounding the homes and other structures lessens the chance that wildfires will destroy buildings.

Prescribed Burns

The continual buildup of surface fuels such as leaves, needles, branches, and low undergrowth and grasses provides the fuel for wildfires. Fire management agencies use prescribed burns to remove these surface fuels so as to lessen the likelihood of future uncontrolled fires. A **prescribed burn** is an intentionally-set fire that takes place under a set of defined parameters and predictable weather conditions. Prior to 2000 the term **controlled burn** was used to describe pre-set fires. Such fires are monitored and actively managed by fire personnel to prevent their getting out of control. Established control lines and some suppression techniques aid in keeping the fire within desired boundaries. In addition to burning material as it lies on the ground, fire managers will also often ignite slash piles consisting of small, rotten, or otherwise undesirable wood discarded during logging (Figure 12.19). There have been times, however, when slash has provided the fuel for devastating fires such as the fires in Michigan in the 19th century and the 1910 fires in northern Idaho.

Source: National Wildfire Coordinating Group.

Figure 12.19 Slash piles are burned to reduce fuel in forests.

prescribed burn
A controlled fire purposefully set to remove fuel from an area.

controlled burn
A fire set to remove fuel. Term is no longer used, as not all such fires were kept under control.

Prescribed burns reduce the fuel side of the fire triangle by removing ground litter and undergrowth. The resulting ash returns nutrients to the soil, helping to sustain the growth of new plants.

There are times, though, when a prescribed burn can turn disastrous, as in the case of the Cerro Grande fire in the summer of 2000. This fire, begun as a controlled burn, quickly got out of hand and inflicted major damage on the city of Los Alamos, New Mexico. Following this catastrophe, the term controlled burn is no longer used by fire management agencies.

Firebreaks and Burnouts

The spread of wildfires is slowed whenever firebreaks are established. **Firebreaks** are areas that have been cleared of surface and low-lying vegetation and ground litter. Clearing small areas involves actual removal of the fuel sources. However, a fire can generate airborne embers that can spread the fire past a narrow firebreak. Larger areas are cleared by setting prescribed burns on dozens or often hundreds of acres.

There are times when fire crews will conduct a **burnout**, a tactic that removes the nearby fuel before a fire gets to it. When a wildfire is threatening a community or if valuables are at risk, firefighters will sometimes deliberately set a backburn, a rather risky decision that is often a last resort. **Backburns** are carried out along a natural break, such as a roadway or near threatened buildings. The burn is drawn into the larger fire because of the convective heat rising from the larger fire.

firebreak
A clearing that provides a gap across which fire should not burn.

burnout
An intentional fire that is lit to burn fuel that lies in the path of a spreading fire.

backburn
A fire that burns back toward the main part of a wildfire.

Source: USFS.

Figure 12.20 Aspen quickly return within 2 years of a fire that burned the predominately ponderosa pine forest.

Source: New Mexico Environment Department.

Figure 12.21 Removal of surface vegetation by wildfires enhances surface erosion.

Results of Wildfires

Rejuvenation

Fires do have a useful purpose. They can help regenerate plants and other flora by opening seeds and by adding nutrients to the soil. This is evident in areas that have sustained severe burns when plants and trees rejuvenate themselves, often within a few months after the fire if moisture is present. In cold climates where there is little chemical breakdown, fire can speed up the process of enriching the soil with nutrients. Nitrogen is released from the burned plants and helps accelerate future plant growth. Large wildfires in Alaska are very beneficial to that state's high-latitude ecosystems because soils are slow to form naturally in cold conditions.

One excellent example of natural rejuvenation is seen in the regeneration of aspen trees following a wildfire. The root systems of aspen are relatively shallow and put out numerous sprouts that penetrate the surface in a fire area. This often occurs very soon after a fire (Figure 12.20).

Erosion

In the time following a large wildfire, the potential for major surface erosion is a primary concern. Steep slopes that were once covered by lush vegetation that could hold water and impede its rapid flow across the surface no longer slow the flow. Rapid runoff removes topsoil and cuts deep channels and rivulets into the surface (Figure 12.21). In regions near highways, road cuts, and other steep slopes, mass wasting is accelerated. California experiences significant slides every year in burned areas. This was also the case with the Thomas Fire that adversely affected the city of Montecito and surrounding communities.

Extreme temperatures exceeding 1500 °C transform surface soils into welded particles that produce an impermeable zone which is termed a hydrophobic layer. The **hydrophobic layer** causes more sheet erosion and flooding. Seasonal rains and snowfall will produce debris flows and landslides, along with flooding that can be catastrophic to low-lying areas. Replanting hillsides with grasses and seedlings helps alleviate the problem, but the landscape will not be the same for many years, especially in arid areas where plant life is slow to grow.

hydrophobic layer
A layer of material that cannot absorb water; usually forms in dry climates or in regions that have experienced intense fires.

Fiscal Costs

There are two main monetary costs related to wildfires: the direct cost of fighting the fires and the secondary cost of their economic impact. Over the past 30 years the cost of fighting fires has risen 500%, reaching $2.3 billion in 2020. This large increase is due to fires being much larger than in the past and often occurring in developed area requiring more resources to put them out.

In forested areas where timber is harvested for sale, the revenue source is lost. The profits from the sale is used to help cleanup and rehabilitated burned areas and to plant and reseed new areas. In previous years, when wildfires were less damaging, the revenue from these timber sales could be used by the affected communities to rehabilitate and strengthen their forests. These steps aid in the recovery process to minimize erosion problem.

Key Terms

backburn *(page 357)*
brush fire *(page 336)*
burnout *(page 357)*
cellulose *(page 337)*
conduction *(page 337)*
controlled burn *(page 357)*
convection *(page 337)*
crown fire *(page 340)*
firebreak *(page 357)*
firestorm *(page 342)*
fire triangle *(page 336)*
forest fire *(page 336)*
fuel *(page 336)*
glucose *(page 336)*
ground fire *(page 339)*
heat *(page 336)*
hydrophobic *(page 339)*
hydrophobic layer *(page 359)*
Incident Command System (ICS) *(page 343)*
Incident Management Team (IMT) *(page 354)*
oxidation *(page 337)*
oxygen *(page 336)*
prescribed burn *(page 357)*
public information officer (PIO) *(page 345)*
pyrolysis *(page 336)*
radiation *(page 337)*
retardant *(page 346)*
Santa Ana winds *(page 340)*
surface fire *(page 339)*
temperature of combustion *(page 336)*
topography *(page 340)*
wildfire *(page 336)*

Summary

Fires have existed for millions of years and will continue to be a natural hazard. Wildfires are unplanned, uncontrolled events that are caused by rapid oxidation of organic matter. Fuel, oxygen, and heat are needed to produce combustion. Removal of any one of these components of the fire triangle causes the fire to be extinguished.

Wildfires are usually caused by humans, either intentionally or accidently or by lightning. These fires can be ground, surface, or crown fires depending on where they occur in the fuel source. Factors that control fire behavior include surface topography, the types of fuels, and weather conditions, particularly wind speed.

Once a fire occurs, fire management personnel decide on how to fight the fire. If it is located in an area that poses no threat to structures or people, the fire is usually allowed to burn itself out, although its progress is monitored. If fires pose a threat, then they are attacked by ground crews and aircraft. There have been many historical fires, including the Great Chicago Fire and those occurring in Yellowstone National Park.

Recently there has been an increase in the number of fires that have affected communities, because urbanization has placed more structures and people closer to fuel sources. A shift in climatic conditions has brought drought to more sections of the United States. Although the number of fires has declined over the past several decades, the average size of fires has increased significantly. As a result, we need to be more aware of the effect fires have on developed areas and the impact major fires have on the landscape and environment. Preparation and planning will minimize the impact of fires in the future.

References and Suggested Readings

Arno, S. F., & Allison-Bunnell, S. (2002). *Flames in our forest: Disaster or renewal.* Island Press.

Egan, T. (2009). *The big burn—Teddy Roosevelt and the fire that saved America.* Houghton Mifflin Harcourt.

Johnson, E. A., & Miyanishi, K. (Eds.). (2001). *Forest fires—Behavior and ecological effects.* Academic Press.

Leschak, P. M. (2003). *Ghosts of the fireground: Echoes of the great Peshtigo Fire and the calling of a wildland firefighter.* Harper Collins.

Maclean, J. N. (1999). *Fire on the mountain—The true story of the South Canyon Fire.* Simon and Schuster.

Maclean, J. N. (2003). *Fire and ashes—On the front lines of American wildfires.* Henry Holt.

Maclean, N. (1992). *Young men and fire.* University of Chicago Press.

Pyne, S. J. (1997). *Fire in America—A cultural history of wildland and rural fire.* University of Washington Press.

Pyne, S. J. (2004). *Tending fire—Coping with America's wildland fires.* Island Press.

Pyne, S. (2016). *California—A fire survey.* University of Arizona Press.

Pyne, S. (2016). *The Southwest—A fire survey.* University of Arizona Press.

Pyne, S., Andrews, P. L., & Laven, R. D. (1996). *Introduction to wildland fire* (2nd ed.). John Wiley.

Wolbach, W. S., Gilmour, I., & Anders, E. (1990). Major wildfires at the Cretacrous/Tertiary boundary. In V. L. Sharpton & P. D. Ward (Eds.), *Global catastrophes in earth history—An interdisciplinary conference on impacts, volcanism, and mass mortality* (pp. 391–400). Geological Society of America.

Wolf, T. J. (2003). *In fire's way—A practical guide to life in the wildfire danger zone.* University of New Mexico Press.

Websites for Further Reference

https://gacc.nifc.gov/links/links.htm
https://landsat.gsfc.nasa.gov/images/image_index.html
https://www.fire.ca.gov
https://www.nifc.gov
https://www.usfa.dhs.gov

Questions for Thought

1. Explain the processes involved in the formation of a fire.
2. What are the three ways in which heat can be transferred?
3. Explain the fire triangle.
4. Why do fires spread more rapidly up a canyon than across a valley floor?
5. What defensive measures would you take to protect a summer resort home located in a forest against wildfires?
6. How do wildfires affect the soil layer and future runoff from rainfall?
7. How safe is your residence in terms of wildfire potential? What could be done to improve the situation?
8. What role do the Santa Ana winds of southern California play in the fires of that area?
9. Following a wildfire, what steps should be taken to begin restoring the ecology of the landscape?
10. How have the characteristics of wildfires changed over the past several decades?
11. What effect did the early policy of putting out all fires have on the wildlands and future fires?

Epidemics, Pandemics, and Infectious Diseases 13

© JOHN WESSELS/Contributor/Getty Images.

Transporting an Ebola victim to the hospital—the Ebola outbreak that hit West Africa in 2014 had an official death toll of 11,310 people, but many deaths went unrecorded and the total could have been two to three times greater.

Still Dangerously Unprepared for Future Deadly Pandemics

Researchers estimate that birds and mammals harbor anywhere from 631,000 to 827,000 unknown viruses that could potentially leap into humans—these are the sources for future epidemics and pandemics. Ebola jumped from bats to humans in Guinea, West Africa, when a boy was infected in December of 2013. Genetic studies found that the virus was spreading quickly and mutated to form a strain that replicated faster, spread faster, and was significantly better at killing its human host. Treating and containing fast-spreading deadly diseases is challenging even with the best medical facilities. In response to just 10 cases of Ebola in 2014, the United States spent $1.1 billion on domestic preparations, including $119 million on screening and quarantine. The World Bank warns that we as a planet are "dangerously unprepared" for future deadly pandemics. According to Jim Yong Kim, president of the World Bank speaking in 2015, the international response to the Ebola crisis was belated and disorganized. We need countries, corporations, and donors working together to prepare for rapid responses to outbreaks and emergencies, with stronger health systems, improved monitoring and supply transportation, and fast-acting medical teams.

hemorrhagic fever

A viral infection that causes victims to develop high fevers and bleeding throughout the body that can lead to shock, organ failure, and death.

biological hazard (biohazard)

A biological substance or condition that poses a threat to the health of living organisms, primarily that of humans.

protozoa

Microscopic one-celled free-living or parasitic organisms that are able to multiply in humans.

parasite

An organism that lives on or in an organism of another species, the host, and benefits by deriving nutrients at the host's expense.

fungus

A single-celled or multicellular organism that gets its food from decaying matter, such as yeast, mold, and mushrooms.

Natural disasters are catastrophic events that have serious health, social, and economic impacts. Deaths related to disasters, especially rapid-onset disasters like earthquakes, tsunamis, or floods are usually due to blunt force trauma, crush-related injuries, or drowning. The number of dead bodies in disaster affected areas increases concerns about disease outbreaks, although human remains do not pose a significant risk for epidemics unless the deaths are from cholera or **hemorrhagic fevers** such as the Ebola virus (Figure 13.1). The main risks for outbreaks after disasters are due to population displacement and overcrowding, and the availability of safe drinking water, sanitation facilities, and health care services. Disasters can greatly increase the risk of epidemics, especially in crowded situations such as refugee camps or temporary living areas for people displaced by natural disasters (Figure 13.2a). Often poor sanitary conditions and the loss of a reliable water supply lead to explosive outbreaks of diarrheal disease, such as a large cholera epidemic that sickened 16,000 people and killed 276 in West Bengal in 1998 after flooding contaminated the water supply. Hepatitis A and E are also associated with lack of access to safe water and sanitation. Overcrowding facilitates the transmission of communicable diseases and is common in displaced populations. After the eruption of Mt. Pinatubo in the Philippines in 1991 more than 18,000 people came down with measles. In addition, standing water from heavy rainfall or river flooding, cyclones, or hurricanes can increase breeding sites for mosquitoes that transmit disease. Malaria outbreaks after flooding are common. Biological hazards can quickly develop and have far reaching, devastating effects at all levels of our population.

A **biological hazard** (also called a **biohazard**) is any organism or substance that has the potential to threaten the health of animals or the environment (Figure 13.2b). These can be in the form of a living and/or replicating pathogen such as a bacterium, virus, **protozoan**, **parasite** or **fungus**. A different type of hazard is material produced by chemical waste or nuclear accidents; this can also become a primary cause of biological disasters (see Chapter 14).

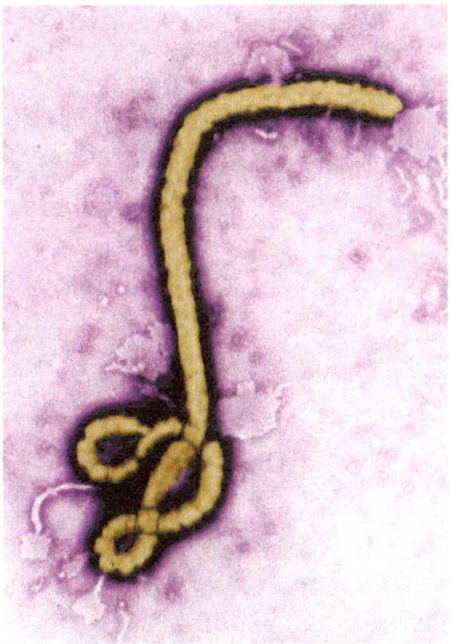

Image created by CDC microbiologist Frederick A. Murphy.

Figure 13.1 This colorized transmission electron micrograph shows the shape of an Ebola virus virion.

© Sk Hasan Ali/Shutterstock.com.

Figure 13.2a Overcrowding, poor sanitation, unsafe drinking water, and lack of medical care all contribute to the spread of disease in displaced populations fleeing natural disasters.

© Kaspi/Shutterstock.com.

Figure 13.2b The biohazard symbol, developed by Dow Chemical Company in 1966, contains four circles representing the chain of infection: agent, host, source, and transmission.

The Centers for Disease Control and Prevention (CDC) was founded in 1992 and is one of the major operating components of the Department of Health and Human Services in the United States. It tracks and responds to global health threats, and has monitored more than 300 outbreaks of different diseases in 160 countries. Infectious illnesses are thought to cause 15% of all deaths worldwide. **Bacteria** are microscopic creatures that usually consist of one cell, contain no chlorophyll, and reproduce by simple cell division. **Viruses** are ultramicroscopic organisms that contain either RNA or DNA, surrounded by a protein case. Viruses can only multiply within living cells.

bacteria

A tiny, single-celled organism that reproduces by cell division, having a shape similar to a rod, spiral, or sphere, and has no chlorophyll.

virus

Acellular, non-living infectious particles of either DNA or RNA associated with various protein coats; these only multiply in living cells.

Nature of Diseases

A **pathogen** is any agent that is capable of causing a disease. Most pathogens are microorganisms. **Microbes** infest a host, which serves as the source of energy and nutrition necessary for them to survive. Microbes have the ability to sustain their existence by not killing the host immediately, which would terminate the particular disease—and the microbes. Many of Earth's 6.7 billion people live in areas that are constantly threatened by outbreaks of highly communicable diseases, such as malaria and tuberculosis (Figure 13.3a).

pathogen

An agent, such as a bacterium or virus, that causes a disease.

microbe

A microscopic living organism, including bacteria, protozoa, fungi, algae, amoebas, and slime molds. Viruses are also called microbes. Some microbes are pathogens, capable of causing disease.

The malaria parasite that infects humans has existed for up to 100,000 years, but increased with the development of settlement and agriculture 10,000 years ago. Both ancient Greece and the Roman Empire suffered from malaria, which means bad air in ancient Italian, and was associated with swamps and standing water (Figure 13.3b). Malaria parasite species infect birds, reptiles, and other mammals, including other primates such as chimpanzees. Bubonic plague has been documented since before the first millennium and has persisted for almost 1,500 years. Several diseases, including tuberculosis (TB), have become resistant to early treatment techniques and are now more difficult to combat.

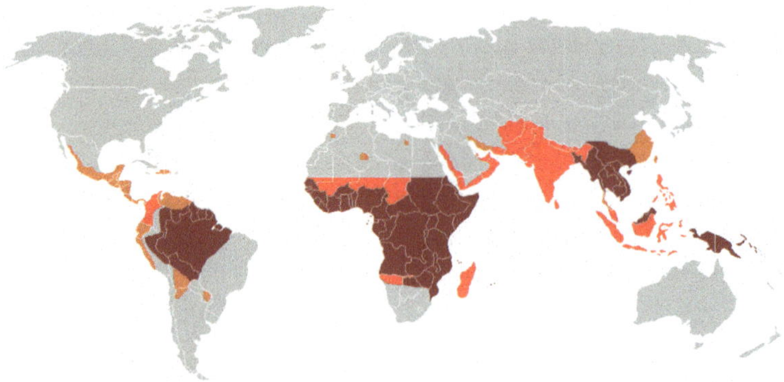

Source: Percherie CHU de Roun.

Figure 13.3a In 2015 malaria infected 350 to 550 million people and killed 438,000, according to the World Health Organization. Death tolls have dropped by 60% since 2000, when almost a million people died, due to the use of insecticide-treated mosquito nets. Malaria is still endemic around the equator in tropical and subtropical regions that have stagnant water that mosquito larvae need to mature. Areas in dark red have malaria resistant to multiple treatments, red areas host chloroquine-resistant malaria, orange areas do not have resistance to treatments, and gray areas do not yet host malaria.

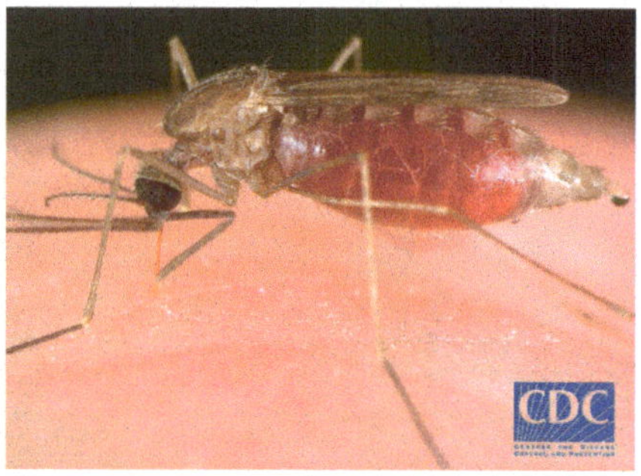

Source: CDC; Global Health, Division of Parasitic Diseases and Malaria.

Figure 13.3b The malaria parasite is transmitted to humans by the host female *Anopheles* mosquitoes which infect people while they are feeding on them for blood.

Epidemics and Pandemics

epidemic
The sudden occurrence of a highly-contagious disease or other event that is clearly in excess of normally expected numbers.

pandemic
A term applied to a highly-contagious disease that spreads throughout the world.

Epidemics are the rapid spread of infectious disease to a large number of people in a given population within a short period of time. In the past, these diseases were usually contained within a region, but with today's mobile society, they can spread rapidly across countries and continents to produce a **pandemic**, which occurs over a wide geographic area such as multiple continents or worldwide, and affects an exceptionally high proportion of the population. An example of an epidemic would be the spread of measles within a local community or a university student body. A pandemic would be much more widespread and create havoc for millions of people across continents. Examples of pandemics include the bubonic plague that overtook Europe in the mid-1300s or the Spanish influenza outbreak of 1918. In the case of the Spanish flu, which started in 1918 and lasted until 1920, up to 40% of the world's population at that time was infected, or 500 million people. The mortality rate was 10 to 20%, with 25 million people dying in the first 25 weeks, and up to 50 million people dying in all. This flu pandemic was particularly bad because it killed healthy young adults instead of those with already weakened immune systems. Diseases such as cancer or congestive heart failure, although killers of many people, are not contagious and hence do not produce epidemics or pandemics.

Historical Outbreaks

Several notable pandemics have affected large civilizations or the entire world. Among the earliest recorded was the Plague of Athens in 430 B.C. through 426 B.C. (Figure 13.4):

> In the winter following the first year of the war, morale had fallen considerably in Athens. It was at the year's public funeral (held annually for men who had fallen in battle in the course of the year)

that Pericles pronounced the famous funeral oration that is so often quoted as summing up the greatness of Periclean Athens (Thuc.2.34–46). Pericles' speech was an encomium on Athenian democracy and it provided the high point of Thucydides' account of the war. It is immediately and dramatically followed in his account by the description of the plague which struck the city in the following summer, as the Spartans again invaded Attica. Crowded together in the city as the result of Pericles' strategy, the Athenians fell victim to the virulent sickness that was spreading throughout the eastern Mediterranean. People died in large numbers, and no preventive measures or remedies were of any avail. It has been estimated that a quarter, and perhaps even a third, of the population was lost. The plague returned twice more, in 429 and 427/6, and Pericles himself died during this time, probably as a result of the disease.

Source: From http://www.indiana.edu/~ancmed/plague.htm.
Used with permission.

In 2006, Manolis Papagrigorakis and others reported in the *International Journal of Infectious Diseases* that analysis of dental pulp extracted from the teeth in bodies in a mass grave in Athens, Greece, contained evidence that the people most likely died from typhoid fever.

The Antonine Plague, which lasted from A.D. 165 to A.D. 180, was most likely smallpox that had been brought back from the eastern Mediterranean. It is estimated that about 25% of those infected died, almost 5 million people. In a second outbreak from A.D. 251 to A.D. 260, as many as 5,000 people died each day in Rome and the surrounding area.

Source: Wellcome Collection.

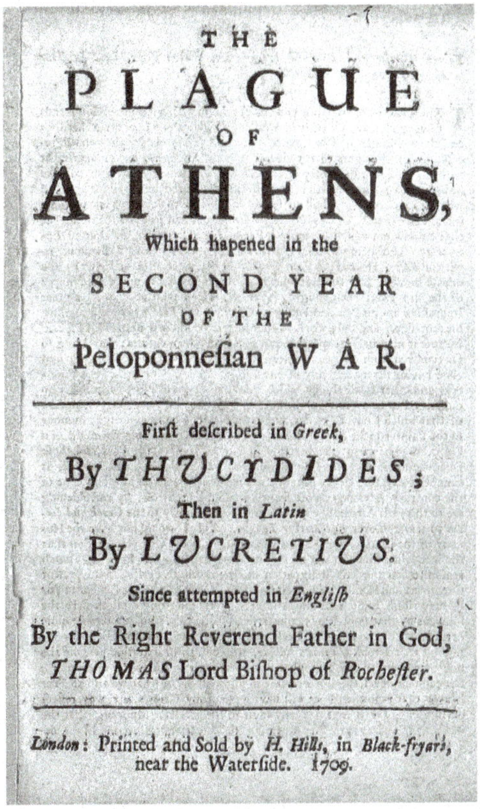
THE
PLAGUE
OF
ATHENS,
Which hapened in the
SECOND YEAR
OF THE
Peloponneſian WAR.

Firſt deſcribed in *Greek*,
By *THUCYDIDES*;
Then in *Latin*
By *LUCRETIUS*.
Since attempted in *Engliſh*
By the Right Reverend Father in God,
THOMAS Lord Biſhop of *Rocheſter*.

London: Printed and Sold by *H. Hills*, in *Black-fryars*, near the Waterſide. 1709.

Figure 13.4 **The Plague of Athens was one of the first epidemics to be recognized in ancient history. It impacted much of the Eastern Mediterranean.**

The bubonic plague first made its appearance when a pandemic outbreak occurred in A.D. 541. The Plague of Justinian began in Egypt and surfaced in Turkey the following year. At its worst in Constantinople, it killed 10,000 people each day and wiped out close to 40% of the population of the city. At the end of the outbreak, 25% of the entire population of the eastern Mediterranean region had died.

Other well known pandemics include the Black Death outbreak of the mid-1300s, a series of influenza pandemics that extended from the late 1800s until the most recent worldwide outbreak in 1968–1969, and typhus, which had its first outbreaks in the late 1400s and recurred intermittently until the Second World War in the 1940s.

Types of Diseases

Bubonic Plague

Bubonic plague primarily affects rodents, but it is transmitted to humans by flea piercings. We incorrectly refer to these as bites, but fleas and mosquitoes do not have teeth (Figure 13.5). The disease is caused by a bacterium called *Yersinia pestis,* is highly communicable, and spreads very rapidly.

bubonic plague

A rare but fast-spreading bacterial infection caused by *Yersinia pestis;* transmitted in humans and rodents by infected flea piercings; symptoms include headaches and painful swelling of lymph nodes.

© Cosmin Manci/Shutterstock.com.

Figure 13.5 The Oriental rat flea is the host for bubonic plague, the plague forms a biofilm in the flea's gut and when the parasitic flea in turn feeds on a host the plague bacteria infects the wound.

Source: Paul Fürst, Der Doctor Schnabel von Rom, 1658.

Figure 13.6 The plague doctor mask had a beak-like front that was filled with herbs and spices because people believed at that time that breathing in contaminated air was the way to transmit disease.

Black Plague (Black Death)
An epidemic of bubonic plague that killed more than 20 million people in Europe during the mid-14th century.

Symptoms include fever, chills, and painful swelling of the lymph glands that are called buboes and are where the plague gets its name. Red spots develop on the skin and these eventually turn black from subdermal dried blood caused by internal bleeding, giving rise to the name **Black Plague**, or the **Black Death**. The Black Death swept through Asia, Europe, and Africa in the 14th century, killing an estimated 50 million people and up to 60% of the European population. Plague doctors were hired by towns to treat the victims; they often lacked medical training and rarely cured their patients. Starting in 1619 plague doctors began to wear beak-like masks filled with aromatic herbs, straw, and spices in the hopes of protecting themselves from the contaminated air which they believed caused the plague (Figure 13.6).

Following the pandemics that were recorded before A.D. 550, including the first recorded cases of bubonic plague, there was a lull in worldwide diseases until 1347, at which time the Black Plague resurfaced in China and found its way to Italy. Sailors on several Italian trade ships were dying from plague as the ships pulled into Sicily. Within a year, the plague had made its way through western Europe and into England (Figure 13.7). Here, it received the name "Black Death" (Box 13.1). Outbreaks slowed in the winter because

BOX 13.1 Ring Around the Rosies Rhyme

Although typically considered to be just a myth, some feel that the song "Ring Around the Rosies" aptly describes the "black plague"; that is, "ring around the rosies" could refer to the red rash rings that those infected would get on their skin; "pocket full of posies" could refer to the fact that people would carry posies (flowers) in their pockets in the belief that this would keep the plague at bay; "ashes, ashes" might have originally been "Achoo, Achoo" to designate sneezing or coughing, a final fatal symptom, and "we all fall down" could refer to the many deaths that the disease caused.

Source: http://www.niehs.nih.gov/kids/lyrics/rosie.htm.

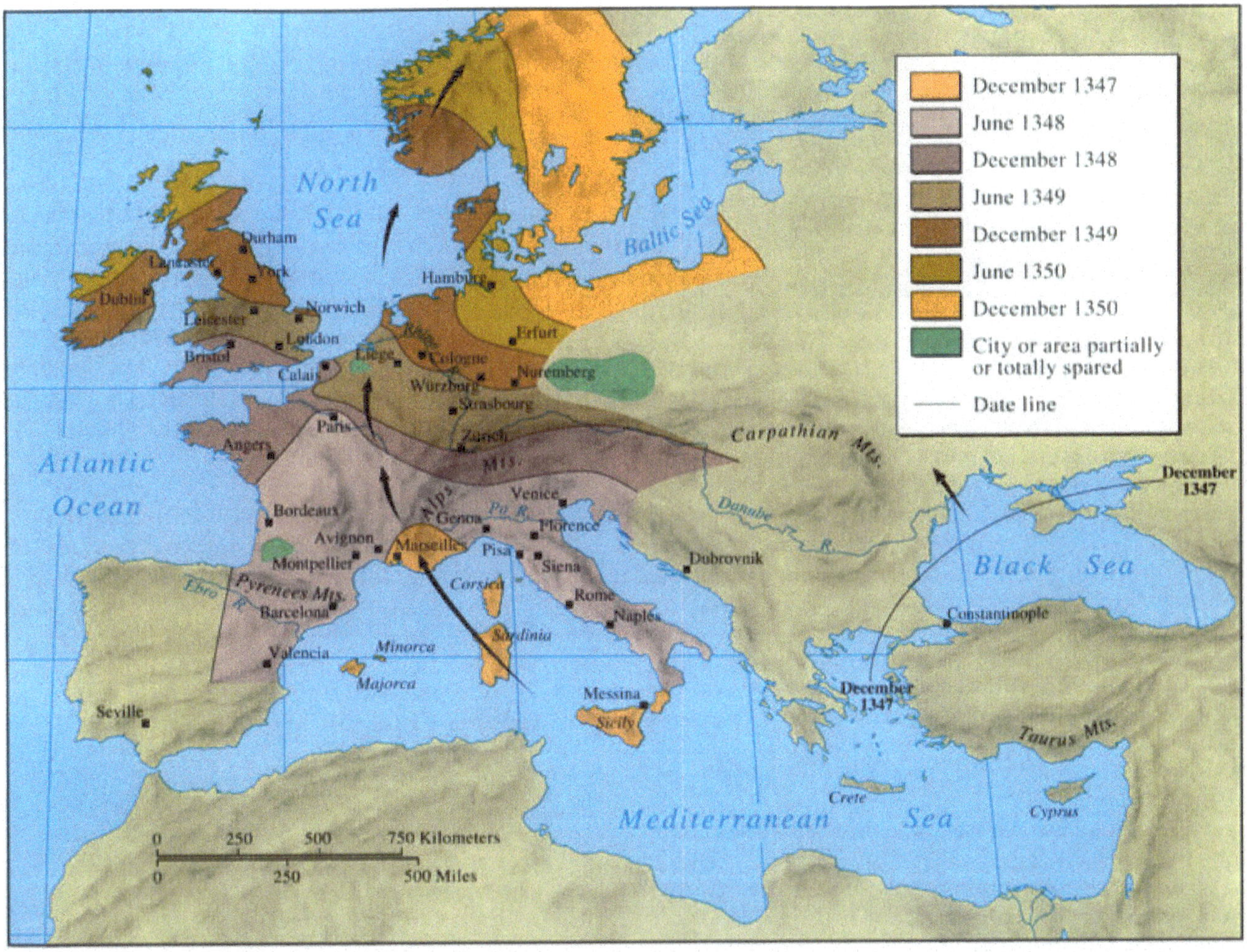

Source: http://www.odu.edu/~mcarhart/hist102/images.htm.

Figure 13.7 Spread of the Black Plague across Europe.

fleas were dormant, but they had a resurgence each spring, when deaths increased dramatically. In *The Black Death* by Philip Ziegler, he chronicles an ongoing debate about the population of Europe and the number of people who died. It is fairly well established that by 1353, more than 20 million Europeans, almost one third of Europe's population, had perished. This loss of life represented the estimated population growth that Europe had experienced in the previous 250 years.

Cholera

Cholera results from an intestinal infection by the bacteria *Vibrio cholerae* that could be present in drinking water or contaminated food (Figure 13.8). It is spread when people come in contact with feces from others who are infected, a problem in areas without proper sanitation facilities or safe water supplies. Extensive breakdown of infrastructure when cholera is present can produce epidemic conditions. Floods and earthquakes easily destroy a community's ability to handle its sanitation needs. There are documented cases of people contracting cholera from ingesting raw shellfish from coastal

cholera
An acute intestinal disease caused by the bacteria *Vibrio cholera;* often found in contaminated drinking water and food.

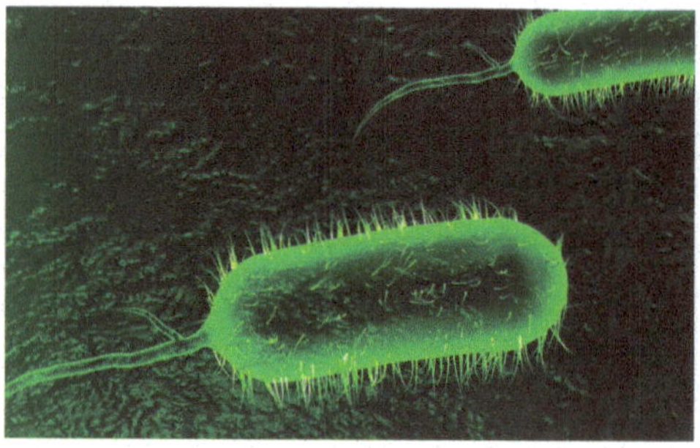

© Creations/Shutterstock.com.

Figure 13.8 **Cholera is a bacteria found in salt water or brackish water that, when ingested, can cause diarrhea and vomiting in the host within several hours to 2 to 3 days of ingestion.**

regions and brackish rivers, areas that are often tainted by raw sewage, especially when cyclonic storms hit populated coastal areas.

Symptoms of cholera range from almost nonexistent to severe, the latter affecting approximately 5% of those who become infected. Vomiting and extremely watery diarrhea create a rapid decrease in body fluids that brings on dehydration and shock. Death is rapid without treatment.

Table 13.1 shows pandemics and concentrated outbreaks of cholera that have occurred since the first outbreak in India in 1816–1826. By the late 1880s, most developed countries understood the need for a clean reliable water supply, which helped reduce the pervasive nature of the disease in most Western nations.

After the 2010 earthquake in Haiti, which displaced approximately 1.5 million people, a cholera epidemic began to ravage the internally displaced persons camps. Studies have traced the outbreak to faulty sanitation in a United Nations Peacekeeping Forces camp. Estimates suggest that more

Table 13.1 **Major Outbreaks of Cholera Throughout the World Since 1816**

Dates	Localities	Comments
1816–1826	First in Bengal, to India in 1820, then to China and eastern Europe	
1829–1851	Began in Europe, then London, Canada, and New York in 1832; West Coast of North America in 1834	
1852–1860	Mainly Russia	More than 1 million deaths, including composer Peter Tchaikovsky
1863–1875	Prevalent in Africa and Europe	
1866	North America	
1892	Hamburg, Germany	City water supply contaminated, killing more than 8,500
1899–1923	Russia	
1961–1966	First in Indonesia, then to Bangladesh, India, and Russia	
1994	Rwanda refugee camps	48,000 cases resulted in almost 24,000 deaths
2010	Haiti earthquake displaced persons camps	>800,000 cases and more than 9,000 deaths

Source: https://www.asm.org/index.php/microbelibrary/367-news-room/iceid-releases/93626-influenza-vaccine-while-not-100-effective-may-reduce-the-severity-of-flu-symptoms

than 9,000 people died of cholera, but a study by Doctors Without Borders suggests that this is severely underreported and it may be up to a factor of three greater than that reported during the most intense early stages of the epidemic. As of 2021, the cholera epidemic is still ongoing in Haiti but the number of cases has dropped dramatically. About 40% of the population in Haiti does not have daily access to clean water and less than 25% has regular use of a toilet, according to Pan-American Health Organization and World Bank, making Haiti vulnerable to recurring strikes of cholera during future disasters such as hurricanes.

Influenza

influenza

An acute viral infection that affects the respiratory system; this can occur as isolated, epidemic, or pandemic in scale; symptoms include fever, headache, nasal congestion, and general lethargy.

Influenza viruses produce the flu, a disease that affects millions of people every year (Figure 13.9). Symptoms include fever, headache, nasal congestion, muscle aches, sore throat, and general loss of energy and appetite. A case of the flu varies from being a mild illness to one that can be quite severe and lead to death in those with low immunity, especially the young and the aged. An annual vaccination is moderately effective against the disease, reducing your risk of developing an influenza-related illness by about 60%, and may also lessen your symptoms if you do get the flu, according to the American Society for Microbiology.

The Centers for Disease Control and Prevention (CDC) report that every year in the United States, on average,

- 5 to 20% of the population get the flu.
- More than 200,000 people are hospitalized from flu complications.
- About 36,000 people die from flu infections.

Pandemics of influenza have struck approximately three times every century since the 1500s. Recurrence intervals range between 10 and 50 years for these major outbreaks, beginning with the one that infected people in Africa and Europe in 1510. A more widespread episode occurred in 1889 and 1890, when influenza was reported in Russia in May 1889 and spread to North America by December of that year. In only 5 months, this pandemic had overtaken India and South America, and it finally reached Australia in the early spring of 1890.

The 1918 influenza pandemic killed more people than World War I, which was just winding down. It is cited as the most devastating epidemic in recorded world history, and more people died of influenza in a single year than in the entire Black Death from 1347 to 1351. Although the strain began in the central United States,

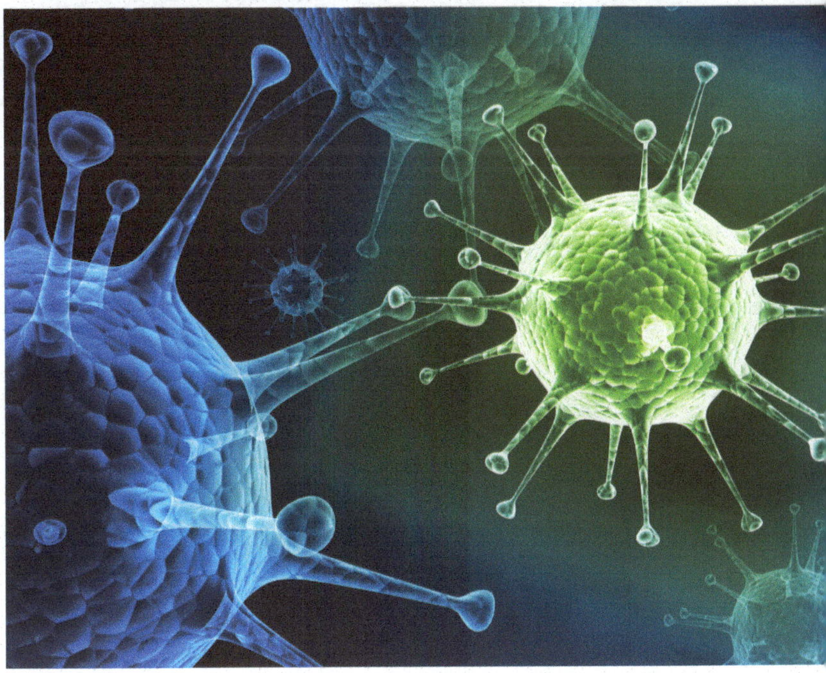

© mathagraphics/Shutterstock.com.

Figure 13.9 Three-dimensional structure of seasonal influenza virus from electron tomography.

this episode became known as the "Spanish flu" or "La Grippe" because the spread of the disease received more press coverage in Spain, which was not involved in World War I and hence news in that country was not censored.

The supposed cause of this episode was from a mutated swine virus that affected soldiers at Camp Funston, Kansas, on March 11, 1918. Within 2 days, more than 500 people were sick, many of whom contracted pneumonia. Within a week, the flu had spread to every state in the United States and in the next 2 months the flu outbreak was worldwide. Interestingly, the disease usually peaked within a few weeks of its initial onset in a given area.

In the end, there were nearly 20 million cases in the United States, from which almost 1 million people died. It infected 28% of all Americans, and a fifth of the world population. The worldwide death toll has been estimated to range between 25 and 40 million people. Overall, most deaths occurred in people between 20 and 45 years of age generally considered the healthiest portion of a population. Besides a shortage of medical workers and medical supplies, there were also shortages of coffins, morticians, and gravediggers. The Red Cross created a special committee to mobilize all resources to fight the flu, and even President Woodrow Wilson suffered from the flu in early 1919 while negotiating the Treaty of Versailles that ended the first World War.

Smallpox

smallpox

An acute viral disease that was once a major killer but has been eradicated; symptoms include headaches, vomiting, and fever, followed by a widespread skin rash that eventually permanently scars the skin.

Smallpox is an acute, contagious, and sometimes fatal disease caused by the variola virus (an orthopoxvirus), and marked by fever and a distinct, progressive skin rash. The only defense is prevention by vaccination, as no adequate treatment exists. Infected people develop raised bumps filled with fluid that appear on the skin covering the body. Historically, there were major episodes of outbreaks in the New World that killed several million people. In the 21st century, the United Nations reported that more than 300 million people worldwide have died from smallpox. The last documented case in the United States was in 1949 and the last case worldwide was in Africa in 1977. In 1980 the disease was declared eradicated following worldwide vaccination programs, and no cases of naturally occurring smallpox have happened since. However, in the aftermath of September 11, 2001, the CDC reported that the US government is taking precautions to be ready to deal with a possible but unlikely bioterrorist attack using smallpox as a weapon. Adequate vaccine supplies exist to inoculate everyone in the United States against the disease.

Typhus

typhus

An acute infectious disease with symptoms of high fever, severe headaches, and a skin eruption; common in wartime, famines, or catastrophes, it is spread by lice, ticks, or fleas.

The CDC reports that **typhus** generally occurs only in communities and populations with poor sanitary conditions and crowding, in which body louse infestations are frequent. This is typically seen in refugee and prisoner populations, particularly during times of wars or famine (**Figure 13.10**). Typhus also occurs sporadically in mountainous regions of Africa, South America, Asia, and Mexico, especially during the colder months when louse-infested clothing is not laundered and person-to-person spread of lice is more frequent.

The four main types of typhus are (1) *epidemic typhus,* (2) *Brill-Zinsser disease,* (3) *endemic* or *murine typhus,* and (4) *scrub typhus.* Scrub typhus was a major problem in World War I as it was associated with trench warfare that was prevalent in Europe. Typhus was a major cause of death during World War II, killing those in POW camps, ghettos, and Nazi concentration camps, including Anne Frank, at the age of 15. The newly discovered insecticide DDT was used as a lice killer and helped avert major post-war epidemics.

Symptoms of epidemic typhus include chills and fever, headache, vomiting, and a rash that generally begins in the trunk area. If these conditions remain untreated, older adults who are infected can experience a death rate as high as 60%. Children usually recover well from epidemic typhus. Death rates for Brill-Zinsser disease, endemic or murine typhus, and scrub typhus are generally less than 1%. The best prevention for any form of typhus is to avoid the carrier insects and to use insect repellents and practice good hygiene.

Source: Centers for Disease Control and Prevention. Photo by James Gathany.

Figure 13.10 Body lice are parasitic insects that live on the body and in the clothing or bedding of infested humans. They spread infections rapidly under crowded conditions where hygiene is poor and there is frequent contact among people. The dark mass inside the abdomen is a previously ingested blood meal.

Yellow Fever

Another viral disease that is transmitted to humans by mosquitoes is **yellow fever**. Symptoms are similar to other tropical diseases in that infected people experience fever, headache, muscle pain, jaundice, and nausea. One's pulse can slow down also, but all these symptoms generally vanish in a few days. There can be a sudden return of more serious conditions that include bleeding from the nose, eyes, mouth, and in the stomach. Kidney failure follows and death usually occurs within 10 to 14 days.

yellow fever

Disease caused by a vector-transmitted virus; symptoms include high fever, headaches, jaundice, and often gastrointestinal hemorrhaging.

An epidemic of yellow fever struck the young United States in the late 1700s. From August to November 1793, between 4,000 and 5,000 people perished from the disease that at the time was unexplained. Philadelphia, then the new nation's capital, was the hardest hit in terms of deaths. The government came to a halt because so many workers were either afflicted with the disease or left the city to seek a safer area. A recurrence in 1798 killed many who were involved with forming the new government of the United States.

Historically, yellow fever was a major disease in equatorial regions and was one of the key impedances to the construction of the Panama Canal in the early 1900s. The research of William C. Gorgas (Figure 13.11), who also played a key role in the quest to eliminate malaria in the region, was paramount in controlling outbreaks of yellow fever. Gorgas contracted yellow fever as a young soldier and became immune to it in his later life. He also contracted typhoid fever in 1898.

Agricultural and forestry workers in South America and Africa are most affected by yellow fever as they are exposed to mosquitoes in their

Source: U.S. Army.

Figure 13.11 **Major General William C. Gorgas (1854–1920), Surgeon General of the US Army, eradicated yellow fever and reduced malaria in Panama.**

workplace, fields, and forests. Moist savanna regions in Central and West Africa are primary areas when the rainy season produces conditions that foster mosquito larva growth.

Yellow fever is now very rarely a concern to travelers but they must have a vaccination to prevent the disease. A single vaccination provides immunity for 10 years or more, with inoculation of a booster dose required every additional 10 years. If people are able to avoid mosquito piercings through the use of protective clothing, mosquito nets, and insect repellent, they can minimize the chance of contracting the disease. As the world's climate becomes warmer, more localities will experience increased infestation of mosquitos. This could result in an increase in yellow fever and other mosquito-borne diseases, especially in developing nations.

HIV Infection and AIDS

Human immunodeficiency virus (HIV) is a retrovirus, or a type of of virus that use RNA as its genetic material, which can lead to **acquired immunodeficiency syndrome (AIDS)** in humans. This disease is considered a pandemic and was first diagnosed in 1981 in the United States, although cases dating back to the late 1950s and 1960s are believed to be the earliest known AIDS infections. There is no cure or vaccine. The disease is thought to originate in non-human primates from Africa and to have been transferred to humans in the early 20th century.

human immunodeficiency virus (HIV)

An RNA (ribonucleic acid) virus that causes an immune system failure and can lead to AIDS.

acquired immunodeficiency syndrome (AIDS)

A serious disease that results from an HIV infection. It is spread through direct contact with contaminated bodily fluids.

The virus attacks the body's immune system, specifically the CD4 or T cells. Over time, so many of these cells are destroyed that the body can't fight off infection or disease. People diagnosed with HIV may become infected with other life-threatening diseases called opportunistic infections, which are caused by microbes such as viruses, bacteria, or fungi that usually do not make healthy people sick. As the HIV infection progresses, it increases the risk of common infections like tuberculosis as well as tumors. These late-stage symptoms, often accompanied by weight loss, are referred to as AIDS.

More than 75 million people have been infected with HIV and 32 million people have died from AIDS-related illnesses since the start of the pandemic (Figure 13.12). As of 2020 about 37.9 million people are living with HIV infections, including 1.7 million children. About 21% of infected people don't know they have the virus. AIDS-related deaths have fallen by 48% since their maximum in 2005, and tuberculosis remains the leading cause of death among people living with HIV. The greatest occurrence of HIV and AIDS is in eastern and southern Africa, where 68% of all HIV cases occur, and it is believed to infect about 5% of the adult population. This is a blood-borne disease that is transmitted through unprotected sexual contact

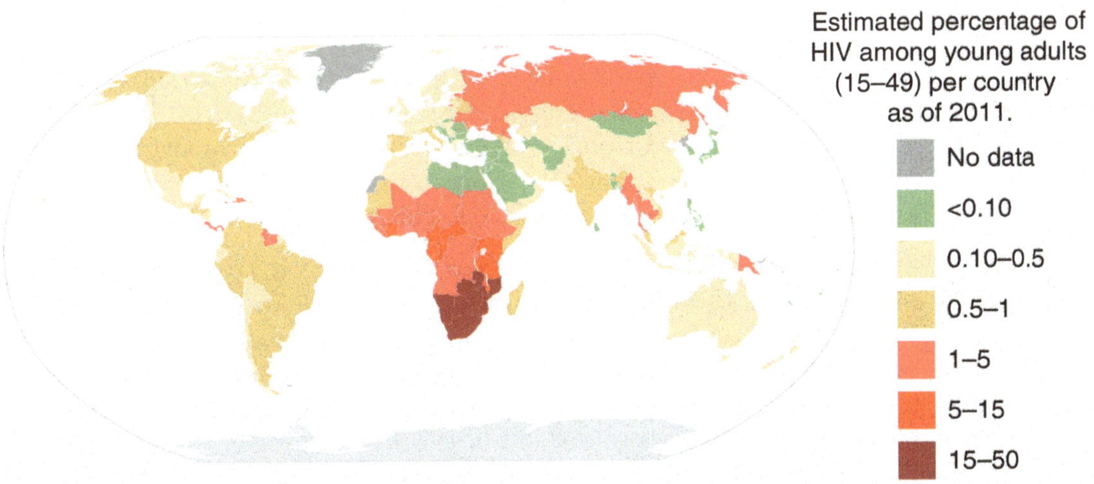

Figure 13.12 **The estimated amount of the population of young adults ages 15 to 49 who were infected with HIV as of 2011, shown per country.**

with infected partners. It is also possible to become infected through the use of nonsterile needles or, on extremely rare occasions, by tainted blood received through transfusions. HIV can be transmitted by women to their babies during pregnancy or birth. Approximately one quarter to one third of all untreated pregnant women infected with HIV will pass the infection to their infants. HIV can also be spread to babies through the breast milk of mothers infected with the virus.

The incidence rate of AIDS is highest among homosexual men, but education programs in high-incidence countries have helped stem the spreading rate. Recent data, however, show an increase in the Americas and in Asian countries.

As reported by the CDC, almost 700,000 cases of AIDS have been reported in the United States since 1981. As many as 1.2 million Americans may be infected with HIV, one fifth of whom are unaware of their infection.

Symptoms of an HIV infection are not evident when someone is first infected. Within 1 or 2 months, flu-like symptoms appear and may include fever, fatigue, headaches, and swollen lymph nodes. Because these generally disappear within a few weeks, they are usually mistaken for some other viral infection. It is during this period that people are highly infectious, especially through unprotected sexual contact.

Medical workers have noticed that the more severe or persistent symptoms may take 10 years or more to be evident, at which time the disease is well-established in the individual. The time it takes for the symptoms to be clearly defined varies with each person.

Much progress has been made toward the treatment of HIV and AIDS during the past decade. Advances and new drug treatments have improved the survival rate. The most current information concerning HIV and AIDS will be found on appropriate internet sites sponsored by major medical institutions such as the CDC and National Institutes of Health (NIH).

Hepatitis

hepatitis
A liver disease caused by a virus; four different forms exist; symptoms include fever, jaundice, fatigue, liver enlargement, and abdominal pain.

Hepatitis is a liver disease and the most common cause worldwide is viruses. Several forms of hepatitis (designated A, B, C, D, and E) exist. Hepatitis A and E are both passed from person to person through contact with fecal material of infected people. These are common in developing countries with poor sanitation, and do not lead to chronic hepatitis.

The most widespread of the five types of hepatitis is Hepatitis B (HBV), which affected about 343 million people worldwide in 2015, compared to 114 million with hepatitis A and 142 million with hepatitis C. In comparison, alcoholic hepatitis affects about 5 million people. Its symptoms include jaundice, fatigue, abdominal pain, nausea, and pain in the joints. One third of the world's population (one out of every three people) is infected with HBV, making it the number one infectious disease, not only of the present time but of all time. Death from chronic liver disease occurs in approximately 25% of all cases, usually as a result of liver cancer or cirrhosis (scarring) of the liver. Most deaths occur in underdeveloped countries.

HBV is spread by contact with the blood of infected people, which enters the bloodstream of the noninfected person. Unprotected sexual contact with infected people and the sharing of needles and other drug paraphernalia will put the uninfected at severe risk of contracting Hepatitis B. HBV can produce a lifelong infection and can cause cirrhosis of the liver, liver cancer, liver failure, and death. A vaccine has existed since 1982 and is recommended for people between birth and age 18 if they have any likelihood of contracting the disease.

The World Health Organization reports that 500 million people are living with chronic Hepatitis B virus or Hepatitis C infection. Many of these people lack access to testing and treatment, and are at risk for liver disease, cancer, and death. Viral hepatitis is a major public health challenge, and causes 1.5 million deaths each year, which is comparable to deaths caused by tuberculosis and HIV. Mortality from hepatitis is increasing. Globally, 84% of children born in 2015 got the three hepatitis B vaccine doses, according to the WHO, and the proportion of children under 5 years of age with the infection has fallen to 1.3%.

Hepatitis C is contracted through contact with blood of an infected person. Hepatitis B and C are the two main types out of the five different infections, and are responsible for 96% of all hepatitis deaths. A primary means is by shared use of needles and other sharp objects related to drug use. Hepatitis D needs HBV present to exist. Hepatitis E, which is transmitted in similar ways to HAV, rarely occurs in the United States. Hepatitis A, B, and D are preventable with immunizations, and chronic viral Hepatitis can be treated medically, although it results in more than a million deaths each year.

Leprosy

Leprosy (Hansen's disease) is an infection caused by slow-growing bacteria called *Mycobacterium leprae* that attacks the skin and peripheral nerves and can have many other manifestations. Two forms of the disease exist: paucibacillary Hansen's disease is the milder form that causes one or more pigmented lesions under the skin. The multibacillary form is recognized by symmetric skin lesions, nodules, thickened dermis, and nasal mucosa which produce nasal congestion and nose bleeds. *Mycobacterium leprae* is a rod-shaped bacterium that spreads very slowly and mainly infects the skin, nerves, and mucous membranes (Figure 13.13).

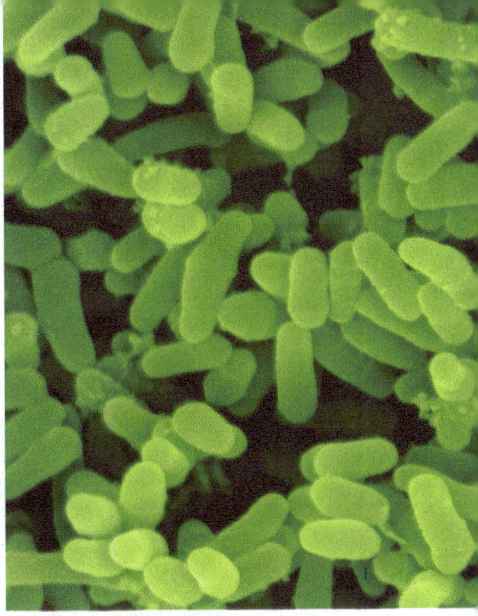

© Dennis Kunkel Microscopy/Science Photo Library.

Figure 13.13 ***Mycobacterium leprae*** **is the bacteria that causes leprosy. A Norwegian scientist, Armauer Hansen, identified the bacteria that causes leprosy in 1873. The first treatment wasn't available until the 1940s.**

The first description of symptoms similar to leprosy comes from writings in India from 600 B.C., and it is mentioned in both the Old and New Testaments in the Bible. It is uncertain how the disease is spread but researchers believe that humans contract the bacillus through respiratory droplets by coughing or other contact with fluid from the nose of an infected person. Contrary to common belief, it is not highly contagious and most of the human population, about 95%, is not susceptible to infection, according to the Health Resources and Services Administration, US Department of Health and Human Services. It is curable with multi-drug therapy.

leprosy (Hansen's disease)

A chronic, mildly infectious disease which affects the nervous system, skin, and nasal regions; it is characterized by skin ulcerations and nodules.

The number of leprosy cases was in the tens of millions in the 1960s. This has dropped dramatically to about 175,000, but there are still about 210,000 new cases diagnosed worldwide every year (Figures 13.14 and 13.15).

Malaria

> Since the beginning of history, malaria has killed half of the men, women, and children who have lived on the planet. It has outperformed all wars, all famines and all other epidemics. Until World War II it still accounted for 50 percent of the business at most cemeteries.
>
> Nikiforuk, 1991

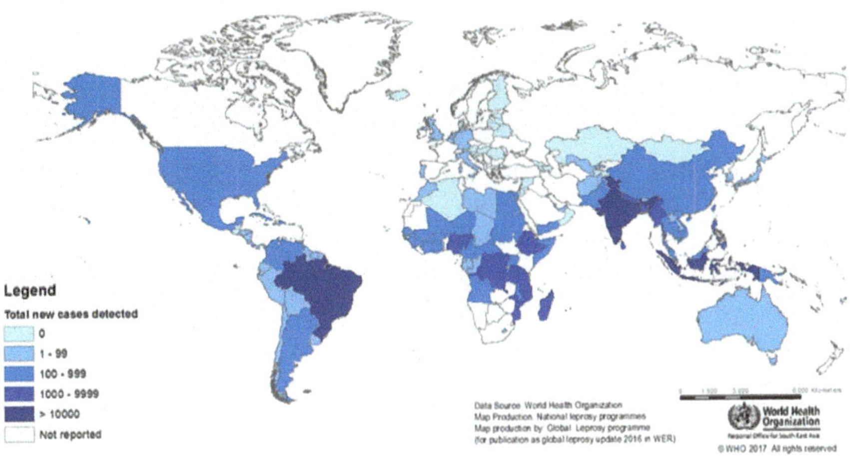

Source: Reprinted from Global Leprosy Programme update 2016 in WER, pg 507, © 2016.

Figure 13.14 **Global map of new cases of leprosy based on World Health Organization data.**

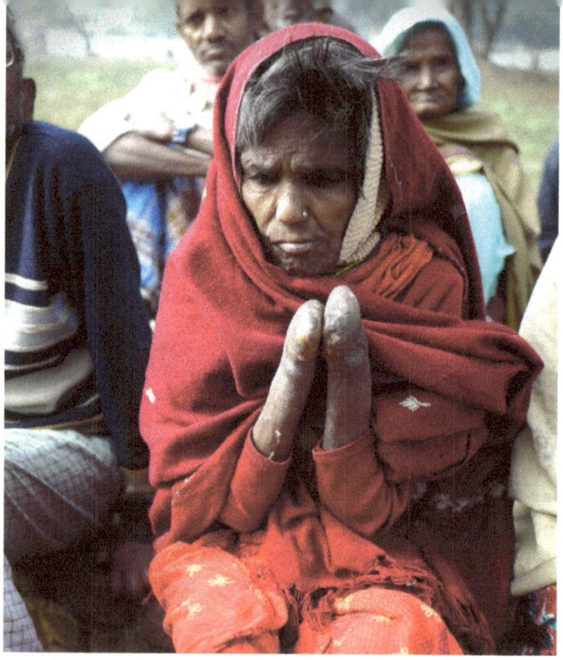

© fivepointsix/Shutterstock.com.

Figure 13.15 Leprosy patient at Munger Leper Colony, Munger, India.

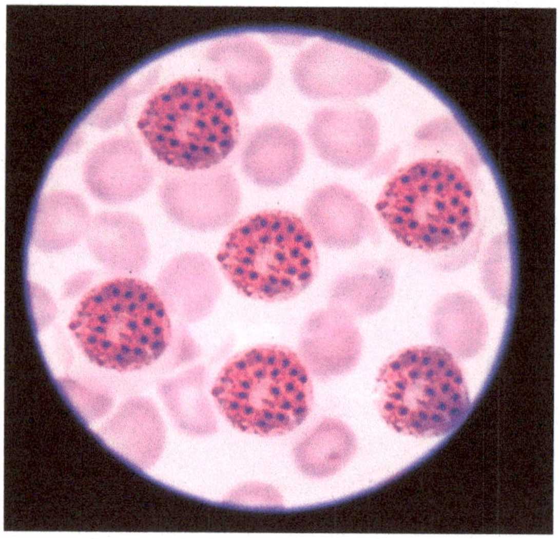

© toeytoey/Shutterstock.com.

Figure 13.16 A *Plasmodium* parasite from an infected mosquito. The mosquito host transmits the parasite into a vertebrate host, the secondary host, and is the transmission vector for malaria.

malaria
An infectious disease caused by the bacterium *Plasmodium falciparum*, which is transmitted by the female *Anopheles* mosquito; symptoms include high fever, chills, and sweating.

vector
An agent such as a flea or mosquito that can spread parasites, virus, or bacteria.

Malaria is derived from the Italian *mala aira*, meaning bad air, and was so named because it was first thought to have been caused by breathing bad air. It is a mosquito-borne infectious disease caused by a parasitic protozoan single-celled micro-organism that has affected humans for more than 50,000 years (Figure 13.16). Malaria has the cyclic symptoms of body chills, muscular pain, flu-like illness, and a high fever, which first appear 10 to 15 days after infection and then may recur months later if not properly treated. Different strains exist, and one type can cause more serious problems, damaging vital organs including the heart, kidneys, lungs, or brain, thereby producing death. The female *Anopheles* mosquito is the reservoir for malaria and is the **vector** that infects people by means of puncturing the skin. Male mosquitoes feed on nectar from flowers and pose no threat of transmitting the disease.

Mosquitoes infected with malaria are less fertile and less long-lived than uninfected mosquitoes, but vector biologist Janneth Rodrigues and colleagues at the National Institute of Allergy and Infectious Diseases in the US has found that many mosquitoes are malaria-resistant and are able to fight off the infection upon exposure. This may be useful in helping fight malaria, which can kill more than a million people every year. Most of them are children and pregnant women, according to the World Health Organization.

Most often found in Central and South America, Africa, and South Asia, malaria is a serious disease in those regions. This was especially true in the 1800s when the British Army had large numbers of soldiers stationed throughout the tropics. The British government recognized that if the cause and source of the disease could be identified, Great Britain would continue to dominate the regions it controlled in South Asia.

Sir Ronald Ross, a former major and surgeon in the British Army (Figure 13.17a), studied the malaria parasite in birds and recognized its connection with mosquitoes, although he did not recognize that malaria was

Figure 13.17a **Sir Ronald Ross (1857–1932).**

Figure 13.17b **Charles Louis Alphonse Laveran (1845-1922).**

transmitted by the female *Anopheles* mosquito. In 1880, Charles L. A. Laveran, a French army surgeon stationed in Algeria (Figure 13.17b), was the first to notice parasites in the blood of a patient suffering from malaria. For his discovery and continuing work, Laveran was awarded the Nobel Prize in Physiology or Medicine in 1907.

In rare instances, the disease can be transmitted if people come in contact with infected blood. It is also possible for an infected pregnant woman to pass malaria to her fetus. It is a myth that malaria can simply be passed to anyone who comes in close contact with a carrier.

Today malaria infects more than 200 million people each year, mostly young children in sub-Saharan Africa. Every 2 minutes malaria kills a child under the age of 5. Over three billion people in 106 countries and territories are at risk of malaria transmission. This depends on climatic factors such as temperature, humidity, and rainfall. Malaria requires tropical and subtropical climates where mosquito hosts can survive and breed, and where the malaria parasite can complete its growth cycle in the mosquito. This can't happen at temperatures less than 20 °C (68 °F). In warm countries close to the equator malaria transmission may be year round and more intense compared to cooler regions, or those with seasonal temperature or moisture fluctuations. No vaccine is currently available for malaria, so preventive drugs must be taken continuously to reduce the risk of infection. Unfortunately, these prophylactic drug treatments are too expensive for most people living in the endemic areas of the tropical and subtropical regions of the world.

The most cost-effective way of preventing malaria is to sleep under an insecticide treated bed net to protect against mosquito bites, which typically happen between 10 p.m. and 2 a.m. Each net costs about $2, lasts for 3 to 4 years, and protects, on average, two people (Figure 13.18). Since the year 2000, more than one billion nets have been given out in Africa alone. However, a WHO study from 2016 suggests that mosquitoes may be developing

© Tuttoo/Shutterstock.com.

Figure 13.18 **Inexpensive, insecticide-treated bed nets help prevent malaria.**

an immunity to the insecticide used in these nets, and malaria increased by five million cases—an estimated 216 million people infected—from the year 2015 to 2016.

Malaria was prevalent in Central and equatorial South America and created a major challenge during the early stages of construction of the Panama Canal. In 1906, more than 26,000 workers were involved with the project. Malaria and yellow fever hospitalized more than 80% of them during some stage of their employment. By the time the workforce had swelled to more than 50,000 people in 1912, fewer than 12% of them were hospitalized. Through the efforts of William C. Gorgas, malaria was significantly reduced as a biological threat in the region.

Malaria is less likely to occur in developed countries where public health measures are better established. Although the United States and Western Europe have eradicated malaria, these areas still harbor *Anopheles* mosquitoes that can transmit it. Reintroduction of the disease is a constant risk. The high levels of deaths from malaria has put the strongest known rates of selective pressure on the human genome since the beginning of agriculture, and humans may be developing a genetic resistance to it. Miguel Soarez and Ana Ferreira of the Gulbenkian Institute of Sciences in Portugal showed, in 2011, that people carrying the gene for sickle-cell anemia disease are protected from malaria, not by protecting from infection, but by preventing the disease from taking hold because of changes to hemoglobin that sickle-cell gene carriers have in their blood.

Tuberculosis

"My sister Emily first declined. The details of her illness are deep branded in my memory, but to dwell on them, either in thought or narrative, is not in my power. Never in all her life had she lingered over any task that lay before her, and she did not linger now. She sank rapidly. She made haste to leave us. Yet, while physically she perished, mentally she grew stronger than we had yet known her. Day by day, when I saw with what a front she met suffering, I looked on her with an anguish of wonder and love. I have seen nothing like it; but, indeed, I have never seen her parallel in anything. Stronger than a man, simpler than a child, her nature stood alone: The awful point was, that while full of ruth for others, on herself she had no pity; the spirit was inexorable to the flesh; from the trembling hand, the unnerved limbs, the faded eyes, the same service was exacted as they had rendered in health. To stand by and witness this, and not dare to remonstrate, was a pain no words can render.

Two cruel months of hope and fear passed painfully by, and the day came at last when the terrors and pains of death were to be undergone by this treasure, which had grown dearer and dearer to our hearts as it wasted before our eyes. Towards the decline of that day, we had nothing of Emily but her mortal remains as consumption left them…

We thought this enough: but we were utterly and presumptuously wrong. She was not buried ere Anne fell ill. She had not been committed to the grave a fortnight, before we received distinct intimation that it was necessary to prepare our minds to see the younger sister go after the elder."

Charlotte Bronte, in a preface for the new addition of *Wuthering Heights* and *Agnes Grey* in 1850, a year after the authors, her sisters Emily and Anne, died, on December 19, 1848, and May 28th, 1849, respectively.

Called "consumption" in Victorian times, tuberculosis (TB) is one of human history's greatest killers, responsible for a billion deaths over 2 centuries. It has been found in human remains from 5000 B.C., Egyptian mummies from 2400 B.C., ancient Greece, and medieval Europe (**Figure 13.19**). Hippocrates identified it as the most widespread disease of his age in 460 B.C. In 1882 Robert Koch discovered that bacterial infection causes TB, and that air and lung fluid secretions such as coughing from victims contains live bacteria and can infect those around them. TB can affect the skin, the bones, or most commonly the lungs. It was called consumption for the wasting away late-stage victims experienced, along with night sweats, fevers, chills, and coughing. TB is estimated to be responsible for 20% of deaths in 17th-century London and 30% in 19th-century Paris. Consumption has been romanticized in music and film, from Mimi dying in Puccini's *La Boheme* to Satine in *Moulin Rouge*. Emily Bronte died at the age of 30 from TB, as

© Andrea Izzotti/Shutterstock.com.

Figure 13.19 **An Egyptian mummy from the British Museum that shows spinal decay from tuberculosis.**

Figure 13.20 **Anne, Emily, and Charlotte Bronte, painted by their brother Branwell in 1834. He initially painted himself into the family portrait but covered himself over with the pillar afterward. Charlotte is the only one of the four who did not die of consumption.**

Detail of Portrait of the Bronte Sisters, c.1834 (oil on canvas), Bronte, Patrick Branwell (1817–48)/National Portrait Gallery, London, UK/Photo © Brian Seed/Bridgeman Images.

did her brother and sister (Figure 13.20). *Wuthering Heights* is filled with characters who develop TB and transmission from one character to another is used as a plot device to show the nature of their relationships.

While fresh air was considered a treatment for TB, and in-patient hospitals called sanatoriums were set up to treat patients, 80% of those who contracted TB died because there was no cure. This changed in 1944 when the first critically ill TB patient was cured with the antibiotic streptomycin. Rates of the disease dropped throughout the world, and in the mid-1980s the American Medical Association predicted that TB would be eradicated by 2010.

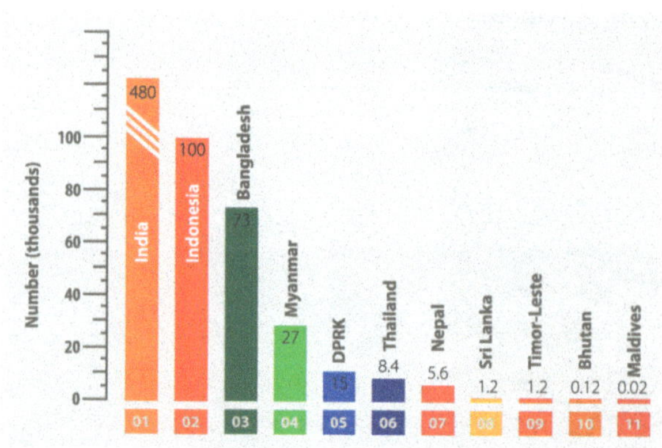

Figure 13.21 **Ranking of South-East Asian countries by tuberculosis incidence, according to the World Health Organization Regional Office for South-East Asia 2017 annual report.**

Source: World Health Organization.

Yet by 1985 the decline in TB cases stopped and then began to rise. The drug regime developed to treat TB needs to be taken continuously, in regular doses, for 6 to 8 months. Most patients do not follow through because of the cost, the inconvenience, or because they feel better and stop taking medication early. This leaves antibiotic-resistant microbes in the host that have turned into treatment-resistant strains of TB. It takes 3 years to develop a new antibiotic, and 3 months to develop a new drug-resistant strain of microbe. Tuberculosis and HIV have formed a new deadly combination and immunocompromised HIV victims represent an additional 1.4 million cases of TB each year. The WHO now calls TB a "fire raging out of control" (Figure 13.21).

Two forms of TB exist: latent and active. One third of the world's population is thought to be infected with TB, but if a person has latent TB, they are asymptomatic and the bacteria are being carried around but are not transmittable (**Figure 13.22**). Approximately 10% of latent TB cases transform into active TB. People with latent infections should seek treatment to prevent them from developing active infections later in life. Active TB spreads through the body of the carrier and can be transmitted to others if the carrier's lungs are infected. TB is the number one cause of death from an infectious disease and people with HIV/AIDS are 16 to 27 times more likely to develop TB than those without.

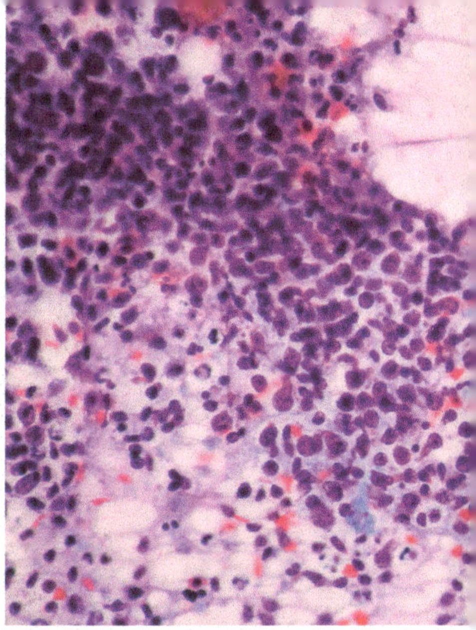

© Komsan Loonprom/Shutterstock.com.

Figure 13.22 **The tuberculosis bacteria in mucus.**

In 2015 the World Health Organization reported more than 4.7 million new TB cases, resulting in 1,945 deaths per day which is equivalent to nine passenger planes crashing each day. Of those new cases, only about 2.5 million were treated. TB death rates are highest in Africa, India, Indonesia, and Bangladesh (**Figure 13.23**). Between 2000 and 2014, improvements in diagnosis and treatment saved 43 million lives worldwide. The current rate of decline in TB is 1.5 to 2% a year, which is not enough to meet the WHO's "End TB" target in the year 2030. TB is not just a biomedical and public health problem, its association with poverty complicates efforts. To meet the WHO target, TB decline has to reach 17% a year to 2025.

Typhoid Fever

The bacterium *Salmonella typhi* is the cause of **typhoid fever**, which is common worldwide, affected 12.5 million people and resulted in 149,000 deaths in 2015. Approximately 400 cases are diagnosed each year in the United States, three quarters of which are attributable to international travelers. The bacteria grow in the intestines and blood, and the disease is spread by eating or drinking food or water contaminated with the feces of an infected person, which can occur through poor sanitation or poor hygiene. Symptoms, which vary from mild to severe, start 6 to 30 days after exposure and include high fever, weakness, abdominal pain, constipation, and headaches. People may develop a skin rash with rose colored-spots. Some people are carriers without having any symptoms, but they can still transmit the disease.

typhoid fever
An illness spread by contamination of water, food, or milk supplies with *Salmonella typhi;* symptoms include fever, diarrhea, stomach aches, and rash.

Although typhoid fever is rare in developed countries, it is still very prevalent throughout Latin America, Asia, and Africa. Typhoid occurs most often in children and young adults between the ages of 5 and 19 years. The rates of typhoid fever in developed countries declined during the first half of the 20th century because of the implementation of vaccines and improvements in public sanitation and hygiene, including the chlorination of public drinking water in 1908. In developed countries typhoid fever rates are low at about five cases per million people per year.

William Wallace Lincoln, son of former US president Abraham Lincoln, died of typhoid in 1862. The English novelist Arnold Bennett drank tap water in a Paris restaurant to prove it was safe, despite being warned by the waiter, and died in 1931 from the typhoid he contracted. Mary Mallon, a cook who worked in New York City between 1900 and 1907, was the most notorious carrier identified for typhoid, even though she was not the most destructive. Three deaths have been directly linked to her but some estimates

Source: https://www.sciencedirect.com/science/article/pii/S147330991730703X#fig2

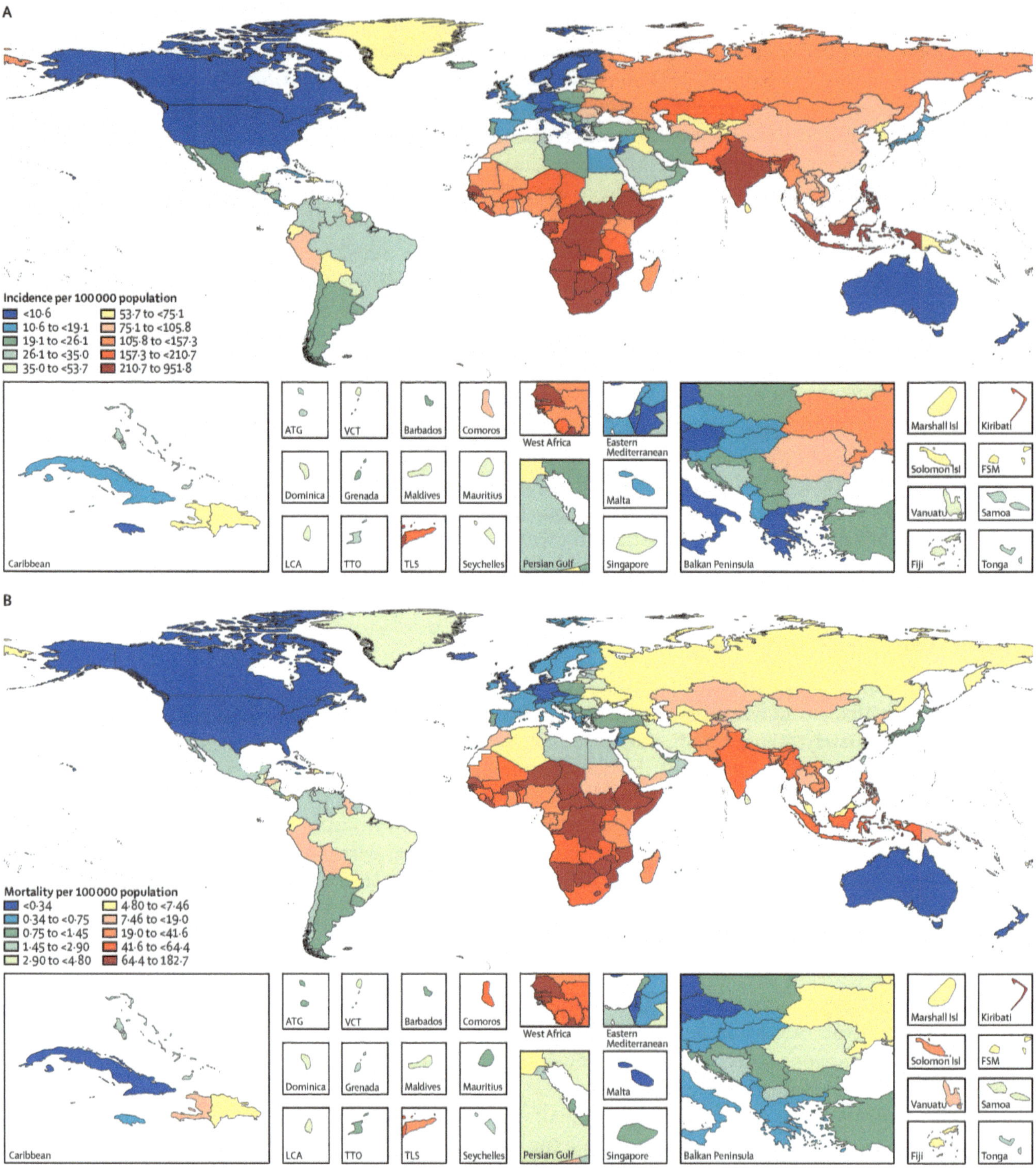

Figure 13.23 For the year 2015, these maps show the age-standardized rates of tuberculosis occurrence in HIV-negative individuals (top map) and the mortality rates (bottom map) per every 100,000 people. This research, published in *The Lancet* by the Global Burden of Disease Tuberculosis Collaborators, was funded by the Bill and Melinda Gates Foundation.

suggest she caused up to 50 fatalities (Figure 13.24). She became the first person to be recognized as a healthy individual who was clearly documented to be a carrier of typhoid fever. As a cook, she ended up infecting 53 people with typhoid fever through her dessert dish—peaches and ice cream. "Typhoid Mary," as she was nicknamed, continually denied her involvement in transmitting the disease and also refused to stop work as a cook, sneaking back under a false name to continue working. She was arrested and permanently quarantined at Riverside Hospital for 26 years total, before dying of pneumonia at age 69.

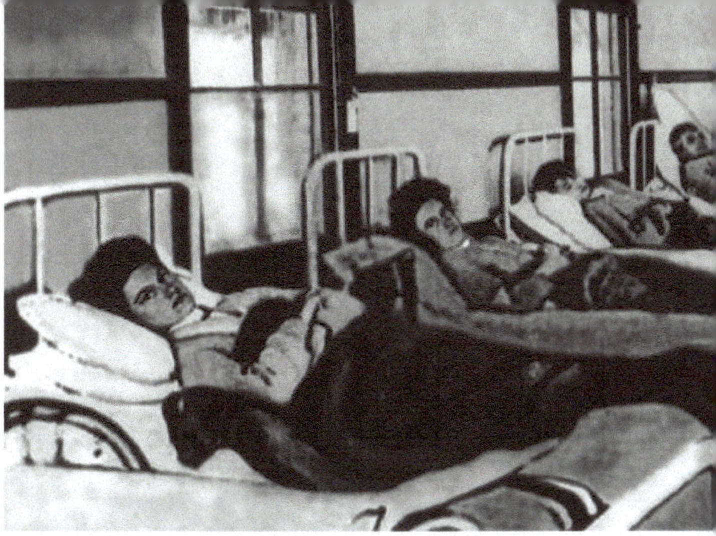

Source: NOVA

Figure 13.24 Mary Mallon (on left) in a hospital after being apprehended by the police in 1907. She was not the only asymptomatic carrier of typhoid, and she wasn't responsible for the largest number of deaths by an asymptomatic carrier, but she was forcibly quarantined and held against her will for more than 26 years.

A vaccine can prevent up to 70% of cases over 2 years, and is recommended for people travelling to areas where the disease is common. Other preventative measures include drinking only clean water, better sanitation, and rigorous handwashing. Food preparers should be checked for infection as they can be asymptomatic but still transmit the disease. The risk of death is as high as 20% if not treated. Antibiotics can treat the infection once it is acquired, but strains of typhoid have been developing resistance which makes treatment and complete recovery more difficult.

For studying several of the diseases discussed in this chapter, five medical researchers won Nobel Prizes for physiology or medicine (Table 13.2). Their work has wide-ranging and beneficial effects for humanity, as do the contributions of countless other scientists and medical personnel who, with less recognition, have greatly expanded our current knowledge of these and other diseases.

Table 13.2 Nobel Prizes for Physiology or Medicine Awarded for Research Into Various Major Diseases

Year	Name	Country	Achievement
1902	Sir Ronald Ross	Great Britain	Recognizing the role of mosquitoes in spreading malaria
1907	Charles L. A. Laveran	France	Discovering the role of parasites in blood as related to the transmission of malaria
1928	Charles Nicolle	France	Researching the causes of typhus
1948	Paul Mueller	Switzerland	Recognizing that DDT killed mosquitoes
1951	Max Theiler	South Africa	Developing a vaccine for yellow fever

Emerging Infectious Diseases

Emerging infectious diseases are infections that have recently appeared in a population, or ones that have had rapid increase in their geographic range or number of cases, or diseases that threaten to increase in the near future. These can be caused by four different factors:

1. New detection of an unknown or previously undetected infectious agent.
2. Known infection agents that have spread to new locations or new populations.
3. Previously known disease agents whose role in specific diseases has just been recognized.
4. Re-emergence of disease agents whose incidence of disease declined in the past, but has reappeared. These are classified as reemergent infectious diseases.

The WHO warns that infectious diseases are emerging at an escalating rate and since the 1970s we have discovered about 40 new infectious diseases. Increased travel, increased population density, and contact with wild animals all increase the potential for emerging infectious diseases to become global pandemics. Many of the following emergent diseases are considered by the WHO to be both a dire epidemic threat and in need of urgent funding into research and development of clinical solutions.

BOX 13.2 SARS-CoV-2 (COVID-19)

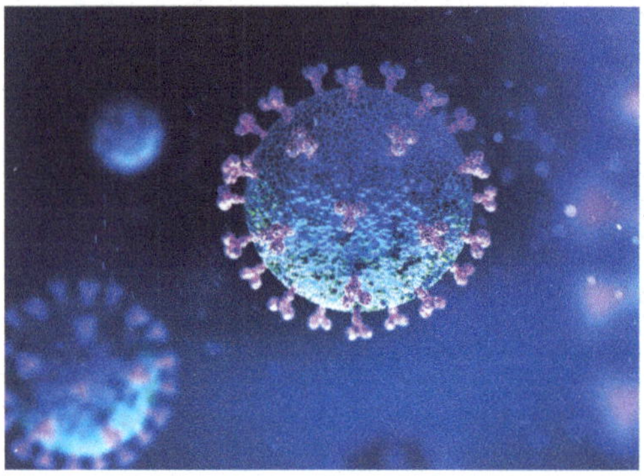

© Andrii Vodolazhskyi/Shutterstock.com.

Figure 13.2.1 The SARS-CoV-2 coronavirus which infects humans, and causes the disease COVID-19.

The first acknowledged human infections of COVID-19, a contagious disease caused by the severe acute respiratory syndrome coronavirus 2 (SARS-CoV-2) were identified in Wuhan, China, on December 1, 2019. Italy had its first confirmed cases on January 31, 2020, when two tourists from China tested positive. By March 13, 2020, the WHO considered Europe to be the active center of a pandemic. Italy overtook China with the most deaths on March 19. By March 26, the US had overtaken both Italy and China with the highest number of confirmed cases. From there, it spread across the planet, with infection fatality rates (the total number of deaths in relation to the total number of infected individuals) of up to 3.4% or higher in some of the hardest-hit places, as estimated by the WHO. New variants began to appear in late 2020, some of which were significantly more transmissible than the original strain. These variants drove record-setting spikes in cases in early 2021, including in the UK, Brazil, South Africa, and India. Johns Hopkins University reported that in 2020 1.8 million people died from the SARS-CoV-2 virus. New estimates by the Institute for Health Metrics and Evaluation suggest that official reports significantly underestimate the global death toll and project that more than 9 million people will die by September 2021.

A number of the first recognized infections in Wuhan were linked to the Huanan Seafood Wholesale Market. This "wet market" is a traditional market selling fresh meat, fish, produce, and other perishable goods, including live animals. Samples from the market showed that the virus was found on surfaces, but tissue samples from the market's animals have not shown any animals that were infected with the virus. In order for the virus to transfer to humans, an animal has to be infected and carrying it.

© Ken Griffiths/Shutterstock.com.

Figure 13.2.2 Scientists believe that a pangolin may have been one of the original hosts of the virus that caused the COVID-19 epidemic. Pangolin, or scaly anteaters, are the only known mammals with scales covering their entire body. Eight species of pangolin are found in Asia and Africa, ranging in size from that of a housecat to over a meter (48 inches), and all are considered to be vulnerable to critically endangered. They are the world's most trafficked non-human mammal, with tens of thousands killed every year for their meat, considered a delicacy in China and Vietnam, and for their scales, which are used in traditional Chinese medicine. More than 1 million pangolin have been poached in the last 10 years.

In March of 2021, the World Health Organization released a report from a month-long mission by a team of Chinese and international experts to uncover the origins of COVID-19. Their results suggest that the virus originated in a bat and may have passed to a pangolin (Figure 13.2.2) and then moved to humans via another intermediary animal that hasn't yet been identified. A growing number of scientists and China experts are theorizing that the coronavirus may have accidentally leaked from the Wuhan Institute of Virology, which is China's only level-4 virology institute with the world's largest collection of bat coronaviruses. Continuing investigations by governments and intergovernmental agencies are ongoing, as there are many unanswered questions. For a similar investigation into the SARS virus, which also originated in bats, the definitive study of its origins wasn't published until 2017, about 15 years after the SARS outbreak occurred.

COVID-19 Cases

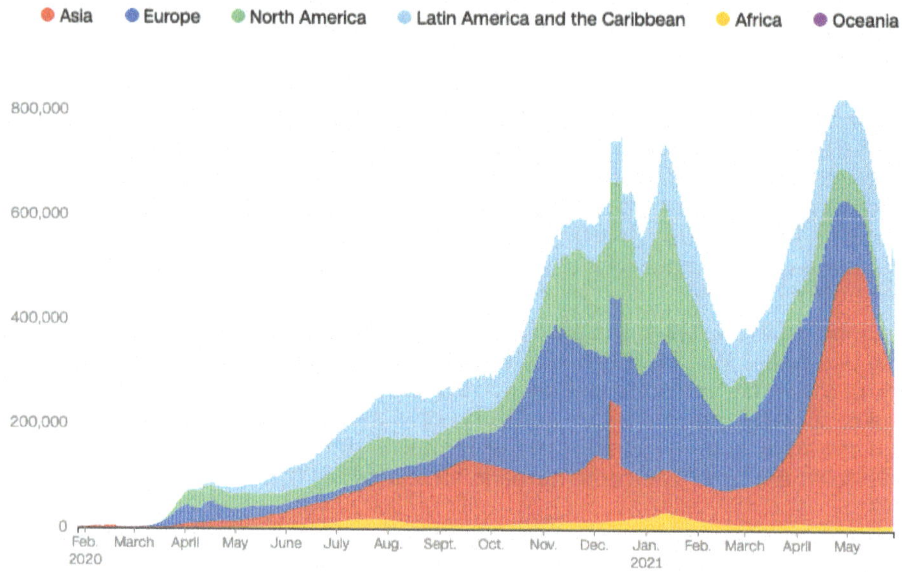

COVID-19 Deaths

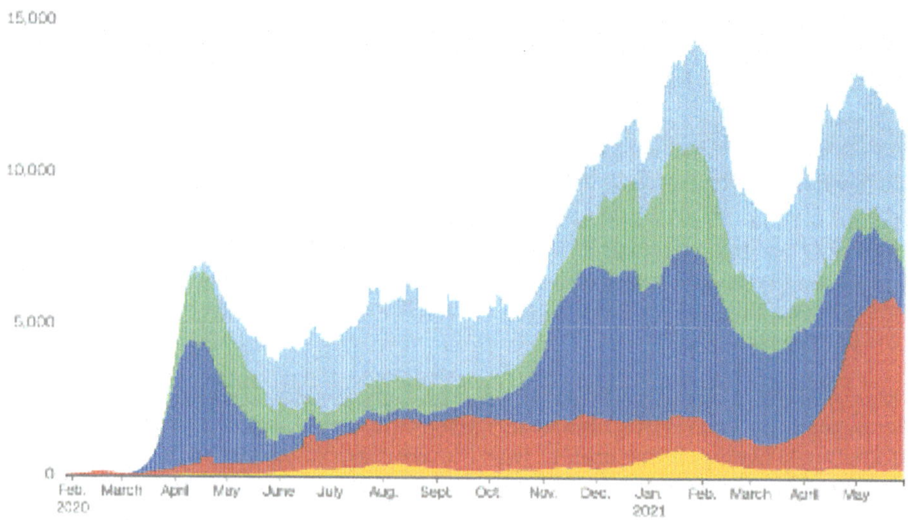

Figures 13.2.3a & b **The global report of daily cases of COVID-19 and deaths attributed to the virus by regions as based on the United Nations classification. The data is a rolling 7-day average to highlight trends and smooth out anomalies like underreporting data on weekends and overreporting data on Mondays.**

Source: Johns Hopkins University Center for Systems Science and Engineering.

Every region on Earth has reported at least one case of the coronavirus, which now infects 240 countries and territories. Confirmed cases rose steeply starting in the middle of 2020, but the true extent of the outbreak is unclear because testing was not widely available. The 100 millionth COVID-19 case was recorded at the end of January 2020, about a year after the first officially diagnosed case. The global economy plummeted, as containment and mitigation efforts interrupt manufacturing, education, and the financial sector.

SARS-CoV-2 is most likely spread between people in airborne respiratory droplets, where the virus moves from an infected person's nose or mouth in small fluid particles when they cough, sneeze, speak, sing, or breathe. A person can become infected when aerosols or droplets containing the virus are inhaled or come into contact with their eyes, nose, or mouth. This usually occurs when people are in close contact, within 1 meter (3 ft) of each other. It can also spread indoors in poorly ventilated or crowded areas, where people spend more time, and the virus can travel longer distances. Symptoms of COVID-19 include fever, cough, difficulty breathing, fatigue, headache, and loss of sense of taste or smell.

Severe COVID-19 can damage the lungs, heart, and brain. It can make blood cells more likely to clot, which in turn causes heart attacks and strokes. It can weaken blood vessels, causing them to leak and leading to long-term liver and kidney problems. People with severe COVID-19 have to be hospitalized and treated in intensive care units, often with ventilators to sustain breathing, which is a traumatic experience that is likely to lead to post-traumatic stress syndrome, depression, and anxiety.

Most people who have COVID-19 recover completely within a few weeks. Some people, even those with mild or no initial symptoms, have symptoms that last for weeks or months, and this is called post-COVID-19 syndrome, or "long COVID." They experience extreme tiredness, shortness of breath, chest pain, brain fog, memory and concentration problems, insomnia, heart palpitations, and other symptoms. Many of the long-term effects of COVID-19 are still unknown.

On January 10, 2020, the genetic sequence for SARS-CoV-2 was published, and by March 19 the global pharmaceutical industry announced a major commitment to develop vaccines intended to provide acquired immunity against the virus. Close to 1 year after the first detected case, almost 20 vaccines are authorized by at least one national regulatory authority for public use. As of mid-2021, close to 2 billion doses of COVID-19 vaccines have been administered. Of the eight main pharmaceutical companies producing vaccines, they anticipate making more than 9 billion doses in 2021. Vaccines can take up to 10 years to develop, and cost up to $500 million US dollars, so the rapid deployment of these vaccines is a scientific and medical achievement of the greatest order.

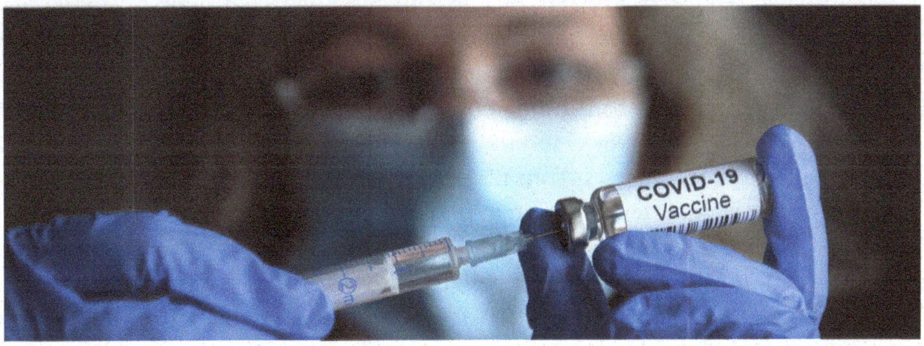

© Viacheslav Lopatin/Shutterstock.com.

Figure 13.2.4 **COVID-19 vaccines have efficacy rates of up to 95%, meaning that you are 95% less likely to become infected than if you did not have the vaccine. They prevent infection in exposed people, reduce the severity of symptoms in those who do become infected, and reduce the likelihood of transmission to others.**

Viral Hemorrhagic Fevers

Viral hemorrhagic fevers (VHF) are a group of illnesses produced by five families of RNA virus and have the common features of causing a severe, multisystem syndrome. They affect many organs, damage blood vessels, and impact the body's ability to self-regulate. They are often accompanied by hemorrhage, or bleeding. Some can cause mild symptoms but many, like Ebola or Marburg, result in severe, life-threatening illness and death. Specific VHF diseases are usually limited to the geographic region of the animal that hosts the virus. Viral hemorrhagic fevers are spread in a variety of ways, some may be transmitted through a respiratory route, and the ability to disseminate by aerosol, along with their severity, means they have the potential to be weaponized for biowarfare.

Crimean-Congo Hemorrhagic Fever

This virus is typically spread by tick bites or contact with livestock carrying the disease, as well as contact with infected people. Infection occurs most frequently in agricultural workers, slaughterhouse workers, and medical personnel. Death rates are up to 40%, and outbreaks occur in Africa, the Balkans, the Middle East, and Asia.

Ebola Hemorrhagic Fever

Fruit bats are believed to be the carrier of the Ebola virus, and can spread the virus without being affected by it (**Figure 13.25**). It spreads by direct contact with body fluids like blood, or items contaminated by body fluids, from an infected person or animal. Eating infected bushmeat (wild animals) may expose people to the virus. Animals can become infected when they eat other infected animals or eat fruit partially eaten by infected bats. People who have recovered from Ebola continue to carry the virus in their semen or breast milk for several weeks and up to 9 months afterward, and can infect others. Death rates from Ebola are extremely high, from 25 to 90% of those who are infected, with an average of about 50% mortality. As soon as 2 days after exposure flu-like symptoms start, followed by vomiting, diarrhea, rash, decreased kidney and liver function, and internal and external bleeding. Victims also suffer with vomiting or coughing up blood, internal bleeding and blood in the stool, as well as bleeding into the eyes.

Two simultaneous outbreaks in 1976 were the first time Ebola was identified, one in South Sudan and one near the Ebola River in the Democratic Republic of the Congo which is where the name comes from. A total of 602 people were infected and 430 died. The largest outbreak currently documented was in West Africa from 2013 to 2016. Guinea, Liberia, and Sierra Leone had 28,661 suspected cases and 11,310 reported deaths, although the WHO believes these numbers were underestimated. Ten percent of the dead were health care workers.

Source: CDC.

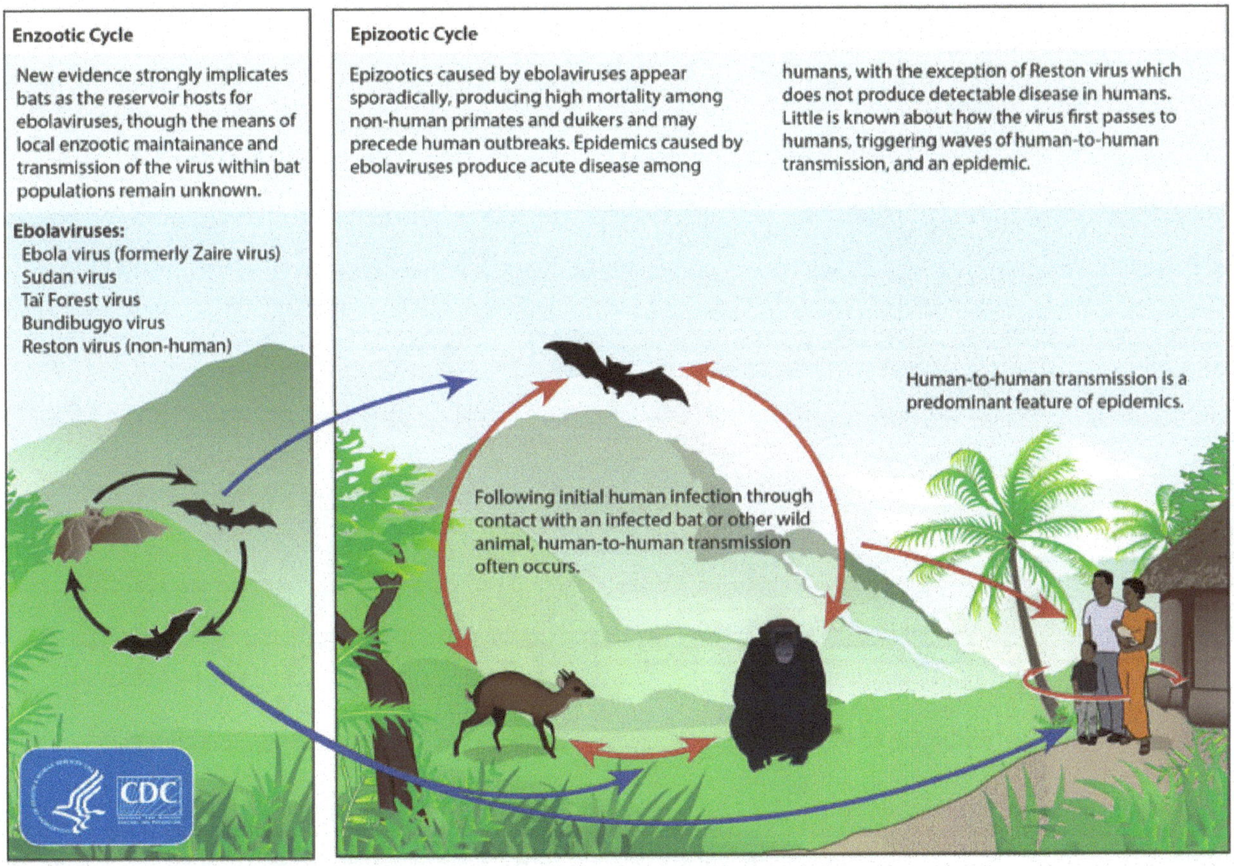

Figure 13.25 The Ebola virus host and transmission cycle from wild animals to humans.

Lassa Hemorrhagic Fever

Lassa fever is relatively common in West Africa, with about 300,000 to 500,000 cases each year. Rats are the disease vector and can spread the virus through their urine or feces. It can then be spread from person to person. Most of those who contract the virus do not develop symptoms, which are similar to Ebola, although a complication from the disease is that about a quarter of those who survive develop deafness, which can improve over time. Death rates are about 1%, with about 5,000 deaths each year, although in epidemics death rates can reach 50%.

Marburg Hemorrhagic Fever

Marburg was first described in 1967 from an outbreak in Marburg and Frankfurt, Germany, and Belgrade, Yugoslavia, where 31 cases were identified and seven people died when workers were exposed to infected tissue from grivet monkeys at a German biopharmaceutical company. The virus is carried, and can be transmitted, by fruit bats and also between infected

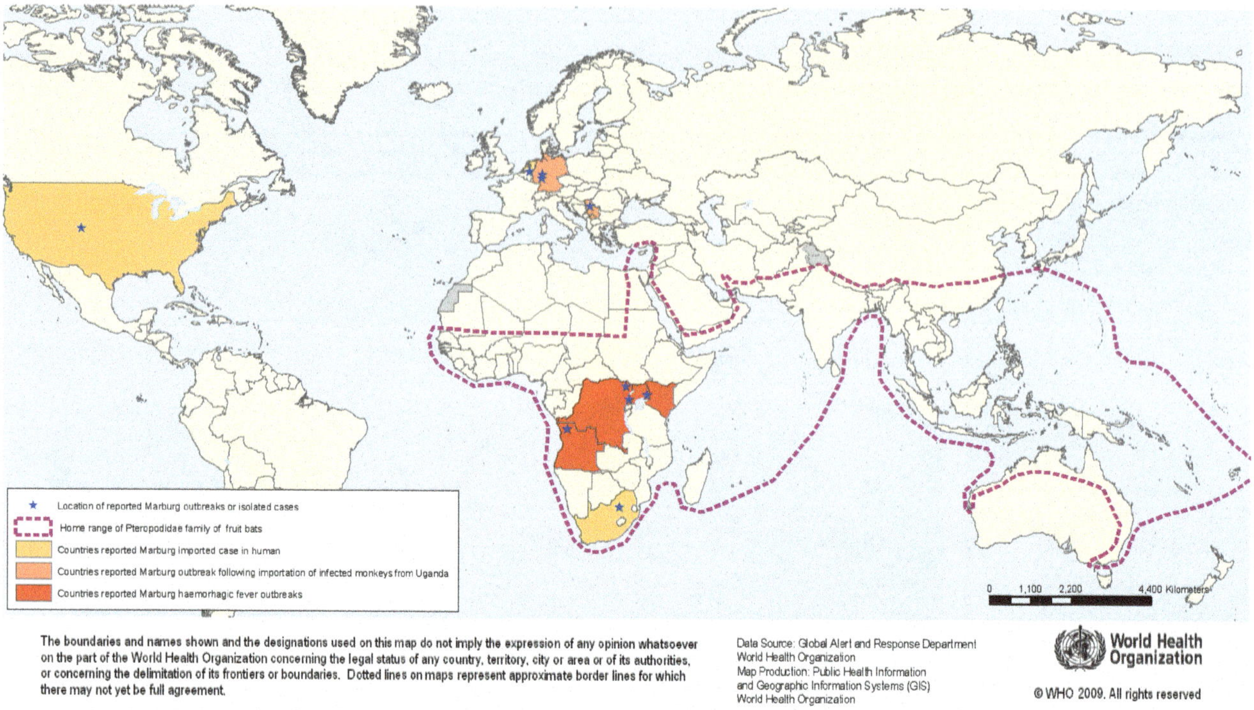

Source: WHO.

Figure 13.26 **Geographic distribution of Marburg hemorrhagic fever outbreaks and fruit bats of Pteropodidae Family.**

people or other animals via body fluids, broken skin, unprotected sex, or exposure to infected tissue. Prolonged exposure to mines or caves inhabited by bats carrying the disease may also result in infection. Disease progression and symptoms are similar to Ebola. The largest outbreak so far was in Angola in 2004 to 2005, with 252 cases and 227 deaths (Figure 13.26). The most recent outbreak, in Uganda in 2017, was successfully controlled within weeks of detection, and resulted in only three deaths, according to the WHO. It was Uganda's fifth Marburg outbreak in 10 years.

Chikungunya Disease and Dengue Fever

Chikungunya disease is caused by a virus that is related to Dengue fever; both are tropical diseases that are transmitted by mosquitoes. In the past, both were restricted to tropical areas around the Indian Ocean, but with modern climate change, tropical diseases such as these are spreading to new areas and can have significantly greater impact (Figure 13.27). In 2007, there was a mysterious outbreak of illness that caused fever, exhaustion, and severe bone pain in Italy, which was later identified as Chikungunya. By 2014 outbreaks were reported from Europe, Asia, Africa, the Caribbean, Central and South America, and North America. Initially only in

Source: CDC.

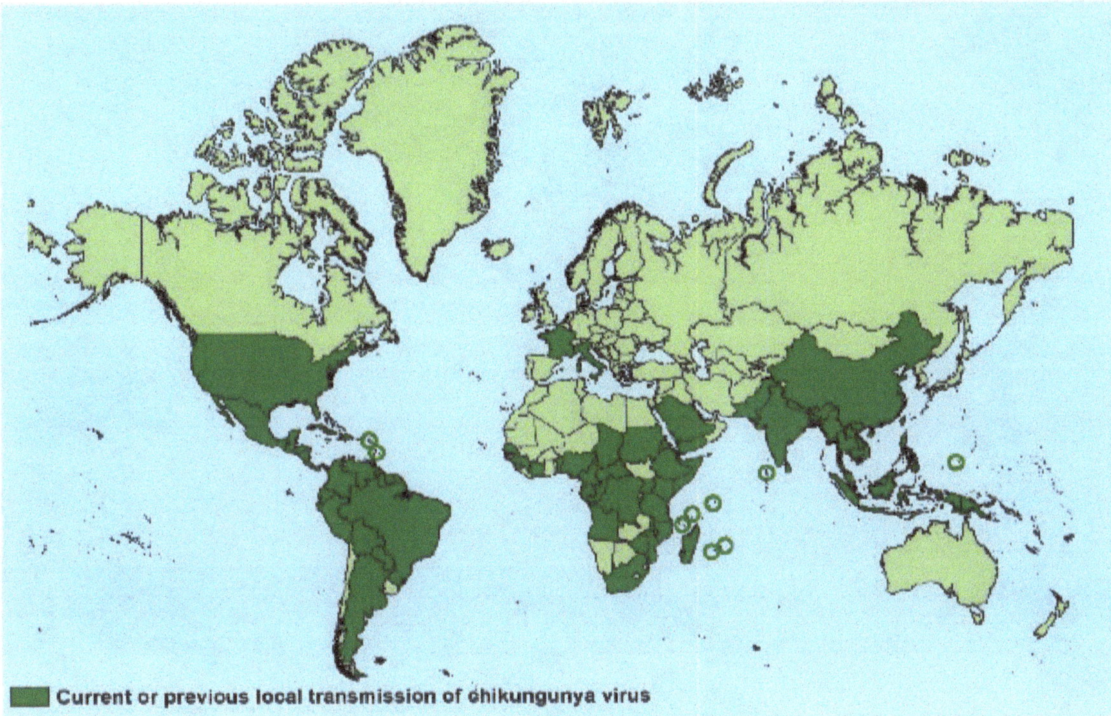

Figure 13.27 **Countries and territories where Chikungunya cases have been reported. This does not include countries or territories where only imported cases have been documented.**

Florida, but as of April 2016 cases have occurred as far north as Alaska. Chikungunya has now been identified in 45 countries and infects three million people each year (Figures 13.28a and 13.28b).

Dengue Fever is also a tropical, mosquito-transmitted viral infection that has epidemic potential, but is not classified as an emergent disease because there are active major disease control and research networks, and mechanisms for intervention, already in place. About half of the world's population is at risk for Dengue Fever, 3.9 billion people in 128 countries. Severe Dengue is a leading cause of serious illness and death among children in some Asian and Latin American countries. Incidences of Dengue have risen dramatically around the world in the last several decades, and according to the WHO up to 390 million people are infected each year, of which 96 million develop mild to severe symptoms of the disease. Not only are the number of cases increasing, but severe outbreaks are also occurring, and the threat of Dengue in places like Europe is real. Outbreaks have been reported in France, Portugal, and Croatia. In the United States cases have been reported from Florida and Hawaii. In 2016 there were large Dengue outbreaks worldwide, with more than 2.38 million in the Americas.

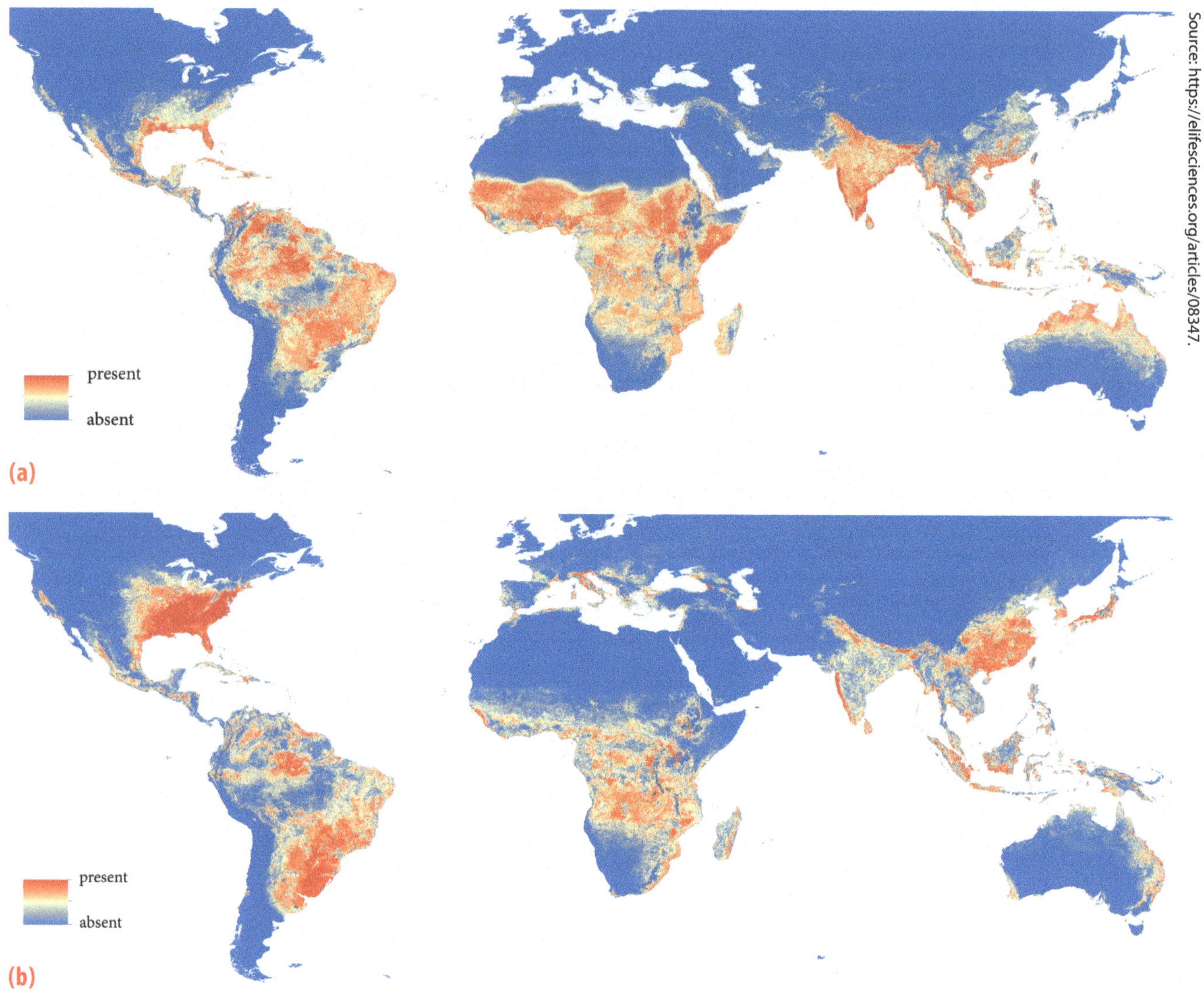

Figure 13.28 Global maps of the distribution of *Aedes aegypti* and *Aedes albopictus* (top and bottom maps, respectively), mosquito species that host and spread Dengue and Chikungunya Fever. This study by Moritz Kraemer and colleagues, published in *Ecology, Epidemiology and Global Health* in 2015, shows the distribution of these mosquito species to be the widest ever recorded. They are now extensive in all continents, including North America and Europe.

Zika Fever

In early 2015, a Zika Fever epidemic which originated in Brazil spread through South and North America. The 2015–2016 outbreak was picked up by social media, and an analysis of tweets on February 2, 2016, showed that there were 50 tweets per minute posted about Zika. The World Health Organization declared this outbreak a public health emergency of international concern.

Zika is a virus hosted by monkeys, and transmission to humans was rare before the current pandemic began in 2007. Zika is spread among monkeys, and to humans, by a mosquito species most active during daytime. Related virus that infect the same mosquito species, such as Dengue, are known to be intensified by urbanization and globalization, and this is thought to be a factor in the current spread. The potential risk of Zika is dependent on the distribution of the mosquito vector that transmits it, and the global distribution of this species is the most extensive ever recorded—across all continents including North and South America and Europe—and expanding due to global travel and trade. This mosquito species may be adapting to colder climates and has been found in Washington, DC, where they can now survive through the winter, according to Dr. David Severson, expanding their range from the tropical and subtropical areas where they are commonly found. Zika can also be transmitted from men and women to their sexual partners, most cases are from infected men to women.

Most people infected with Zika have few or no symptoms, and those who develop symptoms have fever, joint pain, headache, and rash. Symptoms last about a week and no deaths have been reported due to the initial infection. A much greater problem with Zika is the risk of transmission from a pregnant mother who becomes infected to her unborn child. The disease can then cause birth defects such as underdevelopment in the fetal brain leading to microcephaly. Adults who have been infected may have increased risk of developing Guillain-Barre Syndrome, a rapid-onset muscle weakness caused by the immune system damaging the central nervous system. This can be life-threatening in about 8% of victims, especially those who develop weakness of the breathing muscles or abnormalities in heart rate or blood pressure.

Because of the evidence for a link between Zika infection and birth defects and neurological problems, the Centers for Disease Control issued a travel alert in early 2016 advising pregnant women to consider postponing travel to areas with ongoing Zika virus cases. The advice was updated to caution pregnant women to avoid these places entirely, and several countries in Latin America and the Caribbean issued warnings for women to avoid pregnancy until more is known about the virus and its impact on fetal development.

Animal Influenza Viruses (Zoonotic Influenza)

Humans can be infected by avian, swine, and other animal-based influenza viruses. These infections are mainly from direct contact with infected live or dead animals or contaminated environments. Symptoms of animal-based flu virus infections called zoonotic influenza, range from mild flu-like symptoms to death.

Flu viruses infect many different species of animals, such as birds, seals, pigs, cattle, horses, and dogs, in addition to humans. While flu viruses are adapted to the specific species that hosts them, they can jump to other species, such as the **Avian flu** outbreak of **H5N1** in 1997 which has been linked back to a goose (Figure 13.29). Species such as pigs, which can be infected with swine, avian and human viruses allow the virus genes to mix and can facilitate the creation of a new virus which people may have little to no immunity from. The diversity of zoonotic influenza viruses is "alarming" according to the WHO, and necessitates strengthening surveillance and pandemic preparedness planning.

avian flu (H5N1)
An acute viral disease of chickens and other birds (except pigeons) capable of being transmitted to humans; first noticed in the Far East, its symptoms include fever and lack of energy.

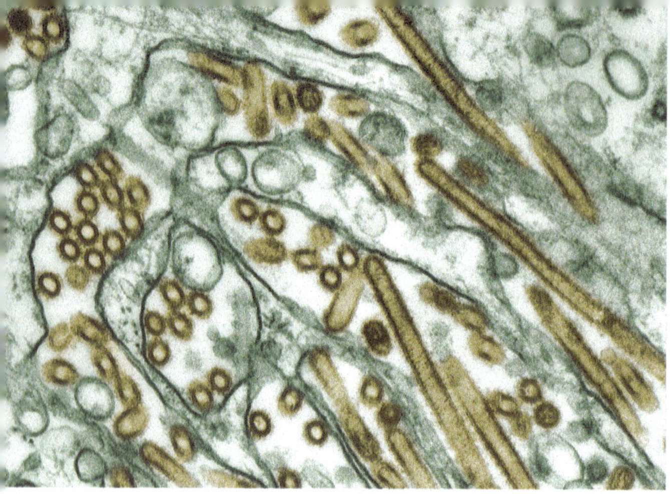

Source: Centers for Disease Control and Prevention's Public Health Image Library image #1841.

Figure 13.29 Transmission electron micrograph of the avian flu virus, shown in yellow.

Prion Diseases

Prion diseases are part of a class of transmissible spongiform encephalopathies (TSE) that are characterized by a malformed protein molecule that produces clumps in the brain. Bovine spongiform encephalopathy (BSE or mad-cow disease) is a TSE, as is scrapie, a nervous system disease in goats and sheep, and chronic wasting disease in deer. Human consumption of a diseased animal can lead to the spread of TSE among humans. Once contracted, prion diseases are always fatal.

prion disease
An infliction that attacks the brain and nervous system, and disrupts normal protein activity within the neural cells.

TSEs are spread by prions, which are only protein material, unlike other diseases which are spread by agents with DNA or RNA, such as virus or bacteria. Transmission occurs when a healthy animal consumes tainted tissue from one with the disease. BSE spread as an epidemic in cattle in the 1980s and 1990s because cattle were fed processed remains of other cattle, a practice which is now banned in many countries. People who ate the infected cattle also contracted prion disease, called variant Creutzfeldt-Jacob disease (vCJD), which is a fatal brain disease with an extremely long incubation period of many years (Figure 13.30). Prions can't be transmitted through the air, by touch, or most other forms of casual contact. They can be transmitted through contact with infected tissue, body fluids, or contaminated medical instruments. Normal sterilization procedures such as alcohol, boiling, acid, or irradiation do not destroy the prions. Infected brains that have been sitting in formaldehyde for decades can still transmit spongiform disease.

Included in the human category of TSEs is Kuru, a rare brain disorder that reached epidemic proportions in the mountains of New Guinea in the 1960s. Natives called Kuru the "laughing disease" and thought it was caused by sorcery. Symptoms include shaking, dementia, and uncontrolled

Figure 13.30 This brain tissue has been magnified 100 times and shows prominent spongy texture in the cortex and loss of neurons—the patient was diagnosed with the prion disease Creutzfeldt-Jakob (vCJD). This was first diagnosed in the UK in 1996 and results from eating cows infected with bovine spongiform encephalopathy (BSE or "mad cow" disease), which is the same agent responsible for the outbreak of vCJD in humans.

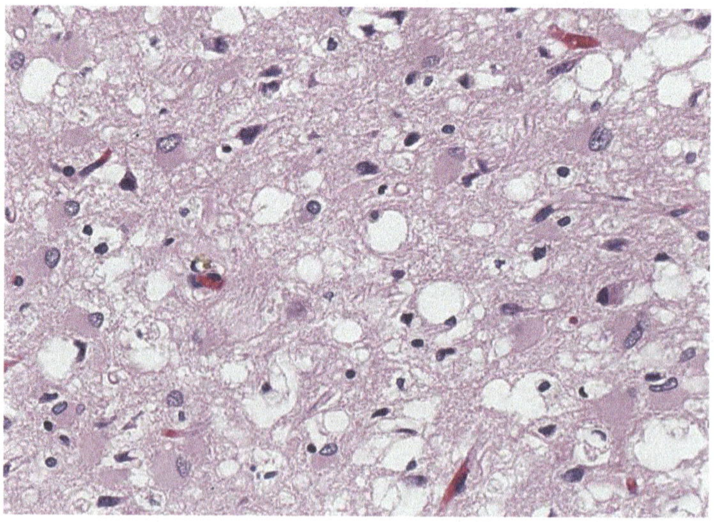

Source: CDC Public Health Image Library #10131.

crying and laughing. It incubates for about 10 years and victims die within 12 months of developing symptoms. People suspected of being sorcerers and inflicting Kuru on others were murdered brutally, by biting their trachea and using clubs or stones to crush their genitals. As a result of rituals involving mortuary cannibalism among some of the indigenous tribes, Kuru spread to a large percentage of the population, but it has now essentially disappeared due to government intervention.

severe acute respiratory syndrome (SARS)

A respiratory disease of unknown source that first occurred in mainland China in 2003; symptoms include fever and coughing or difficulty breathing; it is sometimes fatal.

SARS and MERS

SARS (severe acute respiratory syndrome) and MERS (Middle East respiratory syndrome) are infectious respiratory diseases caused by coronaviruses, which have a crown-like appearance in microscopic view, and can both be deadly to humans (Figure 13.31). SARS is characterized by severe pneumonia-like symptoms, and is transmitted from person to person through respiratory droplets produced by sneezing or coughing and through direct contact with a surface contaminated with respiratory droplets. It was first identified in Guangdong Province, China, in November of 2002, and by July 2003 there were 8,098 cases in 37 countries resulting in 774 deaths. No cases of SARS have been reported since 2004.

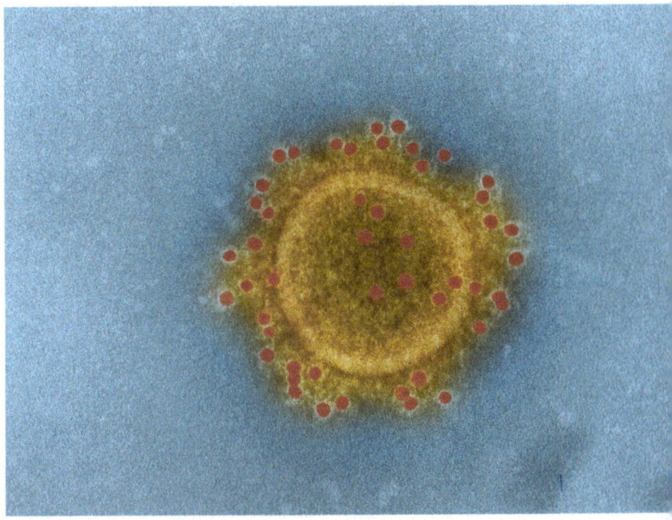

Source: NIH NIAID.

Figure 13.31 Middle East Respiratory Syndrome Coronavirus (MERS-CoV) is a viral respiratory illness first reported in Saudi Arabia in 2012. Both MERS and SARS (Severe acute respiratory syndrome) belong to a family of viruses called coronaviruses.

The SARS epidemic was brought to the public spotlight in February of 2003, when an American businessman flying back from China got extremely sick on the airplane to Singapore. The flight made an emergency landing in Vietnam and the victim died. Several of the medical professionals who treated him in the hospital also died, and Dr. Carlo Urbani identified the risk and alerted the World Health Organization; he later died of SARS also.

The source of the virus was traced to cave-dwelling horseshoe bats in Yunnan Province, China, in late 2017 (Figure 13.32). Antibiotics are ineffective since SARS is a viral disease, and there is no vaccine. Isolation and quarantine are the most effective means of preventing the spread of SARS. Because SARS is most infectious in extremely sick patients, which usually occurs during the second week of illness, quarantine can be highly effective if infected people are isolated in the first few days of their sickness.

© All-stock-photos/Shutterstock.com.

Figure 13.32 Cave-dwelling horseshoe bats like these, living in Yunnan Province, China, are thought to be the source of the SARS epidemic that killed 774 people.

MERS was first identified in the Arabian peninsula in 2012, and MERS is similar to SARS because it is a severe respiratory illness that can be fatal to humans, with a death rate of from 30 to 40%. MERS is also thought to

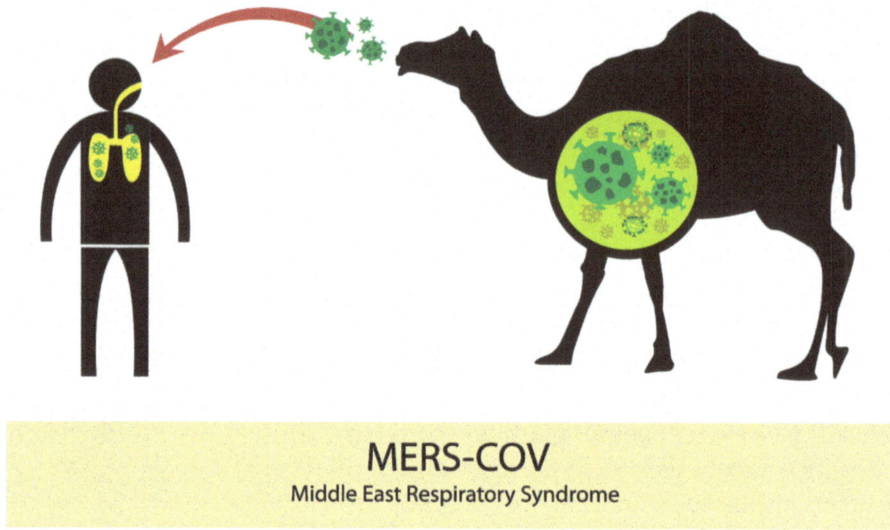

© Crystal Eye Studio/Shutterstock.com.

Figure 13.33 **MERS is transferred from infected camels to humans through close contact.**

originate in bats, but was passed to camels which then infected humans (**Figure 13.33**). Spread between humans requires close contact with an infected person, and because of that, so far the risk to the global population is considered to be relatively low. As of February of 2018, the WHO has identified 2,143 confirmed cases, with 750 MERS-related deaths in 27 countries. The two largest localized outbreaks so far have been in Saudi Arabia in April of 2014, with 688 people infected and 282 deaths, and in South Korea in May of 2015. The first case was a man who had visited Saudi Arabia, and as of June 2015, when it was finally contained, a further 186 cases of infection and 36 deaths occurred there.

Evidence for Epidemics in the Geologic Record

Scientist Jie Fei and colleagues found evidence that the 1600 A.D. Huaynaputina eruption in Peru, the largest volcanic eruption over the last 2,000 years, caused abrupt and unseasonably cold summer temperatures in both China and the Korean Peninsula, based on historical literature and tree ring dendrochronology. Widespread disease outbreaks and epidemics after the volcanic eruption have been linked to these unusually cold conditions. These types of events that can precipitate conditions that support disease outbreaks and epidemics are not uncommon in Earth's history.

It is likely that epidemics have occurred frequently in the geological past, affecting humans, our earlier ancestors, and other animals. Malaria, once thought to be a disease of modern origins, has been traced back and identified in insects that lived 100 million years ago, according to research by

George Poinar, Jr. These early forms of the disease, carried by biting midges, are at least 100 million years old, and probably much older, according to Poinar. Modern malaria vectored by mosquitoes is at least 20 million years old. Poinar used insects preserved in amber to discover this fossil malaria (Figure 13.34). He also suggests that the extinction of the dinosaurs 65 million years ago, which has been linked to catastrophic events such as meteorite impact and flood basalt eruptions, may have been helped along by insects carrying disease. Insects, microbial pathogens, and vertebrate diseases were all evolving at about that time, and may have negatively impacted the dinosaurs as well as had a huge effect on animal evolution.

© Matteo Chinellato/Shutterstock.com.

Figure 13.34 ***Priscoculex burmanicus*** **is a newly identified genus and species of anopheline mosquito, one that carries malaria, and existed 100 million years ago.**

Lessons From the Historic Record and the Human Toll

Many serious diseases have existed in the past and continue to do so now. Throughout human history, countless millions of people have died from diseases, most of them related to pandemic outbreaks. Because the sources of some diseases are difficult to eradicate, such as wiping out all mosquitoes worldwide, it will be virtually impossible to completely eliminate many of these maladies.

However, there is a glimmer of hope. Smallpox is an example of a success story. A disease that was once a dreaded scourge on Earth, it has been put into the extinct category, as it has been eradicated after a very concerted, worldwide effort. The CDC reported that the last recorded case in the United States was in 1949 and the last one in the world was in Somalia in 1977. Although the disease is eradicated, samples of the virus are kept under guard at the Centers for Disease Control for future research and the production of a smallpox vaccine, if needed.

The Anti-Vaccine Movement

Almost 9 million children in the US alone are at risk of contracting measles from under-vaccination. People who are reluctant to vaccinate themselves or their children, called anti-vaccers, are one of the top 10 global health threats, according to the World Health Organization in 2019. Despite overwhelming scientific consensus that vaccines are safe and effective, some people choose to delay or avoid vaccinating their children, which results in outbreaks of preventable diseases such as measles, whooping cough, polio, and mumps. Inoculation or vaccination, also called variolation, first began in North America and England in 1721 as a means to fight the smallpox epidemic. Smallpox was successfully eradicated worldwide in 1980 as a result

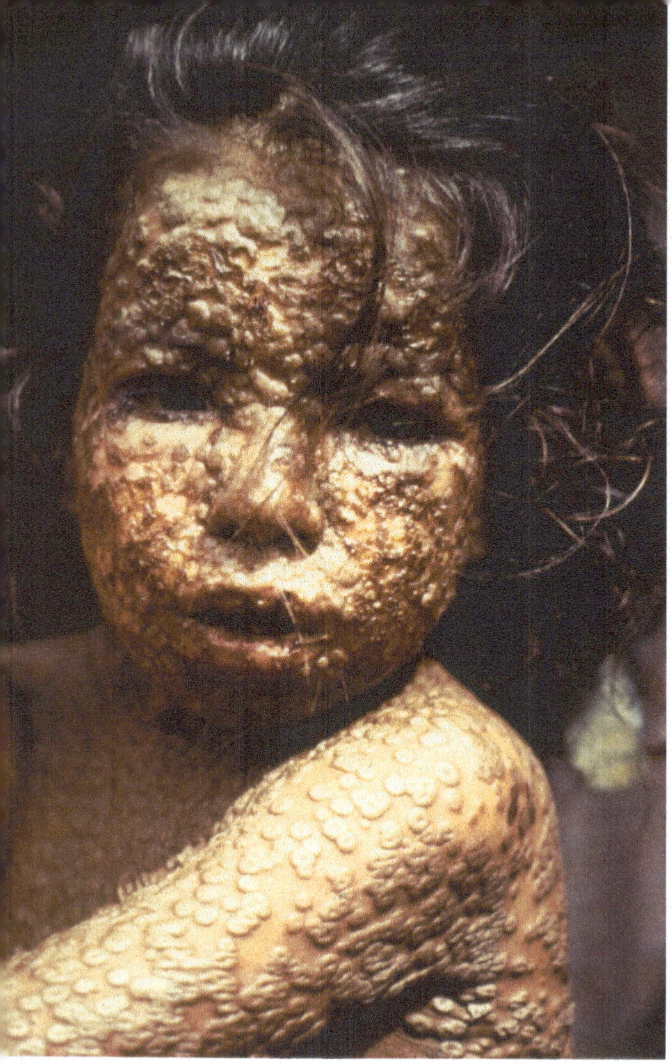

Figure 13.35 A young girl in Bangladesh infected with smallpox in 1973. Bangladesh was officially declared smallpox-free in 1977.

of vaccinations. It had previously killed up to 30% of all victims, with higher rates among babies and the young with up to 1 in 7 children dying (Figure 13.35). Vaccines are effective, they save the lives of 33,000 children born each year and prevent 14 million infections each year in the US alone. Incomplete vaccine coverage increases the risk of disease for the entire population, including those who have already been vaccinated, and threatens those who are too young, too old, immunocompromised, or otherwise unable to be vaccinated for medical reasons such as severe allergies to the ingredients in the vaccine. If enough of a population is vaccinated, **herd immunity** takes effect and the overall risk to all people is decreased.

Minimizing Biological Hazards

One means of reducing the spread of insect-borne diseases is the use of insecticides. Perhaps the most widely used has been DDT (dichloro-diphenyl-trichloroethane), first synthesized in 1874 by a German chemistry student as a thesis project. In 1939, its ability to kill insects was discovered by the Swiss scientist Paul Müller (Nobel Prize in physiology or medicine in 1948). Several countries used it during World War II to rid areas of lice that carried typhus. Following the war, it was used in numerous countries to kill mosquitoes that carried malaria. However, the use of DDT had profound effects on many forms of wildlife, including the embryos of bald eagles and peregrine falcons. Its use in the United States was banned in 1972. However, it is still used in many tropical regions to curtail the spread of mosquitos.

herd immunity

A form of indirect protection from infectious diseases when a sufficient percentage of the population has become immune to an infection through previous infection or vaccination.

An example of the eradication of a disease in a developed country occurred in 1951 when malaria was officially declared no longer a threat in the United States. This happened after a 4-year plan had been carried out to spray millions of homes and their surroundings to kill the carriers. As late as 1947, 15,000 cases of malaria were reported in the United States; by 1951 there were none.

Safe water sources and clean living and sanitary environments will help minimize many diseases that are associated with poverty. Diseases can move with people who are sick whenever they are trying to avoid a pending natural disaster, such as an imminent volcanic eruption. We must all be vigilant to prevent the spread of diseases that can become a challenge to our very existence. Global warming will enhance the spread of diseases that thrive in warmer temperatures. Malaria and yellow fever will spread when mosquitoes are able to expand their habitats across the globe.

Zombie Apocalypse Preparedness

Zombies are reanimated corpses with a taste for brains. Their cause is often attributed to a virus passed through an infected human host by bites, or wounds that come in contact with infected bodily fluids. The Zombie Survival Guide identifies the source as the virus Solanum. Harvard Medical School psychiatrist Dr. Steven C. Schlozman diagnosed Night of the Living Dead zombies as suffering from Ataxic Neurodegenerative Satiety Deficiency Syndrome, attributed to an infectious agent released when a radioactive space probe exploded in Earth's atmosphere. Other (fictional) zombie plagues have been blamed on mutations in prions, mad-cow disease, measles, and rabies. No matter the cause, the preparation is the same—have an emergency kit in your house, ready for any disaster, zombie-based or otherwise (Figure 13.36). Emergency kits, based on CDC guidelines, should include:

- Water (4.5 liters/1 gallon per person per day)
- Food (non-perishable items)
- Medications (7-day supply of prescription and non-prescription)
- Tools and supplies (utility knife, duct tape, battery-powered radio, and spare batteries)
- Sanitation and hygiene (household bleach, soap, towels)
- Clothing and Bbedding (a change of clothes for each person in the household, blankets)
- Important documents (copies of driver's license, passport, and birth certificate, plus other important paperwork)
- First Aid supplies (for basic cuts and lacerations, won't help with zombie bites)

You should also have a household emergency plan, listing who you would call and where you would go in case of emergency. The first step is to identify the types of emergency that might occur in your area, pick a meeting place in case you need to flee your home, one right outside your house for sudden emergencies and another outside your neighborhood in case you are not able to get to, or return to, your home. Make a list of all emergency contacts including police, fire departments, and out-of-state relatives. Have this printed out, in case your phone does not work. Finally, plan an evacuation route in case you need to leave home. Have multiple routes in case one road or highway is blocked.

Source: CDC.

Figure 13.36 CDC guidelines for Zombie Apocalypse preparedness are also useful for other disasters.

Key Terms

acquired immunodeficiency syndrome (AIDS) *(page 374)*
Avian flu (H5N1) *(page 395)*
bacteria *(page 365)*
biological hazard (biohazard) *(page 364)*
Black Plague (Black Death) *(page 368)*
bubonic plague *(page 367)*
cholera *(page 369)*
epidemic *(page 366)*
fungus *(page 364)*
hemorrhagic fever *(page 364)*
hepatitis *(page 376)*
herd immunity *(page 400)*
human immunodeficiency virus (HIV) *(page 374)*
influenza *(page 371)*
leprosy (Hansen's disease) *(page 377)*
malaria *(page 378)*
microbe *(page 365)*
pandemic *(page 366)*
parasite *(page 364)*
pathogen *(page 365)*
prion disease *(page 396)*
protozoa *(page 364)*
severe acute respiratory syndrome (SARS) *(page 397)*
smallpox *(page 372)*
typhoid fever *(page 383)*
typhus *(page 372)*
vector *(page 378)*
virus *(page 365)*
yellow fever *(page 373)*

Summary

Natural disasters often set up conditions that are conducive to the development of disease outbreaks in survivors—they concentrate large numbers of people into small areas with poor sanitation, lack of health services, and safe drinking water. Biological hazards are any organism or substance that threatens health or the environment. They can be living or replicating pathogens—any agent capable of causing disease—such as bacteria, virus, protozoa, parasites, or fungus. Microbes infect the host, which provides them with energy and nutrition. They live off the host, which now becomes a vector for spreading the disease. Epidemics are when a disease spreads rapidly to a large number of people in a given area; pandemics occur over a wide geographical are or worldwide.

Pandemics have occurred in the historical record, and the earliest one recognized was an outbreak of typhoid fever, then called the Plague of Athens, in 430 through 426 B.C. The Antonine Plague occurred in A.D. 165 to 180 (attributed to smallpox), and killed almost 5 million people. The Bubonic Plague started in A.D. 541 and wiped out 25% of the entire population of the Mediterranean region. The Black Death occurred in the 1300s, and influenza pandemics started in the late 1800s.

Bubonic plague is transferred to humans from fleas, and caused the Black Death, which killed an estimated 50 million people in the 1300s. Cholera results from a bacterial infection contracted through contaminated drinking water or contact with other infected people. It generally occurs in overcrowded conditions with poor sanitation, which are common after disasters. The 2010 Haiti earthquake displaced 1.5 million people and refugee camps suffered cholera outbreaks that have killed more than 9,000 people. Influenza virus produces the flu, which affects millions of people every year. Influenza pandemics result from unusually strong strains and have struck on average three times each century since the 1500s. The 1918 influenza pandemic killed more people than World War 1 and is the most devastating epidemic in recorded world history. Smallpox is a viral disease that has killed more than 300 million people in the 21st century—it has now been globally eradicated through intensive vaccinations. Typhus is spread by body lice and occurs in areas of poor sanitation, and was a major cause of death during World War II. Yellow fever is transmitted by mosquitoes, and was a major disease in tropical regions—there is currently a vaccine that provides immunity for 10 years. HIV infection is caused by a retrovirus and can lead to AIDS in humans. This pandemic was first diagnosed in 1981 in the United States, although earlier cases are believed to have occurred as far back as the late 1950s. More than 78 million people have been infected with HIV and 35 million people have died from AIDS-related illnesses in the last 35 years. Hepatitis is a liver disease caused by viral infection, and type B affected 343 million people and caused 1.34 million deaths in 2015 alone. Leprosy is a bacterial infection that attacks the skin and nerves, first described in India in 600 B.C. and mentioned in the Bible. Malaria is a mosquito-borne infection that has killed half of all people who have lived on Earth. Today it still infects more than 200 million people each year, and kills a child under the age of 5 every 2 minutes. Tuberculosis has killed over a billion people in the last 200 years, and has been found in Egyptian mummies from 5000 B.C. The World Health Organization considers this to be a "fire raging out of control"—1,945 people die from TB every day. Typhoid Fever is caused by bacteria and while a vaccine can prevent infection, it is still common in developing countries and antibiotics-resistant strains make treatment complicated.

Emerging infectious diseases are those that were previously unknown, have spread to new locations, have just been identified in causing disease, or represent disease that has re-appeared and threaten to increase. Viral hemorrhagic fevers are produced by RNA virus and cause severe, multisystem syndromes. These include Ebola, Lassa and Marburg Hemorrhagic Fevers. Chikungunya Disease and Dengue Fever are both spread by mosquitoes, and as climate change increases the range of tropical mosquitoes carrying these diseases, their occurrence has spread across the globe. Zika fever is another mosquito-borne disease that became an international concern during an outbreak in 2015–2016. Zoonotic Animal Influenza Viruses such as Avian flu are transmitted from animals such as swine, birds, and dogs to humans, and so far have not had sustained transmission among humans. Prion disease attacks the brain and nervous system, and is transmitted by eating infected tissue. Once contracted, prion disease is always fatal. BSE or mad-cow disease can be transmitted to humans, and Kuru is a prion disease contracted through cannibalism. SARS and MERS are respiratory syndromes caused by coronaviruses transmitted to humans from bats and camel, respectively.

Epidemics are likely to have occurred throughout the geologic record, and there is evidence for malaria among dinosaurs more than 100 million years ago. Vaccines help protect us through "herd immunity" and provide protection for those that are too young or too old to be vaccinated, those who are immune compromised, or those unable to get vaccinated due to severe allergies to the ingredients in vaccines.

References and Suggested Readings

Brief History of Tuberculosis. (2001, November). www.umdnj.edu/~ntbcweb/history.htm

Brothers, M. (2003). *The zombie survival guide.* Broadway Books.

Bryder, L. (1988). *Below the magic mountain: A social history of tuberculosis in the twentieth century.* Oxford University Press.

Ewald, P. W. (1994). *Evolution of infectious disease.* Oxford University Press.

Feldberg, G. D. (1995). *Disease and class: Tuberculosis and the shaping of modern North American society.* Rutgers University Press.

Giesecke, J. (2002). *Modern infectious disease epidemiology* (2nd ed.). Arnold.

Herlihy, D. (1997). *The black death and the transformation of the West.* Harvard University Press.

Hurster, M. M. (1997). *Communicable and non-communicable disease basics—A primer.* Bergin and Garvey.

Lashley, F. R., & Durham, J. D. (Eds.). (2002.) *Emerging infectious diseases—Trends and issues.* Springer.

Nikiforuk, A. (1991). *The fourth Horseman—A short history of epidemics, plagues, famines and other scourges.* M. Evans.

Rifkind, D., & Freeman, G. L. (2005). *The Nobel Prize winning discoveries in infectious diseases.* Elsevier Academic Press.

Ryan, F. (1992). *Tuberculosis: The greatest story never told.* Swifth Publishers.

Watson, J. T., Gayer, M., & Connolly, M. A. (2007). Epidemics after natural disasters. *Emerging Infectious Diseases, 13*(1), 1. https://doi.org/10.3201/eid1301.060779

Ziegler, P. (1991). *The black death.* Alan Sutton.

Websites for Further Reference

https://www.cdc.gov/

https://www.cdc.gov/ncidod/dbmd/diseaseinfo/

https://www.cyndislist.com/medical.htm

https://www.infoplease.com/ipa/A0903696.html

https://www.mla-hhss.org/histdis.htm

https://virus.stanford.edu/uda/

Johns Hopkins:

https://gisanddata.maps.arcgis.com/apps/dashboards/bda7594740fd40299423467b48e9ecf6

http://www.healthdata.org/

Questions for Thought

1. What is a biological hazard and what form can it take?
2. Distinguish between a bacterium and a virus.
3. Explain the difference between an epidemic and a pandemic.
4. Why was the Black Plague so named?
5. List the major diseases that are spread by insects.
6. How does a warming climate affect the potential spread of diseases?
7. Why is the "Spanish flu" outbreak in 1981 so named, and where did it begin?
8. List major diseases that are readily passed along from one person to another by means of direct contact.

Environmental Sustainability and Our Role in Earth's Future

14

© Jag_cz/Shutterstock.com.

Earth's natural environment is a complex and multifaceted living organism.

Biological Annihilation and the 6th Mass Extinction: Us

Planet Earth is in the middle of the sixth wave of mass extinction of plants and animals in the last half-billion years—this one caused almost entirely by humans. Right now, 99% of threatened species are at risk from human activities, mainly those causing habitat loss, global warming, and from invasive species. We are presently living through the highest extinction rates since the dinosaurs 65 million years ago. Extinctions are part of the evolution of our planet, and "background" rates are from one to five species each year. We are currently losing species at 1,000 to 10,000 times this background rate. More than 37,400 species are threatened with extinction, and that is from an analysis of only 130,000 of the 1.8 million species on Earth, according to the International Union for Conservation of Nature. The dodo, the great auk, the thylacine, Chinese river dolphin, passenger pigeon, imperial woodpecker, all have been driven extinct, some deliberately. Using cloning to "de-extinct" species, or bring them back, sounds like science fiction but could become a reality. However, putting species back after they've gone extinct is loaded with its own problems. The Earth is a delicately balanced system and our actions have a profound impact on all aspects of our planet.

© The Art of Pics/Shutterstock.com.

Figure 14.0.1 **The dodo was a large (1 m, 3.3 ft) flightless bird from Maritius that was last seen alive in 1662. It was driven to extinction within less than a century of its discovery by hunting and habitat loss.**

Human Population Growth

In Chapter 1 we learned that global population has been estimated at eight million inhabitants 10,000 years ago. With the domestication of animals and the development of agriculture that allowed people to move away from hunter-gatherer societies, the number of people on Earth began to rise. Growth rates are estimated to have been fairly low for the next 11,000 years, but by the Middle Ages from the years 1100 to 1500 A.D. almost 500 million people were living on the planet (**Figure 14.1**). A setback occurred between 1348 and 1350 A.D. when the Black Plague struck Europe, killing tens of millions of people (refer to Chapter 13). The world population rebounded in the 1400s and by 1800 it was slightly less than one billion.

The world is expected to reach a global population of 8 billion people by 2023. This represents 2 centuries of rapid population growth with a sevenfold increase in world population over that time. Population growth peaked at 2.1% in 1968 and has slowed to just over 1% per year since then.

- It reached 1 billion in 1804;
- It took 123 years to reach 2 billion in 1927;
- 33 years to reach 3 billion in 1960;
- 14 years to reach 4 billion in 1974;

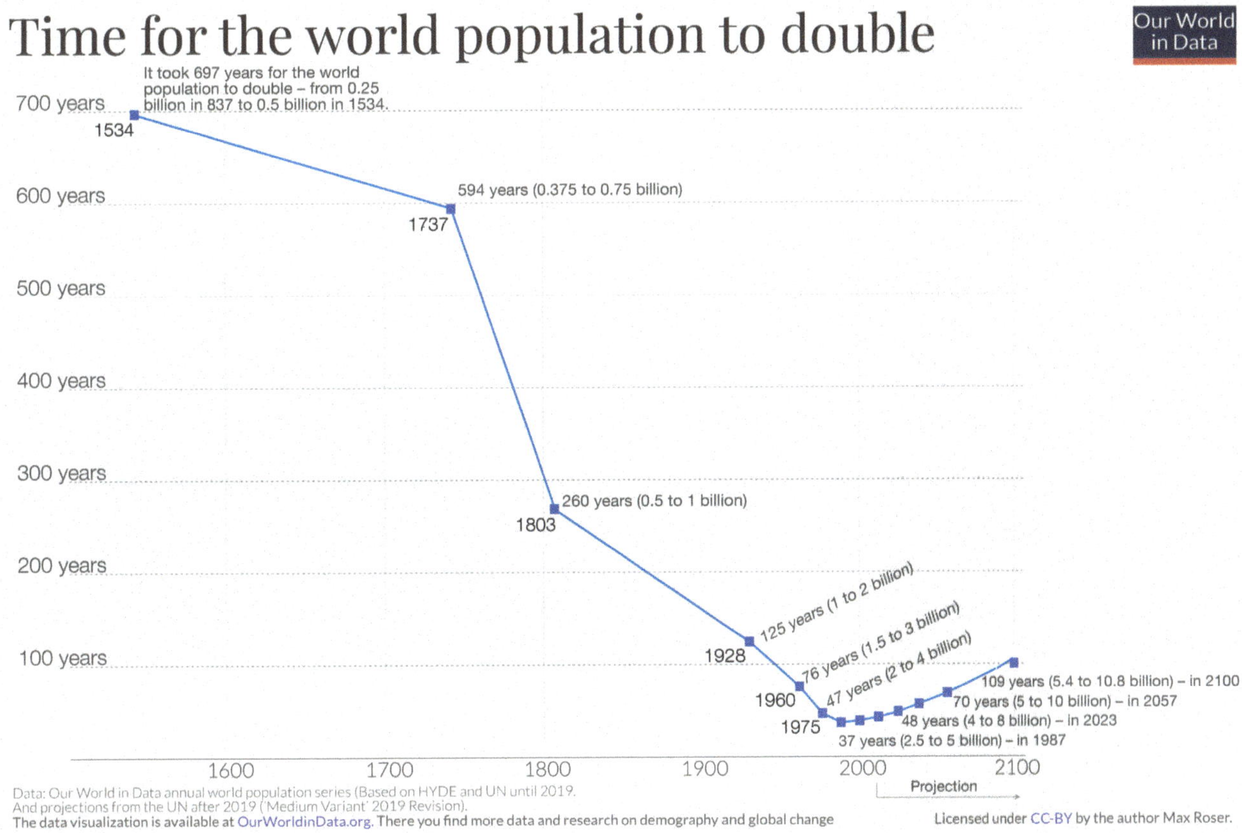

Source: Max Roser, Hannah Ritchie, & Esteban Ortiz-Ospina. (2013). World Population Growth. *Published online at OurWorldInData.org.* Retrieved from https://ourworldindata.org/world-population-growth.

Figure 14.1 The fastest doubling of the planet's population happened between 1950 and 1987.

- 13 years to reach 5 billion in 1987; and
- 12 years to reach 6 billion in 1999.

The United Nations estimated that global population would reach 7 billion in 2013, but according to National Geographic, the planet reached 7 billion people on October 31, 2011 (Figure 14.2). There has been a decline in the annual growth increment from about 2% in 1968 to 1.1% in the early 21st century. However, it is still a positive growth rate. Current global human population growth is about 1.1% per year, or almost 83 million people annually.

Contributors to the exponential growth in population worldwide include increased industrialization in Europe in the 18th century; increased food sources and improved nutrition; the eradication of certain diseases; and development of medicines to improve longevity. Unfortunately diseases such as cholera, influenza, malaria, and tuberculosis continue to kill hundreds of millions of people each year, mainly in underdeveloped countries that lack sanitation and good health services. Childhood diseases also take a huge toll in poor nations, where infant mortality exceeds 10%.

Life expectancy is a key metric that captures the mortality of a population along the entire life course, and tells us the average age of death in a population (Figure 14.3).

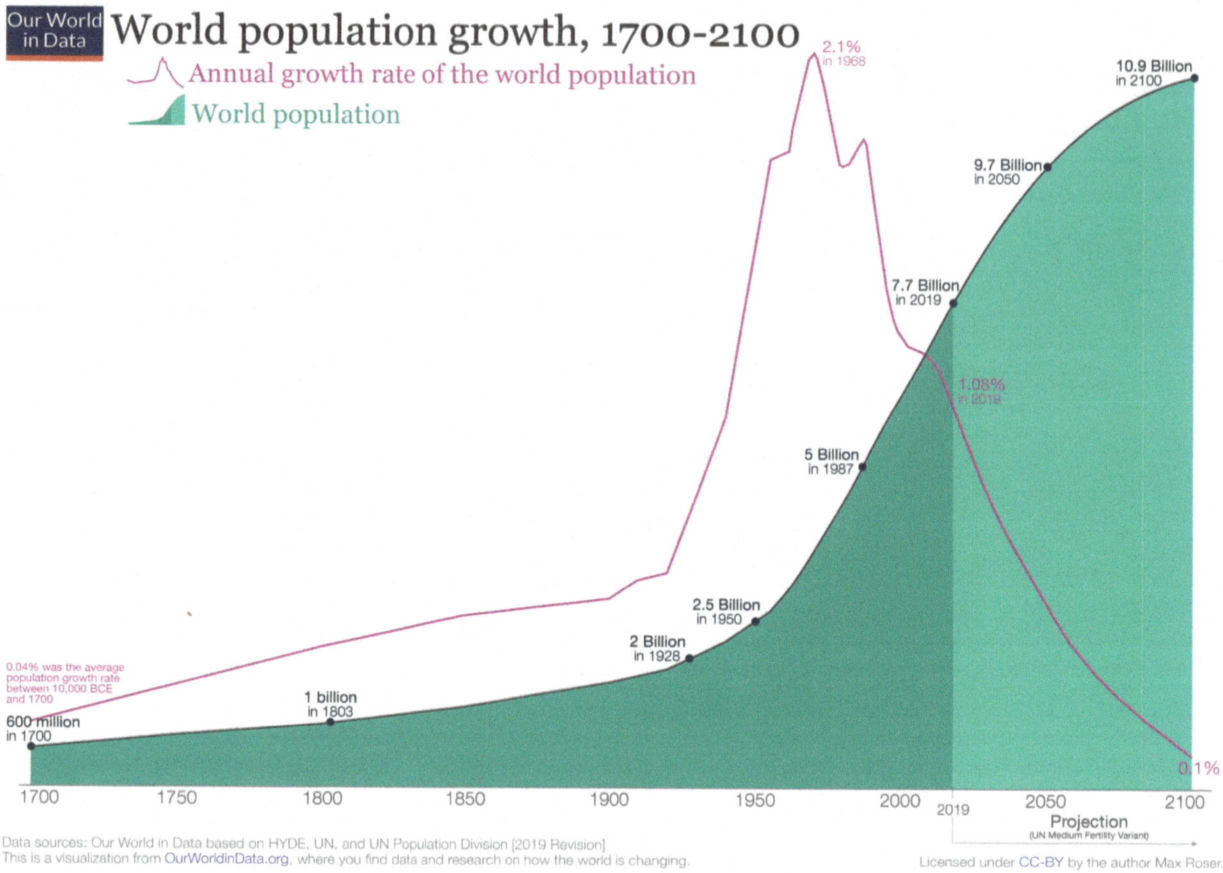

Source: Max Roser, Hannah Ritchie, & Esteban Ortiz-Ospina. (2013). World Population Growth. *Published online at OurWorldInData.org*. Retrieved from https://ourworldindata.org/world-population-growth.

Figure 14.2 **Every year 140 million people are born and 58 million die, which means that we add 82 million people to our planet's population.**

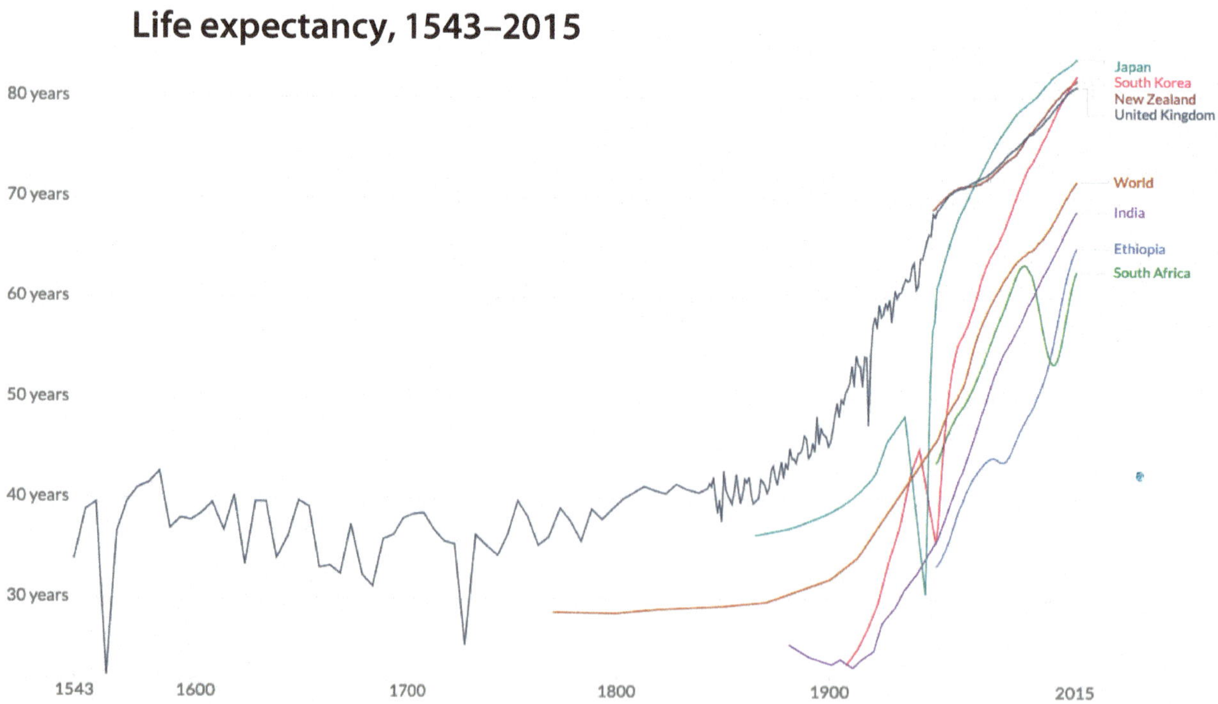

Source: Max Roser, Hannah Ritchie, & Esteban Ortiz-Ospina. (2013). World Population Growth. *Published online at OurWorldInData.org*. Retrieved from https://ourworldindata.org/world-population-growth.

Figure 14.3 **Life expectancy is a key metric for assessing population health. In a pre-modern world, life expectancy was about 30 years everywhere. It has improved dramatically since then, but there are still huge differences between countries today.**

The Effects of Overpopulation

Land Use

The continued exponential population growth worldwide has placed additional demands on Earth's resources, both those that occur naturally such as water and oil, and on resources that are produced, such as agricultural goods. Destruction of the landscape has had a profound effect on the environment. Our natural resources have not increased, only our ability to find them. Essentially no new land is created (tens of thousands of years are often necessary for volcanic material to form productive soils).

© igorstevanovic/Shutterstock.com.

Figure 14.4 Urbanization means that people are becoming concentrated in small areas, more than half of the world's population is currently living and working in cities. This results in huge social, economic, and environmental transformations.

Urbanization has placed a higher percentage of people into compact cities (Figure 14.4). In the United States only 3% of the area of the country is categorized as urban (60 million acres). Often these municipalities are built at the expense of farmland, areas once used to produce crops and raise animals.

The Department of Agriculture (USDA) reports that 97% (2.2 billion acres) of the land in the United States is classified as agricultural, forest, or other use land. In the United States the population explosion in the Southwest has transformed once **desert** areas into bedroom communities. In recent years the overbuilding in cities such as Phoenix and Las Vegas has resulted in thousands of unsold homes and a significant decrease in home values due to a glut of housing. Parts of southern California and southern Arizona that were once rich agricultural areas have been reshaped into suburbs populated by large numbers of people.

desert
A region that receives less than 25 cm (10 in) of annual precipitation.

Agriculture

The Earth's surface is dominated by oceans, with only 29% land (Figure 14.5). Of that, 10% is covered by glaciers and 19% is made of deserts, salt flats, beaches and sand dunes, and barren rock—all of this is barren and uninhabitable land. This leaves 71% of the land surface as habitable areas. For almost all of human history, most of the Earth was wilderness made up of forests, grasslands, and shrub lands. One thousand years ago less than 4% of the planet's ice-free and non-barren land area were used for farming. Today that value is half—50% of all habitable land is used for agriculture. The remaining half of our planet's surface is 37% forests, 11% shrub and grasslands, 1% is freshwater areas and wetlands, and the remaining 1 percent is built-up urban areas which includes all of our cities, towns, villages, roads, and other human infrastructure.

If we classify agricultural land based on use, 77% is dedicated to livestock farming, yet livestock produces only 18% of the world's calories and 37% of the total protein consumed. The remaining 23% of agricultural land used for farming crops produces 82% of the planet's calories and 63% of the world's protein supply. This expansion of agriculture, especially for livestock, is our greatest impact on the environment.

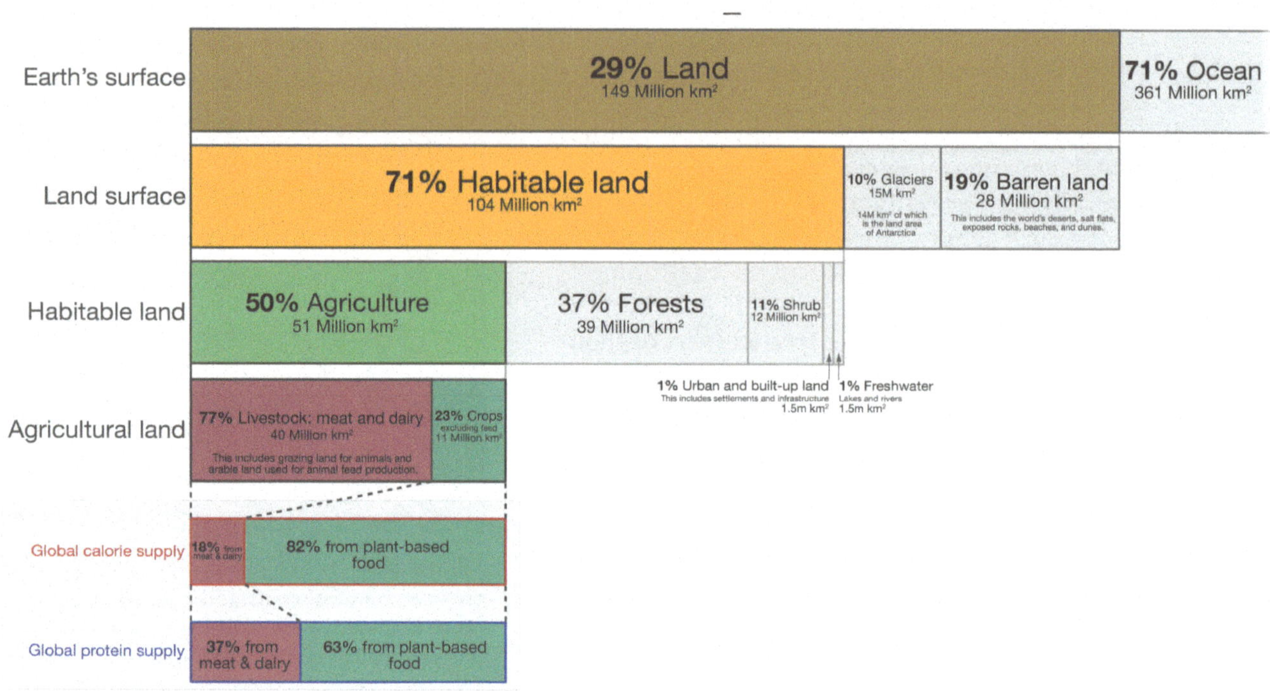

Source: UN Food and Agriculture Organization.

Figure 14.5 Global land use for food production—the expansion of agriculture is humanity's largest impact on the environment and the greatest pressure on biodiversity.

The total biomass of the human race accounts for just 0.01% of all life on Earth, and yet we have caused a global transformation. Of all the mammals on Earth, 36% are human beings, 60% are our food livestock, and only 4% are wild animals. Almost 4 out of every 10 animals on Earth are people, and the remaining 6 out of 10 are our cows and pigs. Chicken and other poultry make up 70% of all birds on Earth, which means that 7 out of every 10 birds on our planet are raised for food. Just one sixth of wild mammals remain compared to pre-agricultural times, according to a recent study in the Proceedings of the National Academy of Sciences. Humanity has caused the loss of 83% of all wild mammals. Of the 37,400 species that are currently considered to be threatened with extinction by the International Union for Conservation of Nature, almost all of them are under threat from agriculture.

© Jeff Whyte/Shutterstock.com.

Figure 14.6 An increase in the urban population has led to the construction of many new homes in areas that were once used for agricultural purposes. This housing development outside Calgary, Canada, is built on land previously used for farming.

Factors Contributing to Land Use Change

Land has shifted into crop production from other nonurban uses in response to rising commodity prices. However, land-use changes are gradual, due to conversion costs.

Global population doubled between 1950 and 1987, from 2.5 to 5 billion people. More land was converted to urban uses, especially for homes (Figure 14.6). New residential uses also require land for schools, office buildings, shopping sites, and other commercial and industrial uses. Urban areas represented 17.6 million hectares in 1950 and increased to 59 million hectares by 2015. These increases came mostly from pasture, range, and forest land, a shift that decreased the amount of land available for farming.

Effects of Human Population Growth on the Environment

Rapid increases in global population during the past 2 centuries have affected cycles related to the environment and ecosystems. These alterations and disruptions have changed the amounts of important elements and components of the environment, such as carbon, oxygen, nitrogen, phosphorus, and water. The disruption and pollution of the atmosphere and hydrosphere have long-ranging effects on ecosystems and **biodiversity**.

biodiversity
All the diversity of life on Earth, ranging from single cell organisms to the most complex ecosystems.

biogeochemical cycle
Natural processes that recycle nutrients in various chemical forms from the environment, to organisms, and then back to the environment. Examples are the carbon, oxygen, nitrogen, phosphorus, and hydrologic cycles.

ecosystem
A community of plants, animals, and other organisms that interact together within their given setting.

carbon cycle
The movement of carbon between the biosphere and the nonliving environment. Carbon moves from the atmosphere into living organisms that then move it back into the atmosphere.

The Biogeochemical Cycles

Besides water cycling through the biosphere in plants and animals, other components, such as carbon and nitrogen, have high concentrations in the biosphere. Cycles that include the interactions between the biosphere and other reservoirs involve biological processes including respiration, photosynthesis, and decomposition (decay), which are referred to as biogeochemical cycles. A **biogeochemical cycle** is a pathway by which a chemical element or molecule moves through both biotic and abiotic components of an **ecosystem** (Figure 14.7). All chemical elements occurring in organisms are part of biogeochemical cycles and, in addition to being a part of living organisms, these chemical elements also move through the hydrosphere, atmosphere, and lithosphere. All the chemicals, nutrients, or elements, such as carbon, nitrogen, oxygen, and phosphorus, used by living organisms are recycled. An important example is the **carbon cycle**, which is important in regulating our global climate by acting as a greenhouse gas within the atmosphere.

© Jacob Lund/Shutterstock.com.

Figure 14.7 Mountain environments provide a complex and diverse array of habitats for a large range of plants and animals.

Carbon Cycle

Carbon (C) is the basic building block of life. As the fourth most abundant element in the universe (after hydrogen, helium, and oxygen), carbon occurs in all organic substances, including DNA, bones, coal, and oil. Carbon moves through the hydrosphere, geosphere, atmosphere, and biosphere, each of which serves as a reservoir of carbon and carbon dioxide (CO_2). The carbon cycle (Figure 14.8) involves the biogeochemical movement of carbon as it shifts between living organisms, the atmosphere, water environments, and even solid rock. The least amount of carbon is contained in the atmosphere, while the largest amount is found in the lithosphere.

Carbon dioxide occurs in all spheres on Earth and, as a gas, is readily mobile. Our use of fossil fuels has increased the amount of CO_2 in the atmosphere, with some of the gas being taken up in the hydrosphere and biosphere. Marine and fresh water organisms use CO_2 in their life cycles and

Figure 14.8 The carbon cycle shows where carbon is contained in the geosphere, hydrosphere, biosphere, and atmosphere. Each year more than 3 gigatons of carbon are added to the cycle.

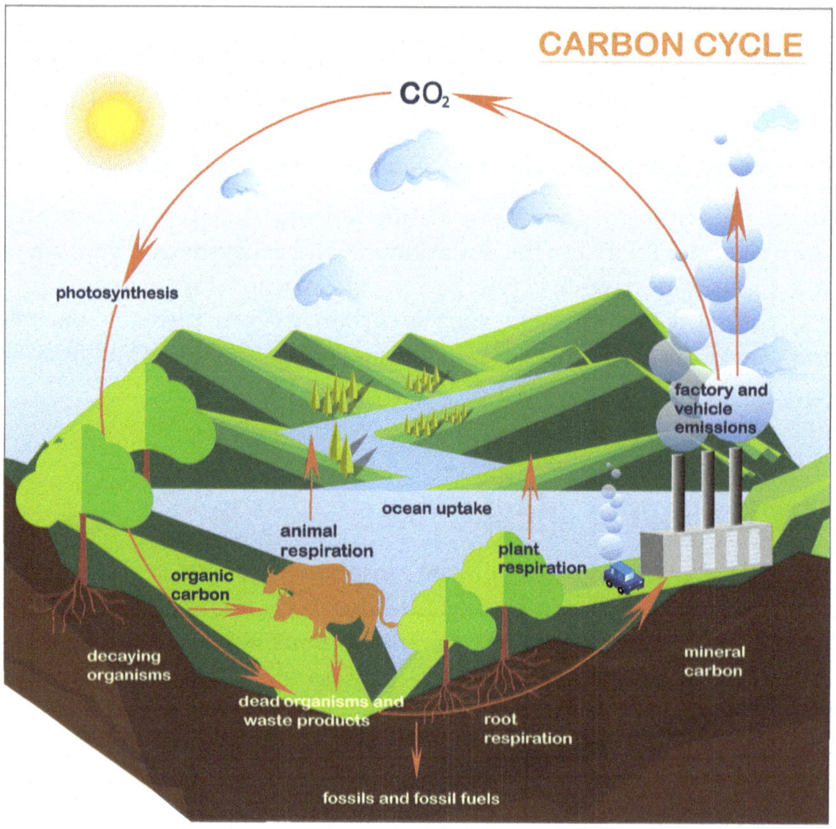

© danylyukl/Shutterstock.com.

can increase the amount of CO_2 in the geosphere. Coral reefs form by corals extracting dissolved CO_2 from seawater and creating their colonies in warm water environments.

Generally there is a dynamic equilibrium between and among the four reservoirs of carbon. However, these are often changed by natural processes that contribute to the production and hence increased abundance of CO_2. Volcanic eruptions and wildfires are two natural hazards that can upset the carbon equilibrium.

Carbon sequestration is the long-term removal or capture of carbon dioxide from the atmosphere. By removing and storing the carbon dioxide, we are attempting to slow or reverse atmospheric CO_2 pollution in order to slow or reverse global warming. Natural processes that occur in peatlands and wetlands capture and store carbon dioxide, acting as sinks by accumulating partly decayed biomass that would otherwise continue to decay. Establishing forests in places where trees currently don't grow is another natural means of accumulating carbon dioxide in biomass. Trees that live over 100 years will store carbon for substantial periods of time. The planet has enough space to plant more than a trillion new trees, and these would offset 10 years of carbon emissions. To store the carbon more permanently, we must use the wood itself to keep the carbon from being recycled into the atmosphere. Recent studies from researchers at the Potsdam Institute for Climate Impact Research in Germany calculated that if all new construction over the next 30 years uses 90% wood products, 700 million tons of carbon would be sequestered. This is equivalent to 7 days worth of global

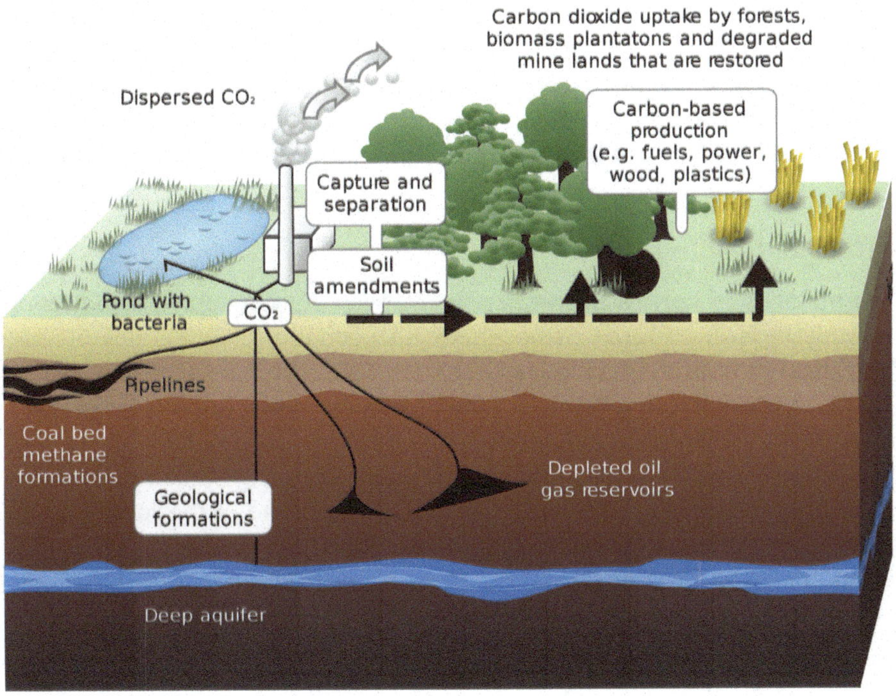

Source: Oak Ridge National Laboratory, www.ornl.gov.

Figure 14.9 **Natural and artificial mechanisms capture and store carbon dioxide, removing it from the atmosphere.**

carbon emissions in 2019. Currently only 0.5% of new buildings are made of wood.

We can accelerate the natural biological, chemical, and physical processes that capture carbon from the atmosphere through changes in land use and agricultural practices. We can also use artificial processes, including large-scale capture and storage of industrial carbon dioxide in subsurface aquifers, the ocean, or old oil fields. Geological sequestration stores carbon dioxide underground in old oil and gas reservoirs, saline water aquifers, and deep coal beds. Carbon dioxide is captured and compressed into a fluid, and that fluid is then injected deep below the surface where it would be stable for hundreds to millions of years. Iron fertilization in the ocean encourages the growth of phytoplankton that use carbon, and when they die some sinks to the deep ocean and takes the carbon with it. Artificial iron fertilization is controversial due to our limited understanding of its impact on the marine ecosystem.

Oxygen Cycle

Oxygen is one of the most abundant elements on Earth and represents a large part of all of Earth's reservoirs. The largest reservoir of oxygen is in the silicate and oxide minerals of the crust and mantle, which has 99.5% of all oxygen by weight on Earth. In the earliest stages of Earth's formation, no free O_2 existed in the atmosphere, and very little was present in the oceans. Approximately one billion years after Earth formed, bacteria began to produce O_2 and the amounts of this gas increased significantly. The atmosphere is 20.9% oxygen by volume. As an anion, oxygen can actively combined with other cations and is a key constituent in numerous organic compounds. Oxygen is abundant in the lithosphere and is often freed up

oxygen cycle
The biogeochemical transitions of oxygen atoms between different oxidation states in ions, oxides, and molecules through redox reactions within and between the reservoirs of Earth.

through chemical reactions to move into the atmosphere, hydrosphere, and biosphere (**Figure 14.10**). The main source of atmospheric oxygen is photosynthesis, which produces oxygen from carbon dioxide and water. The main way atmospheric oxygen is consumed is through respiration and decay, where animal and plant life consume oxygen and release carbon dioxide.

Nitrogen Cycle

nitrogen cycle
Biogeochemical cycle in which nitrogen is converted into many geochemical forms as it circulates among the atmosphere, terrestrial, and marine ecosystems.

The majority of Earth's atmosphere is nitrogen, making up 78% and representing the largest source. Atmospheric nitrogen is difficult to integrate into biological processes, making it scarce in many ecosystems despite its abundance. Although it is a key part of proteins and other organic substances, nitrogen is difficult for most living organisms to assimilate. Soil bacteria must alter N_2, which is then taken up by plants such as peanuts, peas, beans, soybeans, and alfalfa. These bacteria, which are found in the roots of these plants, generate a chemical reaction that allows N_2 to be transformed into ammonia (NH_3) and nitrate compounds (those containing NO_3). Animals and other organisms that use these plants as a food source then take in the N_2 and expel it as organic waste. Waste products rich in N_2 are used as fertilizers which return the N_2 to the soil and the **nitrogen cycle** begins again (**Figure 14.11**)

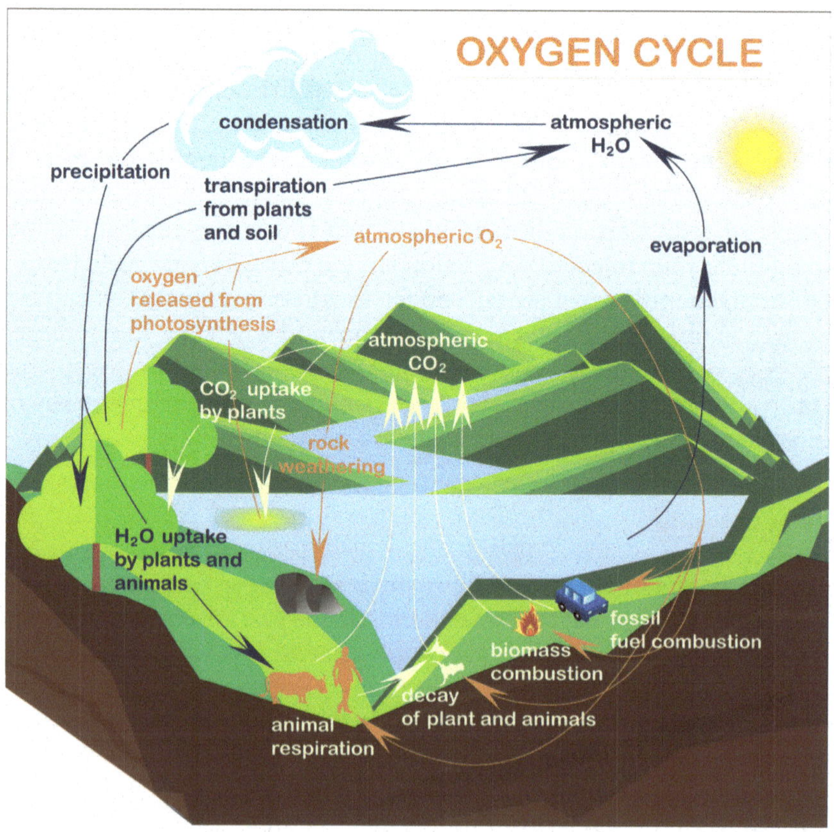

© danylyukk1/Shutterstock.com.

Figure 14.10 Storage and movement of oxygen occur between the atmosphere, biosphere, hydrosphere, and lithosphere. Photolysis is the chemical decomposition of the atmosphere caused by sunlight.

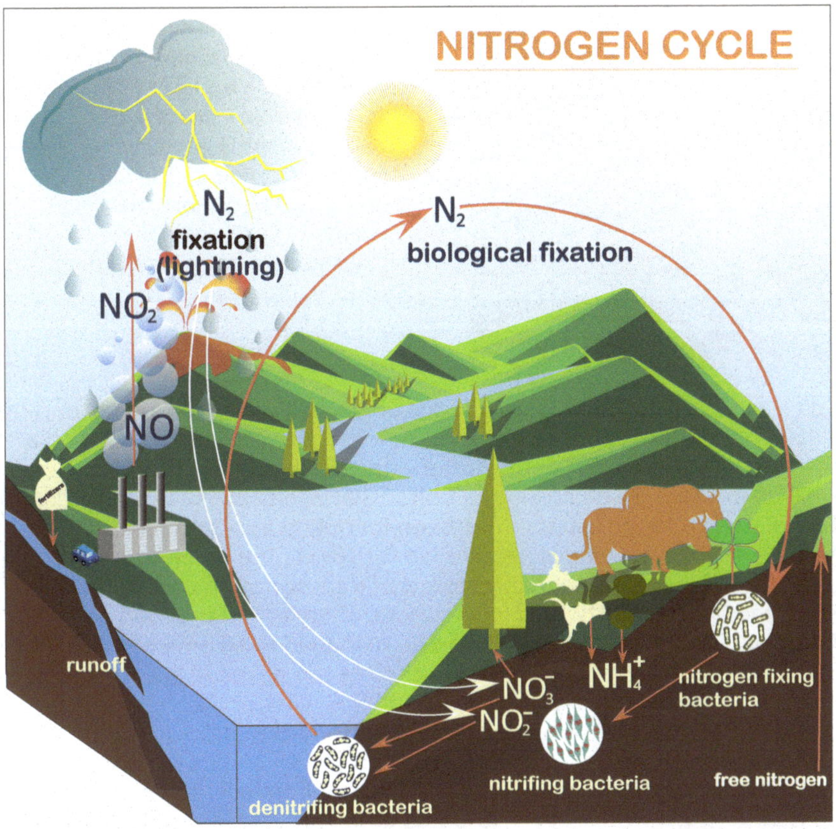

© danylyukk1/Shutterstock.com.

Figure 14.11 The Nitrogen Cycle, similarly to the carbon cycle and oxygen cycle, moves through the lithosphere, biosphere, atmosphere and hydrosphere.

BOX 14.1 Cows: Pee, Poop, and the Nitrogen Cycle

© Labellepatine/Shutterstock.com.

Figure 14.1.1 Cow urine is a significant contributor to global warming and nitrogen pollution.

One molecule of nitrous oxide (N_2O) released into the atmosphere contributes almost 300 times more to climate change than a single molecule of carbon dioxide. Half of the world's nitrogen is from industrialization, and 70% of man-made N_2O comes from agriculture. Nitrous oxide is released from urine patches in paddocks, dung produced during grazing, manure spread on pasture, and from nitrogen fertilizer. The average cow weighs 545 kg (1,200 lb), drinks 133 liters (35 gallons) of water and eats 45 kg (100 lb) of grass every day. That grass and water turns into 13 liters (3.5 gallons) of urine and 29.5 kg (65 lb) of feces every day. That feces contains about 2% phosphorous and the urine has 10 grams of nitrogen per liter. When pasture is too cold, too wet, too dry, or bare, plants can't take up the nitrogen and it washes into rivers and groundwater. The concentrated deposition of cow urine and feces during grazing also provides more nitrogen than plants can use—the bright green patches in pastures are ultra-rich overfertilized spots from uneven dispersal.

Figure 14.1.2 The average annual global consumption of beef per person is 8 kg (17.4 lb), but ranges from a low of 0.3 kg (0.6 lb) in the Democratic Republic of Congo, 2 kg (4 lb) in India, 17.6 kg (39 lb) in New Zealand, 36 kg (79 lb) in the United States, and up to 56 kg (124 lb) in Uruguay.

New Zealand's 5 million dairy cows put 200 tons of leached nitrogen into groundwater every day, and that's not counting the input from 3.6 million beef cattle. India has 305 million cattle, about 30.5% of the entire world population, with 232 million in Brazil, 99 million in China, 94 million in the United States, and 88 million in the EU. The increasing global cattle population has been identified as a significant threat to climate, wildlife, and forests, and is responsible for 18% of the greenhouse gases that produce global warming, more than all forms of transportation combined. This issue is complicated, as it is intertwined with culture, lifestyle, and the food choices we make every day. Farming and livestock can be managed in an environmentally progressive way, promoting healthier soil. Food choices can be made in a conscious and educated way, supporting industries that employ environmentally sustainable practices. It is an unavoidable fact that it takes less land to grow food directly for humans, than to feed animals which humans then eat. Over 25% of the world's entire land area is used to graze livestock or grow feed for livestock, and if everyone ate a plant-based diet, we would need 75% less farmland than we currently use. Every person plays a part, and every choice has an impact. We can all take a minute and consider our choices, and make better ones when and where we are able.

The nitrogen cycle is very important to ecologists because nitrogen directly impacts plant growth and decomposition. Human use of nitrogen fertilizer has had a profound impact on the global ecosystem and human health. Through our agricultural practices, we have more than doubled the annual transfer of nitrogen into biologically active forms. The countries that use the most fertilizer are Singapore, Qatar, Hong Kong, New Zealand and Malaysia, based on WorldBank data from 2002–2015. All of these countries use more than 1,500 kg per hectare of arable land, with the most fertilizer used in Singapore at 33,000 kg per hectare. Nitrogen can be directly toxic to certain plant species, can lead to soil acidification, and can lead to water acidification and eutrophication of freshwater and saltwater ecosystems dead zones.

Blue baby syndrome, also known as infant methemoglobinemia, is a condition that limits the amount of oxygen in the blood. When babies drink water contaminated with nitrates the body converts them to nitrites which binds with hemoglobin and prevents it from carrying oxygen. Infants under 3 months old are at highest risk, but it can also occur in adult populations who have preexisting conditions such as kidney failure or ulcers. In addition, current research by Stanford University shows that premature births increased almost 50% when nitrates were present in drinking water at levels of 5 mg per liter, which is less than half of the current safe level in New Zealand of 11.3 mg. Nitrate contamination is most common in farming communities that use well water, and comes from fertilizer and manure.

Phosphorus Cycle

Phosphorus (P) is an important nutrient for plants and animals. It moves through the biosphere, hydrosphere, and geosphere but is lacking from the atmosphere, because it is in a solid state at normal temperature and pressure. Phosphorus is part of DNA molecules and is an essential nutrient for plants, animals, and humans. It makes up a significant portion of bones and teeth. Phosphorus is also a component of ATP, the energy molecule for all life forms. Since phosphate is slowly lost in water runoff, it becomes gradually less abundant in soils over time. This reduces plant growth and slows down microbial activity. The **phosphorus cycle** has been significantly impacted by modern agricultural practices, and 80% of of mined phosphorous is used to make fertilizer. Synthetic fertilizers are not fully utilized by plants, and the excess phosphate is carried away in water runoff. Over-fertilization and misuse of animal waste fertilizer can also add phosphate to runoff. Water pollution from excess phosphorous results in uncontrolled growth of cyanobacteria (Figure 14.12), which dies, rots, and removes oxygen from the water. Excess phosphorous can also generate a toxic algal bloom that makes the water poisonous to animals and humans.

© igorstevanovic/Shutterstock.com.

Figure 14.12 Blue-green algae or cyanobacteria thrive in phosphorous-rich aquatic environments, growing into a toxic algal bloom which depletes the water of oxygen when it dies and decomposes.

phosphorus cycle

Biogeochemical cycle that shows the movement of phosphorous through the lithosphere, hydrosphere, and biosphere. Because phosphorous and phosphorous-based compounds are usually solid, they do not play a significant role in the atmosphere.

Water Cycle and Water Resources

The **water cycle**, which is also referred to as the hydrologic cycle, shows where water is located and how it moves through the atmosphere, hydrosphere, lithosphere, and biosphere (Figure 14.13). The movement of water is related to the three states or conditions in which water can exist—as a liquid, solid, or a gas. More than 75% of Earth's surface is covered by water in either a liquid or solid form. Initially, water formed on the surface through

water cycle

The water cycle describes the continuous movement of water on, above, and below the surface of the Earth.

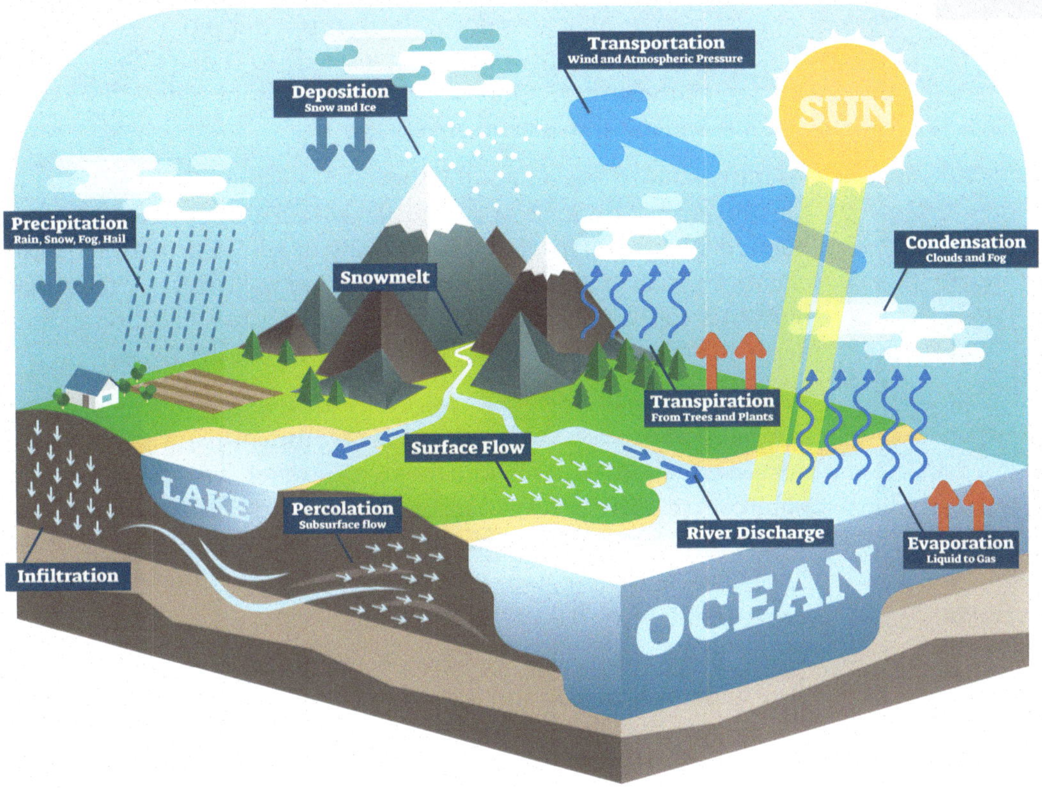

© VectorMine/Shutterstock.com.

Figure 14.13 The water cycle (also referred to as the hydrological cycle) shows how water in its various phases moves or rests on Earth.

its being expelled by volcanic eruptions from water-bearing magmas and also by comets from outer space bringing frozen ice to Earth.

The water cycle involves the transfer of energy as water evaporates, removing energy from the environment and reducing the temperature. When it condenses as rain, it releases energy and warms the environment. These heat exchanges drive our atmosphere and climate system. Evaporation purifies water by leaving other dissolved ions behind, and this replenishes the hydrosphere with freshwater. Erosion and sedimentation from the energy of flowing liquid water and moving ice reshapes the geological features of the surface of our planet. The water cycle is responsible for the existence of almost all life and ecosystems on Earth.

The average residence time of water in the different reservoirs on Earth ranges from 9 days in the atmosphere to a few months in rivers, soils, and seasonal snow cover, to 10s to 100s of years in glaciers, lakes, and shallow groundwater, to 1,000s of years in the oceans. Most water is being stored for much longer time periods than it is actually moving through the water cycle, and the biggest storage area is the oceans, which house about 97% of all water on Earth, and they supply about 90% of the evaporated water that goes into the water cycle. Without evaporative cooling from the oceans, natural greenhouse warming would result in a surface temperature of 67 °C (153 °F).

Water Pollution

Water is essential for plant and animal growth. Water must be clean enough for its consumer to use it without adversely affecting the health and well-being of the plant or animal.

Most individuals do not think about where the water they drink comes from or where the water that goes down the drains winds up. If asked, their responses range from—"the city provides my water" to "I don't know, it just flows when I open the faucet." Only when we are without water do we consider where it is and ask if it is drinkable.

© alexmisu/Shutterstock.com.

Figure 14.14 **Irrigation of fields uses large amounts of ground and surface water. The water must be of good quality to provide the moisture needed to raise crops and other agricultural products.**

"Is it drinkable?" becomes an important question. Because water is the universal solvent, most chemicals will eventually become incorporated. None of us would intentionally drink water with harmful chemicals such as arsenic, cadmium, or lead in it. Thus we trust that our water suppliers, be it a city or a bottler of water, have some standards in what can and cannot be in the water we drink.

When you flush anything down the drain, for example, some expired prescription drugs, the water treatment plant may or may not remove those drugs before the treated water is discharged into a watercourse. Downstream another city takes in that water and treats it to destroy harmful bacteria and viruses, but what about the other substances in the water? Where would you like to live? Upstream or downstream of such a community?

Much of the freshwater we have on the planet is used in agricultural endeavors to feed and nourish our ever-increasing population. Pumps run 24/7/365 to move water from deep aquifers to the surface for irrigation (Figure 14.14). These aquifers are not recharging at anywhere near their depletion rate. Thus the water is being mined in a permanent sense. In some areas of our country where there is a preponderance of agricultural activities and associated irrigation, saltwater encroachment is occurring in the aquifers.

Source: Special Collections, Michael Schwartz Library at Cleveland State University.

Figure 14.15 The 1952 fire on the Cuyahoga River caused over $1,000,000 US in damage when the 15 cm (6 inches) of oil and chemicals floating on the water surface ignited.

The pollution of water in aquifers beneath cities and military installations occurs as a result of leaks and accidental spills into the ground, which eventually enters the water table. For example, petroleum pipelines connecting refineries and distribution centers around the country occasionally break. Certainly if the leak is detected mediation (cleanup) measures are instigated. But what about undetected leaks, or leaks from years ago when regulations were lacking or ignored, many times out of ignorance? Today we have monitoring wells around sites of known hazardous leaks. These wells are regularly sampled to ensure that the chemical plume's vicinity to groundwater is known. But again these are the sites we know about.

When the Cuyahoga River in Cleveland, Ohio, caught on fire for the 12th time in June 1969, a wakeup call was sounded to the nation and the world (Figure 14.15). The Cuyahoga River first started burning in 1868, when industrial pollution from steel mills, paint and chemical works, and at least 20 oil refineries all drained into the river system. In the early days of oil refining, it was common practice to dump unusable refined crude oil straight into the waterways for disposal. Gasoline was essentially useless until cars were developed, so that was dumped too. The water itself wasn't flammable, but the floating chemical slicks certainly were. The 1969 Cuyahoga River fire took on mythic status as an ecological crisis, resulting in the formation of the United States Environmental Protection Agency and the establishment of the first Earth Day, both in 1970.

Runoff from agricultural and livestock feeding facilities represents another source of water pollution. Fertilizers, insecticides, fungicides, and herbicides that are sprayed on fields to improve crop yields find their way into streams, rivers, lakes, and eventually oceans. Fertilizers, while enriching the soil, can end up in the water supply and alter the biota by accelerating the growth of certain algal components. Bacteria feed on the decaying algae, depleting the dissolved oxygen levels, causing fish kills with their associated ramification to the environment. It takes one domino falling into another to cause the remaining ones in line to fall.

Usually not mentioned but certainly on the list of water pollution is thermal pollution. Downstream outflows from power generating stations, which use river water to cool the machinery of power production, alter temperature regimes, which alter habitat and thus the biota. Some organisms flourish while others die off in the heated water.

Oceans are the final receiver of continental runoff (Figure 14.16). Oceans and lakes seem vast and bottomless, but in some areas, usually at the mouths of rivers, they may be virtual cauldrons of toxins. These toxins may be suspended in the water column or accumulate in the sediments. Ocean food chains are linked to the water column and the sediments.

© Gary Webber/Shutterstock.com.

Figure 14.16 Surface drainage and sewage can enter the ocean without any monitoring or treatment for water quality.

Organisms feeding in these areas accumulate toxic compounds that are then later accumulated at higher levels in larger organisms (biomagnification). Biomagnification was introduced by Rachel Carson in her book, *Silent Spring.* Considering that many species of ocean life are consumed by humans around the world, and depending on where in the world we are, ocean life is sometimes a staple in the diet and sometimes a delicacy. Either way, the toxins placed by us, either accidently or by intent, return to us in an often harmful way.

Air Pollution

Clean air is necessary for life to survive and thrive on the planet. We have seen that the atmosphere near the surface consists of 78% nitrogen, 21% oxygen, and about 1% of many lesser gases. Oxygen is the component critical for animal life. However, when the air (and atmosphere) is contaminated with pollutants such as nitrous and sulfurous oxides and dangerous hydrocarbon volatiles, the air is rendered harmful (Figure 14.17). Pollution can come from numerous sources and can include particulate matter such as dust and ash from volcanoes and wildfires. Anthropogenic sources include coal-fired power plants, factories, vehicles that operate on fossil fuels, and even dry-cleaning establishments.

A secondary effect of air pollutants is the formation of haze that can reduce visibility around large cities and in wilderness areas and in national parks. In northern Arizona a large coal-fired generating station is located only 25 km (15 miles) from the northeastern boundary of Grand Canyon National Park. Until several years ago, a similar type of station located to the west of the park contributed significantly to lower visibility in the regions of national parks in northern Arizona and southern Utah. The Mohave Generating Station in Laughlin, Nevada, was shut down in December 2005 rather than have its operators face the expense of installing scrubbers and other pollution control equipment to reduce its emissions.

The Clean Air Act, originally passed in 1963, has been amended several times to insure healthy air in the United States. The EPA is entrusted to insure that clean air standards are met by monitoring the amounts of air pollutants across the country.

© alexmisu/Shutterstock.com.

Figure 14.17 Pollutants from industrial smokestacks enter the atmosphere without any reduction in the particulates or gasses generated through the manufacturing process.

BOX 14.2 Case Study: Fracking

Fracking is a process for extracting natural gas from shale layers thousands of feet below the surface of the Earth. These shale layers contain hydrocarbons, but they are not very permeable, so the gas is trapped in the rock. The process of extracting that gas is called hydraulic fracturing, or fracking, and allows industry to extract natural gas from rock reservoirs that otherwise wouldn't be economically viable resources. The process involves drilling a well to the target rock and pumping down a fracking fluid made up of water and thickening agents with a solid material like sand suspended in it, which pressurizes the rock and cracks it. The fluid flows in and the sand grains or other solid material, called a proppant, moves into the fractures, holding them open (or propping them apart). Once the rock is fractured and the fractures are kept open with the proppant, the trapped gas can more easily flow into the fractures and be extracted for commercial use (Figure 14.2.1).

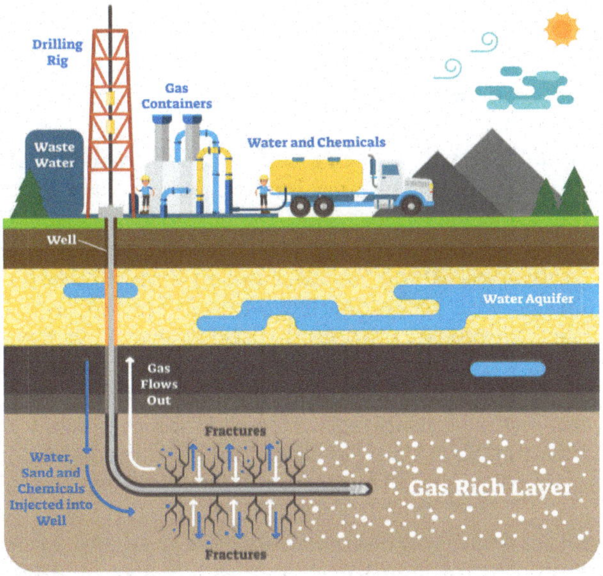

© VectorMine/Shutterstock.com.

Figure 14.2.1 **Hydraulic fracturing is a process that allows petroleum companies to extract hydrocarbons from shale reservoirs that would not otherwise be economically viable resources. However, it is highly controversial, with proponents arguing for the economic benefits of the petroleum resources and opponents arguing that there are significant and long-term environmental impacts.**

Fracking began commercially in 1950, and current estimates of how many wells have been fracked are difficult to make because reporting requirements are variable in different places, and data are difficult to access. A 2017 analysis of petroleum production in the United States suggests there are as many as 1.3 million fracked wells in 34 states that are actively producing oil or gas (Figure 14.2.2). The largest number of these are in Texas, which has almost 400,000 facilities alone. Other states with significant numbers of fracked wells are Pennsylvania (more than 100,000), California, Kansas, and Ohio (more than 90,000 each), New Mexico (almost 60,000), and Colorado and Illinois (about 50,000 each).

In late October of 2018, the energy firm Cuadrilla began drilling for gas reservoirs in shale rocks in Lancashire, United Kingdom. Two days later, three earthquakes occurred in the area of the drilling, and after 9 days there had been 17 earthquakes. These earthquakes measured up to 0.8 on the Richter Scale and were too small to be felt on the surface but can be detected by seismic monitoring equipment. Any earthquake over 0.5 magnitude requires an immediate stop to drilling for 18 hours. Environmentalists tried to block the fracking in court but were not successful. After the earthquakes, Manchester's metro mayor Andy Burnham said, "The Earth is telling us something. Stop this process now." These induced earthquakes are one of the reasons fracking is so controversial; they can be generated by any over-pressurization of layers under the surface of the Earth because pumping fracking fluid into shale reservoirs pressurizes these layers. Fracking wastewater can have significant public health and environmental impacts, such as ground and surface water contamination.

© Anton Foltin/Shutterstock.com.

Figure 14.1.2 After the well is drilled, it is connected to pumps that push the fracking fluid into the well to pressurize it.

In 2008, people living in Pavillion, Wyoming, began to notice a bad taste and smell in their drinking water. They live in the middle of a natural gas basin that was being drilled and fracked for petroleum resources. Wastewater from the fracking wells was being stored in unlined pits dug in the ground, where it leaked into surface waters and contaminated the drinking water. Fracking fluids can also contaminate subsurface water reservoirs if the wells are drilled through those layers and the fracking fluids aren't sealed off from the water reservoirs. Documenting groundwater contamination from fracking is difficult because a lot of the compounds used for hydraulic fracturing are not commonly analyzed in commercial labs. Chemicals like methanol can be part of the fracking fluid and can trigger permanent nerve damage and blindness when consumed at high enough levels. However, it degrades rapidly and virtually disappears from samples over a few days. New testing methods confirmed that groundwater wells from Pavillion contained methanol, high levels of diesel compounds, salts, and other compounds used in fracking fluids. New tests have detected 19 concerning chemicals in the water supply, half of which are unstudied and do not have known safe exposure levels. Residents of Bradford County in Pennsylvania have methane contamination in their drinking water, which is attributed to fracking that began in the area in 2008. Bradford has more wells than any other county in Pennsylvania, and some residents near the wells have noticed chemical smells and tastes in their drinking water, including methane levels high enough to light their water on fire (**Figure 14.1.3**). Chesapeake Energy, who drilled in the area, reached a $1.6 million settlement with three families in 2012 after they sued the company for contaminated drinking water from their wells. According to an EPA study: "The most important findings are that drilling, fracking, and the use of hazardous chemicals necessary to frack have caused groundwater contamination."

Christian Science Monitor/Contributor/Getty Images.

Figure 14.1.3 Water wells in Bradford County, Pennsylvania, contain high levels of methane attributed to fracking contamination from drilling that began in 2008. This woman lives 30 meters from a gas well, and methane levels in her drinking water are so high she can light her tap water on fire.

Soil Contamination

The introduction of hazardous materials either on the surface or into pore spaces in soil can produce harmful conditions for humans, animals, and plant life. Before the EPA was established in 1970, numerous instances of dumping of hazardous chemicals and other materials caused the soil and groundwater to be filled with substances that had a dangerous effect of life. There was no concern or knowledge that materials placed in dump areas would produce negative effects on future generations.

Dangerous contaminants quickly work their way into the food chain as plants take up the material into their structure and then animals ingest the plants. Eating of the animals by humans and other animals then passes the substances along, where they can have a very deleterious effect on life.

Heavy minerals and liquids can contaminate the soil and lead to life-threatening conditions (Figure 14.18). The cleaning up of contaminated areas involves: (a) treating the soil in place with chemical and other procedures that clean the soil, (b) leaving the soil in place and containing it to a small area, thereby preventing the contaminants from reaching animals, humans, or plants, or (c) removing the soil and treating it or disposing of it in a way that removes the harmful material.

© Vladimir Melnik/Shutterstock.com.

Figure 14.18 Soil contamination is caused by the presence of human-made chemicals that can be from industrial activity, agricultural chemicals, or improper disposal of waste material. The most common chemicals are hydrocarbons, solvents, pesticides, lead, and other heavy metals.

Worldwide there are millions of sites that contain some amount of contamination. The proliferation of harmful chemicals and the thoughtless manner in which industrial wastes were handled in the early and middle 20th century have created numerous harmful conditions. Two such examples in the United States are the incident at Love Canal in the State of New York and the mining of uranium in the Western United States.

The Love Canal Environmental Crisis

To quote Eckardt Beck, former Administrator of Region 2 of the EPA, "Love Canal is one of the most appalling environmental tragedies in American history." Construction of the canal first began in the 1890s as the plan of William T. Love, a developer who had the idea of constructing a canal between the upper and lower Niagara Rivers in New York. The idea was to generate inexpensive hydroelectric power for the homes and industries in the area. In the early 20th century digging on the canal began but was abandoned shortly thereafter. The excavated area was left as an eyesore, which became a dumping ground for at least 20,000 tons of chemical waste used in the manufacture of pesticides, plasticizers, and caustic soda in the 1920s and 1930s. In 1953 a chemical company sold the land, which had been buried in soil, to the City of Niagara Falls for one dollar.

© Andy Levin/Science Source.

Figure 14.19 **The Love Canal neighborhood near Niagra Falls, NY, was built on top of a toxic waste site.**

Several years later, as the region began to expand, the area was developed as homes and a school were constructed on the land that rested on decaying chemical containers and other hazardous waste (Figure 14.19). The winter of 1976 and spring of 1977 brought record precipitation to the region, and the percolating water was enough to release toxic chemicals that had been stored in the subsurface. Within months residents began developing major health problems that were traced to a myriad of extremely harmful chemicals. Although the affected area only covered about 16 acres, the 21,000 tons of buried toxic waste led to miscarriages and birth defects in unusually high proportions.

Assistance soon came to the area. Intervention on the part of the State of New York and the federal government through a Presidential Declaration by President Carter in 1978 provided funds to initiate cleanup in the area. This event resulted in the establishment of the Comprehensive Environmental Response, Compensation, and Liability Act (CERCLA), or the Superfund Act. More than 200 homes were purchased, and implementation of a major environmental restoration project led to restoration of the area.

© John Carnemolla/Shutterstock.com.

Figure 14.20 **Uranium mines create large waste piles as well as waste ponds full of slurry resulting from processing of the rock to extract the uranium ore. When these tailings piles are located near rivers, they can leach pollutants like ammonia into the water system.**

Uranium Mining in the West

Uranium is a naturally occurring element that is processed to provide fuel for nuclear power plants. In the 1950s and 1960s uranium was refined to use in the manufacturing of nuclear bombs as part of the arms race with the Soviet Union. The main source for uranium in the United States is the Western States, where the geology provided numerous sources from deposits found in sedimentary rocks. The mining and refinement of the source material created large piles of mine tailings that contained small amounts of uranium and other dangerous elements (Figure 14.20).

Very little attention was paid to the massive piles of waste material. There are many of these piles that have leached harmful substances into the environment, especially into the hydrosphere. The city of Moab, Utah, was an active center for mining, as well as Grand Junction, Colorado. Each of these cities lies along the Colorado River. A major cleanup site is located within a few kilometers of Moab as a massive pile of material is located on the banks of the Colorado River. The EPA has designated this location as a Superfund site. An estimated 16 million tons of mine tailings are being moved about 50 km north to the town of Crescent Junction, Utah, in order to get the toxic materials away from the river. Completion of the project is estimated for the year 2022. In the meantime, the waters of the Colorado River will continue to carry some of the toxic materials downstream.

The cost of hauling away the Moab tailings could exceed $1 billion, according to the latest estimate by the U.S. Energy Department, the agency managing the cleanup. Congress ordered updated cost projections based on a cleanup timetable that is nearly a decade shorter than DOE's. The department had been planning to spend about $30 million a year for the next 20 years to remove the leftover uranium waste piled up on the banks of the Colorado River outside of Moab. But securing annual funding is likely to be much tougher when the annual costs are between $79 million and $103 million from 2010 to 2019, as the DOE estimates. The total cost would be between $844.2 million and $1.1 billion, under the projections DOE submitted in a report to Congress on July 1, 2008. (Reprinted by permission of the *Salt Lake Tribune.*)

Ecosystems Effects and Impacts

Ecology is the study of the interactions between organisms and their environment (Figure 14.21). Organisms such as plants (producers), consumers (squirrels, foxes), and decomposers (fungi, bacteria) are the biotic components. Examples of the abiotic components are water, temperature, soil nutrients, pH of the substrate, elevation, latitude, aspect, slope, and amount of solar radiation received. If both the biotic and abiotic components are combined and studied, this discipline is referred to as ecosystem ecology.

ecology

The study of the interaction between organisms and their environments.

A grassland ecosystem, for example, would include all of the grasses, insects, birds, rodents, earthworms, as well as the soil nutrients, water availability, soil and ambient temperatures, and parent rock.

All of the ecosystems on the planet form the biosphere. With few exceptions, ecosystems are named after the dominant vegetation that occurs within the system. Some examples of ecosystems are grasslands, savannahs, tundra, tropical rain forests, deciduous forests, deserts, coral reefs, oceans, and estuaries.

Each ecosystem has its own flora and fauna. Some of these are endemic, while others are found across several ecosystems. Ecosystems suffer from the effects caused by humans, directly and indirectly. An example of this would be the loss of a section of tropical rain forest. When the trees are removed, such as by burning or logging activities, animals and plants that used the trees as their habitat are immediately without a residence. The nutrients that were stored in the trees are removed entirely from that ecosystem. The soil, which is already very poor in nutrients, will not be replenished. When the seasonal rains come, there is no vegetation to intercept its flow, and soil erosion becomes rampant. If the surface has significant slope, landslides can result. Next the unchecked water with the eroded soil enters streams and rivers, causing sediment loads, which interfere and greatly reduce primary production

© sarayut_sy/Shutterstock.com.

Figure 14.21 All organisms on Earth are related to one another and their physical surroundings.

© Anton Foltin/Shutterstock.com.

Figure 14.22 A desert ecosystem is made up of all the living and non-living components of a climate that gets less than 25 cm of rainfall a year. These harsh systems usually have poor soil as well, but organisms are adapted to survive and thrive with the limited supply of water.

of alga and aquatic vegetation. This loss of photosynthetic activity causes collapse of food chains within the associated aquatic habitats.

Deserts are another ecosystem that has been greatly affected over the past century by humans. Deserts have their own very unique life forms that have evolved through adaptations to endure the hot, dry days and cold evenings (Figure 14.22). Deserts, especially in the Southwestern United States, have suffered dramatic changes due to development of cities. Houses, roads, buildings, and infrastructure now rise up from the same place that cacti and other desert flora once flourished. Furthermore, the desert temperatures, coupled with atmospheric pollution from cars, trucks, and planes, plus the dust from unpaved desert roads, create a brown pall over many Southwestern cities.

Acidification, the process of lowering the pH of terrestrial and aquatic environments, begins with the combination of gases and water vapor in the atmosphere. Examples of three gases released into the atmosphere are CO_2, SO_2, and NO_2. Volcanoes, internal combustion engines, and various industries produce these oxides. Geological activities such as volcanoes emit oxides of sulfur, and carbon dioxide is released from catastrophic forest fires. Once these gases are released and combined with water vapor the resulting acidic precipitation falls onto terrestrial and aquatic habitats. At first the buffering effects of various ions of the soil and water prevent the radical shift of pH. However, over time the acidification process gains momentum, resulting in losses of vegetation resulting in the loss of food and cover for the various organisms. In some lakes, acidification has rendered the water sterile of all life. If this sounds like the "Domino Theory"—it is. To paraphrase John Muir, everything is tied together in an ecosystem and therefore changing one aspect changes everything else within the system.

Since 75% of Earth's surface is covered by oceans and lakes, it is logical to some of our species (*Homo sapiens*) that the oceans are limitless with respect to being able to swallow the sewage, garbage, sludge, and other pollutants so frequently and directly shoved into the water. There is an indirect absorption by oceans of agricultural runoff from rivers that are fed by surface runoff, which are fed by farms. This agricultural "cocktail" contains fertilizers, herbicides, fungicides, insecticides, and in some cases fecal materials from feedlots. The oceans and their associated shores and intertidal zones are the largest ecosystem and as such touch every human in some way. The most obvious way is the utilization of the numerous foods that are harvested from the oceans. The pollutants, for example mercury, that have entered the oceans do come back to shore in the fish we eat. This is because of a process called **biomagnification**. Small amounts of a residual pollutant that are absorbed in the lower links of a food chain are increased in the tissues of the larger members of the food chain.

biomagnification
The increase in amounts of toxins in animals higher in the food chain as a result of ingesting organisms that contain toxic materials.

Changes to Biodiversity

Biodiversity is the full spectrum of all living species on Earth, including plants, animals, insects, bacteria, and fungi. It can be used to refer to all of the species in one region or ecosystem. Earth's biodiversity is so rich that many species still remain to be discovered. Scientists estimate that there are about 8.7 million species of just plants and animals, but only about 1.2 million have been identified and described so far. The World Wildlife Federation found an average 60% decline in global populations of mammals, fish, birds, reptiles, and amphibians since 1970. The Intergovernmental Platform on Biodiversity and Ecosystem Services reports that 1,000,000 plant and animal species are now threatened with extinction. Human activity impacts three fourths of land-based environments and about 66% of ocean environments, resulting in significant alteration. More than one third of the Earth's land surface and 75% of freshwater resources are devoted to agriculture in the form of crop or livestock production.

© Peter La Fontaine/Shutterstock.com.

Figure 14.23 Orangutans are great apes that spend most of their time in trees, can live for over 30 years, and are among the most intelligent primates. They are considered critically endangered from poaching, habitat destruction, and the illegal pet trade.

Some of the most important biodiversity locations on Earth, biodiversity hotspots like Borneo, are under pressure. Borneo is home to more than 15,000 plant species and 1,400 different animal species, including more than 3,000 species of orchids, clouded leopards, rhinos, pygmy elephants, and orangutans (Figure 14.23). Borneo's forests are being cut down for timber, palm oil plantations, wood pulp, rubber, and mineral mining. Satellite images show that protected lowland tropical rain forests have been deforested significantly since 1950, and with up to 98% of the orangutan habitat destroyed by 2022.

© Leonardo Mercon/Shutterstock.com.

Figure 14.24 This bromeliad in the Guarapari rain forest of Brazil is the home of a white-banded tree frog and a bromeliadas tree frog.

There are tiny frogs that spend their entire life in the pool, from egg to adult (Figure 14.24). So who cares if they are lost; after all, out of sight out of mind, as the saying goes. We do assign labels to some species like endangered, threatened, species of concern, and once a label is appended to that species some protection is provided. Good examples would be bald eagles and peregrine falcons; populations of both have recovered sufficiently to have their labels removed.

A common way for biodiversity to be reduced in an area is remove the habitat or alter it so significantly that there is no chance for certain species to hang on. The Dusky seaside sparrow fits into this category. Between flooding and then draining its habitat in Florida and with the addition of numerous pesticides in its environment, the "duskies" departed, going extinct in the late 1980s. The dodo, which was thought to have only existed on Mauritius in the Indian Ocean became extinct in the late 17th century (Figure 14.25). On rare occasions a species is "rediscovered in the wild." An example is the Banggai crow (Corvus unicolor), thought extinct from the early 1900s, was rediscovered in 2007 on Peleng Island in eastern Indonesia.

© Daniel Eskridge/Shutterstock.com.

Figure 14.25 **An illustration of two dodo birds, now extinct, on the island of Maurita.**

© Luke Ferguson/Shutterstock.com.

Figure 14.26 **Kudzu has overgrown an abandoned building.**

Another reason for losses of endemic biodiversity, organisms that only live in a particular area, is the invasion of non-native biota, meaning organisms that did not evolve within the same ecosystem and therefore have no natural controls on their spread or population levels. There is a National Invasive Species Information Center in the United States. Some examples of invasive species are:

- The majestic American chestnut tree grew to great heights and diameters in the Eastern United States until the chestnut blight (a fungus) was accidentally released in the United States on some imported wood from Asia. A few American chestnut trees remain, perhaps because of immunity to the blight.
- The European gypsy moth, brought into the United States in the 1860s for silk production is a notorious defoliator of hardwood trees across North America.
- The European starling was introduced into the United States in the 1890s for the sole purpose of having all the birds mentioned in any of William Shakespeare's writings to have a habitat in this country. These birds have displaced many native songbirds and have spread to all 48 contiguous states and Alaska. From the original 100 starlings released in New York's Central Park, hundreds of millions of these birds have resulted.
- Kudzu, a member of the bean family, was introduced into the United States around 1880 from Japan. Kudzu was originally touted as a good plant to control erosion. It found the Southern United States to its liking and took off, covering everything in its path. Kudzu has grown around roads by traveling across electrical lines from one side of the road to another (**Figure 14.26**). Some organisms came in accidentally, other introduced on purpose, but either way each organism was not native and has disrupted native species as a consequence. This is one reason the United States Customs Service asked arriving travelers into this country if they have any plants or animals in their possession.

More than 99% of all species that ever lived on Earth are estimated to be extinct. Of the 2 million to 1 trillion species currently living on Earth, we have only described one thousandth of 1% of them, according to scientific estimates. Extinction is a normal part of the evolutionary process, but current levels of extinction are 1,000 times higher than normal background rates, and are likely to be 10,000 times higher in the future. Current levels of extinction over the last 100 years would have taken between 800 and 10,000 years to occur naturally. We are in a period of exceptionally rapid loss of biodiversity, leading to a new period of mass extinction, this one driven by us. Averting this loss of biodiversity and its impact on ecosystems is still possible through intensive conservation efforts, but that window of opportunity is closing rapidly.

Key Terms

biodiversity *(page 411)*
biogeochemical cycle *(page 411)*
biomagnification *(page 426)*
carbon cycle *(page 411)*
desert *(page 409)*
ecology *(page 425)*
ecosystem *(page 411)*
nitrogen cycle *(page 414)*
oxygen cycle *(page 413)*
phosphorus cycle *(page 417)*
water cycle *(page 417)*

Summary

Just 10,000 years ago the population of all people on Earth was 8 million. It reached 1 billion in 1804, 2 billion 123 years later in 1927, and 3 billion 33 years after that in 1960. Since then, 1 billion people have been added to the global population about every 12 years—4 billion in 1974, 5 billion in 1987, 6 billion in 1999, and 7 billion in 2011. Population increase has a significant impact on the Earth's systems. Land use and agriculture change dramatically with urbanization, resulting in land that was previously used for agriculture or other resources converted to urban use.

Carbon, oxygen, nitrogen, and phosphorous are important chemical cycles in the biosphere, hydrosphere, geosphere, and atmosphere. These are important for all living organisms. The water cycle is important for life on Earth and is easily impacted by human activities. Water pollution can occur as a result of accidental or deliberate release of chemicals onto the surface or in the subsurface. Air pollution can come from chemicals or particulate matter. Soil contamination results from chemicals added to the surface or into the pore spaces in soils. The Love Canal neighborhood near Niagara Falls, New York State, is the site of catastrophic environmental contamination from toxic waste and resulted in high levels of miscarriages and birth defects in people living in this neighborhood. Uranium mining in the Southwestern United States generates large volumes of tailings piles that can leach compounds like ammonia into nearby water systems. Chemical changes can have major effects on ecosystems and impact the biodiversity on Earth. When species are pushed out of their habitat or forced to compete with invasive species they can decline in productivity and may die off. Sometimes these impacts are accidental and unintentional, sometimes they are deliberate. Our actions as a species are now shaping our planet and its ecosystems.

References and Suggested Readings

Botkin, D. B., & Keller, E. A. (2009). *Environmental science: Earth as a living planet.* John Wiley & Sons.

Campbell, N. A., Reece, J. B., & Simon, E. J. (2007). *Essential biology with physiology* (2nd ed.). Pearson Publishing Company.

Jordan, E. (2016). *Small footprint, big impact—Introduction to environmental science* (4th ed.). Kendall Hunt Publishing Company.

Keller, E. A. (2012). *Introduction to environmental geology* (5th ed.). Pearson Publishing Company.

Levkov, J. (2008). *As the Earth turns: Perspective on environmental issues in the 21st century.* Kendall Hunt Publishing Company.

Mader, S., & Windelspecht, M. (2018). *Essentials of biology* (5th ed.). McGraw-Hill Publishing Company.

Websites for Further Reference

https://biology.usgs.gov

https://www.census.gov/

https://www.epa.gov/

https://www.gbif.org/

https://www.iucn.org/theme/global-policy/our-work/convention-biological-diversity-cbd/post-2020-global-biodiversity-framework

https://www2.usgs.gov/envirohealth/cbp/index.php

Questions for Thought

1. Explain in general terms how global population has changed over the past 12,000 years.
2. Cite data to show how rapidly the world's population has changed in the past 2 centuries.
3. What conditions have caused the population to increase during the past 2 centuries?
4. Explain how changes in land use in the United States have occurred in the past 50 years.
5. What is the biogeochemical cycle? (Give two examples.)
6. What role do the oceans play in the water cycle?
7. Research an example of water pollution in an area near your hometown.
8. Explain the effects of coal-fired power plants on the atmosphere.
9. Although grassland ecosystems tend to occur in regions with a lower amount of rainfall, explain the effect the grasses have on animal life in the area.
10. Explain the concept of extinction in terms of a plant or animal.
11. How widespread is water pollution?

Climate Change 15

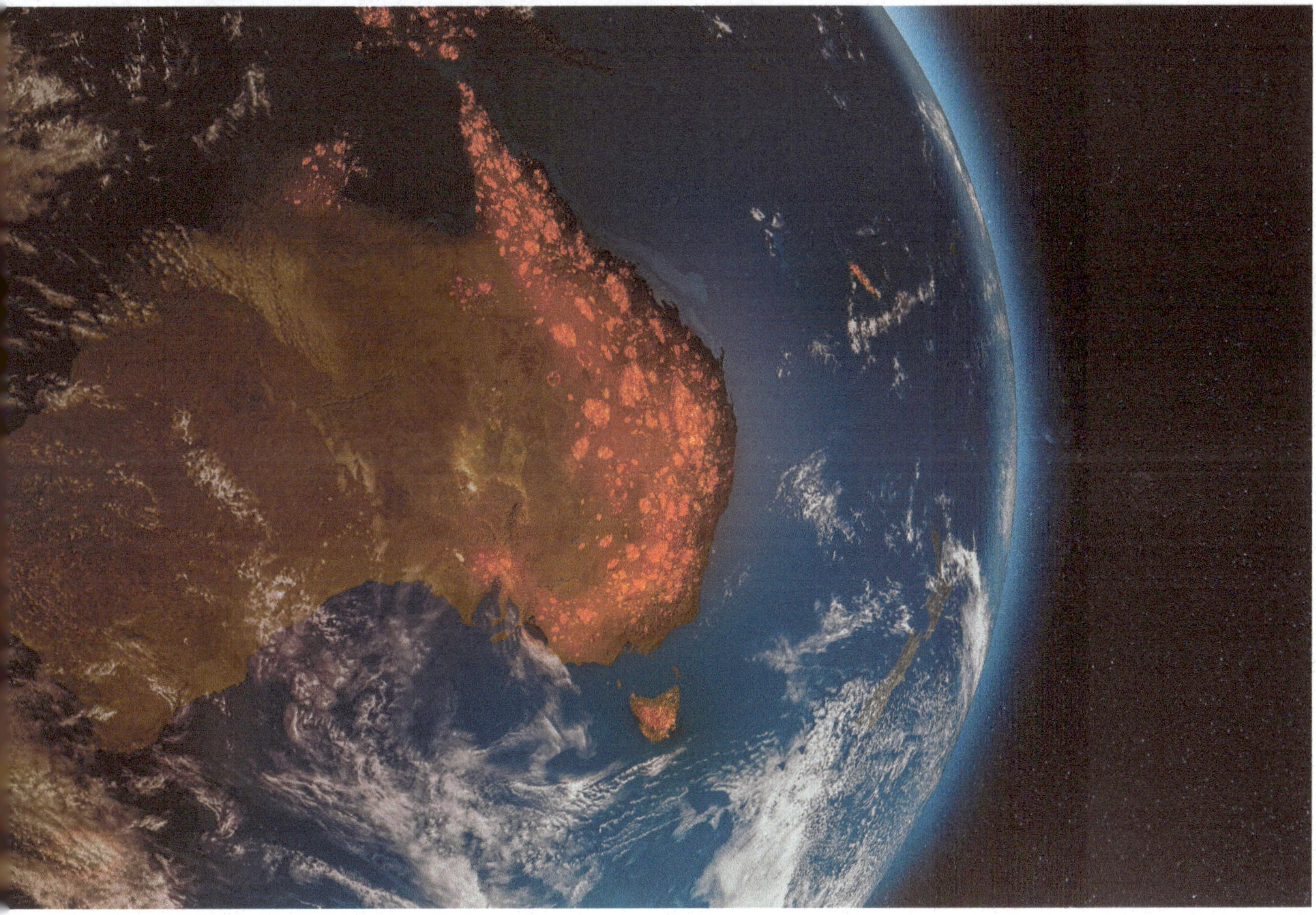

© OSORIOartist/Shutterstock.com.

Global warming increases the average wildfire season by several months a year and hotter weather increases the incidence of lightning, which is the major cause of naturally occurring biomass burning. The 2019–2020 wildfires in Australia, known as Black Summer, were some of the most catastrophic wildfire events on Earth and killed nearly 3 billion animals. Twenty-seven people were killed by the fires directly, and smoke is thought to be a cause of more than 445 additional deaths.

Melting Glaciers Are Changing Earth's Rotation

Since 1980, the planet's poles have moved about 4 meters in distance—the North Pole in Canada heading toward Russia at a rate that is 17 times faster than the previous 15 years. This shift in Earth's axis is due to melting of the world's glaciers, especially from the polar regions, and was observed by scientists at the Institute of Geographic Sciences and Natural Resources Research at the Chinese Academy of Sciences, led by Dr. Sanshan Deng. Hundreds of billions of tons of ice melted each year is redistributing water to the oceans. Pumping of groundwater also eventually moves water to the sea, and humans have removed 18 trillion tons from subsurface reservoirs over the last 50 years. The extra weight of water in the oceans, combined with the mass deficit of melting ice, changes the Earth's moment of inertia and gives it a new wobble, shifting the planet's axis. While the movement may not have a practical effect on humans, it is indicative of the profound changes that human-induced climate disruption can have.

© Risto Raunio/Shutterstock.com.

Earth's climate is not fixed, and has had significant variations over the history of the planet. In the Jurassic Period about 200 to 146 million years ago atmospheric levels of carbon dioxide were about 1900 ppm, compared to pre-industrial levels of 280 ppm. The Jurassic was significantly warmer than today, and was an ice-free world. At the height of the last glacial maximum about 21,000 years ago, Earth's carbon dioxide levels were 180 ppm and the average global temperature was about 6 °C (11 °F) colder than today.

Over the last thousand years climate has varied naturally due to solar influences and volcanic eruptions (**Figure 15.1a**). The Medieval Climate Anomaly was a warm period that lasted from 950 to 1250, with average global temperatures thought to be similar to the early 1900s. During this time, the Vikings colonized Greenland, Iceland, and North America. Prolonged

Figure 15.1a **Northern Hemisphere temperature changes over the last 1,000 years up to about 1900 shows warm and cool periods that were dependent on orbital parameters changing the amount of solar energy reaching Earth's surface. Present-day temperatures are significantly warmer than anything observed in the pre-industrial climate record.**

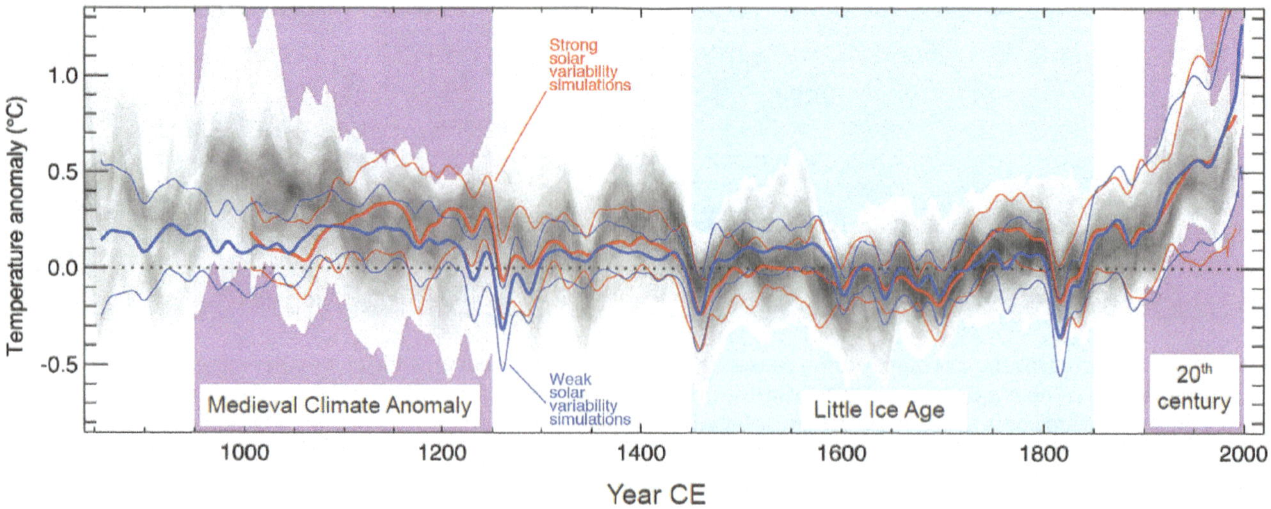

Source: NASA.

drought during this time in the Western United States is thought to have impacted Native American cultures, causing settlement disruption and population movement. Following the Medieval Warm Period, the coldest decade of the last 1,000 years was the 1430s, at the beginning of a time known as the Little Ice Age. From about 1300 or 1400 to 1850 cold climate conditions resulted in crop failures, heavy snowfall, and famine in Europe and North America (**Figure 15.1b**). In New Zealand, the Franz Josef glacier in the Southern Alps advanced rapidly and reached its maximum extent in the early 18th century, one of the few examples of a glacier front ending in a rain forest.

© Everett-Art/Shutterstock.com.

Figure 15.1b The painting *Ice Skating in a Village,* by Hendrick Avercamp, shows Dutch villagers ice skating on a frozen Dutch canal. Cold conditions during the Little Ice Age may have been due to a period of low solar activity.

Weather and Climate

People often confuse the terms *weather* and *climate.* **Weather** is the short-term condition of the atmosphere and how it behaves, mainly with respect to its impact on life and human activities. Often measured over periods of hours, days, or even weeks, these conditions are usually localized or confined to a small region. Most people think of weather in terms of temperature, precipitation, humidity, cloudiness or sunniness, and wind. There are a lot of components that make up weather, and when someone asks "What's the weather like today?" you're likely to answer with information about sunshine or clouds, rain, hail, sleet or snow, wind, thunderstorms, blizzards, steady rains from cold fronts or warm fronts, cold or warm temperatures, heat waves, and other types of short-term events (**Figure 15.2**).

weather

The composite condition of the near Earth atmosphere, which includes temperature, barometric pressure, wind, humidity, clouds, and precipitation. Weather variations over a long period create the climate.

Climate is the long-term average of weather, usually over periods of 30 years or more. It represents the average and variability of conditions such as precipitation, temperature, humidity, and wind velocity that occur in a particular place over a long period of time. Just like weather, climate can be affected by the latitude, terrain, and altitude of the place, as well as any nearby lakes or oceans. It is most commonly classified by temperature and precipitation, and the commonly used Köppen classification has five main climate types: tropical, dry, mild mid-latitude, cold mid-latitude, and polar.

climate

The long-term average pattern of weather, usually taken over a period of years or decades, for a particular region and time period.

© LiliGraphie/Shutterstock.com.

Figure 15.2 An easy way to remember the difference between weather and climate is to think about what you expect, like a hot sunny summer, which is climate. If you compare that to what you actually get during those summer days, like occasional thunderstorms and some days that are very hot and others that are cooler, that is weather.

Those main climate types can be further subdivided into more restricted climate zones such as a rain forest within a tropical climate zone, which has high rainfall between 1.75 and 2.0 m (69–79 inches) and mean temperatures above 18 °C (64 °F) all year round.

Weather and climate are intertwined with the atmosphere and the hydrosphere in complex ways. Varying amounts of water in the oceans change the climate and thus affect weather. Changes in solar radiation patterns over long periods of time can increase or decrease global temperatures, which can cause changes in the amount of ice held in polar regions. Changes in the concentration of different atmospheric gases such as carbon dioxide can also alter global temperatures. People who study past climate over Earth's history are called paleoclimatologists. They use evidence from ice sheets, tree rings, ocean and lake sediments, rocks and fossils, and ancient corals to construct a record of past climate, often using chemical measurements to reconstruct past temperature, rainfall, and other climate conditions.

The Atmosphere

Earth is the only planet in the solar system with an atmosphere that can sustain life and it is composed of a layer of gases, which we call air, that surrounds the planet and is kept from drifting into space by Earth's gravity. In the lower atmosphere, the troposphere extends up to about 7 to 10 km (4 to 6 mi) above Earth's surface at the poles and 17 to 18 km (10.5 to 11 mi) high above the equator, and consists primarily of nitrogen (78%) and oxygen (21%). Argon makes up slightly less than 1% of the atmosphere. Other gases present in very small amounts include carbon dioxide, methane, and nitrous oxides. Depending on atmospheric conditions, water vapor can make up as much as 4% of the atmosphere when the air is saturated. Water content varies with air temperature.

relative humidity

The percentage of moisture present in the air as measured against the amount it can hold at a given temperature and pressure to be saturated.

We measure the amount of water present in air in terms of the saturation level. **Relative humidity** (measured as a percent) indicates the amount of saturation present as measured against total saturation (100%) at a given temperature. Total saturation results in condensation of moisture as dew and precipitation (either rain or snow). A relative humidity of 50% tells us that the air is holding one half of the amount it would hold if totally saturated at a given temperature.

A cross section of the atmosphere shows that it is layered (Figure 15.3a). Heat in the lowest part of the atmosphere helps drive the weather that occurs here. Turbulence is generated as wind blows over the Earth's surface, and by warm air rising from the land as it is heated by the Sun. This turbulence drives atmospheric circulation and helps redistribute heat and moisture from areas on the Earth that have an excess, like the equator, to areas that have a deficit, like the poles. Air becomes less dense with increasing altitude, as there are fewer gas molecules present. There is also less gravitational attraction to hold the molecules close to the surface.

The atmosphere is divided into six different layers based on the changes in physical properties as you move up from the surface of the Earth into space.

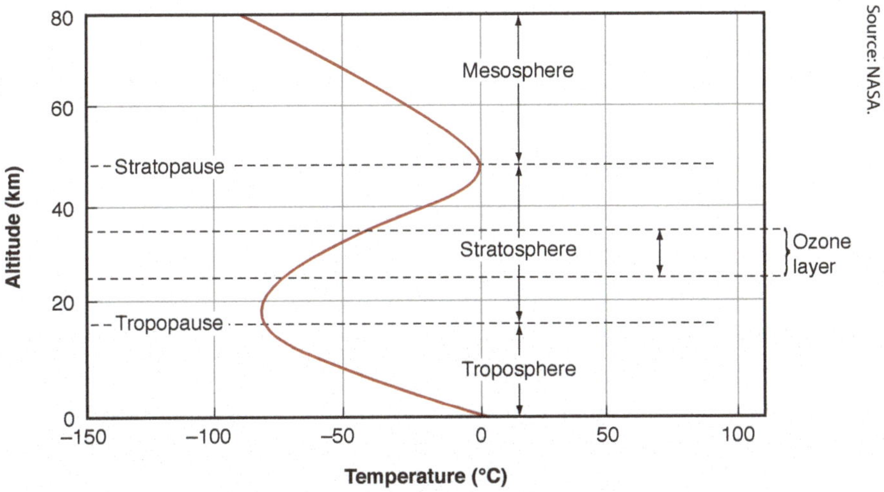

Figure 15.3a A simplified cross section of Earth's atmosphere up to 60 km. Human activity and weather are confined to the troposphere. The solid, curved line shows the change in temperature as altitude changes.

Figure 15.3b A photo of the Earth's troposphere (orange-red), stratosphere, and mesosphere from the International Space Station.

The troposphere (Figure 15.3a) starts at Earth's surface and extends up to 6 to 18 km high (3.7 to 11 mi). It is thickest at the tropics and thins toward the poles. Almost all the weather we experience occurs here. This is the most dense part of the atmosphere and contains 75% of the total mass and 99% of the total water vapor in all of the atmosphere. The composition of the troposphere is what we think of as air, and it contains 78.08% nitrogen, 20.95% oxygen, 0.93% argon, 0.04% carbon dioxide, and very small amounts of other gases. The temperature of the troposphere generally decreases as altitude increases. The top of the troposphere is a boundary region called the tropopause, which is identified based on temperature change as you move into the overlying atmosphere that begins to warm up with increasing height. The troposphere does not mix with any overlying atmospheric layers.

The stratosphere, the second major layer of Earth's atmosphere, lies above the troposphere and tropopause boundary layer. It starts at about 7 km above the poles to 20 km above the equator (4.3 to 12 mi) and extends up to about 50 km (31 mi) above the surface of the Earth. This layer contains Ozone (O_3) which forms here from UV radiation from the Sun splitting oxygen molecules. When oxygen atoms combine with oxygen molecules, they form ozone and release heat. The stratosphere increases in temperature with altitude due to that heat release and ozone absorbing solar energy. Temperatures range from about –51 °C (–60 °F) near the tropopause to an average of about –15 °C (5.0 °F) near the top of the stratosphere. Commercial airlines typically cruise in the lower parts of the stratosphere, mainly to conserve fuel in the low temperatures and low air density.

The **mesosphere** is the third highest main layer of Earth's atmosphere, extending from an altitude of about 50 km (31 mi) to 80–85 km (50–53 mi) above sea level. Temperatures drop with increasing altitude and have an average temperature of –85 °C (–120 °F), making it the coldest place on Earth. This is also the layer where meteors burn up on entering Earth's atmosphere.

mesosphere
The third-lowest layer of Earth's atmosphere, this is the coldest place on Earth and where meteors burn up on entering the atmosphere.

thermosphere
The second-highest layer of Earth's atmosphere that has extremely high temperatures but near-vacuum conditions.

The **thermosphere** is the second-highest main layer of the atmosphere and extends from about 80 km (50 mi) to 500–1,000 km (310–620 mi). The height varies depending on solar activity, as does the temperature, which can reach 2500 °C (4530 °F) or more. Despite the high temperatures, the extremely low density of atmospheric gases here means that molecules can't conduct heat and a person or object will actually experience cold temperatures. The International Space Station orbits in the thermosphere, and the aurora borealis and aurora australis occur in the upper parts of this layer.

exosphere
The highest layer of Earth's atmosphere that thins and merges with space, with extremely low amounts of atmospheric gases. This is the uppermost boundary of Earth.

The outer layer of Earth's atmosphere before we enter space is the **exosphere**, which extends up to about 10,000 km (6,200 mi). This layer has so few atoms that individual hydrogen and helium can travel hundreds of kilometers without colliding with another gas atom. The outer edge of this layer merges into space, and is the outermost part of planet Earth.

Solar Radiation and the Atmosphere

In Chapter 2 we saw that electromagnetic radiation from the sun has a wide range of wavelengths. Part of this spectrum of energy includes visible light. However, a much wider range of energy bombards Earth, including many wavelengths that are harmful. Certain gases in the atmosphere interact with selected wavelengths in different ways. Ozone in the stratosphere shields the Earth's surface from 97 to 99% of the Sun's UV radiation, which damages the genetic material of DNA and is related to some types of skin cancer. Destruction of the ozone layer has occurred by the release of man-made organic compounds, including chlorofluorocarbons (CFCs), into the atmosphere. Ozone reacts readily with CFCs and is destroyed in the chemical reaction. Ultraviolet radiation can then penetrate the atmosphere and reach the surface.

United States Geological Survey.

Figure 15.4 **Volcanic ash and gas, especially sulfur, from volcanic eruptions can reach the stratosphere and cause global cooling.**

Effects of Volcanic Activity

Volcanic eruption blast ash and gas into the atmosphere, and large eruptions may reach the stratosphere (Figure 15.4). Sulfur gases are a component in explosive volcanic eruptions and are able to reflect short wavelength solar radiation, which can cause global cooling for years after large eruptions. The combination of sulfur gases and volcanic ash can produce a significant drop in temperature that is equivalent to a nuclear winter. In the 1991 eruption of Mount Pinatubo in the Philippine Islands, more than 20 million tons of sulfur dioxide were thrown into the atmosphere. This caused global temperatures to drop 0.5 °C for a 2-year period following the eruption. When measured in the context of geologic time, major eruptions similar to these can occur frequently. One major eruption every 100 or 200 years adds volcanic gases to the atmosphere, and water to the hydrosphere.

In the geologic past, several episodes of massive flood basalt eruptions sent millions of tons of gases into the atmosphere. The flood basalts of Siberia that formed about 250 million years ago and the Deccan Traps in India that erupted 66 million years ago both had a large global climate impact, resulting in major climate change and mass extinction events. The continual eruption of basalts associated with seafloor spreading adds heat and gases directly into ocean waters.

Greenhouse Gases

Once solar radiation penetrates the atmosphere and strikes the surface of the Earth, some of the energy is absorbed by the land and water, while some of it is reflected back into space. Gases and dust particles also reflect radiation back into the upper atmosphere, as well as clouds and ice.

When this energy is reflected, there is a change in the wavelengths of the radiation. Longer wavelength infrared waves are absorbed by moisture and gases in the middle and upper troposphere (Figure 15.5). This absorption of heat by these gases causes the temperature of the atmosphere to rise. Radiation is also re-reflected back toward the surface. This trapped energy causes a global warming effect, similar to what takes place in a greenhouse, and results in a land and ocean surface temperature that is on average 14 °C warmer than it would be without this process. It is important to note that this greenhouse effect is a natural process and helps contribute to the habitability of Earth; it is the increase in **greenhouse gases** since the Industrial Revolution that has caused a change in the composition of the atmosphere and increased the magnitude of the greenhouse effect.

greenhouse gases
Atmospheric gases, primarily carbon dioxide, methane, and nitrous oxide restricting some heat-energy from escaping directly back into space.

greenhouse effect
The heating that occurs when gases such as carbon dioxide trap heat escaping from the Earth and radiate it back to the surface.

Several gases contribute to this **greenhouse effect**. The primary gases are carbon dioxide, water vapor, methane, and ozone. Because so much water is present on Earth's surface, it is easily heated, and rises to become a

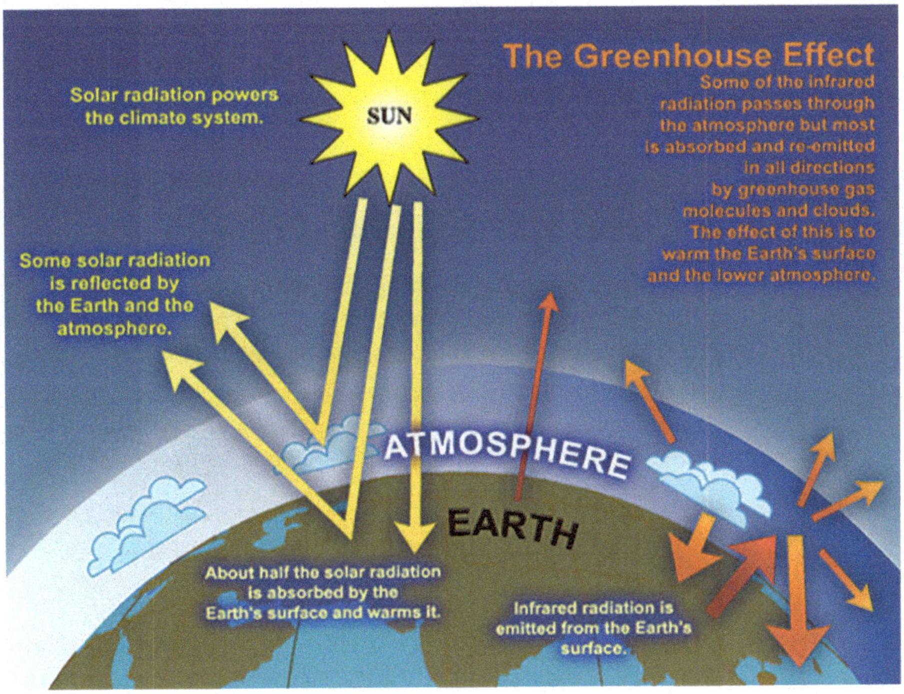

Source: IPCC—Intergovernmental Panel on Climate Change.

Figure 15.5 Energy from solar radiation is divided between reflected and reradiated energy.

significant contributor to the warming process. However, carbon dioxide, a gas that is generated by the oxidation or burning of fossil fuels which contain carbon, is also a key greenhouse gas. Over the past 200 years, there has been a 45% increase in the amount of carbon dioxide in the atmosphere. This has contributed to the warming of the atmosphere.

Methane (CH_4) is the simplest of the hydrocarbons, and is also a greenhouse gas. While carbon dioxide stays in the atmosphere for 100 years, methane only lasts for about 20. However, while it is in the atmosphere, it is about 30 times stronger than carbon dioxide in causing warming. It is a byproduct of the decay of organic material, and it also forms during the digestive process of organisms. Estimates show that as much as 20% of all methane is produced by cattle. Methane is also released during the processing and transportation of petroleum products.

Reducing the Presence of Carbon

anthropogenic impact
Human-caused changes in the environment, ecosystems, natural resources, or biodiversity.

carbon sequestration
The storage or removal of carbon from the environment or the reducing or elimination of its presence.

An increase in the concentration of greenhouse gases like carbon dioxide and methane traps more solar energy and results in warming of the Earth. With current emissions rates, our planet's temperature could increase by 2 °C (3.6 °F) over the next few decades. The majority of these gases come from **anthropogenic** sources such as burning fossil fuels, deforestation, and other changes in land use. If the amount of carbon can be reduced, the end result would be an environment that is closer to the baseline levels of greenhouse gases that existed before the start of the Industrial Revolution. International scientific agencies and many governments, in conjunction with private industries, are working on the process of **carbon sequestration**, which involves removing carbon from the environment and storing it below ground in order to temporarily lower atmospheric carbon dioxide levels over the next several decades to a hundred years. The increased use of hybrid vehicles decreases the demand for petroleum and natural gas and reduces the amount of carbon dioxide put into the atmosphere. The rapidly increasing use of wind turbine technology to generate electricity reduces the need for coal-fired power plants (**Figure 15.6**) and reduces the production of carbon dioxide.

Source: David M. Best.

Figure 15.6 This wind turbine farm in southern California is part of a larger plan nationwide to increase the production of electricity from clean, existing resources.

Is Global Warming Occurring?

For centuries people have recorded temperatures in cities and other locations throughout the world. Annual data for the past 125 years show that average global temperatures have been increasing significantly (**Figure 15.7**), with an increase of more than 1 °C in the past century. When viewed over the past 15,000 years, temperatures have risen more than 4 °C.

Effects of Global Warming

Increased warming causes sea level to rise as glaciers and ice sheets melt. Because water has a high capacity to retain heat, the oceans absorb more heat, causing the molecules

Source: NOAA.

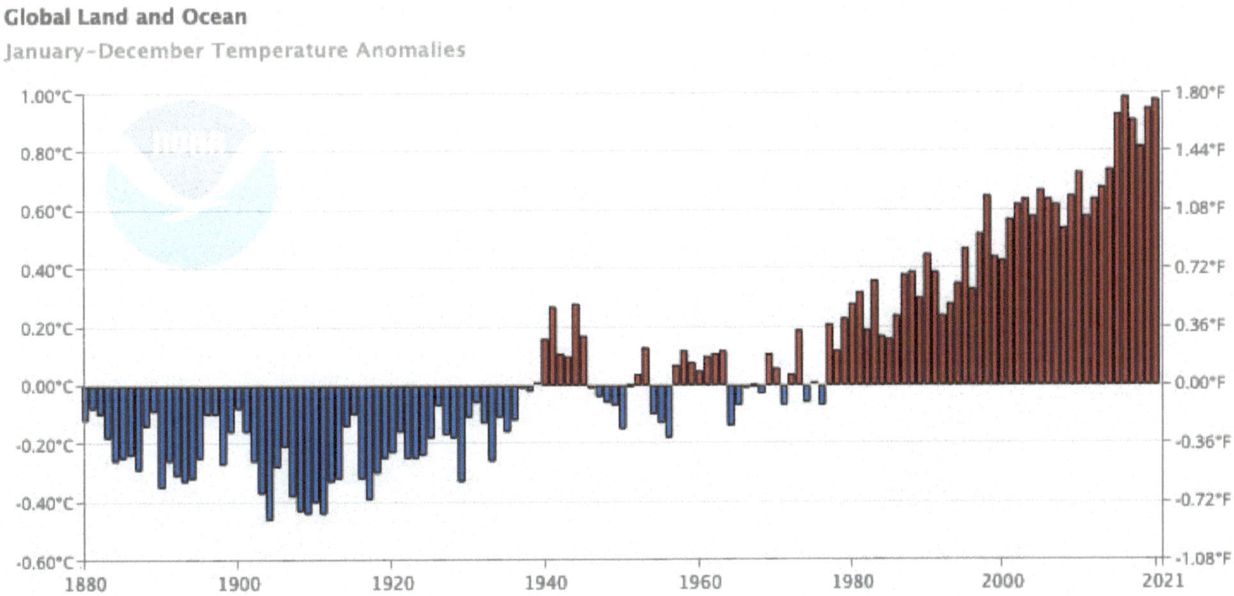

Figure 15.7 Annual global temperatures have been increasing over the past 125 years, although most of the increase has occurred since 1980.

to expand, thereby raising sea level further. Large bodies of water are slow to heat up, but they are also slow to cool down.

Warmer ocean waters have increased thermal energy, which leads to more tropical storms. Higher ocean temperatures allow cyclonic storms to intensify as they pass over these heat sources. This was the case with Hurricane Katrina as it moved into the Gulf of Mexico before striking the Louisiana-Mississippi coastline in August 2005 (Figure 15.8). Temperatures in the Gulf of Mexico were well above average, providing more thermal energy to intensify the hurricane.

Continental areas are affected by global warming as changes in the hydrologic cycle alter the distribution of precipitation, which can affect vegetation patterns. Droughts can destroy once productive agricultural areas, causing famine and the possible need to relocate people.

Increased temperatures in high latitudes melt permafrost, ground that is continuously below 0 °C (32 °F) for 2 or more years, either on land or under the ocean. About 11% of the planet's surface is underlain by permafrost. Because these ecosystems contain large amounts of organic material, decay processes are accelerated, which release trapped methane into the atmosphere as the permafrost melts. This methane then becomes part of the greenhouse gases.

Source: NASA.

Figure 15.8 Hurricane Katrina is located in the Gulf of Mexico several days before striking the Gulf Coast region of the southern United States. Orange to red colors in the Gulf of Mexico reflect the warmer seawater temperatures providing energy to the storm.

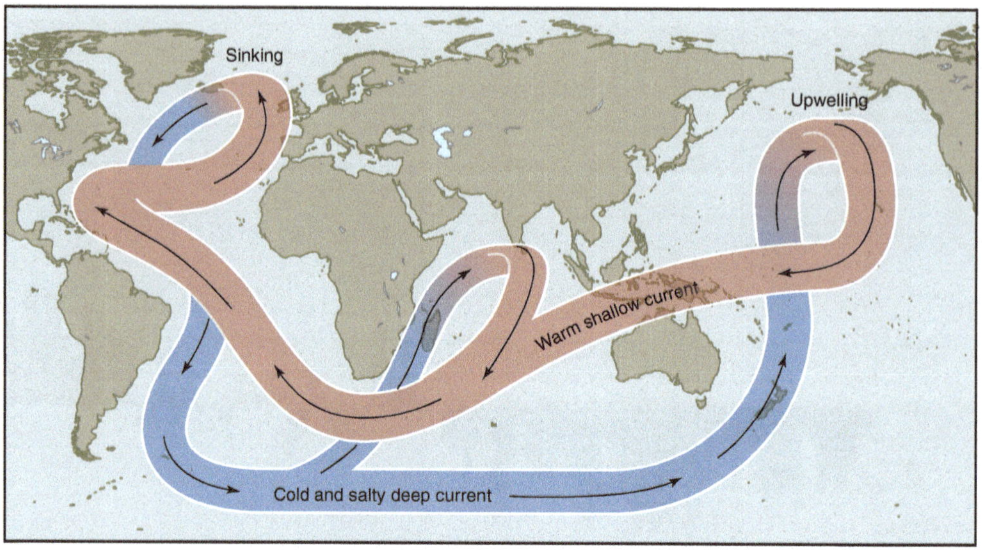

Figure 15.9 Movement of waters in the ocean can be thought of as a conveyor belt transporting deep, cold water to the surface where the warm, shallow water moves near the surface.

© Kendall Hunt Publishing Company.

This is an example of a positive feedback cycle, warming causes methane release that causes more warming.

climate change

The long-term fluctuation in temperature, precipitation, wind, and other aspects of the Earth's climate.

Long-Term Climate Changes

Variations in climate have occurred since the formation of the atmosphere several billion years ago. In addition to the role the oceans play in these variations, continents also contribute in less obvious ways. The positions of the continents have changed drastically through geologic time. The location of these large land masses has a direct effect on large-scale circulation patterns in the oceans. Today, many ocean currents flow near the edges of continents, moving massive amounts of water and heat across Earth's surface. Near the end of the Permian Period about 250 million years ago, one large land mass or supercontinent existed called Pangaea. As it began to break up, the earliest shapes of the present-day continents began to form. This breakup altered oceanic circulation. Today's configuration of the continents allows water to move readily between the poles and the open oceans (**Figure 15.9**).

Figure 15.10 Arctic tundra in the Brooks Range of Alaska, US.

© Troutnut/Shutterstock.com.

The rearrangement of the continents also puts the land masses in different locations. When located near the equator, an area would experience warmer temperatures and more precipitation than a land mass situated at a high latitude. Think of portions of central Africa and southern and central Alaska today. The warm equatorial regions of Africa have a wide range of life forms, while the harsh, colder climate of Alaska limits the varieties of plants and animals that can survive there (**Figure 15.10**).

Short-Term Climate Changes

The cause of short-term changes in Earth's climate is generally controlled by the amount of incoming solar radiation. Both atmospheric and astronomical factors contribute to these changes. Atmospheric factors include the concentration of greenhouse gases and the amount of dust and other aerosols in the air. Researchers have examined ice cores from the polar regions and found that periods of increased glaciation correspond to increased amounts of dust in the ice core. The additional atmospheric dust reflected solar energy back into space, producing a drop in surface temperatures. Short-lived decreases in global temperatures can also result from large volcanic eruptions injecting ash and aerosols into the air.

Astronomical influences are mainly related to Earth's position in space relative to the Sun. Variations in Earth's orbit around the Sun produce changes in the distance between the two bodies. The Earth is approximately 150 million kilometers (93 million mi) away from the Sun, which is defined as 1 Astronomical Unit. The Earth's orbit is not perfectly circular, so this distance can vary from about 147.5 million kilometers (91.6 million mi) when it is closest to the Sun—called perhelion—to 152.6 million kilometers (94.8 million mi) when it is furthest away—called aphelion. Eccentricity is a measure of how circular versus elliptical the Earth's orbital path is. As Earth revolves around the Sun, its path changes due to the gravitation attraction of the Sun, Moon, and other more distant objects. When the orbit become more circular, the distance between the Earth and the Sun is less than when the path is more elliptical (Figure 15.11a). These variations occur over periods of about 100,000 years and the ellipticity changes about 5%.

As the Earth rotates on its axis, there are times when the angle of tilt of the axis ranges from 22.1° to 24.5°. Currently Earth is tilted 23.5° from the vertical. The period of change is roughly 41,000 years. The planet's tilt reached a maximum in 8,700 B.C.E., we are currently about halfway between the maximum and minimum, and will reach a minimum in 11,800 C.E. Increased tilt means that each hemisphere's summer will get more incoming solar radiation as the pole is tipped toward the sun, and less solar radiation in winter as the pole is tipped away from the sun. Our current trend of decreasing tilt and the changes in solar energy reaching the poles promotes warmer winters and cooler summers. As Earth rotates, it also tends to wobble on its axis, in the same way a spinning top begins to wobble as its rotation

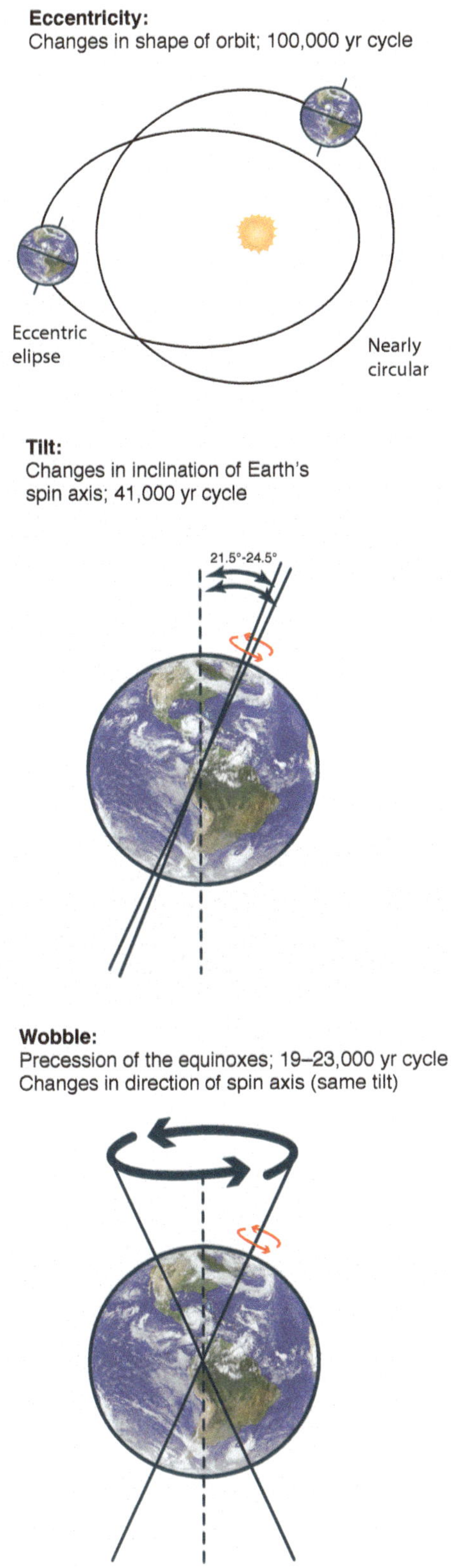

© Kendall Hunt Publishing Company.

Figure 15.11a The eccentricity, tilt, and wobble of the Earth as it moves around the Sun contribute to long-term changes in the climate.

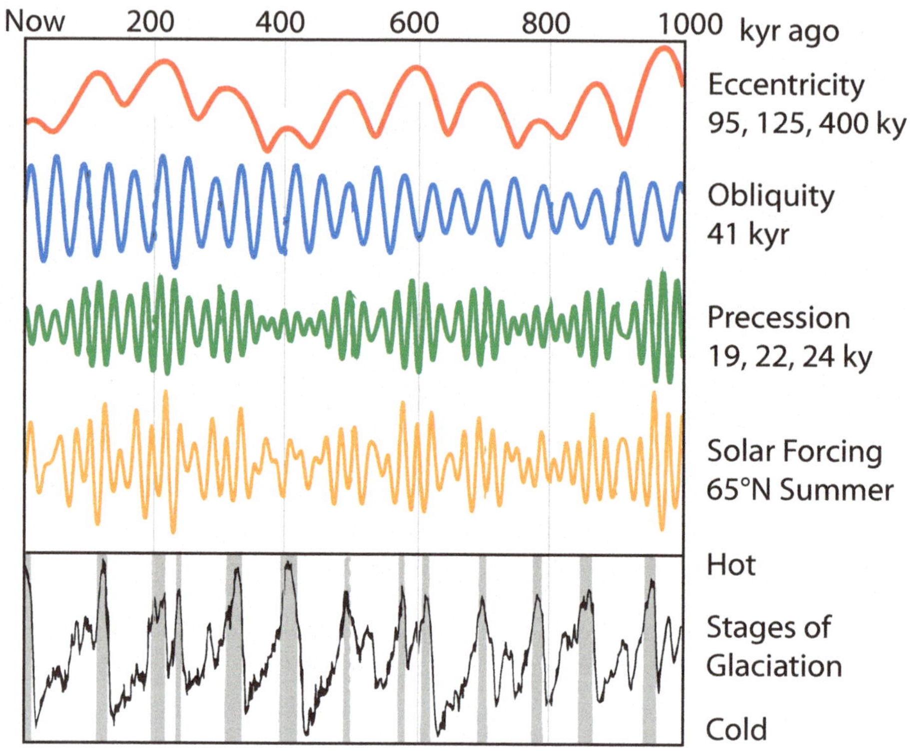

Figure 15.11b
Milankovitch cycles show how changes in the Earth's orbit, its tilt and distance from the Sun, affect the climate on the scale of thousands to tens of thousands of years.

Source: NOAA National Centers for Environmental Information. Modified by Kendall Hunt Publishing Company.

axial precession
The wobble that occurs when a spinning object slows down.

rate decreases. This wobble, also termed **axial precession**, has a period of about 26,000 years. It is caused by the tidal forces from the Sun and the Moon on the solid Earth. The net effect of all these astronomic factors occurring together is that their period, together with their maximum influences, corresponds to times when active glaciations happened, due to major changes in the amount of solar radiation striking Earth over time (**Figure 15.11b**).

The Effects of Global Climate Change

According to the World Economic Forum, global climate change is the "most impactful risk" facing us worldwide in the coming decades. Climate change has a cascading effect, as it transforms ecosystems it affects everything from where we live, to what we eat, to the water we drink, to the air we breathe. As our air temperature rises, pollutants increase, and allergenic substances like mold and pollen increase too. Warmer ecosystems increase the range for disease-carrying insects like mosquitoes which can spread malaria, dengue fever, yellow fever, and the Zika virus. The World Health Organization predicts that climate change will be responsible for an additional 250,000 deaths per year between 2030 and 2050, due to increases in heat stroke, and cardiovascular and kidney diseases. A warmer atmosphere is capable of holding more moisture, fueling bigger, more intense, and more frequent thunderstorms and cyclonic storms. Higher temperatures increase the frequency of storms, floods, droughts, and wildfires. The likelihood of being displaced by a disaster is 60% higher than it was 40 years ago, and the largest increases in displacement are driven by weather- and climate-related events. Rising seas threaten coastal ecosystems and low-lying areas across the globe, including some of the world's largest cities. The ocean's absorption of carbon dioxide has changed the composition of seawater to make it more acidic, by roughly 30% compared to preindustrial times. This threatens ocean life and ecosystems. On land, wildlife are forced to change their habitat and migration patterns to find cooler climates and higher altitudes. Increasing temperatures can fundamentally transform entire ecosystems, putting many species at increased risk of extinction. Mammals, fish, birds, and reptiles are disappearing more than 100 times faster than they should, due to climate change, pollution, and deforestation. Despite this overwhelming impact of climate change, we can mitigate and help lessen its effects on our planet. Doing so will require tackling its root cause, carbon dioxide pollution from burning fossil fuels. This is a huge task, and requires a global effort. We must all step up.

Key Terms

anthropogenic impact *(page 438)*
axial precession *(page 442)*
carbon sequestration *(page 438)*
climate *(page 433)*
exosphere *(page 436)*
greenhouse effect *(page 437)*
greenhouse gases *(page 437)*
mesosophere *(page 435)*
relative humidity *(page 434)*
stratosphere *(page 435)*
thermosphere *(page 436)*
troposphere *(page 435)*
weather *(page 433)*

Summary

More than 70% of Earth's surface is covered by water, which strongly influences the planet's temperature regulation in the troposphere. Weather and climate are two different but related ideas—weather is the short-term condition of the atmosphere in a particular place whereas climate describes the atmospheric conditions over a long period of time. Earth's atmosphere is layered and made of the troposphere, stratosphere, and mesosphere, and is dominantly made of nitrogen, oxygen, and small amounts of argon, carbon dioxide, methane, and nitrous oxides. Volcanic activity contributes sulfur gases to the atmosphere, producing a cooling effect that can be large if the volcanic eruption is big enough. Solar radiation interacts with the gasses in the atmosphere to generate a natural warming process called the greenhouse effect. Anthropogenic increase in greenhouse gasses, especially carbon dioxide, have increased this warming effect significantly. Increased warming causes ice to melt, warmer ocean waters to expand, sea levels to rise, and increased energy fueling tropical storms.

Long-term climate change has been impacted by geological processes such as plate tectonics and the location of continents with respect to the planet's poles and equator. Shorter-term climate change is influenced by orbital parameters such as the eccentricity, tilt, and wobble of Earth's orbit around the sun. Milankovitch cycles recorded in sediments and periods of glaciation result from these orbital variations.

References and Suggested Readings

Fagan, B. (2002). *The great warming.* Bloomsbury Press.

Houghton, J. (2004). *Global warming: The complete briefing.* Cambridge University Press.

Mayewski, P. A., & White, F. (2002). *The ice chronicles.* University Press of New England.

Peters, E. K. (2012). *The whole story of climate—What science reveals about the nature of endless change.* Prometheus Books.

Romm, J. J. (2016). *Climate change—What everyone needs to know.* Oxford University Press.

Schmidt, G., & Wolfe, J. (2008). *Climate change—Picturing the science.* W. W. Norton & Company.

Schneider, B. (2012). *Extreme weather: A guide to surviving flash floods, tornadoes, hurricanes, heat waves, snowstorms, tsunamis and other natural disasters.* St. Martin's Press.

Silver, J. (2008). *Global warming and climate change demystified.* McGraw Hill.

Spencer, R. W. (2010). *The great global warming blunder—How Mother Nature fooled the world's top climate scientists.* Encounter Books.

Websites for Further Reference

https://www.noaa.gov/weather

https://350.org/

Questions for Thought

1. Explain the difference between weather and climate.
2. What are the three layers of the atmosphere closest to Earth's surface, and what is one key characteristic of each?
3. What effects do volcanic eruptions have on the makeup of the atmosphere, and how can they alter solar radiation reaching Earth?
4. What is the main cause of long-term climate change, and explain how it changes the climate?
5. Explain how changes in Earth's eccentricity affect global climate.

acquired immunodeficiency syndrome (AIDS) A serious disease that results from an HIV infection. It is spread through direct contact with contaminated bodily fluids.

acre foot The amount of water that covers one acre to a depth of one foot; equivalent to 325,851 gallons of water.

active A term applied to a volcano that has erupted in recorded history or is currently erupting.

advisory An announcement that is issued when a tsunami could result from a large earthquake that occurs near a coastal region.

air mass A large body of air of considerable depth which are approximately homogeneous horizontally. At the same level, the air has nearly uniform physical properties, especially temperature and moisture.

amino acids Organic compounds composed of nitrogen, carbon, hydrogen, and oxygen, along with an additional alkyl side chain group.

angle of repose The natural angle of a slope that forms in a pile of unconsolidated material.

anthropogenic impact Human-caused changes in the environment, ecosystems, natural resources, or biodiversity.

anticyclone An area of high atmospheric pressure having clockwise circulation in the Northern Hemisphere and counterclockwise motion in the Southern Hemisphere.

ash Pyroclastic material that has an average particle size less than 2 mm (0.08.

asteroid A rocky or metallic body ranging in size from a meter to almost 1,000 km (621 mi) in size.

asthenosphere (1) The uppermost layer of the mantle, located below the lithosphere. This zone of soft (plastic), easily deformed rock lies at depths between 80 and up to 800 km (50 to 500 mi). (2) The semi-plastic to molten zone beneath the rigid lithosphere.

astronomical unit (AU) The distance of the Earth from the Sun, averaging 150 million km or 93 million miles; used to describe large distances between bodies in space.

atmosphere The air surrounding the Earth, from sea level to outer space.

avian flu (H5N1) An acute viral disease of chickens and other birds (except pigeons) capable of being transmitted to humans; first noticed in the Far East, its symptoms include fever and lack of energy.

axial precession The wobble that occurs when a spinning object slows down.

backburn A fire that burns back toward the main part of a wildfire.

backshore zone The part of the beach extending landward from the high tide level to the area reached only during storms.

backwash The flow of water down the beach face toward the ocean from a previously broken wave.

bacteria A tiny, single-celled organism that reproduces by cell division, having a shape similar to a rod, spiral, or sphere, and has no chlorophyll.

barrier island A long, usually narrow accumulation of sand, that is separated from the mainland by open water (lagoons, bays, and estuaries) or by salt marshes.

basalt An extrusive, fine-grained, dark volcanic rock that contains less than 50% silica by weight and a relatively high amount of iron and magnesium.

beach An aggregation of unconsolidated sediment, usually sand, that covers the shore.

beach drift The movement of sand along a zigzag path along the beach parallel to shore due to successive waves on the beach.

beach nourishment A soft stabilization technique primarily accomplished by adding sediment to the shoreline.

berm A low, incipient, nearly horizontal or landward-sloping area, or the landward side of a beach, usually composed of sand deposited by wave action.

Big Bang Theory The idea that the Universe formed from an initial point mass that exploded about 13.7 billion years ago and moved outward in an ever-expanding fashion.

biodiversity All the diversity of life on Earth, ranging from single cell organisms to the most complex ecosystems.

biogeochemical cycle Natural processes that recycle nutrients in various chemical forms from the environment, to organisms, and then back to the environment. Examples are the carbon, oxygen, nitrogen, phosphorus, and hydrologic cycles.

biological hazard (biohazard) A biological substance or condition that poses a threat to the health of living organisms, primarily that of humans.

biosphere The living and dead organisms found near the Earth's surface in parts of the lithosphere, atmosphere, and hydrosphere. The part of the global carbon cycle that includes living organisms and biogenic organic matter.

Black Plague (Black Death) An epidemic of bubonic plague that killed more than 20 million people in Europe during the mid-14th century.

blizzard A severe winter storm that has the following conditions that are expected to prevail for a period of 3 hours or longer: sustained wind or frequent gusts to 35 miles an hour or greater and significant falling and/or blowing snow (i.e., reducing visibility frequently to less than one fourth mile).

block A large angular fragment of lava measuring more than 64 millimeters (6.4 cm or 2.5 inches) in diameter.

body wave A seismic wave that is transmitted through the Earth. P-waves and S-waves are body waves.

bomb A large lava fragment larger than 64 millimeters (6.4 cm or 2.5 inches) in diameter that is erupted as a liquid and becomes rounded as it is aerodynamically shaped while traveling through the air. This bomb is from the 2014–2015 eruption of Bárðarbunga, Iceland, and is called a "cowpie" bomb for its flat patty shape that forms as the fluid bomb impacts the ground.

break When a wave steepness exceeds 1/7, the wave becomes too steep to support itself and it breaks, or spills forward, releasing energy and forming whitecaps often observed in choppy waters or the surf area along a beach.

breaker A wave in which the water at the top and leading edge falls forward producing foam.

breakwater Structure built offshore and parallel to shore that protects a harbor or shore from the full impact of waves.

brecciated: Mesosiderites broken into angular fragments.

brush fire A fire that burns primarily brush and material low to the ground.

bubonic plague A rare but fast-spreading bacterial infection caused by Yersinia pestis; transmitted in humans and rodents by infected flea piercings; symptoms include headaches and painful swelling of lymph nodes.

burnout An intentional fire that is lit to burn fuel that lies in the path of a spreading fire.

caldera A large basin-like depression that is many times larger than a volcanic vent.

carbon cycle (1)The biogeochemical cycle by which carbon is exchanged among the biosphere, geosphere, hydrosphere, and atmosphere of the Earth. (2) The movement of carbon between the biosphere and the nonliving environment. Carbon moves from the atmosphere into living organisms that then move it back into the atmosphere.

carbon sequestration The storage or removal of carbon from the environment or the reducing or elimination of its presence.

carrying capacity The maximum population size that can be regularly sustained by an environment.

cause A controlling factor that makes a slope vulnerable to failure.

cellulose The primary substance that makes cell walls in plant tissue.

central peak structure Part of a complex crater that has an uplifted center area formed by rebound of the crater floor during a meteorite impact event.

cholera An acute intestinal disease caused by the bacteria *Vibrio cholera*; often found in contaminated drinking water and food.

cinder cone A conical-shaped hill created by the buildup of cinders (lapilli) and other pyroclastic material around a vent.

circular orbital motion The movement of a particle by moving in a circle.

climate The long-term average pattern of weather, usually taken over a period of years or decades, for a particular region and time period.

climate change The long-term fluctuations in temperature, precipitation, wind, and other aspects of the Earth's climate.

closed system A system in which no matter or energy can leave or enter from the outside.

coastal submergence The permanent flooding of coastal areas by global sea level rise or land subsidence.

coastline Unique boundary where the geosphere, atmosphere, and hydrosphere meet and the systems interact.

collapse The sudden sinking of the surface into a subsurface void.

comet Body of icy and rocky material that moves around the Sun; most comets originate from the Oort Cloud.

complex crater A type of large impact crater that forms at diameters greater than 2 km on Earth, where the collapse and modification of the transient crater is extensive and may result in a central uplift and ring structures.

composite volcano A volcano that forms from alternating layers of lava and pyroclastic debris; also known as a stratovolcano.

conduction (1) Heat transfer directly from atom to atom in solids. (2) Transmission of heat through a material.

continental crust The rocky, relatively low density, outermost layer of the Earth ranging in thickness from 30 to 50 km (19 to 31 mi).

controlled burn A fire set to remove fuel. Term is no longer used, as not all such fires were kept under control.

convection Transfer of heat by movement in the atmosphere caused by density differences produced by heating below. (1) (Physics) Heat transfer in a gas or liquid by the circulation of currents from one region to another; also fluid motion caused by an external force such as gravity. (2) (Meteorology) The phenomenon occurring where large masses of warm air, heated by contact with a warm land surface and usually containing appreciable amounts of moisture, rise upward from the surface of the Earth.

convection cell Within the geosphere it is the movement of the asthenosphere where heated material from close to the Earth's core becomes less dense and rises toward the solid lithosphere. At the lithosphere-asthenosphere boundary heated asthenosphere material begins to move horizontally until it cools and eventually sinks down lower into the mantle, where it is heated and rises up again, repeating the cycle.

convergent plate An area where two or more lithospheric plates are coming together, such as along the west coast of South America.

convergent boundary The collision can be between two continents (continental collision), a relatively dense oceanic plate and a more buoyant continental plate (subduction zone) or two oceanic plates (subduction zone).

core The innermost layer of the Earth, made up of mostly iron and nickel metal. The core is divided into a liquid outer core and a solid inner core. The core is the densest of the Earth's layers.

Coriolis Effect An imaginary force that appears to be exerted on an object moving within a rotating system. The apparent force is simply the acceleration of the object caused by the rotation. Along the equator, there will be no such rotation.

crest The highest point of a wave.

crown fire A fire that moves through the upper portions of trees; these are very difficult to extinguish.

cryosphere The frozen water part of the Earth system, including ice caps and glaciers.

cyclone A rotating mass of low pressure in the atmosphere that covers a large area; the warm air mass rotates counterclockwise in the Northern Hemisphere and clockwise in the Southern Hemisphere. The term is strictly applied to large low pressure storms in the Indian Ocean and South Asia.

cyclonic storm A generic term that covers many types of weather disturbances that are typified by low atmospheric pressure and rotating, inwardly directed winds.

decompression melting Melting of hot rocks in the subsurface caused by a reduction in overlying pressure as the material rises toward the surface.

deep-focus earthquake An earthquake that has its focus located between a depth of 300 kilometers and roughly 700 kilometers.

derecho A derecho is a widespread, long-lived, straight-line wind storm that is associated with a fast-moving group of severe thunderstorms with wind speeds equivalent to tornadoes and hurricanes.

desert A region that receives less than 25 cm (10 in) of annual precipitation.

discharge The volume of water flowing through a stream channel in a given period of time, usually measured as cubic feet per second or cubic meters per second.

dissolution The natural dissolving of a substance by chemical reactions.

divergent boundary An area where two or more lithospheric plates move apart from each other, as along a mid-oceanic ridge.

Doppler radar Radar that can measure radial velocity, the instantaneous component of motion parallel to the radar beam (i.e., toward or away from the radar antenna).

dormant A term applied to a volcano that is not currently erupting but has the likelihood to do so in the future.

drainage basin An area that drains water to a given point or feature, such as a lake.

drawdown A lowering of water level at the beach when the trough of a tsunami comes ashore.

driving force A force that produces downslope movement caused by gravity.

dynamic equilibrium The state in which the action of multiple forces produces a steady balance, resulting in no change over time.

Earth system Composed of the geosphere, hydrosphere, atmosphere, and biosphere, and all of their components, continuously interacting as a whole.

Earth system science Study of our planet as a system composed of numerous interconnecting subsystems governed by natural laws.

ebb tide That period of tide between a high water and the succeeding low water; falling tide.

ecology The study of the interaction between organisms and their environments.

ecosystem A community of plants, animals, and other organisms that interact within their given setting.

emergent coastline A coastline which has experienced a fall in sea level, because of global sea level change, local land uplift, or isostatic rebound.

epicenter The point on Earth's surface directly above the focus. This is the position that is reported for the occurrence of an earthquake as a latitude and longitude value can be assigned to the point.

epidemic The sudden occurrence of a highly-contagious disease or other event that is clearly in excess of normally expected numbers.

eustatic Global changes in sea level.

exosphere The highest layer of Earth's atmosphere that thins and merges with space, with extremely low amounts of atmospheric gases. This is the uppermost boundary of Earth.

exponential growth Growth in which some quantity, such as population size, increases by a constant percentage of the whole during each year or other time period; when the increase in quantity over time is plotted, this type of growth yields a curve shaped like the letter J.

extinct A term applied to a volcano that no longer is expected to erupt.

eye The central core of a cyclonic storm, normally relatively small and lacking clouds, moisture, and wind.

eye wall The boundary between the eye of a cyclonic storm and the innermost band of clouds.

Factor of Safety (FoS) The ratio of resisting force to driving force.

fetch The distance over which the wind blows across open water.

fire triangle Consists of three parts that cause fire to occur: heat, fuel, and oxygen.

firebreak A clearing that provides a gap across which fire should not burn.

firestorm A widespread, intense fire that is sustained by strong winds and updrafts of hot air.

flood stage The point in time when a body of water, such as a river, rises to a level that causes damage to the adjacent areas.

flood tide The incoming or rising tide; the period between low water and the succeeding high water.

floodplain The flat area alongside a stream that becomes flooded when water exceeds the banks of the stream.

flux melting Melting of a solid rock that is caused by adding water or other volatiles to the hot rock, which lowers the melting temperature.

focus The point in the subsurface where an earthquake first originates due to breaking and movement along a fault plane; sometimes referred to as a hypocenter.

foreshore zone The area located between the normal low and high tide levels.

forest fire A fire that occurs in a forest or stand of trees.

frequency Occurrence of specific events. Used in interpreting the past record of events to predict occurrences of that event in the future.

fuel The part of the fire triangle that actually burns.

fungus A single-celled or multicellular organism that gets food from decaying matter, such as yeast, mold, and mushrooms.

geologic time scale A relative time scale based upon fossil content. Geological time is divided into eons, eras, periods, and epochs.

geosphere The soil, sediment, and rock layers of the Earth including the crust, both continental and oceanic.

glucose Is an organic substance with the chemical formula $C_6H_{12}O_6$.

gravitational energy The force of attraction between objects due to their mass and is produced when an object falls from higher to lower elevations.

greenhouse effect The heating that occurs when gases such as carbon dioxide trap heat escaping from the Earth and radiate it back to the surface.

greenhouse gases Atmospheric gases, primarily carbon dioxide, methane, and nitrous oxide restricting some heat-energy from escaping directly back into space.

groin Solid structure built at an angle from a shore to reduce erosion from long shore currents, and tides.

ground fire Fire that moves along the ground and rises several feet above the surface.

hail Solid, spherical ice precipitation that has resulted from repeated cycling through the freezing level within a cumulonimbus cloud.

headland A steep-faced irregularity of the coast that extends out into the ocean.

heat The part of the fire triangle that provides the energy to create fire.

heat transfer Heat moving from a hot body to a cold one through processes of conduction, convection, or radiation, or any combination of these.

heat wave A prolonged period of excessively hot weather and high humidity.

hemorrhagic fever A viral infection that causes victims to develop high fevers and bleeding throughout the body that can lead to shock, organ failure, and death.

hepatitis A liver disease caused by a virus; four different forms exist; symptoms include fever, jaundice, fatigue, liver enlargement, and abdominal pain.

herd immunity A form of indirect protection from infectious diseases when a sufficient percentage of the population has become immune to an infection through previous infection or vaccination.

human immunodeficiency virus (HIV) An RNA (ribonucleic acid) virus that causes an immune system failure and can lead to AIDS.

hurricane Term applied to cyclonic storms that occur in the North Atlantic Ocean, Caribbean Sea, Gulf of Mexico, or eastern Pacific Ocean. Minimum wind velocity is 74 miles per hour. Refer to cyclone and typhoon.

hydrologic cycle (1)The cyclic transfer of water in the hydrosphere by water movement from the oceans to the atmosphere and to the Earth and return to the atmosphere through various stages or processes such as precipitation, interception, runoff, infiltration, percolation, storage, evaporation, and transportation. (2) The cycle that moves water and water vapor among the oceans, land, and atmosphere through evaporation, condensation, precipitation, transpiration, and respiration.

hydrophobic Incapable of absorbing water.

hydrophobic layer A layer of material that cannot absorb water; usually forms in dry climates or in regions that have experienced intense fires.

hydrosphere The part of the Earth composed of water including clouds, oceans, seas, lakes, rivers, underground water supplies, and atmospheric water vapor.

hypothesis A supposition or proposed explanation made on the basis of limited evidence as a starting point for further investigation.

ice-jam flood A flood, usually in the spring, that results from broken pieces of river ice blocking the flow of a river, thereby flooding areas adjacent to the river.

igneous rock A rock formed when molten rock (magma) has cooled and solidified (crystallized). Igneous rocks can be intrusive (plutonic) or extrusive (volcanic).

impact energy Cosmic impacts with a larger body convert their energy of motion (kinetic energy) to heat.

Incident Command System (ICS) A system that is used to establish a chain of command for multiple agencies fighting fires.

Incident Management Team (IMT) A rapid response team that oversees the operations of large-scale firefighting.

influenza An acute viral infection that affects the respiratory system; this can occur as isolated, epidemic, or pandemic in scale; symptoms include fever, headache, nasal congestion, and general lethargy.

information bulletin Notification to scientists and researchers that an earthquake has occurred, but not necessarily a tsunami was generated.

inner core The solid central part of Earth's core.

intermediate-focus earthquake An earthquake that has its focus located between a depth of 70 kilometers and 300 kilometers.

island arc A chain of volcanic islands produced where an oceanic plate is sinking beneath another oceanic plate.

jet stream A high-speed, meandering wind current, generally moving from a westerly direction at speeds often exceeding 400 kilometers (250 mi) per hour at altitudes of 15 to 25 kilometers (10 to 15 mi).

jetty A structure extending into the ocean to influence the current or tide in order to protect harbors, shores, and banks.

kiloton 1 kiloton or metric kiloton (unit of mass) is equal to 1,000 metric tons. A metric ton is exactly 1,000 kilograms (SI base unit) making a kiloton equal to 1,000,000 kilograms. 1 kt = 1,000,000 kg.

kinetic energy The energy inherent in a substance because of its motion, expressed as a function of its velocity and mass, or $MV^2/2$.

Kuiper Belt A region beyond the orbits of Neptune and Pluto that is a source of short-period comets.

lahar A volcanic mudflow or landslide that contains unconsolidated pyroclastic material.

lamellae Thin layers, sheets, or plates.

landslide A general term used to describe the downslope movement of material under the force of gravity.

lapilli Pyroclastic material that ranges in size from 2 mm to 64 mm (0.08 to 2.5 inches).

latent heat of condensation Heat released when water vapor absorbs heat to be transformed to water.

latent heat of fusion Heat released when water freezes to form ice.

latent heat of vaporization Heat stored in water vapor as it changes states from a liquid to a vapor.

lava Molten rock that flows onto the surface and cools.

lava dome A dome-shaped mountain formed by very viscous lava flows.

law (or principle) The unvarying sequence of a set of naturally occurring events.

leprosy (Hansen's disease) A chronic, mildly infectious disease which affects the nervous system, skin, and nasal regions; it is characterized by skin ulcerations and nodules.

lightning A visible electrical discharge produced by a thunderstorm. The discharge may occur within or between clouds, between the cloud and air, between a cloud and the ground or between the ground and a cloud.

liquefaction A condition that exists when an overabundance of liquid, usually water, is present.

lithosphere The outer layer of solid rock that includes the crust and uppermost mantle. This layer, up to 280 km (174 mi) thick, forms the Earth's tectonic plates and rests on the mobile, ductile asthenosphere.

lithospheric plate A series of rigid slabs of cold crust plus a thin layer of cold underlying mantle that make up the Earth's outer shell. These plates float on top of a softer, more plastic layer in the Earth's mantle known as the asthenosphere.

local magnitude A term that describes the size of an earthquake in an area near the epicenter; see Richter scale.

longshore current A current that flows parallel to the shore just inside the surf zone. It is also called the littoral current.

longshore drift The net movement of sediment parallel to the shore.

Love waves These move particles with a twisting, side to side motion.

magma Molten rock that lies below the surface.

magnitude The size or scale of an event, such as an earthquake.

Main Asteroid belt A region around the Sun lying between Mars and Jupiter that is the most common source of asteroids.

malaria An infectious disease caused by the bacterium *Plasmodium falciparum*, which is transmitted by the female *Anopheles* mosquito; symptoms include high fever, chills, and sweating.

mantle The layer of the Earth below the crust and above the core. The uppermost part of the mantle is rigid and, along with the crust, forms the 'plates' of plate tectonics. The mantle is made up of dense iron and magnesium rich (ultramafic) rock such as peridotite.

mass movement (mass wasting) A term used more by geologists to describe the downslope movement of material.

megaton 1 megaton or metric megaton (unit of mass) is equal to 1,000,000 metric tons. A metric ton is exactly 1,000 kilograms (SI base unit) making a megaton equal to 1,000,000,000 kilograms. 1 mt = 1,000,000,000 kg.

mesosophere The third-lowest layer of Earth's atmosphere, this is the coldest place on Earth and where meteors burn up on entering the atmosphere.

metamorphic rock A rock that has been altered physically, chemically, and mineralogically in response to strong changes in temperature, pressure, shearing stress, or by chemical action of fluids.

meteor A rapidly moving body that passes into the atmosphere and begins to burn up, producing what is sometimes referred to as a "shooting star."

meteorite A meteor that hits Earth's surface.

meteorite fall This is also called an observed fall, and is a meteorite collected after its fall from space was observed by people or automated devices.

meteorite find Any meteorite found by people, but whose fall was not observed.

meteoroid A relatively small object that moves through space; they range in size from dust to 1 m in size; composed of rock, metal, or ice.

microbe A microscopic living organism, including bacteria, protozoa, fungi, algae, amoebas, and slime molds. Viruses are also called microbes. Some microbes are pathogens, capable of causing disease.

mid-ocean ridge An uplifting of the ocean floor that occurs when mantle convection currents beneath the ocean force magma up where two tectonic plates meet at a divergent boundary. The ocean ridges of the world are connected and form a global ridge system that is part of every ocean and form the longest mountain range on Earth.

mineral Any naturally occurring inorganic substance found in the Earth's crust as a crystalline solid.

mitigation The act of making less severe or intense; measures taken to reduce adverse impacts on humans or the environment.

moment magnitude A measure of an earthquake that is based on the area affected, the strength of the rocks involved, and the amount of movement along the primary fault.

natural catastrophe A massive natural disaster often affecting a large region and requiring significant amounts of time and money for recovery.

natural disaster The loss of life, injuries, or property damage as a result of a natural event or process, usually within a more local geographic area.

natural hazard An event or phenomenon that could have a negative impact on people and their property resulting from natural processes in the Earth's environment.

neap tide A tide that occurs when the difference between high and low tide is least; the lowest level of high tide. Neap tide comes twice a month, in the first and third quarters of the moon. Contrast with spring tide.

nitrogen cycle Biogeochemical cycle in which nitrogen is converted into many geochemical forms as it circulates among the atmosphere, terrestrial, and marine ecosystems.

nor'easter A strong low-pressure system with winds from the northeast that affects the mid-Atlantic and New England states between September and April. These weather events are notorious for producing heavy snow, copious rainfall, and tremendous waves that crash onto Atlantic beaches, often causing beach erosion and structural damage.

normal fault A plane along which movement has occurred such that the upper block overlying the fault has moved down relative to the lower block.

normal force A force that is acting perpendicular to a surface.

oceanic crust That part of the Earth's crust of the geosphere underlying the ocean basins. It is composed of basalt and has a thickness of about 5 km.

oceanic trench Deep, linear, steep-sided depression on the ocean floor caused by the subduction of oceanic crustal plate beneath either other oceanic or continental crustal plates.

offshore zone The portion of beach that extends seaward from the low tide level.

Oort Cloud A far-reaching, spherically shaped body of material located between 50,000 and 100,000 AU from the Sun; the major source of comets.

outer core The liquid outer layer of the core that lies directly beneath the mantle.

overwash A deposit of marine-derived sediments landward of a barrier system, often formed during large storms; transport of sediment landward of the active beach by coastal flooding during a tsunami, hurricane, or other event with extreme wave action.

oxidation A process in which oxygen combines with a substance.

oxygen A basic element that allows oxidation to take place; a necessary part of the fire triangle.

oxygen cycle The biogeochemical transitions of oxygen atoms between different oxidation states in ions, oxides, and molecules through redox reactions within and between the reservoirs of Earth.

P-wave The primary wave, which is a compressional wave; this is the fastest moving of all the seismic waves, and arrives first at a recording station. P-waves are a type of body wave.

pandemic A term applied to a highly-contagious disease that spreads throughout the world.

parasite An organism that lives on or in an organism of another species, the host, and benefits by deriving nutrients at the host's expense.

pathogen An agent, such as a bacterium or virus, that causes a disease.

peak-ring structure A complex impact crater with a roughly circular ring or plateau, possibly discontinuous, that surrounds the center of the crater and is inside the outer rim.

phosphorus cycle Biogeochemical cycle that shows the movement of phosphorous through the lithosphere, hydrosphere, and biosphere. Because phosphorous and phosphorous-based compounds are usually solid, they do not play a significant role in the atmosphere.

plate Slab of rigid crust and uppermost mantle that forms the outermost brittle layer of the planet.

plate boundary According to the theory of plate tectonics, the locations where the rigid plates that comprise the outer surface of the Earth meet.

plate tectonics The theory that the Earth's lithosphere consists of large, rigid plates that move horizontally in response to the ductile deformation of the asthenosphere beneath them, and that interactions among the plates at their boundaries cause most major geologic activity, including the creation of oceans, continents, mountains, volcanoes, and earthquakes.

plunging breaker Forms on shorelines with more steep offshore slopes and have a curling crest that moves over an air pocket.

potential energy The energy available in a substance because of position (e.g., water held behind a dam) or chemical composition (hydrocarbons). This form of energy can be converted to other, more useful forms (for example, hydroelectric energy from falling water).

prescribed burn A controlled fire purposefully set to remove fuel from an area.

prion disease An infliction that attacks the brain and nervous system, and disrupts normal protein activity within the neural cells.

protozoa Microscopic one-celled free-living or parasitic organisms that are able to multiply in humans.

public information officer (PIO) A person who is designated to provide official information regarding disasters.

pyroclastic Related to material that is thrown out by a volcanic eruption.

pyrolysis The process in which a series of reactions break down complex chemical compounds into simpler ones.

radiation (1) Energy emitted in the form of electromagnetic waves. Radiation has differing characteristics depending upon the wavelength. Because the radiation from the Sun is relatively energetic, it has a short wavelength (ultra-violet, visible, and near infrared) while energy radiated from the Earth's surface and the atmosphere has a longer wavelength (infrared radiation) because the Earth is cooler than the Sun. (2) The process in which heat energy is sent through space in the form of rays.

radioactive decay Natural spontaneous decay of the nucleus of an atom where alpha or beta and/or gamma rays are released at a fixed rate.

rapid-onset hazard A hazard that develops with little warning and strikes rapidly. They expend their energy very quickly, such as volcanic eruptions, earthquakes, floods, landslides, thunderstorms, and lightning.

Rayleigh waves These move particles in a rolling, up and down sense that travel in the direction of propagation of the energy.

recurrence interval A statistical expression of the average time between events equaling or exceeding a given magnitude. It also refers to the average time interval, usually in years, between the occurrence of events of a given magnitude or larger.

regolith The layer of varied material that overlies unaltered bedrock and includes unconsolidated and fragmental particles.

relative humidity The percentage of moisture present in the air as measured against the amount it can hold at a given temperature and pressure to be saturated.

resisting force A force that tends to prevent downhill movement.

retardant A chemical substance used to impede the spread of a fire.

reverse fault A plane along which movement has occurred such that the upper block overlying the fault has moved up relative to the lower block.

Richter scale A measurement scale developed by Charles Richter to determine the size of earthquakes in California. This is commonly used to describe many types of earthquakes but it is more correctly used for small, localized events.

rip current Movement of water back into the ocean in narrow zones through the surf zone.

rift valley A depression formed on the surface caused by the extension of two adjacent blocks or masses of rock.

rip current Movement of water back into the ocean in narrow zones through the surf zone.

river A large stream.

rock A naturally occurring aggregate of minerals. Rocks are classified by mineral and chemical composition, the texture of the constituent particles, and also by the processes that formed them. Rocks are thus separated into igneous, sedimentary, and metamorphic rocks.

rock cycle The sequence of events in which rocks are formed, destroyed, and reformed by geological processes. Provides a way of viewing the interrelationship of internal and external processes and how the three rock groups (igneous, sedimentary, and metamorphic) relate to each other.

rogue wave Large solitary wave caused by constructive wave interference that usually occurs unexpectedly amid waves of smaller size.

runoff Water that flows across a surface and into a stream or other body of water.

runup The height to which water comes up onto the shoreline, either from natural wave action or the occurrence of a tsunami.

S-wave The secondary wave, which is a shear wave that moves material back and forth in a plane perpendicular to the direction the wave is traveling. S-waves do not travel through liquids. S-waves are a type of body wave.

Santa Ana winds Seasonal winds that originate from high pressure over the Great Basin of Nevada, pushing dry air from east to west across southern California.

scientific method A systematic way of studying and learning about a problem by developing knowledge through making empirical observations, proposing hypotheses to explain those observations, and testing those hypotheses in valid and reliable ways.

sea arch An opening through a headland caused by erosion.

sea stack An isolated rock island that is detached from a headland by wave action.

seawall Massive structure built along the shore to prevent erosion and damage by wave action.

sedimentary rock A sedimentary rock is formed from pre-existing rocks or pieces of once-living organisms. They form from deposits that accumulate on the Earth's surface. Sedimentary rocks often have distinctive layering or bedding.

seismic gap An area in a faulted region where no seismic activity has occurred in recent time. These areas are undergoing stress buildup and could be the site of future earthquakes.

seismic sea wave A large ocean wave that is produced by a major disturbance in the ocean, such as an earthquake, volcanic eruption, or a landslide.

seismogram The written or electronic record of ground motion detected by a seismograph.

seismograph A device that records the ground motion of an earthquake.

severe acute respiratory syndrome (SARS) A respiratory disease of unknown source that first occurred in mainland China in 2003; symptoms include fever and coughing or difficulty breathing; it is sometimes fatal.

shallow-focus earthquake An earthquake that has its focus located between the surface and a depth of 70 kilometers.

shear force A force that acts parallel to a surface.

shield volcano A broad volcano that has gentle slopes consisting of low viscosity basaltic lava flows.

silicate Refers to the chemical unit of the silicon tetrahedron SiO_4 which is the fundamental building block of silicate minerals. Silicate minerals represent about one third of all minerals and hence make up most rocks we see at the Earth's surface.

simple crater A large, bowl-shaped cavity in the ground or on a planetary object, typically caused by an explosion or the impact of a meteorite. Simple craters are generally up to 2 km (1.2 mi) in diameter.

sinkhole A circular depression on the surface that forms from the collapse of material into an underlying void.

slide The movement of material along a curved or flat plane.

slope failure The downslope movement of an unstable area.

slow-onset hazard A hazard that takes years to develop such as drought, insect infestations, disease epidemics, and global warming and climate change.

slump A mass movement in which generally unconsolidated material moves down a hillside along a curve, rotational subsurface plane.

smallpox An acute viral disease that was once a major killer but has been eradicated; symptoms include headaches, vomiting, and fever, followed by a widespread skin rash that eventually permanently scars the skin.

spilling breaker Forms on shorelines with gentle offshore slopes and are characterized by turbulent crests spilling down the front slope of the wave.

spring tide The highest high and the lowest low tide during the lunar month. The exceptionally high and low tides that occur at the time of the new moon or the full moon when the Sun, Moon, and Earth are approximately aligned. Contrast with neap tide.

storm surge An onshore flood of water created by a low pressure storm system.

strain The response of an object which is being stressed; strain can be elastic, in which case the object returns to its original shape or inelastic, when the object does not recover its shape, thereby being deformed.

stratosphere The second lowest level of the atmosphere, above the troposphere, which extends to about 55 km (34 mi) above the Earth's surface and contains the ozone layer.

stream A body of water that flows downhill under the influence of gravity and lies within a defined channel.

stress The force being applied to a surface; forces can be compressional, extensional, or shearing.

strike-slip fault A fault in which the motion of the two adjacent blocks is horizontal, with little if any vertical movement.

subduction Process of one crustal plate sliding down and below another crustal plate as the two converge.

subduction zone Also called a convergent plate boundary. An area where two plates meet and one is pulled beneath the other.

sublimation The process that changes a solid into a gas, bypassing the liquid phase.

submarine landslide The collapse of land material either underwater or from the land that slides into water, producing a massive wave.

submergent coastline A coastline which has experienced a rise in sea level, due to a global sea level change or local land subsidence.

subsidence A relatively slow drop in the ground surface caused by the removal of rock or water located under the surface.

surf zone The nearshore zone of breaking waves.

surface fire Fire that moves along the ground and consumes ground litter.

surface wave A seismic wave that moves along a surface or boundary.

surging breaker Forms when the offshore slopes abruptly and the wave energy is compressed into shorter distance and the wave surges forward right at the shoreline.

swash A turbulent sheet of water that rushes up the slope of the beach following the breaking of a wave at shore.

swell Wave of uniform wavelength moving away from a storm center. They can travel great distances before the energy of the wave is released by breaking and crashing onto the coast.

talus The natural pile of material that builds up at the base of a cliff or hillside by material falling from above.

temperature of combustion The temperature at which a specific type of material will ignite.

tephra A general term for all types of pyroclastic material produced by a volcano.

theory A comprehensive explanation of a given set of data that has been repeatedly confirmed by observation and experimentation and has gained general acceptance within the scientific community.

thermal energy The amount of energy in a system that is related to the temperature of its constituents; for hurricanes this is heat originally taken from the oceans.

thermosphere The second-highest layer of Earth's atmosphere that has extremely high temperatures but near-vacuum conditions.

thunderstorm A local storm produced by a cumulonimbus cloud and accompanied by lightning and thunder.

tide The periodic rising and falling of the water that results from the gravitational attraction of the moon and sun acting on the rotating Earth.

topography The general configuration of the land surface.

tornado A rotating column of air usually accompanied by a funnel-shaped downward extension of a cumulonimbus cloud and having a vortex several hundred yards in diameter whirling destructively at speeds of up to 600 kilometers per hour (350 miles per hour).

tornado warning This is issued when a tornado is indicated by the WSR-88D radar or sighted by spotters; therefore, people in the affected area should seek safe shelter immediately.

tornado watch This is issued by the National Weather Service when conditions are favorable for the development of tornadoes in and close to the watch area. Their size can vary depending on the weather situation.

transform fault A strike-slip fault with side to side horizontal movement that offsets segments of a continental or oceanic plate.

transform boundary An area where two plates move past each other in a horizontal, sliding motion.

transform plate boundary An area where two plates meet and are sliding side to side past each other.

transient crater The initial approximately bowl-shaped meteorite impact crater with a topographically elevated crater rim that has reached its maximum size.

tributary A stream that flows into another larger stream.

trigger An event that disturbs the equilibrium of a slope, causing movement to occur.

tropical cyclone A low-pressure system having a warm center that developed over tropical (sometimes subtropical) water and has an organized circulation pattern. The magnitude of its winds defines it as a disturbance, depression, storm, or hurricane/typhoon.

tropical depression A slow-forming cyclonic storm with sustained surface winds of 38 miles per hour (62 kph) or less.

tropical disturbance The beginning stage of a cyclonic storm lasting at least 24 hours, originating in the tropics or subtropics; clouds and moisture become organized and a vertically rotating wind mass creates atmospheric instability.

tropical storm A cyclonic storm with sustained surface winds between 39 miles per hour to 73 miles per hour (62 kph to 117 kph). At this level of activity the system is assigned a name to identify and track it.

troposphere The lowest part of the atmosphere that is in contact with the surface of the Earth. It ranges in altitude from the surface up to 10 to 12 km (6 to 7.5 mi). This is where nearly all weather takes place.

trough The low spot between two successive waves.

tsunami A series of giant, long wavelength waves produced by the displacement of large amounts of ocean water by an earthquake, volcanic eruption, landslide, or meteorite impact.

typhoid fever An illness spread by contamination of water, food, or milk supplies with *Salmonella typhi*; symptoms include fever, diarrhea, stomach aches, and rash.

typhoon A cyclonic storm that forms in the central and western Pacific Ocean. Refer to cyclone and hurricane.

typhus An acute infectious disease with symptoms of high fever, severe headaches, and a skin eruption; common in wartime, famines, or catastrophes, it is spread by lice, ticks, or fleas.

undercutting A process whereby a slope or hillside has supporting material removed by erosion.

updraft A small-scale current of rising air. If the air is sufficiently moist, then the moisture condenses to become a cumulus cloud or an individual tower of a towering cumulus.

vector An agent such as a flea or mosquito that can spread parasites, virus, or bacteria.

virus Acellular, non-living infectious particles of either DNA or RNA associated with various protein coats; these only multiply in living cells.

viscosity A measure of the internal resistance of a substance to flow; a lower viscosity means the material flows easily.

volcanic arc A chain of volcanoes formed above a subducting plate. The arc forms where the downgoing plate becomes hot enough to release fluids that rise into the overlying mantle and cause it to melt.

volcanic explosivity index (VEI) A measure of the intensity of a volcanic eruption, with a value of 0 representing the quietest and 8 being the most explosive.

warning The highest alert level that is given when a tsunami has been formed and its arrival is highly possible.

watch An alert that a tsunami could potentially occur in an area.

water cycle The water cycle describes the continuous movement of water on, above, and below the surface of the Earth.

watershed See drainage basin.

wave Energy in motion that is the result of some disturbance that moves over or through a medium with speeds determined by the properties of the medium. Ocean waves are usually generated by wind blowing across the water surface.

wave base Depth equal to one half the wavelength where there is no movement associated with surface waves.

wave-cut platform A gently sloping surface produced by wave erosion, extending far into the sea or lake from the base of the wave-cut cliff.

wave height The vertical distance between the crest and adjacent trough of a wave.

wave period The time it takes for one full wavelength to pass a given point.

wave refraction The process by which the part of a wave in shallow water is slowed down, causing it to bend and approach nearly parallel to shore.

wave speed The velocity of propagation of a wave through a liquid, relative to the rate of movement of the liquid through which the disturbance is propagated.

wave steepness The measured ratio of wave height to wavelength.

wavelength The horizontal distance between any two similar points of the wave, such as two crests or two troughs.

weather The composite condition of the near Earth atmosphere, which includes temperature, barometric pressure, wind, humidity, clouds, and precipitation. Weather variations over a long period create the climate.

weathering A process that includes two surface or near-surface processes that work in concert to decompose rocks. Chemical weathering involves a chemical change in at least some of the minerals within a rock. Mechanical weathering involves physically breaking rocks into fragments without changing the chemical makeup of the minerals within it.

wildfire An uncontrolled fire in a natural setting.

yellow fever Disease caused by a vector-transmitted virus; symptoms include high fever, headaches, jaundice, and often gastrointestinal hemorrhaging.

Index

F

I

J

K

L

M

N

T

U

V

W

Y

Z